Supramolecular Chemistry

Supramolecular Chemistry

AN INTRODUCTION

FRITZ VÖGTLE

with the collaboration of

F. Alfter, W. Calaminus, S. Grammenudi, V. Hautzel, M. Hecker, R. Hochberg, P. Knops, W. Orlia, A. Ostrowicki, K. Saitmacher, Ch. Seel, P. Stutte, E. Weber

University of Bonn, Germany

Translated by Michel Grognuz

JOHN WILEY & SONS

Chichester · New York · Brisbane · Toronto · Singapore

Other Wiley Editorial Offices

John Wiley & Sons, Inc., 605 Third Avenue,
New York, NY 10158-0012, USA

Jacaranda Wiley Ltd, G.P.O. Box 859, Brisbane,
Queensland 4001, Australia

John Wiley & Sons (Canada) Ltd, 22 Worcester Road,
Rexdale, Ontario M9W 1L1, Canada

John Wiley & Sons (SEA) Pte Ltd, 37 Jalan Pemimpin #05-04,
Block B, Union Industrial Building, Singapore 2057

Library of Congress Cataloging-in-Publication Data:

Vögtle, F. (Fritz), 1939–
[Supramolekulare Chemie. English]
Supramolecular chemistry ; an introduction / Fritz Vögtle ; with the collaboration of F. Alfter . . . [et al.] ; translated by Michel Grognuz.
p. cm.
Translation of; Supramolekulare Chemie.
Includes bibliographical references and index.
ISBN 0 471 92802 X
1. Macromolecules. I. Alfter, F. II. Title.
QD381.V6413 1991
547.7—dc20 90-48199
CIP

British Library Cataloguing in Publication Data:

Vögtle, Fritz
Supramolecular chemistry.
1. Molecules. Chemical reaction
I. Title II. Supramolekulare Chemie. *English*
541.39

ISBN 0 471 92802 X

Printed and bound in Great Britain by Biddles Ltd, Guildford, Surrey

Contents

Preface

This book emerged from a series of lectures, held at Bonn University (Germany), which were entitled 'Modern Methods, Reactions and Structures in Organic Chemistry', 'Recent Results and Problems in Organic Chemistry' and 'New Molecules and Reactions in Organic, Bioorganic and Supramolecular Chemistry'.

Originally, the book was to contain two equal parts, the first being a description of 'Fascinating Molecules in Organic Chemistry' and the second a discussion of the 'Aggregation of Molecules to Supramolecular Structures'. In 1987, researchers on supramolecular chemistry were awarded the Nobel Prize and, therefore, the latter half of the book had to be expanded. Subsequently, the size of the book exceeded the limits set within the German textbook series, of which it was to be part. Both parts of the original book were therefore expanded into two separate volumes, entitled 'Fascinating Molecules in Organic Chemistry' and 'Supramolecular Chemistry'. Even so, owing to the limited number of pages available, several planned chapters had to be abandoned.

The discussion of the phthalocyanines at the end of the volume on 'Fascinating Molecules in Organic Chemistry' offers a transition to this volume, while the section on bipyridine in Chapter 2 of 'Supramolecular Chemistry' picks up at the end of 'Fascinating Molecules in Organic Chemistry' and leads on to supramolecular structures.

Whereas in the volume on 'Fascinating Molecules in Organic Chemistry' the individual molecule is very much in the foreground, the present book discusses the interactions of several molecules, i.e. the properties, functions and applications of molecular aggregates and molecular assemblies. Molecular recognition, the directed interaction between host and guest, is discussed in some detail in many different places.

In this volume, supramolecular structures are highlighted in the context of their common field, the discussion extending to the latest results available in the primary literature.

The book consists of comparatively short, nearly self-contained chapters on supramolecular structures. Using numerous examples of 'aggregates', the chemistry involved is described from different angles. Here, the multidisciplinary aspect of this domain of chemistry becomes apparent, as the close relationships with physics, physical chemistry and biochemistry (material sciences and life sciences) are emphasized. This introduction to supramolecular chemistry can by no means be exhaustive. The choice of structures is debatable, as their numbers are almost unlimited. Molecular aggregates with predominantly biochemical character (e.g. nucleic acids), and also polymers, had to be left out; the very large domain of the membranes would also have exceeded the limits of this volume. Related questions could only be touched upon here and there. The various subjects are not all discussed to the same extent and depth. After all, the amount and quality of the literature available in particular areas are variable. The number of references given depends on the type of literature found: whenever recent reviews existed, the number of original papers cited was kept low. Conversely, where no recent overviews were available, a list of original papers was compiled, e.g. in the areas of the crown ethers, liquid crystals, clathrates, host–guest chemistry and organic switches.

Supramolecular chemistry has been subject to such a rapid development that, as yet, it is difficult to divide it into different areas and to give precise definitions. The book starts with simple complex-forming compounds and ends with complex 'supramolecules' and 'superstructures' ('supermolecules', 'supercomplexes' and 'superorbitals' are terms which have mostly been avoided).

The introductory part of the book can be understood with just a basic knowledge of chemistry, and undergraduates should have no great difficulty in understanding large parts of the book. For postgraduates and research chemists, the opportunity is given to learn about new developments and, with the help of the cited literature, to extend some of their knowledge.

A particularity of this volume is the inclusion of many stereoscopic views of supramolecular structures, most of them based on X-ray crystallography. To date, no such collection of three-dimensional structures, allowing direct comparison, has been put together; what is more, they are often of great aesthetic beauty.

Let me thank all the people who worked so enthusiastically on this book with me. In addition to Prof. Dr E. Weber and those co-workers named on the title page, the following have also helped: R. Baginski, S. Billen, J. Dohm, W. Jaworek, H.D.W. Losensky, W.M. Müller, L. Rossa, A. Schröder, H.D.P. Schwenzfeier, A. Wallon and D. Worsch. I would also like to thank Prof. Dr J.F. Stoddart for the original drawings of kohnkene, trinacrene and the catenanes. I am endebted to Dr M. Grognuz for his excellent translation and his willingness to oblige with my requests for changes and additions while working on it. I am grateful to Teubner Verlag, and in particular to Dr P. Spuhler there, and to John Wiley & Sons for their kind guidance throughout.

Bonn, Autumn 1990 F. Vögtle

1 Supramolecular, Bioorganic and Bioinorganic Chemistry

1.1 INTRODUCTION

The area of **supramolecular chemistry** is still a young one. It is based both on the development of the chemistry of crown ethers and cryptands, and on the progress made in the study of the self-organization of molecules (e.g. membranes and micelles) and of organic semiconductors and conductors. So far, the area has not been given a formal structure. Strictly, we look at (intermolecular) interactions between at least two, if not several, molecular species and also ions in solution.

In the solid state (e.g. crystals), molecules can be ordered even without the aid of intermolecular interactions. In this book, these situations will be associated with the term *supramolecular* only when it is used in its broadest sense. If, on the other hand, the molecules in the crystal interact with each other, e.g. through hydrogen bonding, then we shall refer to *supramolecular structures* in a more particular way.

The boundaries, however, are not well defined: In Chapter 5 on 'Clathrate Inclusion Compounds' and in Chapter 6 on 'Directed Crystal Formation with Tailored Additives' we shall see some interesting borderline cases, where it is not easy to define the situation.

1.2 SUPRAMOLECULAR, BIOORGANIC, BIOINORGANIC AND BIOMIMETIC CHEMISTRY

Supramolecular chemistry is closely related to bioorganic and bioinorganic chemistry. In the following discussion, we shall point out common features and differences between these three areas:

> **Bioorganic chemistry** is the biomimetic, *in vitro* chemistry of natural products and analogous compounds. 'Mimicking' Life amounts to simplifying the processes involved; one restricts oneself to selected, well defined properties of reactions in or on biological cells. These bioorganic properties, as compared to conventional natural product chemistry, include e.g. the type of reaction, the reagent or the reaction conditions... [1].
>
> Biologically, the most important type of reaction is the selective aggregation of natural products to form molecular complexes and membranes. Formation of cells, catalysis in enzyme–co-enzyme–substrate complexes, photochemical charge separation in organized redox chains, and interactions of hormones and

> drugs with biological receptors ensue, in a first approximation, from selective weak interactions between natural products. Because bioorganic chemistry is about understanding and mastering these processes, one finds that analysis, synthesis and reactions of molecular aggregates ('supramolecular chemistry') have become (possibly) its most important topics. [1]

> A second special feature of biological reactions, compared with organic chemical ones, is their reaction environment. To put it simply, it is composed of three components, i.e. water, polar surfaces and the less polar cavities of membranes and biopolymers. This multi-component system is structurally organized and allows sequential reaction chains, as well as highly regioselective and stereoselective reactions. [1]

In both bioorganic and supramolecular chemistry, the 'synthesis' and the use of organized spaces for reactions are attempted.

We find that in addition to organic functional groups, nearly all transition metals in the fourth row of the Periodic Table have been identified as essential components of enzymes and proteins. The search for active centres, the elucidation of their structures and the understanding of the elementary reactions that occur there are the object of an interdisciplinary area of research in both inorganic chemistry and biochemistry, which is called *bioinorganic chemistry*. As seen by an inorganic chemist, these metal complexes are Werner-type *coordination compounds* with very large ligands. It is possible for the bioinorganic chemist, using modern physico-chemical methods (ESR, UV–Visible, Mössbauer, Raman spectroscopy; magnetic susceptibility, magnetic circular dichroism; EXAFS spectroscopy) to gather a great amount of varied information with which to describe the way the metal coordinates to its ligands.

Once the structure and the coordination sphere are known, the next step consists in synthesizing model complexes on a lower molecular size level; studies on the stereochemistry, the reactions, the biomimetic behaviour and the (industrial) applications in catalysis then follow.

1.3 FROM MOLECULAR MATERIALS TO SUPRAMOLECULAR STRUCTURES

Molecules are made up of atoms bound by strong covalent and other bonds and can be described unambiguously with stereochemical terms such as constitution, configuration and conformation. Supramolecular structures may be formed by the aggregation of such molecules [2]. The 'molecular subunits' can be prepared selectively by a large number of reactions and can be characterized by physico-chemical properties and criteria [melting point, boiling point, chemical behaviour, redox potential, polarity, polarizability, spin status, position of HOMO and LUMO orbitals, colour, chirality (optical activity)], and by dynamic parameters (vibrational states, half-life of excited states, etc.).

Macroscopically, they appear as either plasmas, gases, liquids or solids. The solid phases are of particular interest, as the molecular building blocks order themselves and condense to form higher structures in a specific manner (molecular materials). While the properties of the condensed phases can be derived from the characteristics of the isolated molecular building block, it is difficult to relate the structure of the molecular subunit with the type of organization found in the condensed phase. This is mainly the result of the large number of directing parameters that affect the organization.

In Figure 1, a classification of the most important types of molecular building

INORGANIC:

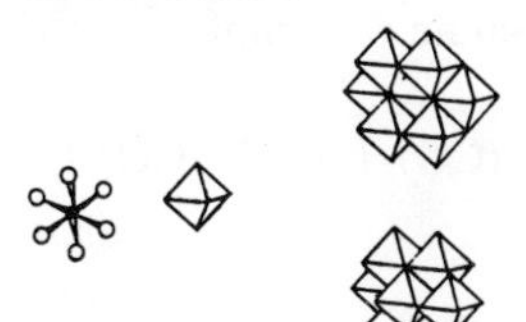

Organizational parameters

— Type of order
— Strength and anisotropy of interactions between subunits
— Symmetry of packing
— Intermolecular vibrations

ORGANIC:

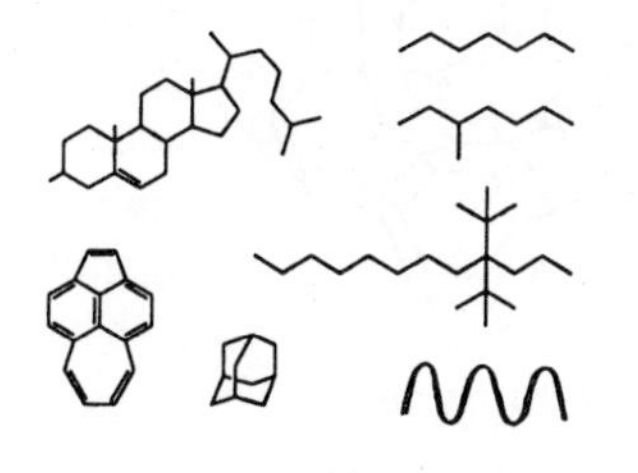

Organizational parameters

— Symmetry of molecules
— Polarizability of a molecule's backbone
— Type, number and polarity of substituents
— Degree of flexibility of submolecular components
— Hydrophilic and hydrophobic properties

METALLOORGANIC:

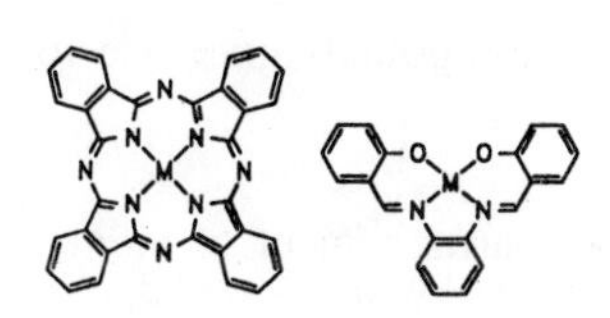

Organizational parameters

— Strength of the covalent part of the bond between the metal ion and the coordination atoms of the ligand
— Electrostatic forces
— Oxidation state of the metal, usually high
— Charge density
— Weak polarizability of the metal ion

ORGANOMETALLIC:

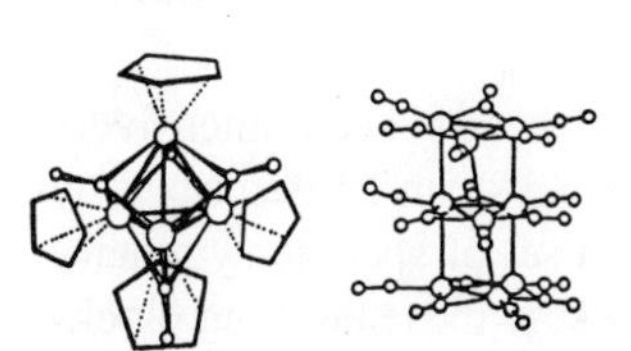

Organizational parameters

— Strong covalent bonding
— Low oxidation state of the metal ion
— High polarizability of the metal ion
— d–d; d–f; f–f orbital interactions

Figure 1. Molecular building blocks and organizational parameters for the corresponding 'molecular materials' [3]

blocks for 'molecular materials' is undertaken; the examples come from areas such as inorganic, organic, metalloorganic and organometallic chemistry, and the corresponding organizational parameters are included [3].

Accordingly, the main types of molecular building blocks that can be used for the building of 'molecular materials' are as follows:

— polymolybdate and polytungstate units, or the β-cage shown in Figure 2 (in *inorganic chemistry*);
— the steroid skeleton, linear and branched hydrocarbon chains, polymers, aromatic systems, adamantane (in *organic chemistry*);
— metallophthalocyanine, bis(salicylidene)ethylenediamine complexes (in *metalloorganic chemistry*);
— transition metal haptocomplexes such as $[(\eta^3\text{-}C_5H_5)Fe(CO)_4]$ or $[Pt_9(CO)_9(\mu^2\text{-}CO)_9]^{2-}$ [3] (in *organometallic chemistry*).

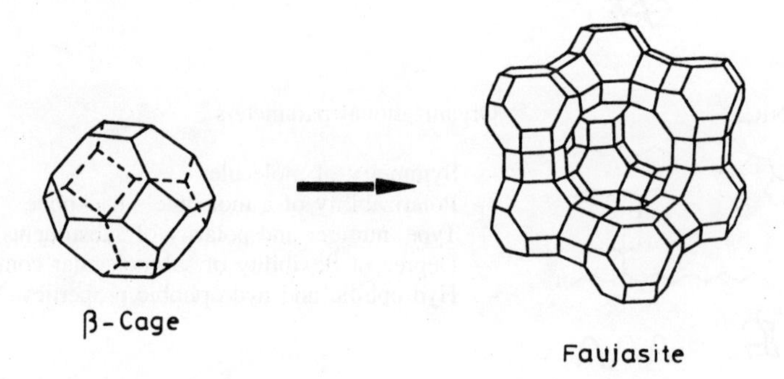

Figure 2. Transition from the β-cage building block to the structure of faujasite

The criteria for the organization of different types of compounds give rise to three separate forms of condensation into the solid state:

— The *single crystal* with its three-dimensional periodic arrangement of atoms, ions and molecules;
— the *polycrystalline form*, where smaller single crystals are ordered in relation to each other;
— the *amorphous state*, where the positions of basic building blocks correlate only in small areas.

Above all, it is the *solid state* of organic matter that has been intensively investigated. Accordingly, the different basic building blocks can be described by a total of 32 point groups. Each point group consists of a set of specific symmetry operations, which can be applied to a given building block. If these building blocks form a three-dimensional solid phase, it will be described by one of 230 space groups. Of the 30 000 organic compounds that have been investigated, 80% belong to only six of these space groups. Following a common model, the principle of highest packing density is applied. On the other hand, the use of symmetry elements should not lead to the assumption that solid phases are rigid structures. Hence, even in the solid state, vibrational processes and rearrangements may occur.

The rotation of methyl groups, with an energy barrier between 1 and 50 kJ/mol, and of aromatic molecules (along their main axis) also belong to the properties of the solid phase, even if X-ray structure analysis does not indicate any disorder. The time constants related to these dynamic processes are limited by the lattice relaxation time (intermolecular vibrations) and cannot therefore exceed 10^{-11} to 10^{-12} s. However, if, through heating, an abnormal coordination of, say, five instead of six neighbours is created, then thermal motion will generate a similar disorder in this area of the crystal and the disorder will then spread throughout. This phenomenon is also used to explain the process of melting.

In addition to these dynamic processes, which appear to leave the crystal generally undisturbed, we also find permanent disorders, such as differences in distances, substitutional disorders or defects in the packing. The investigation of these structural characteristics and the properties that arise from cooperative effects in solid condensed matter belongs to the area of solid-state physics. In this context, think of piezoelectricity, pyroelectricity and semiconductors (cf. Chapter 10).

However, solid materials composed of organic building blocks also remain interesting objects for research. Particularly in the field of liquid crystals, decisive progress has been made. Without it, the numerous applications of liquid crystals would not have been possible (cf. Chapter 8). Liquid crystalline properties occur as a result of intermolecular interactions and orientations of the same molecular building block alone.

If we extend the cooperative effects between molecules to interactions between two or more different species (molecules, ions), we enter the field of *supramolecular chemistry* [2–9]. During the last 20 years, research and development on a molecular level has yielded substantial progress in many areas of chemistry. A natural consequence of this is today's trend by chemists to explore more complex systems and interactions. Representative examples are the developments in areas such as macropolycyclic and host–guest chemistry, the study of active centres of metalloenzymes (bioorganic chemistry), electronic systems on a molecular level, organic superconductors and supramolecular effects in photochemistry and electrochemistry.

In contrast to molecular chemistry, which is predominantly based on the covalent bonding of atoms, supramolecular chemistry is based on intermolecular interactions, i.e. on the association of two or more building blocks, which are held together by intermolecular bonds. Intermolecular (supramolecular) interactions are the foundation for highly specific biological processes, such as the substrate binding by enzymes or receptors, the formation of protein complexes, the intercalation complexes of nucleic acids, the decoding of the genetic code, neurotransmission processes and cellular recognition (immunology).

The exact knowledge of the energetic and stereochemical characteristics of these non-covalent, multiple intermolecular interactions (electrostatic forces, hydrogen bonding, van der Waals forces, etc.) within defined structural areas should allow the design of artificial *receptor molecules*, which bind the substrate strongly and selectively by forming (tailored) supramolecular structures, 'supramolecules', of defined structure and function. In Figure 3, the formation of such a 'supramolecule' (bottom) is contrasted with that of a single molecule (top). Subsequently, we shall use the example of bipyridine to illustrate the transition from a single molecule to a supramolecular structure. This theme will be developed in a different context at

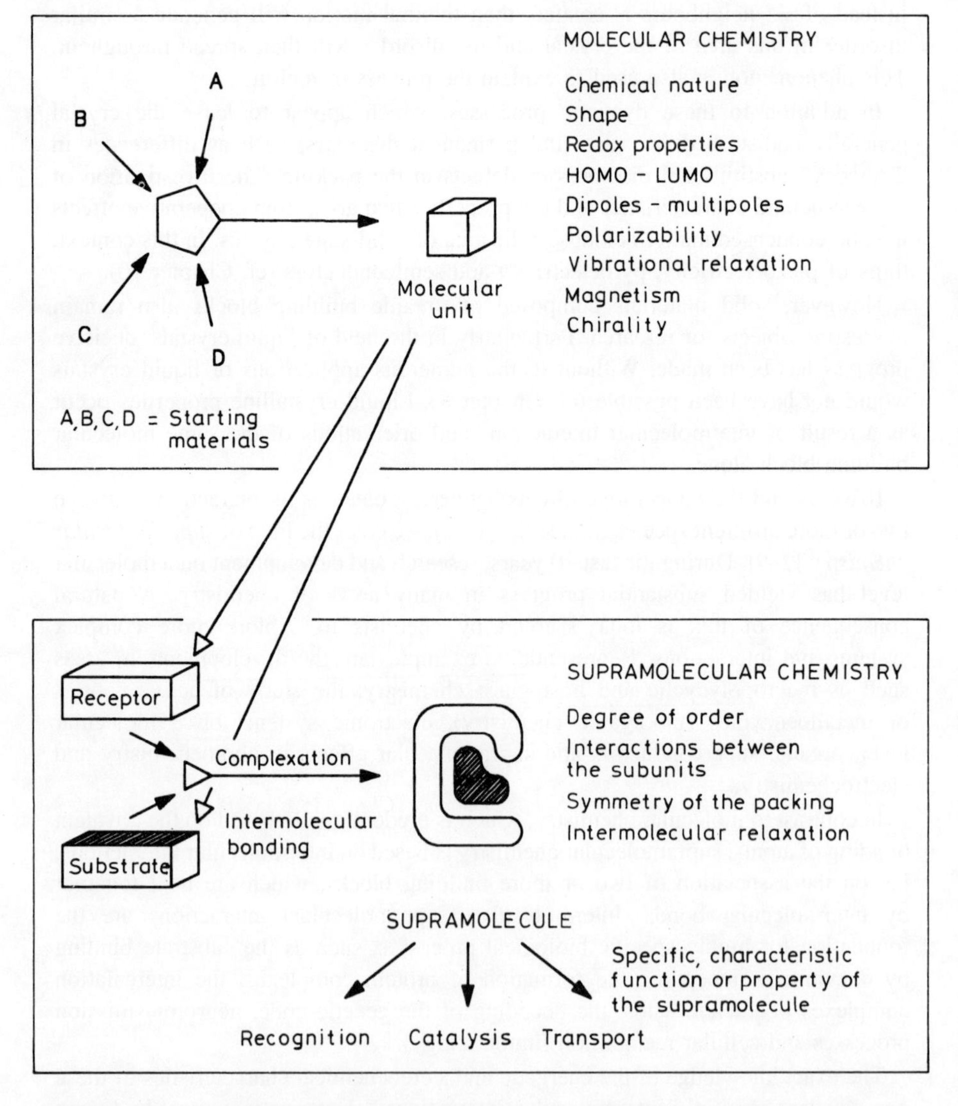

Figure 3. 'Supramolecules, therefore, are to molecules and their intermolecular bonds what molecules are to atoms and their covalent bonds' [2,4]

the end of the volume on 'Fascinating Molecules in Organic Chemistry' using the example of the phthalocyanines.

Thus, we shall go from considering the classical complexation of transition metals to subsequent large chapters describing crowns, cryptands, spherands, etc. There, as in the proposed volume on 'Fascinating Molecules in Organic Chemistry', we shall discuss briefly some topics including the synthesis and stereochemistry of the supramolecular structure, in addition to the individual ligand molecules; the transition from molecule to supramolecule is ever present. Only with this background can supramolecular structures, supramolecular complexes [10], supramolecular catalysis [11] and supramolecular photochemistry [10a] be understood and developed further [12].

To sum up at the end of this first chapter, *Supramolecular Chemistry* is defined as chemistry 'beyond the molecule', as chemistry of tailor-shaped inter-molecular interaction. In '*supramolecules*' information is stored in the form of structural peculiarities. Moreover, not only the combined action of molecules is called *supramolecular*, but also the combined action of characteristic *parts* of one and the same molecule.

References

1. With minimal alteration from J. Fuhrhop, *Bioorganische Chemie*, Thieme, Stuttgart, 1982.
2. J.-M. Lehn, *Science*, **227**, 849 (1985).
3. J. Simon, J. André and A. Skachios, *Nouv. J. Chim.*, **10**,, 295 (1986).
4. J.-M. Lehn, *Pure Appl. Chem.*, **52**, 2441 (1980).
5. Comments on the Nobel Prize for Chemistry 1987: E. Weber and F. Vögtle, *Nachr. Chem. Tech. Lab.*, **35**, 1149 (1987); H.J. Schneider, *Chem. Unserer Zeit*, **6**, A60 (1987); J.F. Stoddart, *Ann. Rep. Progr. Chem. Sect. B*, **85**, 353 (1988–9).
6. J.-M. Lehn, *Angew. Chem.*, **100**, 92 (1988); see also H. Inokuchi, *Molecular Assemblies; Higher-Order Structures and Organizations, and their Functions, Summary Report of Special Research Project 1983–1985*, Institute for Molecular Science, Japan, 1986; Y. Murakami (Ed.), *Supramolecular Assemblies.* Collective Report on Special Research Project, Mita Press, Tokyo; D.H. Busch and N.A. Stephenson, *Coord. Chem. Rev.*, **100**, 119 (1990); G.W. Gokel (Ed.), *Advances in Supramolecular Chemistry*, JAI Press, Tokyo, 1990.
7. D. Karlin, *J. Am. Chem. Soc.*, **109**, 2668 (1987).
8. D.J. Cram, *Angew. Chem.*, **98**, 1041 (1986); **100**, 1041 (1988).
9. H. Ringsdorf, B. Schlarb and J. Venzmer, *Angew. Chem.*, **100**, 117 (1988).
10. (a) V. Balzani (Ed.), *Supramolecular Photochemistry*, Reidel, Dordrecht, 1987; (b) A. Benzini, A. Bianchi, E. Garcia-Espana, M. Giusti, S. Mangani, M. Micheloni, P. Oriollo and P. Paoletto, *Inorg. Chem.*, **26**, 3902 (1987); M. Gubelmann, A. Harriman, J.-M. Lehn and J.L. Sessler, *J. Chem. Soc., Chem. Commun.*, **77** (1988).
11. M.W. Hosseini and J.-M. Lehn, *J. Chem. Soc., Chem. Commun.*, 397 (1988).
12. J.-M. Lehn, *Angew. Chem.*, **102**, 1347 (1990); M. Ahlers, W. Müller, A. Reichert, H. Ringsdorf and J. Venzmer, *Angew. Chem.*, **102**, 1310 (1990); D. Seebach, *Angew. Chem.*, **102**, 1363 (1990).

2 Host–Guest Chemistry with Cations and Anions

2.1 BIPYRIDINE

2.1.1 INTRODUCTION

2,2′-Bipyridine (**1**) and its derivatives are renowned for their ability to form coordination compounds with metal ions of almost all groups in the Periodic Table. The description, based on MO theory, of the coordinative bonds in these complexes requires that the central ion and the ligands be able to form σ- and π-bonds.

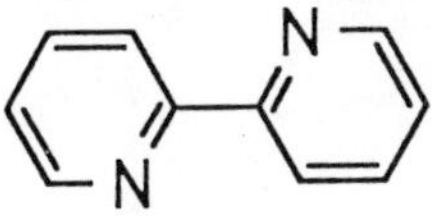

1

It is no wonder, therefore, that bipyridine is a molecular building block *par excellence* (cf. Chapter 1) for a wide variety of types of molecular and ionic aggregates ('supramolecules,' cf. Chapters 2.2, 2.5 and 12). This is a good reason to concern ourselves with bipyridine and its chemistry: in the next few sections, we shall give a general discussion, whilst occasionally focusing on certain topics.

In 1888, Blau [1] synthesized the first complexes of Fe(II) salts with 2,2′-bipyridine and isolated a series of salts with the composition $Fe(bipy)_3X_2$.

When solutions of Fe(II) salts are mixed with 2,2′-bipyridine (bipy or bpy for short), intensely red compounds are formed, which are also characterized by a very high stability. Their colour is attributed to the complexed Fe(II). On treatment with chlorine or potassium permanganate in acidic solution, the red colour changes to blue. This suggests that the blue substance contains trivalent iron, which was confirmed by the isolation of the complex $[Fe(bipy)_3]_2[PtCl_6]_3 \cdot 5.5H_2O$. This behaviour of bipyridine complexes is used in analytical chemistry for the determination of metals, especially Fe(II) [2].

2,2′-Bipyridines can also influence biological systems [3]. Their activity is usually a consequence of their ability to complex those metals which are jointly responsible for the enzymatic activity in a living organism. Moreover, they are able to stimulate the activity of some enzymes, probably by removing the metal which inhibits them [3b].

Apart from the antibiotic caerulomycin (**2**), the most important biologically active

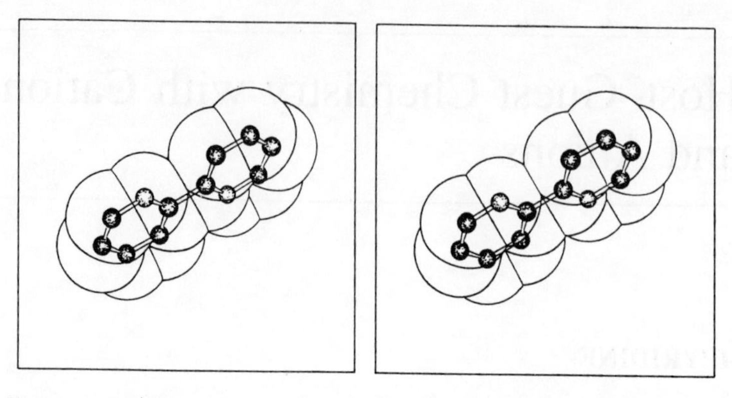

Figure 1. 2,2′-Bipyridine (stereoview based on X-ray diffraction analysis)

OCH_3 N N CH=NOH

2

N N $2Br^{\ominus}$

3

derivative of 2,2′-bipyridine is the bis(pyridinium) salt **3** ('diquat' dibromide), which is produced industrially and is used as a herbicide [4].

Diquat is sprayed on the fields in spring. It is first absorbed by the growing weeds and then reduced to the radical cation. During its reoxidation in the presence of light, a high concentration of hydrogen peroxide develops, which poisons the plant. The green plant loses colour (chlorosis) as a radical chain reaction destroys the chlorophyll. Because its radical cation has only a short life-span in the presence of light and oxygen, and in the fields degrades rapidly to non-poisonous, water-soluble compounds, diquat offers the possibility of restraining the growth of weeds in a timed and environmentally friendly manner. In addition, **3** is used as a redox indicator and as an electron transfer reagent.

2,2′-Bipyridine (**1**) is also of importance in non-biological areas. It acts as an activator for the polymerization of a number of alkenes [5] and as a catalyst for certain reactions [6]. Moreover, it enhances the quality of galvanization processes [7] and is used as an additive for copying materials [8]. As a consequence of their versatile and important applications, the 2,2′-bipyridine compounds and their complexing abilities are still being investigated.

2.1.2 SYNTHESIS OF 2,2′-BIPYRIDINE

Pure 2,2′-bipyridine (**1**) was synthesized and analysed by Blau [9] in 1889, when he obtained it by distillation of copper picolinate. Oxidation of 1,10-phenanthroline (**4**)

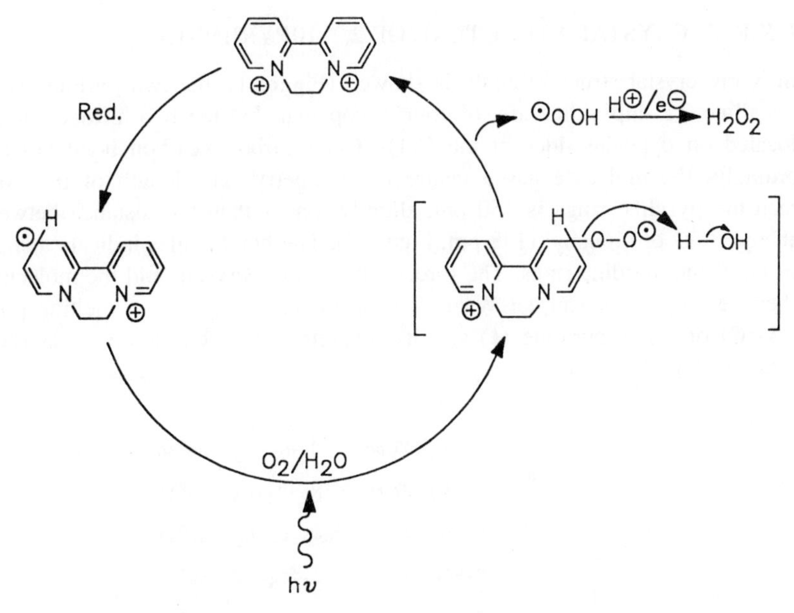

Scheme 1. Reoxidation of diquat (**3**) in the presence of light

4
$KMnO_4$
$OH^{\ominus}$
6
HO_2C CO_2H
5
$-CO_2$
1

with alkaline permanganate produces only some 2,2′-bipyridine-3,3′-dicarboxylic acid (**5**), in addition to diazafluorenone (**6**); a subsequent decarboxylation of **5** yields 2,2′-bipyridine [10].

The conversion of 2-halopyridines with copper, following the Ullmann procedure, also yields 2,2′-bipyridine (**1**) [11]. This reaction can also be used to produce symmetrically substituted 2,2′-bipyridines [12], 2,2′ : 6′,2″-terpyridines and polypyridines [11a].

In 1956, a more convenient synthesis of 2,2′-bipyridine was achieved by Badger and Sasse, using Raney nickel as a catalyst [13]. Substituted 2,2′-bipyridines were prepared from α-picoline, 4-ethylpyridine, nicotinic acid, ethyl nicotinate and other related compounds.

2.1.3 X-RAY CRYSTAL STRUCTURE OF 2,2′-BIPYRIDINE

As an X-ray crystal structure analysis showed (Figure 1), the two pyridine rings of crystalline 2,2′-bipyridine are, of course, coplanar, but the two nitrogen atoms are located on opposite sides of the C(1)—C(1′) carbon—carbon bond [14,15]. Accordingly, the molecule has a centre of symmetry. The length of the bond between the pyridine rings is 150 pm, slightly longer than the distance between the phenyl rings of biphenyl (148 pm, Figure 2). The bond angles indicate a slight distortion of the pyridine rings. The forces which in the crystal hold the molecules together are only weak van der Waals forces; this explains the low melting point (70–73 °C) of 2,2′-bipyridine (**1**) and its propensity to sublime slowly at room temperature (b.p. 273 °C).

C_1–N_2	135 pm	C_1–N_2–C_3	116°40′
N_2–C_3	137 pm	N_2–C_3–C_4	124°17′
C_3–C_4	137 pm	C_3–C_4–C_5	118°32′
C_4–C_5	137 pm	C_4–C_5–C_6	119°40′
C_5–C_6	140 pm	C_5–C_6–C_1	118°17′
C_6–C_1	141 pm	C_6–C_1–N_2	122°28′
		C_6–C_1–C_1'	121°25′
C_1–C_1'	150 pm	N_2–C_1–C_1'	116°08′

Figure 2. Bond lengths and bond angles in the 2,2′-bipyridine molecule (**1**)

For 2,2′-bipyridine (**1**), dipole moments between 0.61 and 0.91 D have been measured [16]. The molecule has no dipole moment while in its planar transoid conformation, but has a dipole moment of 3.8 D when in its cisoid conformation. In benzene solution a nearly planar, transoid-like conformation predominates; the angle between the planes of the pyridine rings is only approximately 20°. Mesomerism and dipole–dipole interactions are too weak to stop any rotation and a completely planar transoid conformation cannot be stabilized in solution.

2.1.4 REACTIONS OF 2,2′-BIPYRIDINE

The reduction of 2,2′-bipyridine (**1**) with sodium in boiling alcohol [17] or by catalytic hydrogenation [18] yields 2,2′-bipiperidine (**7**). In contrast, reduction with tin and HCl [18b,19], controlled catalytic hydrogenation [20] and electrochemical hydrogenation [21] all yield 1,2,3,4,5,6-hexahydro-2,2′-bipyridine (**8**).

By oxidation of 2,2′-bipyridine (**1**) with hot permanganate solution, picolinic acid is obtained [1].

The chlorination of 2,2′-bipyridine (**1**) in the gas phase at 555 °C [22] and the reaction with PCl_5 at 300 °C [23] yield almost quantitatively octachloro-2,2′-

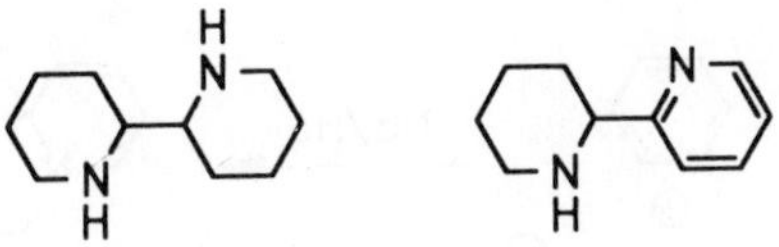

7 8

bipyridine; at lower temperatures (200–400 °C), the reaction in the gas phase between 2,2′-bipyridine and chlorine affords 6-chloro-2,2′-bipyridine and 6,6′-dichloro-2,2′–bipyridine [24]. Corresponding reactions with bromine in the gas phase at 500 °C yield 6-bromo-2,2′-bipyridine and 6,6′-dibromo-2,2′-bipyridine, but at lower temperatures (250 °C) the products are 5-bromo-2,2′-bipyridine and 5,5′-dibromo-2,2′-bipyridine.

Sulphonation reactions with sulphuric acid at 300 °C lead to the formation of 2,2′-bipyridine-5-sulphonic acid and 2,2′-bipyridine-5,5'-disulphonic acid [25], whereas with sulphur trioxide at 200–225 °C [26] or with oleum and a catalyst [27] the product is 2,2′-bipyridine-5-sulphonic acid only. Depending on the conditions, aryllithiums and alkyllithiums react with 2,2′-bipyridine to yield 6-aryl-2,2′-bipyridine and 6,6′-diaryl-2,2′-bipyridine [27,28], and 6-alkyl-2,2′-bipyridine and 6,6′-dialkyl-2,2′-bipyridine [28c,29], respectively.

On mixing 2,2′-bipyridine and an alkyl halide in a 1:1 ratio, the corresponding *N*-alkyl-2,2′-bipyridinium halide is obtained (cf. **9**) [30]. With excess of alkyl halide or dialkyl sulphate, the double salt is formed [9,31]. Fe(III) oxidizes **9** to pyridone **10**, which in turn can react with a mixture of phosphorus pentachloride and phosphorus oxychloride to yield 6-chloro-2,2′-bipyridine.

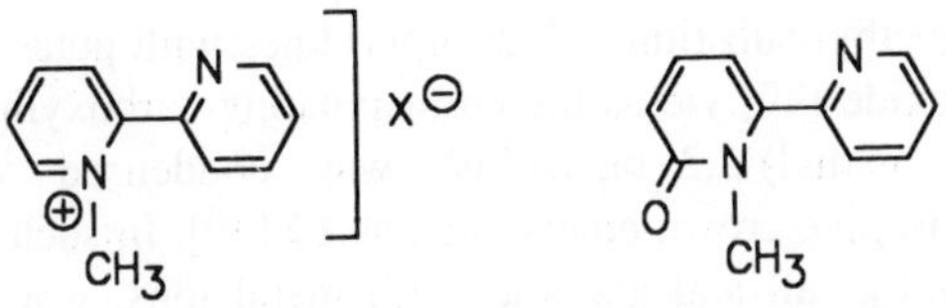

9 10

By reacting 2,2′-bipyridine (**1**) with 1,2-dibromoethane, the bridged bis(pyridinium) salt **3** was obtained. As already mentioned, it is known as diquat dibromide, an important herbicide. After the discovery of its herbicidal properties, substantial research was carried out and more efficient syntheses were sought [4b].

An interesting reaction is the intramolecular coupling of the bis(pyridinium) salt **11** to form **12** [32].

Dehydrogenation of the 6-hydroxy derivative of **3** with thionyl chloride affords salt **13** [33]. Some other homologues of the same type (**14**–**16**) can also be mentioned [34–36].

When 2,2′-bipyridine is made to react with excess of hydrogen peroxide in glacial acetic acid, the *N*,*N*'-bis-*N*-oxide **17** [37] is obtained, in addition to the mono-*N*-oxide **18** [37b,d,e].

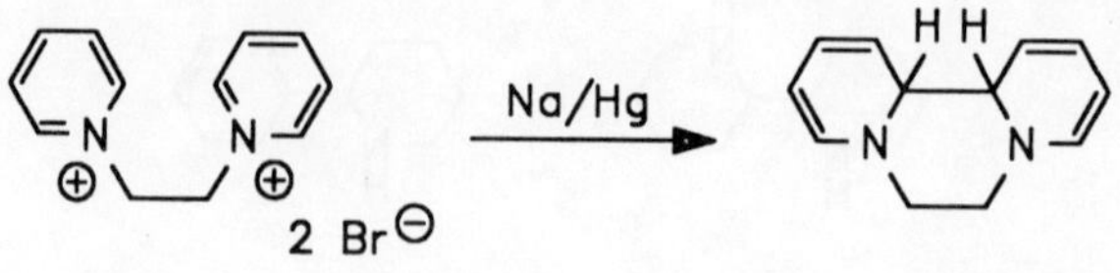

11 Na/Hg → 12

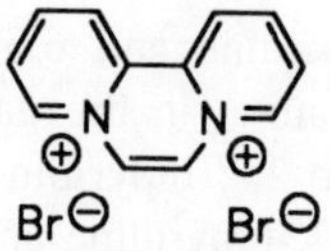

13 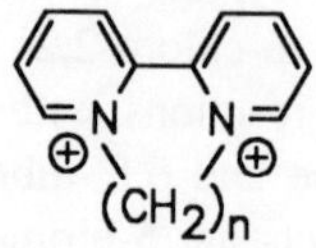 14 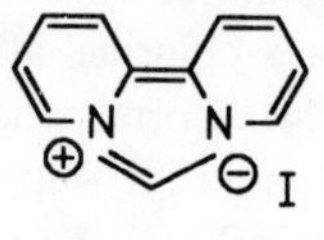 15 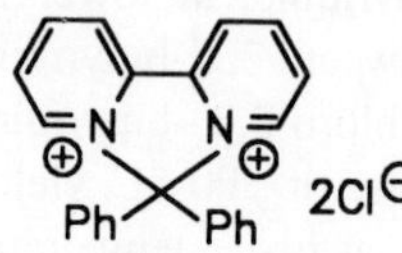 16

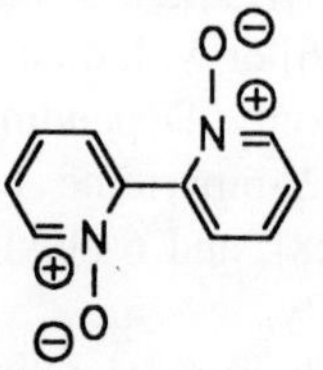

17 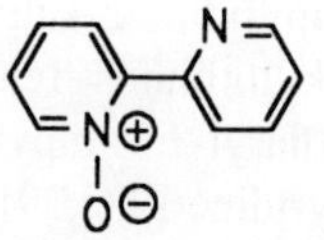 18

2.1.5 REACTIONS OF SUBSTITUTED 2,2′-BIPYRIDINES

The oxidation of methyl-substituted 2,2′-bipyridines with potassium permanganate [38] or selenium oxide [39] yields the corresponding carboxylic acids.

3,3′-Bis(hydroxymethyl)-2,2′-bipyridine was condensed with polyethylene glycol ditosylates to give crown ethers such as **19** [40]. In such ligands, the crown ether part will bind alkali and alkaline earth metal ions, whereas the bipyridine nitrogen atoms will bind transition metal cations; supramolecular structures may also be formed in the process (cf. Sections 2.2 and 2.3).

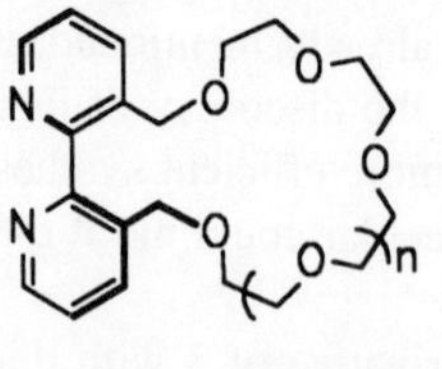

19 a: n = 0
b: n = 1
c: n = 2

2,2′-Bipyridine-6,6′-dicarboxylic acid dichloride reacts with long-chain diamines to form *ansa*-type rings, such as **20** [41].

On treatment with an excess of methyllithium, 2,2′-bipyridinium-3,3′-dicarboxylic acid yields the diol **21**, which then can be cyclized to **22** with hot sulphuric acid [42].

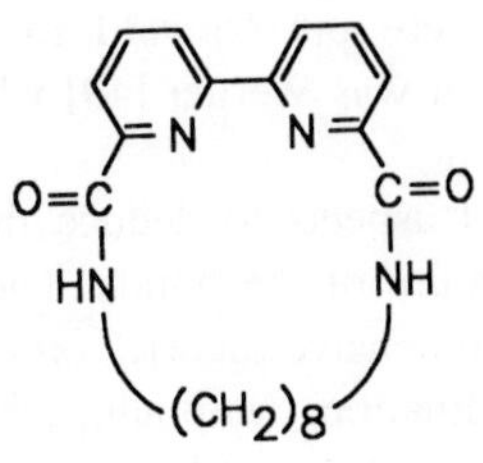

20

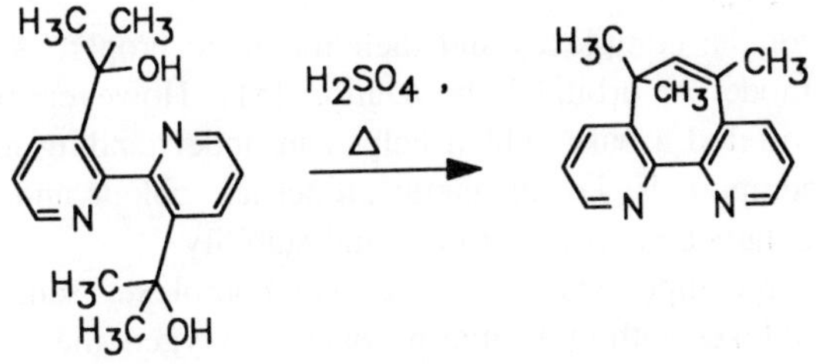

21

22

2.1.6 LIGANDS WITH 2,2′-BIPYRIDINE AS A DONOR CENTRE

2.1.6.1 History

The earliest description of a coordination compound dates back to 1704 and is due to Diesbach, who reported the preparation of Prussian blue from iron sulphate and potash in which organic nitrogen-containing substances had been boiled [43].

With the research done by Tassaert [44], who was working on cobalt–amine complexes, interest in the chemistry of complexes increased, as underlined by work done by Thénard [45] and Fremy [46]. Having measured the conductivity of the complexes and precipitated individual components, scientists still remained puzzled as to what the structure of these compounds might be.

Owing to the lack of knowledge about the nature of bonds in complexes, those were taken to be similar to those in carbon chains, which led to rather adventurous structure formulae [47].

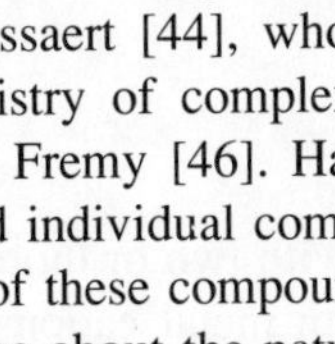

Figure 3. Historical representation of the cobalt complex $[Co(NH_3)_6]Cl_2$ according to Blomstrand [48]

Soon, a number of theories were put forward to explain the properties of coordination compounds [47], but it was Werner [49] who in 1893, gained a better insight.

Werner used phenomenological aspects to deduce the spatial arrangements in the complexes, rather than the nature of the bonds. The increasing knowledge in chemistry led to ever more comprehensive theories on complexes. Having applied Lewis's octahedron rule to coordination compounds, Sidgwick [50] was able to explain how the stability of a complex would change with different oxidation states of the central cation.

The geometry of the complexes and their magnetic properties were explained using Pauling's model of orbital hybridization [51]. However, only the ligand-field theory [52] offered a basis which helped in understanding most phenomena in coordination compounds, i.e. magnetic properties, colour and thermodynamic quantities, such as lattice energy, ion radii and stability.

With coordination compounds such as carbonyl complexes, sandwich and alkene complexes or complexes with molecular nitrogen or oxygen, the ligand-field theory fails; only the MO theory offers solutions [53]. At first, only inorganic coordination compounds were considered, but soon organic molecules also proved to be able to form stable complexes.

2.1.7 COMPLEXING ABILITY OF 2,2′-BIPYRIDINE AND ITS DERIVATIVES

Since the discovery of the usefulness of 2,2′-bipyridine as a probe for iron [1], many of its derivatives have been tested for their ability to complex heavy metals. Their capacity to build strong complexes stems from the chelating effect and the special structure of these ligands.

2.1.7.1 The chelating effect [54]

Ligands such as 2,2′-bipyridine with two or more donor centres which are suitable for coordination may cyclize with metal cations in such a way as to incorporate the ion into the ring. Five- and six-membered rings are the most stable. This form of complexation is called a chelate (Greek for crab-pincers), describing the way in which the ligand 'attacks' the metal ion. Chelating ligands are among the most important organic coordination ligands.

Although the two nickel complexes **23** and **24** possess the same number of analogous donor centres, their complexation constants (K) differ considerably. On going from **23** to **24**, a remarkable 'chelating effect' is observed, which can be explained thermodynamically, since K, the complexation constant, is related to the free enthalpy ΔG by

$$\ln K = -\Delta G/RT \qquad (1)$$

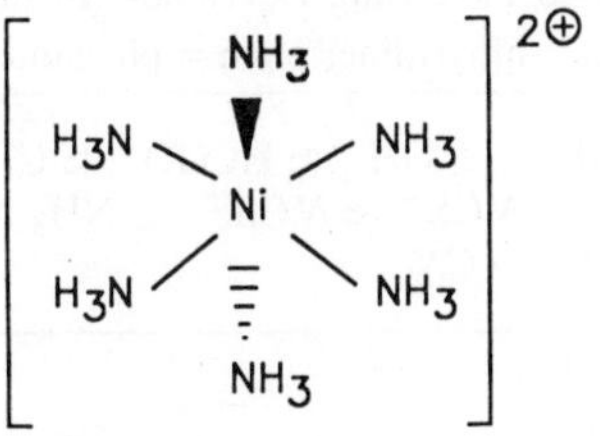

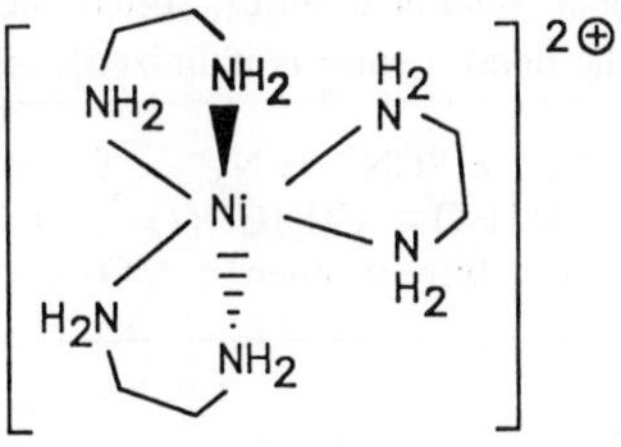

23: $K_1 = 10^{8.7}$ **24**: $K_2 = 10^{18.7}$

ΔG is related to both the enthalpy ΔH and the entropy ΔS:

$$\Delta G = \Delta H - T\Delta S \qquad (2)$$

(Gibbs–Helmholtz equation). Hence, we obtain

$$\ln K = -\Delta H/RT + \Delta S/R \qquad (3)$$

The entropy, a measure of the disorder in a system, grows after the coordination of a bidentate ligand, as two solvent molecules are displaced, and thus the number of independently mobile particles increases. From the point of view of thermodynamics, the chelating effect is one of entropy. Polydentate ligands produce a particularly strong chelating effect.

2.1.7.2 The nature of bonding

The description of coordinative bonding with the help of the MO theory is based on the assumption that both the central ion and ligand are able to form σ and π-bonds.

2,2′-Bipyridine (**1**) is both a σ-donor and a π-acceptor. The lone pair of electrons on the nitrogen is able to form a σ-bond with an unoccupied s-orbital of the metal ion and, in doing this, will also provide both electrons. Using the aromatic system of the bipyridine, occupied orbitals of the metal ion having the appropriate geometry, e.g. d-orbitals, can overlap with the unoccupied π^*-orbitals of the bipyridine.

Both parts of the bonding support each other: as the σ-donor bonding increases the electron density on the metal ion, its ability to form a π-bond with the ligand is enhanced, i.e. back-bonding occurs.

The diimine part, —N=C—C=N—, of bipyridine effects a delocalization of the electrons in the chelate ring. This produces a greater π-acceptor strength and a place higher up on the 'spectrochemical scale' [55] (Table 1), in addition to the appearance of 'low-spin' complexes ($^1A_1$), e.g. trichelates of Cr(II) and Fe(II) cations with 2,2′-bipyridine. In contrast, the corresponding trichelates of Mn(II), Co(II), V(II), Ni(II) and Cu(II) are all of the 'high-spin' type ($^5T_2$) [56].

Table 1. Spectrochemical series, sorted according to increasing D_q-values (in multivalent molecules, the donor centre is italicized); bipy = 2,2′-bipyridine; phen = phenanthroline

$I^- < Br^- < Cl^- \approx SCN^- \approx N_3^- < F^- < \mathit{O}C(NH_2)_2 < OH^- \approx HCOO^- \approx CH_3CO\mathit{O}^-$
$< (CO\mathit{O})_2^{2-} \approx H_2O \approx CH_2(CO\mathit{O})_2^{2-} < \mathit{N}CO^- < \mathit{N}CS^- \approx \mathit{N}CSe^- < NH_3 \approx C_5H_5N$
$< CH_2(\mathit{N}H_2)_2 <$ bipy $\approx$ phen $< \mathit{N}O_2^- < \mathit{C}NO^- < C\mathit{N}^-$

2.1.7.3 Basicity of bipyridine

Bipyridine is a weakly basic ligand and in aqueous solution it is usually monoprotonated (LH^+). The diprotonated molecule has also been detected, but only in acidic solution (UV measurements). The stability constants for LH^+ and LH_2^{2+} are $pK_1 = 4.3$ and $pK_2 = 0.2–0.5$ [57].

According to Baxendale and George [58], monoprotonated bipyridine, LH^+, is stabilized by hydrogen bonding; this would hinder the formation of the dication LH_2^{2+}. However, other workers, argue that in solution, and also in the solid hydrochloride, hydrogen bonding is of no real importance [59].

2.1.7.4 Stability constants of bipyridine complexes

Bipyridine acts as a soft σ-base and as a soft π-acid [60]. Precisely this is shown by the fact that it stabilizes soft metal ions, especially transition metals in low oxidation states, such as Fe(II) and Cu(I) [61]. The reason for this is the higher $d\pi–p\pi^*$ interaction between metal (M) and ligand (L) [62]. A summary of the stability constants K_1, K_2 and K_3 of bipyridine is given in Table 2 [63], where

Table 2. Stability constants of 2,2′-bipyridine complexes with bivalent metal cations and the stoichiometries M^{2+}bipy, $M^{2+}(bipy)_2$ and $M^{2+}(bipy)_3$

Cation	Log K_1	Log K_2	Log K_3
V^{2+}	4.9	4.7	3.9
Cr^{2+}	4.0	6.4	3.5
Mn^{2+}	2.6	2.0	1.0
Fe^{2+}	4.3	3.7	9.5
Co^{2+}	5.9	5.5	4.7
Ni^{2+}	7.1	6.8	6.3
Cu^{2+}	9.1	5.5	3.4
Zn^{2+}	5.2	4.4	3.8
Cd^{2+}	4.3	3.5	2.6
Hg^{2+}	9.6	7.1	2.8

K_1, K_2 and K_3 relate to the complexes with the stoechiometries $M^{2+}L$, $M^{2+}L_2$ and $M^{2+}L_3$, respectively.

2.1.8 SPECTROSCOPY OF BIPYRIDINE COMPLEXES

2.1.8.1 IR spectroscopy

Compared with the IR spectrum of the free ligand, the spectra of 2,2′-bipyridine complexes show a marked shift towards longer wavelengths, especially in the region between 1600 and 1000 cm^{-1} [64]. A bathochromic shift of the peaks at 995 and 759 cm^{-1} is also observed [65]. The region between 600 and 200 cm^{-1} ('far infrared' [66]) has also been investigated, as information about the metal–ligand bonding can be obtained from it.

2.1.8.2 UV spectroscopy

The free bipyridine base (**1**) absorbs at 280 nm (band I) and at 235 nm (band II). These are π–π^* transitions [67]. Kiss and Csaszar assumed that the n–π^* transitions are below the π–π^* transitions [68]. After complexation with metal ions, the UV absorption shifts to longer wavelengths [69]. The UV spectra of bipyridine complexes with metal ions in the same oxidation state, e.g. $Fe(bipy)_3{}^{2+}$ and $Zn(bipy)_3{}^{2+}$, are similar [70]. Monoprotonated bipyridine absorbs at 303 nm (band I) and at 243 nm (band II), while for the dication only one band at 290 nm is observed. The occurrence of colour in the complexes is due to an allowed t_{2g}–π^* transition [71].

2.1.8.3 ^1H NMR spectroscopy

The ^1H NMR spectrum of bipyridine is dependent on the solvent used [72]: in non-polar solvents the molecule adopts a transoid planar conformation (interplanar angle = 0°, see above), while in polar solvents a transoid skew conformation was found for mono- and diprotonated bipyridine (interplanar angle between 25° and 30°). The ^1H NMR spectra of the complexes show almost no difference, when compared with **1**, although the proton α to the nitrogen is shifted by approximately 1 ppm upfield. The shift depends on the length of the metal—nitrogen bond [72].

2.1.9 NEW DEVELOPMENTS

The bipyridines linked with crown ethers, e.g. **25**, synthesized by Rebek *et al.*, possess two different coordination sites: the crown ether cavity to bind alkali, alkaline earth and ammonium ions and the bipyridine moiety to bind transition metal cations [40]. This leads to 'allosteric effects:'

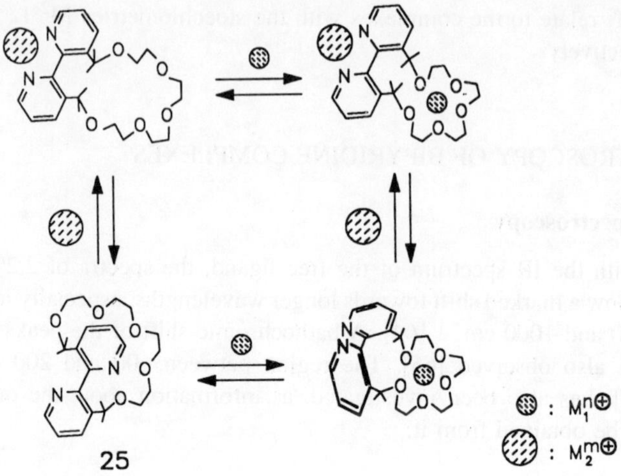

25

The complexation of bipyridine by transition metal cations forces the pyridyl rings into a planar conformation. This restricts the conformational mobility of the crown part, which in turn reduces its complexation ability (cf. Section 2.2). Another type of 2,2′-bipyridine crown ethers (**26**), which binds $CoCl_2$, was described by Newkome *et al*. [73].

The bipyridine cryptand **27** is a strong ligand for alkali metal ions [41]. In it, the crown ether moieties were later replaced with bipyridine and phenanthroline units [74]. Chambron and Sauvage successfully synthesized ligand **28**, which contains two coordination sites with different selectivities [75].

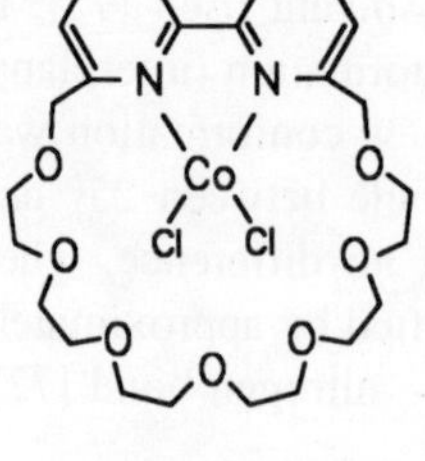

26 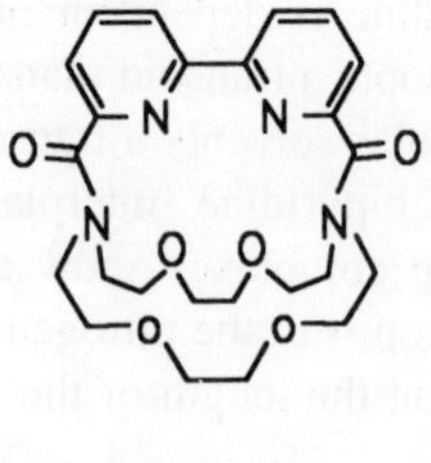 27 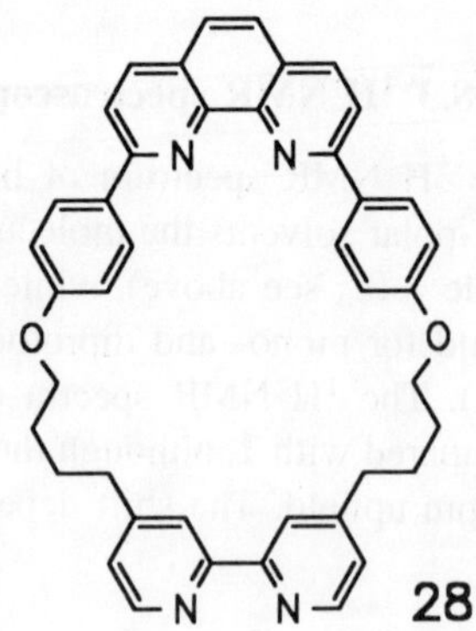 28

Multiple-site ligands of that type may exhibit allosteric properties (see above) if one side of the molecule is able to influence the complexation behaviour of the other side:

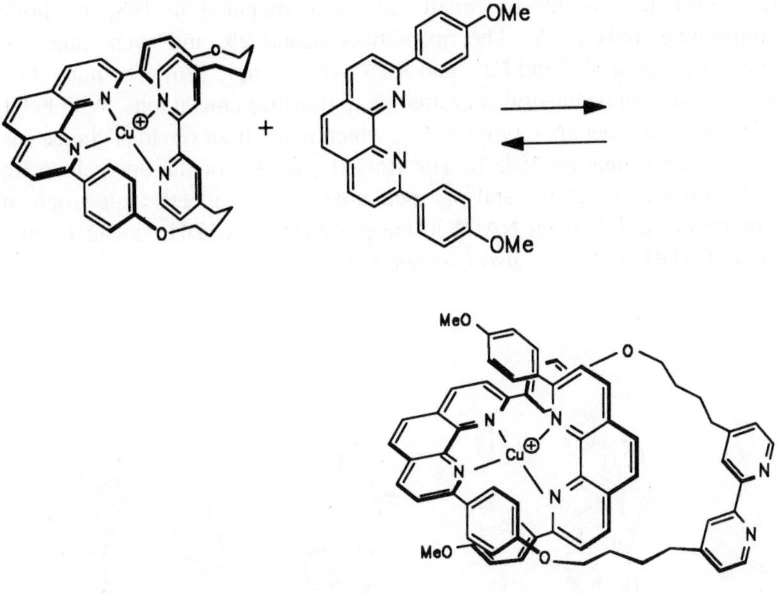

The binding of Ru(II) in an octahedral environment allows the complexation of Cu(I) in a tetrahedral arrangement:

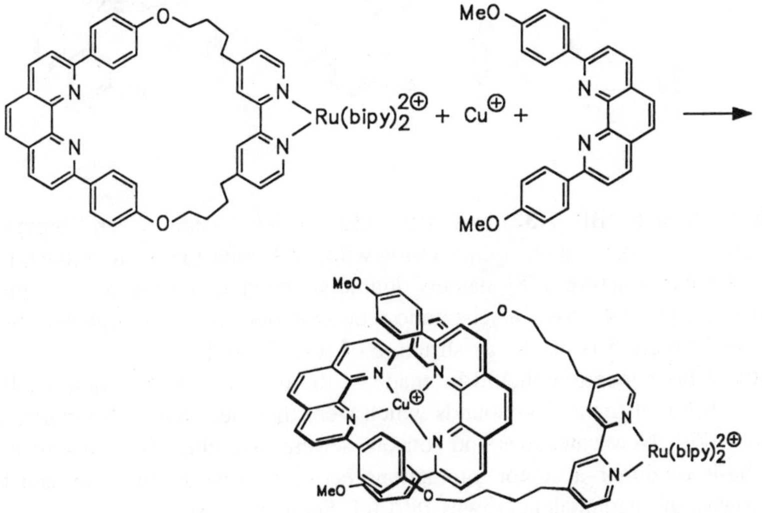

Moreover, it is possible to conduct electron-transfer studies on photo- and electroactive centres fixed within the molecule, studies that are important for the understanding of redox processes, including photosynthesis.

Large molecular cavities for small ions were prepared in 1987 by bridging three bipyridine units twice. The open-chain ligand **29**, and even more so the macrobicyclic ligand **30**, bind Fe^{2+} and Ru^{2+} very strongly [76]. The macrobicycle **30** is able to offer an octahedral coordination to transition metal ions. With Fe(II) an intensely red colour occurs (complex **31**), which made it an obvious choice for the colorimetric determination of iron. The Rh(III) complex of the open-chain ligand **29** can be used as a redox catalyst (mediator) for the selective electrochemical regeneration of NADH from NAD^+ in the presence of NADH-dependent enzyme systems (HLADH) [77] (see also Chapter 12).

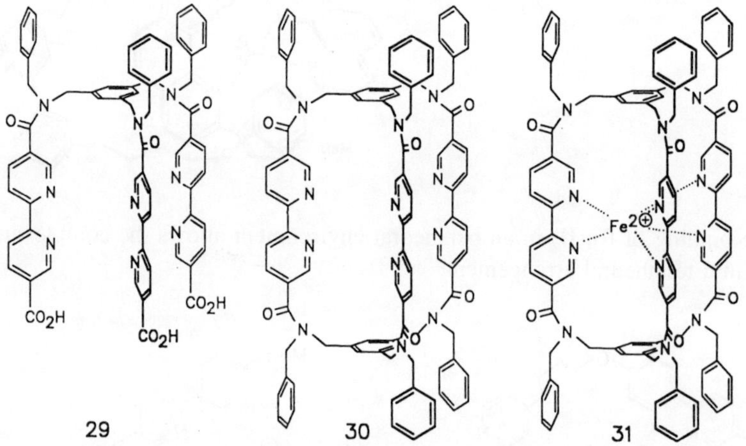

29 30 31

With the ligands 'BP_2' (**32**) and 'BP_3' (**33**), which contain several bipyridine units, Lehn *et al*. obtained on complexation with Cu(I) ions a very attractive type of 'supramolecular structure' [78], namely double-stranded, doubly helical complexes ('helicates'). Figure 4 shows a schematic representation of the arrangement of the ligands, and Figure 5 is the X-ray structure of $[Cu_3(BP_3)_2]^{3+}$.

Finally, a brief mention should be made of the use of certain [30]crown-10 as a hosts to different 'quat' compounds which were then detected with ion-selective electrodes [79]. New catenanes and rotaxanes were assembled by Stoddart *et al*. on the basis of donor–acceptor interactions between bipyridinium ions and both benzocrowns and naphthalenocrowns [80] (cf. Section 2.5.3).

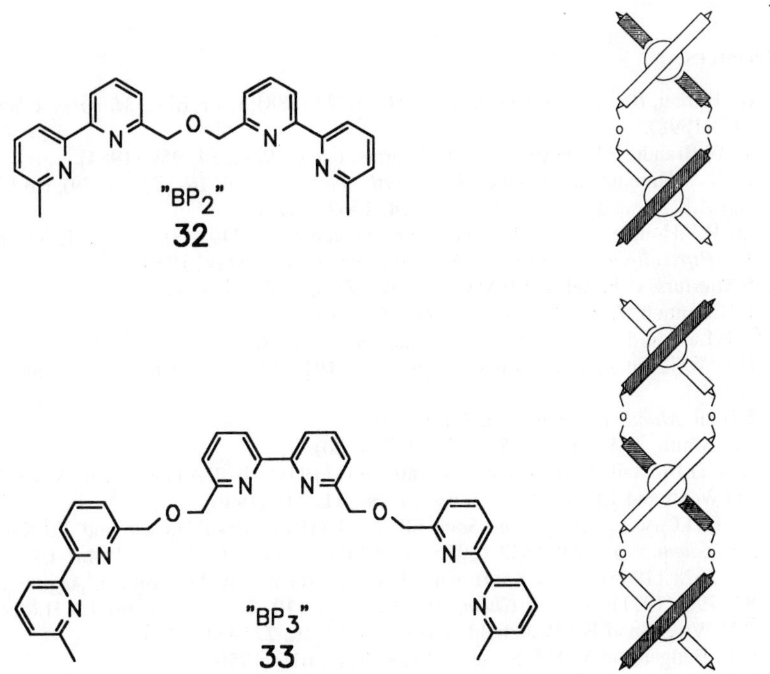

Figure 4. Schematic representation of the double-stranded 'helicates', which result from the complexation of Cu(I) with BP_2 and BP_3 (cf. Figure 5)

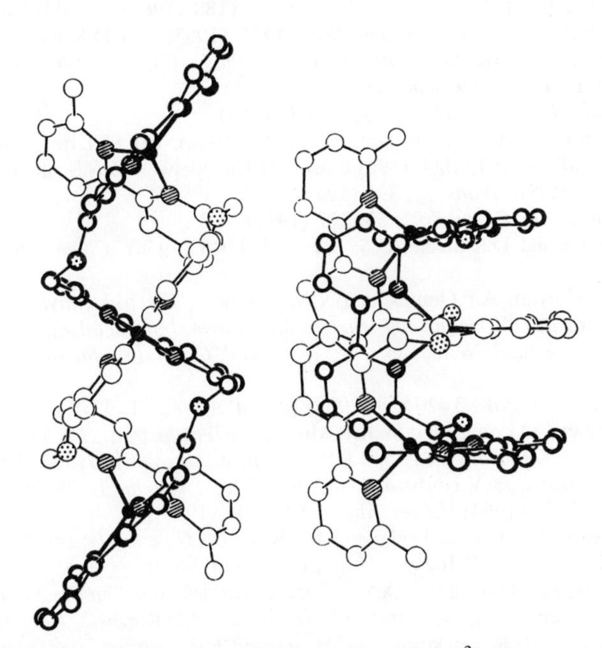

Figure 5. X-ray crystal structure of the complex $[Cu_3(BP_3)_2]^{3+}$. Two views sideways; one of the two helix strands is drawn in bold face. ● = Cu; ◍ = N; ◌ = O (cf. Figure 4)

References

1. (a) F.Blau, *Ber. Dtsch. Chem. Ges.*, **21**, 1077 (1888); (b) F.Blau, *Monatsh. Chem.*, **19**, 647 (1898).
2. W.W.Brandt, F.P.Dwyer and E.C.Gyartas, *Chem. Rev.*, **54**, 959 (1954).
3. (a) K.V.Thimann and S.Satler, *Proc. Natl. Acad. Sci. USA*, **76**, 2770 (1979); (b) J.W.Earl and R.K.Kennedy, *Phytochemistry*, **14**, 1507 (1975).
4. (a) R.F.Homber and T. E. Tomlinson, *J.Chem.Soc.*, 2498 (1960); (b) L.A.Summers, *The Bipyridinium Herbicides*, Academic Press, New York, 1980.
5. S.Mierturs, O.Kysel and P.Majek, *Chem. Zvesti*, **33**, 153 (1979).
6. L.G.Cannell, *J. Am. Chem. Soc.*, **94**, 6867 (1972).
7. K.E.Langford, *Electroplat. Met. Finish.*, **9**, 39 (1956).
8. H.Ishii, T.Tanaka and A.Yamada, *Jpn. Pat..*, **101**, 111 (1977) [*Chem. Abstr.*, **88**, 172078 (1978)].
9. F.Blau, *Monatsh. Chem.*, **10**, 375 (1889).
10. C.R.Smith, *J. Am. Chem. Soc.*, **52**, 397 (1930).
11. (a) F.H.Burstall, *J. Am. Chem. Soc.*, **60**, 1662 (1938); (b) T.A.Geissmann, M.I.Schlatter, I.D.Webb and J.D.Roberts, *J. Org. Chem.*, **11**, 741 (1946).
12. (a) F.H.Case, *J. Am. Chem. Soc.*, **68**, 2574 (1946); (b) F.H.Case and T.J.Kasper, *J. Am. Chem. Soc.*, **78**, 5842 (1956); (c)J.J.Porter and J.C.Murray *J. Am. Chem. Soc.*, **87**, 1628 (1965); (d) T.Kauffmann, E.Wienhofer and A.Woltermann, *Angew. Chem.*, **83**, 796 (1971); *Angew. Chem., Int. Ed. Engl.*, **10**, 741 (1971); (e) J.A.H.MacBride, P.M.Wright and B.J.Wakefield, *Tetrahedron Lett.*, **22**, 4545 (1981).
13. G.M.Badger and W.H.F.Sasse, *J. Chem. Soc.*, 616 (1956).
14. F.W.Cagle, *Acta Crystallogr.*, **1**, 158 (1948).
15. L.L.Merritt and E.C.Schroeder, *Acta Crystallogr.*, **9**, 801 (1956); K.Nakatsu, H.Yoshioka, M.Matsui, S.Koda and S.Ooi, *Acta Crystallogr.*, **28**, 24 (1972).
16. C.A.Goethals, *Recl. Trav. Chim. Pays-Bas.*, **54**, 299 (1935); (b) C.W.N.Cumper, R.F.A.Ginman and A.I.Vogel, *J. Chem. Soc.*, 1188 (1962); (c) P.H.Cureton, C.G.Le Fevre and R.I.Le Fevre, *J. Chem. Soc.*, 1736 (1963); (d) D.Kranbuhl, D.Klug and W.Vaughan, *Natl. Acad. Sci. Natl. Res. Counc. Publ.*, 1705, 22 (1969); (e) P.E.Fielding and R.I.W.Le Fevre, *J. Chem. Soc.*, 1811 (1951).
17. P.Krumholz, *J. Am. Chem. Soc.*, **75**, 2163 (1953).
18. (a) C.R.Smith, *J. Am. Chem. Soc.*, **50**, 1936 (1928); (b) M.T.Beck and M.Halmos, *Nature (London).*, **191**, 1090 (1961); (c) G.O.Khandelwal, G.A.Swan and R.B.Roy, *J. Chem. Soc., Perkin Trans. 1.*, 891 (1974).
19. C.R.Smith, *J. Am. Chem. Soc.*, **53**, 277 (1931).
20. P.N.Rylander and D.R.Steele, *U.S. Pat.*, 3 387 048 (1968) [*Chem. Abstr.*, **69**, 26968 (1968)].
21. (a) Y.N.Forostyan, A.P.Oeilnik and V.M.Artemora, *Elektrokhimiya.*, **7**, 715 (1971); (b) H.Erhard and W.Jaenicke, *J. Electroanal. Chem. Interfacial Electrochem.*, **81**, 79 (1977); (c) H.Erhard, W.Jaenicke, *J. Electroanal. Chem. Interfacial Electrochem.*, **81**, 89 (1977).
22. W.H.Taplin, *U.S. Pat.*, 3 420 833 (1969) [*Chem. Abstr.*, **71**, 3279 (1969)].
23. R.D.Chambers, O.Lomas and W.K.R.Musgrave, *Tetrahedron.*, **24**, 5633 (1968).
24. S.H.Ruetman, *U.S. Pat.*, 3 819 538 (1974) [*Chem. Abstr.*, **81**, 91363 (1974)].
25. O.S.Otroshchenko, Y.V.Kurbatov and A.S.Sadykov, *Nauchn. Tr. Tashk. Gos. Univ. V.I. Lenina.*, **263**, 27 (1964) [*Chem. Abstr.*, **63**, 4248 (1965)].
26. Y.V.Kurbatov, O.S.Otroshchenko and A.S.Sadykov, *Nauchn. Tr. Tashk. Gos. Univ. V.I. Lenina.*, **263**, 36 (1964) [*Chem. Abstr.*, **63**, 8309 (1965)].
27. N.H.Pirzada, P.M.Pojer and L.A.Summers, *Z. Naturforsch. Teil B.*, **31**, 115 (1976).
28. (a) T.A.Geissman, M.I.Schlatter, I.D.Webb and J.D.Roberts, *J. Org. Chem.*, **11**, 741 (1946); (b) G.R.Newkome, D.C.Hager and F.R.Fronczek, *J. Chem. Soc., Chem.*

Commun., 858 (1981); (c) R.F.Knott and J.G.Breckenridge, *Can. J. Chem*., **32**, 512 (1954).
29. T.Kaufmann, J.Koenig and A.Woltermann, *Chem. Ber*., **109**, 3864 (1976).
30. F.H.Westheimer and O.T.Benjey, *J. Am. Chem. Soc*., **78**, 5309 (1956).
31. (a) P.Krumholz, *J. Am. Chem. Soc*., **73**, 3487 (1951); (b) S.Hünig, J.Gross and W.Schenk, *Liebigs Ann. Chem*., 324 (1973); (c) F.G.Mann and J.Watson, *J. Org. Chem*., **13**, 502 (1948).
32. J.F.Cairns and J.A.Corran, *Ger. Pat*., 1 801 365 (1969) [*Chem. Abstr*., **71**, 81424 (1969)].
33. I.C.Calder and W.H.F.Sass, *Tetrahedron Lett*., 1465 (1965).
34. F.D.Popp and D.K.Chesney, *J. Heterocycl. Chem*., **9**, 1165 (1972).
35. I.C.Calder, T.M.Spotswood and W.H.F.Sasse, *Tetrahedron Lett*., 95 (1963).
36. I.C.Calder and W.H.F.Sasse, *Aust. J. Chem*., **18**, 1819 (1965).
37. (a) J.Haginiwa, *J. Pharm. Soc. Jpn*., **75**, 733 (1955); (b) I.Murase, *Nippon Kagaku Zasshi*., **77**, 682 (1956); (c) P.G.Simpson, A.Vinciguerra and J.V.Quagliano, *Inorg. Chem*., **2**, 282 (1963); (d) N.S.Bazhenova, Y.C. Kurbatov, O.S. Otroshchenko and A.S. Sadykov, *Tr. Samark. Gos. Univ. Alishera Navoi*., **206**, 226 (1972) [*Chem. Abstr*., **80**, 95678 (1974); (e) I.Antonini, F.Claudi, G.Cristalli, P.Franchetti, M.Grifantini and S.Martelli, *J. Med. Chem*., **24**, 1181 (1981).
38. C.P.Whittle, *J. Heterocycl. Chem*., **14**, 191 (1977).
39. (a) F.H.Burstall, *J. Chem. Soc*., 1662 (1938); (b) M.Pitman and P.W.Sadler, *J. Chem. Soc*., 759 (1961).
40. (a) J.Rebek, J.F.Trend, R.V.Wattley and S.Chakravorti, *J. Am. Chem. Soc*., **101**, 4333 (1979); (b) J.Rebek and R.V.Wattley, *J. Am. Chem. Soc*., **102**, 4853 (1980).
41. E.Buhleier, W.Wehner and F.Vögtle, *Chem. Ber*., **111**, 200 (1978).
42. J.Rebek and J.E.Trend, *J. Am. Chem. Soc*., **100**, 4315 (1978).
43. *Gmelins Handbuch der anorganischen Chemie*, 8.Aufl., Band Eisen **59**, Teil B, Verlag Chemie, Weinheim, 1932, p.671.
44. B.M.Tassaert, *Ann. Chim*., **28**, 92 (1799).
45. L.J.Thénard, *Ann. Chim*., **42**, 210 (1803).
46. E.Fremy, *Liebigs Ann. Chem*., **83**, 227 (1853).
47. S.M.Jörgensen, *J. Prakt. Chem*., [2], **39**, 1 (1889).
48. C.W.Blomstrand, *Die Chemie der Jetztzeit vom Standpunkt der elektrochemischen Auffassung aus Berzelius Lehre entwickelt*, K.Winters Universitätsbuchhandlung, Heidelberg, 1869, p.280.
49. A.Werner, *Z. Anorg. Allg. Chem*., **3**, 267 (1893).
50. N.V.Sidgwick, *J. Chem. Soc*., **123**, 725 (1923).
51. L.Pauling, *J. Am. Chem. Soc*., **53**, 1367 (1931).
52. H.Hartmann and H.L.Schläfer, *Angew. Chem*., **66**, 768 (1954).
53. H.B.Gray, *Coord. Chem. Rev*., **1**, 2 (1966).
54. F.Kober, *Grundlagen der Komplexchemie*., Salle, Sauerländer, Frankfurt am Main, Aarau, 1979.
55. E.König, *Coord. Chem. Rev*., **3**, 471 (1968).
56. (a) F.H.Burstall and R.S.Nyholm, *J. Chem. Soc*., 3570 (1952); (b) F.Hein and S.Herzog, *Z. Anorg. Allg. Chem*., **267**, 337 (1952).
57. G.F.Condike and A.R.Martell, *J. Inorg. Nucl. Chem*., **31**, 2455 (1969).
58. J.H.Baxendale and P.George, *Trans. Faraday. Soc*., **46**, 55 (1950).
59. (a) I.R.Beattle and M.Webster, *J. Phys. Chem*., **66**, 115 (1962); (b) F.H.Westheimer and O.T.Benley, *J. Am. Chem. Soc*., **78**, 5309 (1956).
60. (a) R.G.Pearson, *J. Am. Chem. Soc*., **85**, 3533 (1963); (b) B.Saville, *Angew. Chem*., **79**, 966 (1967); (c) R.G.Pearson, *Science*., **151**, 172 (1966).
61. F.Kröhnke, *Synthesis*., 1 (1976).
62. W.A.E.McBryde, *A Critical Review of Equilibrium Data for Proton and Metal*

Complexes of 1,10-Phenanthroline, 2,2′-Bipyridyl and Related Compounds., IUPAC Chem. Data Ser. No. 17, Pergamon Press, Oxford, 1978.
63. W.R.McWhinnie and J.D.Miller, *Adv. Inorg. Radiochem*., **12**, 135 (1969); see also G.Nord, *Comments Inorg. Chem*., **4**, 193 (1985).
64. P.Besler and A.V.Zelewsky, *Helv. Chim. Acta*., **63**, 1675 (1980).
65. (a) B.Martin, W.R.McWhinnie and G.M.Waind, *J. Inorg. Nucl. Chem*., **23**, 207 (1961); (b) S.P.Sinka, *Spectrochim. Acta*., **20**, 879 (1964).
66. R.G.Inskeep, *J. Inorg. Nucl. Chem*., **24**, 763 (1962).
67. (a) G.Favini and A.Gamba, *Gazz. Chim. Ital*., **96**, 391 (1966); (b) L.Gil, E.Moraga and S.Bunel, *Mol. Phys*., **12**, 333 (1967).
68. A.Kiss and J.Csaszar, *Acta Chim. Acad. Sci. Hung*., **38**, 405, 421 (1963).
69. (a) K.Stone, P.Krumholtz and H.Stammreich, *J. Am. Chem. Soc*., **77**, 777 (1955); (b) B.Martin and G.M.Waind, *J. Chem. Soc*., 4284 (1958).
70. H.L.Schläpfer, *Z. Phys. Chem*., **8**, 373 (1956).
71. S.M.Al, F.M.Brewer, J.Chadwick and G.Garton, *J. Inorg. Nucl. Chem*., **9**, 124 (1959).
72. S.Castellano, H.Günther and S.Ebersole, *J. Phys. Chem*., **69**, 4166 (1965).
73. G.R.Newkome, J.K.Kohli and F.Fronczek, *J. Chem. Soc. Chem. Commun*., 9 (1980).
74. B.Alpha, J.-M.Lehn and G.Mathis, *Angew. Chem*., **99**, 259 (1987); *Angew. Chem., Int. Ed. Engl*., **26**, 266 (1987).
75. J.-C.Chambron and J.-P.Sauvage, *Tetrahedron Lett*., **27**, 865 (1986).
76. S.Grammenundi and F.Vögtle, *Angew. Chem*., **98**, 1119 (1986); *Angew. Chem., Int. Ed. Engl*., **25**, 1122 (1986); for the photophysics and photochemistry of luminescent ruthenium complexes of **30**, see L.De Cola, F.Barigelletti, V.Balzani, P.Belser, A.von Zelewsky, F.Vögtle, F.Ebmeyer and S.Grammenudi, *J. Am. Chem. Soc*., **100**, 7210 (1988); P.Belser, *Chimia*, **44**, 226 (1990); V.Balzani, F.Barigelletti and L.De Cola, *Top. Curr. Chem*., **158**, 31 (1990).
77. S.Grammenudi, F.Vögtle, M.Franke and E.Steckhan, *J. Inclus. Phenom*., **5**, 695 (1987).
78. J.-M.Lehn, A.Rigault, J.Siegel, J.Harrowfield, B.Chevrier and D.Moras, *Proc. Natl. Acad. Sci. USA*., **84**, 2565 (1987); J.-M.Lehn and A.Rigault, *Angew. Chem*., **100**, 1121 (1988); *Angew. Chem., Int. Ed. Engl*., **27**, 1095 (1988).
79. J.F.Stoddart *et al*., *Tetrahedron Lett*., **29**, 1575 (1988); and references quoted therein.
80. J.-Y.Ortholand, A.M.Z.Slavin, N.Spencer, J.F.Stoddart and D.J.Williams, *Angew. Chem., Int. Ed. Engl*., **28**, 1394 1396 (1989); *Angew. Chem., Adv. Mat*., **28**, 1405 (1989).

2.2 CROWN ETHERS, CRYPTANDS, PODANDS AND SPHERANDS

2.2.1 CROWN ETHERS AND ANALOGUES: CRYPTANDS, PODANDS AND SPHERANDS

The discovery of the crown ethers, approximately 20 years ago [1], has given fundamentally new impulses to chemistry in general and to the chemistry of complexes with organic ligands, especially neutral ones, in particular. A considerable number of original papers, review articles and monographs have been published in this field [2–95]. Research on crown ether-type compounds is not limited to preparative organic chemistry alone but is, as demonstrated by numerous sections in this book, an interdisciplinary task.

The first crown ether was a 'chance discovery': The industrial chemist C.J.Pedersen (DuPont, Delaware, USA; Nobel Prize 1987) wanted to prepare bisphenol **3** from mono-protected catechol **1** and bis(2-chloroethyl) ether **2** (Figure 1). As a slightly impure sample of **1** was used (it contained some unprotected catechol), a very small amount of hexaether **4** (0.4%!) was also obtained [16].

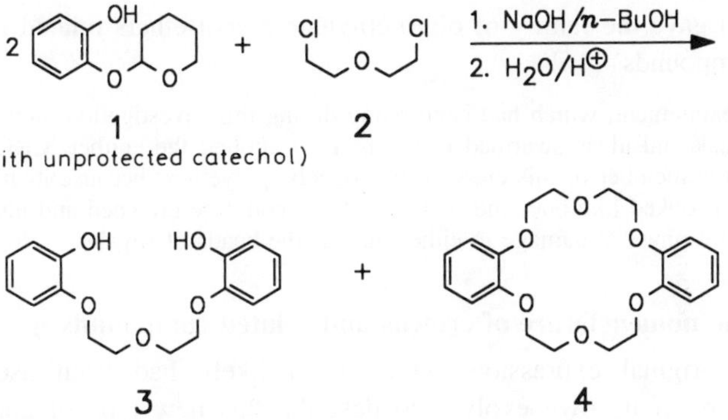

2 OH O O + Cl O Cl 1. NaOH/n–BuOH 2. $H_2O/H^{\oplus}$

1 (with unprotected catechol) 2

OH HO O O O + O O O O O O

3 4

Figure 1. Discovery of crown ethers

Its good crystallizability, and even more so its unusual solubility behaviour, aroused Pedersen's interest: the hexaether **4** was only slightly soluble in methanol, but on addition of sodium salts, it dissolved surprisingly easily. This, and the fact that with it potassium permanganate could be dissolved in benzene or chloroform (giving a purple colour), prompted him to make this, at the time, bold statement [16]:

> It seemed clear to me now, that the sodium (potassium) ion had fallen into the hole at the center of the molecule.

A little earlier, it had been discovered that certain ionophores [e.g. nonactin (**5**), valinomycin (**6**); Figure 2] are able to incorporate alkali metals (Na^+, K^+) into their interior, and thus to transport them into biological systems, e.g. through membranes [96]. This is of particular interest in relation to the stimulation of neural membranes [97]. The analogy to **4** is obvious.

5 6

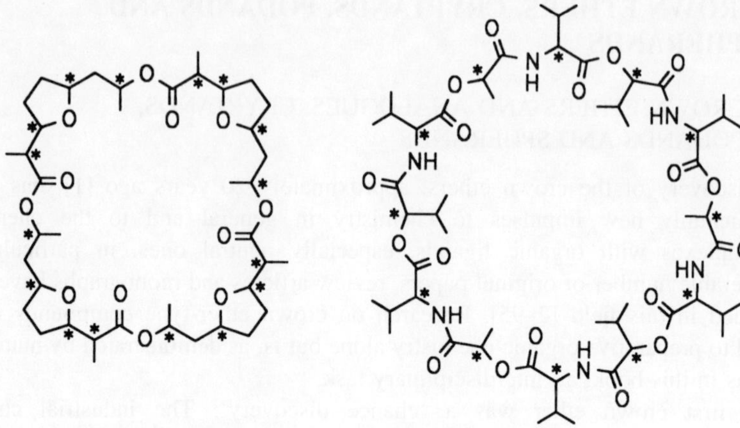

Figure 2. The natural-product ionophores nonactin (**5**) and valinomycin (**6**) (centres of chirality are marked with asterisks

Pedersen gave the family of oligo-ethylene glycol ethers related to 4 the name 'crown compounds' [16]:

> My excitement, which had been rising during this investigation, now reached its peak and ideas swarmed in my brain. I applied the epithet 'CROWN' to the first member of this class of macrocyclic polyethers because its molecular model looked like one, and with it, cations could be crowned and uncrowned without physical damage to either, just as the heads of royalty.

2.2.1.1 The nomenclature of crowns and related compounds

From the original expression, which most likely had been used in jest, a nomenclature of its own evolved to describe this new type of compound [29]. Figure 3 gives a few examples.

The ring size of the crown is given in square brackets, e.g. '[18]' as in **4**. The family name 'crown' then follows. Finally, the number of donor sites is given, in this particular case (**4**), the six oxygen atoms. Additional sub-

7	4	8
[18]Crown–6	Dibenzo[18]crown–6	Dicyclohexano[18]crown–6
18C6	DB18C6	DCH18C6

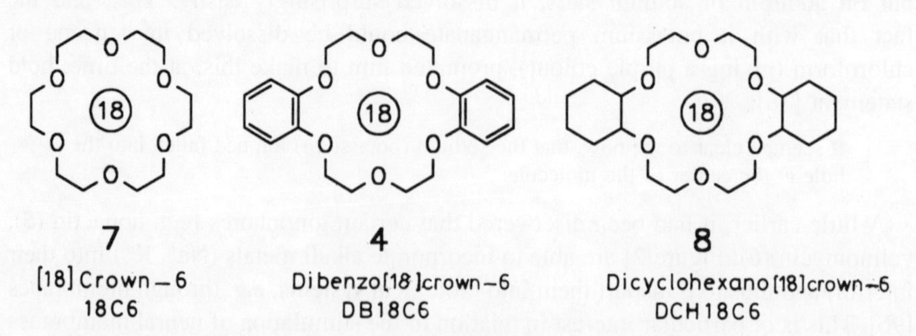

Figure 3. Classical crown ethers (nomenclature)

stituents or fused rings are affixed at the beginning of the whole name, e.g. 'dibenzo' (**4**) or 'dicyclohexano' (**8**). The advantages of this nomenclature become easily apparent when 'dibenzo[18]crown-6' for **4** is compared with its full IUPAC name, 2,5,8,15,18,21-hexaoxatricyclo[20.4.0.$0^{9,14}$]hexacosa-1(22),9,11,13,-23,25-hexaene. Often, acronyms such as '18C6' or 'DB18C6' are also used.

Strictly, Pedersen's crown nomenclature is unambiguous only together with a structural formula, or if certain moieties are taken for granted, e.g. the exclusive use of ethylene glycol units. For this reason, a new classification and nomenclature for neutral organic ligands of any structure has been proposed (e.g. $18<O_6 2_6$coronand-6> for **4**) [98]. However, for the sake of simplicity we shall use Pedersen's nomenclature. The crown ethers shown in Figure 3 are the most popular and most widely used; they are also among the first to be discovered [1].

Almost immediately, a correlation between structure and properties was sought, and all parameters in the classical crown ether ring were varied: size, molecular flexibility and type and number of donor sites in the ring. Bridges were also added and, on other occasions, the ring was cleaved to an open chain (see below). Today, approximately 5000 different crown compounds are known [5]. From a topological point of view, neutral ligands of this type can be classified into three groups (Figure 4) [98]: open-chain compounds, known as *podands*, monocyclic systems, called *coronands*, and oligocyclic (spherical) ligands with the name *cryptands*. For historical reasons, those coronands which contain ether oxygens exclusively have retained the name *crown ether*. Figures 5–12, give a good idea of the great variety of possible structures.

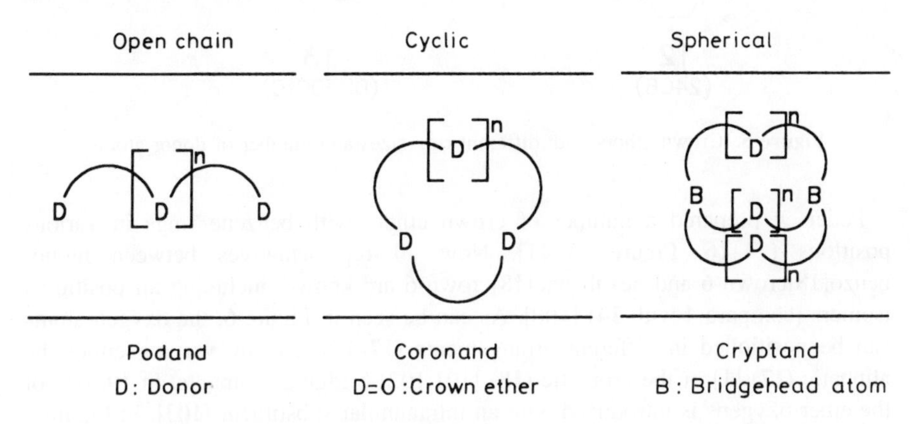

Figure 4. Topology and classification of organic neutral ligands

2.2.1.2 Important crown ethers and neutral ligands

2.2.1.2.1 Crown compounds

Apart from the numerous ether oxygens, condensed rings are the hallmark of classical crown ethers (Figure 3). Any variation of the ring size usually also changes the number of donor atoms.

In Figure 5, crown ethers with ring sizes varying between 12 and 30 and with four to ten donor atoms are shown. [12]Crown-4 (**9**) is the smallest crown ether used; the even smaller crown [9]crown-3 has also been prepared (with some difficulty), but is of no use as its cavity is too small for any strong complexation. Even though dibenzo[30]crown-10 is the largest shown here, it by no means represents the limit in large sizes [5].

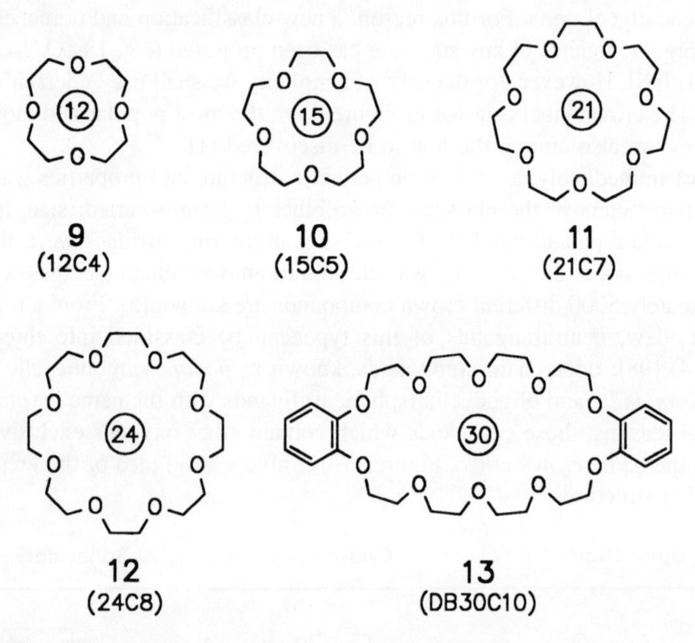

Figure 5. Crown ethers with different ring sizes and number of donor atoms

Pedersen prepared a number of crown ethers with benzene rings in various positions (14–16, Figure 6) [1]. Now all representatives between monobenzo[18]crown-6 and hexabenzo[18]crown-6 are known, including all positional isomers (compare **4** with **14**) [100]. As can be seen in Figure 6, the oxygen atoms can be positioned in different arrangements (**17–19**), e.g. by varying either the aliphatic (**17**) [1] or the aromatic (**18**) [101,102] bridging segments. In **19**, one of the ether oxygens is introduced with an intraannular substituent [103]. In Figure 7 the chemical nature of the donor sites is changed.

Compound **20** is a crown compound in which the ether oxygens have been replaced with sulphur atoms. In other 'thiacoronands', only individual positions may have been replaced [24]. Correspondingly, in 'azacoronands' the oxygens have been replaced with nitrogen (**21**, **22**) [75,104]. Systems in which oxygen, sulphur and nitrogen are all present in the same ring are also possible (**23**) [20,24]. There are also crowns that contain phosphorus (**24**) [4] and arsenic atoms (**25**) [105].

Heteroaromatic moieties may serve as donor sites as well: the furano-, pyridino- and thiopheno-crowns **26–28** (Figure 8) are examples of this type [41,77].

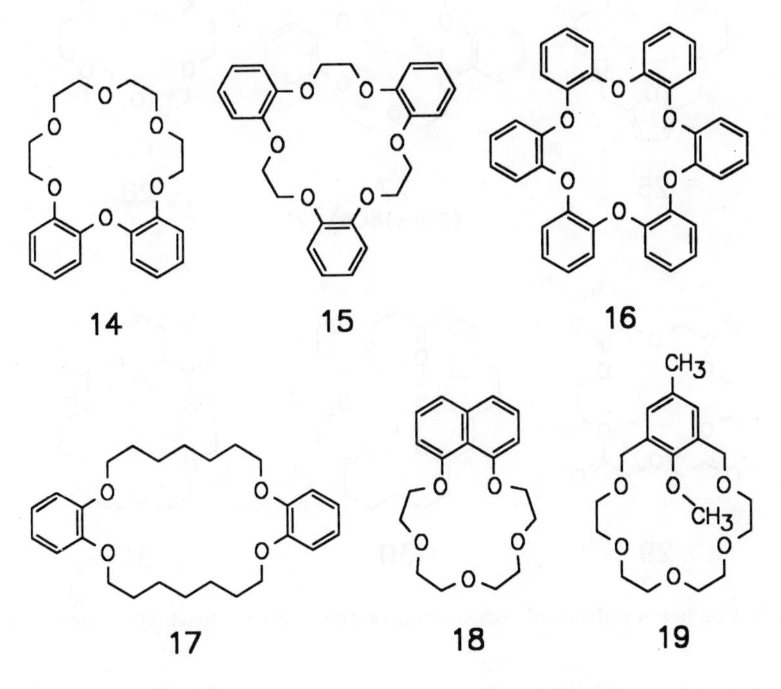

Figure 6. Crown ethers with varying ring flexibilities and layout of donor sites

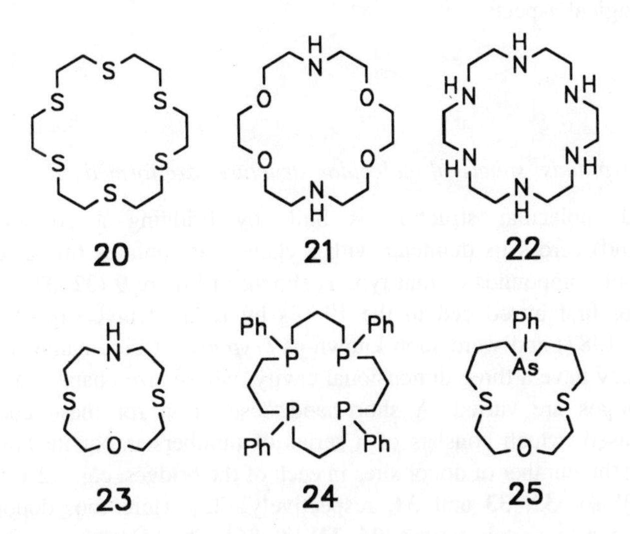

Figure 7. Coronands containing S, N, P and As (hetero-crowns)

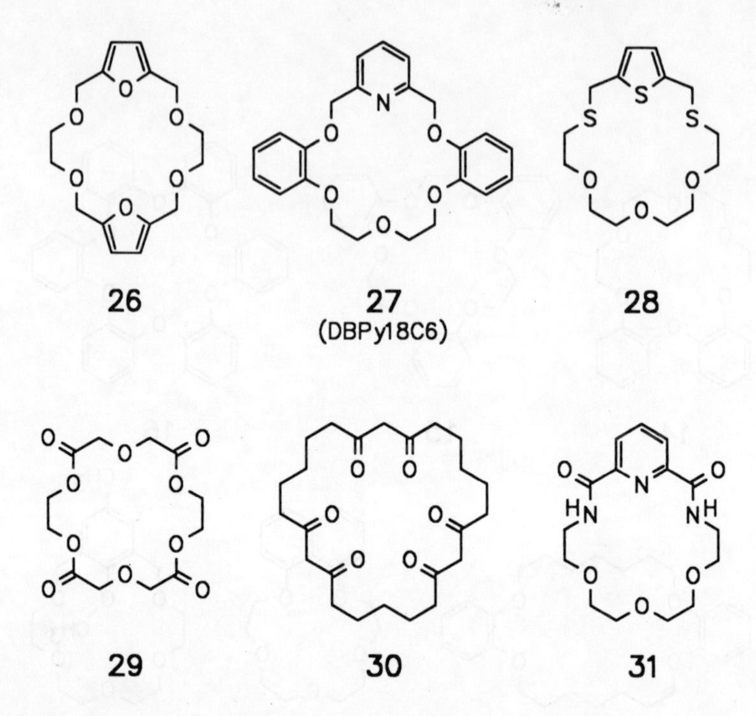

Figure 8. Coronands with heteroaromatic structural elements and functional groups as donor sites

Finally, functional groups such as esters [52], amides and ketones can be incorporated into crown rings to serve as donor sites (**29-31**) [5]. Let us consider a few topological aspects.

2.2.1.2.2 Cryptands: spherical molecular structures are formed

A spherical molecular structure is built by bridging a coronand (e.g. a diazacoronand) across its diameter with a chain that contains further donor sites. A selection of compounds of that type is shown in Figure 9 (**32–37**).

They were first introduced in the 1970's by Lehn (University of Strasbourg; Nobel Prize 1987) and were soon known as *cryptands* (trade name Kryptofix®) [47,106]. They have a three-dimensional cavity, whose size changes as the lengths of their bridges are varied. A short-hand description for these compounds is commonly used, which consists of a series of numbers in square brackets, each representing the number of donor sites in each of the bridges, e.g. '[2.1.1]', '[2.2.1]' and '[2.2.2]' for **32**, **33** and **34**, respectively [22]. Here too, donor sites and substituents can be widely varied (**35–37**) [48,86]. The bridges may also originate on carbon atoms [107,108]. Additional bridges lead to further ligand topologies, such as the cylindrical cryptands (e.g. **38**) and the 'soccerball cryptand' **39** [40].

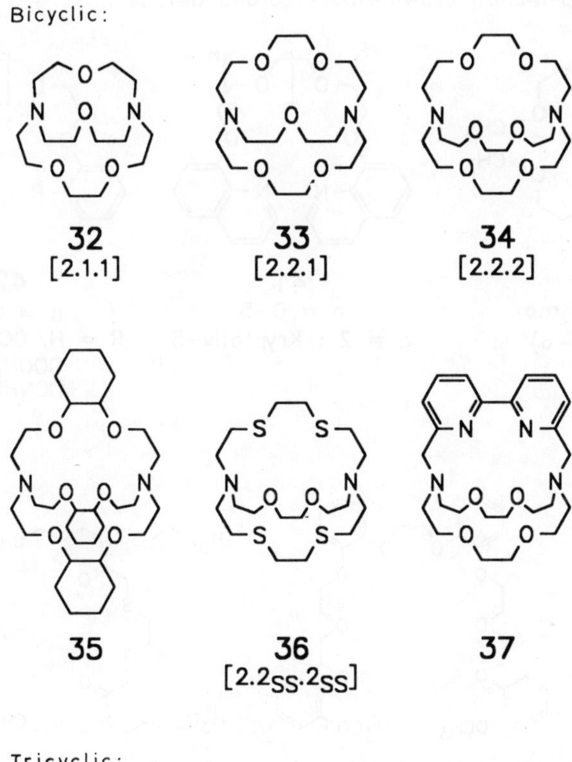

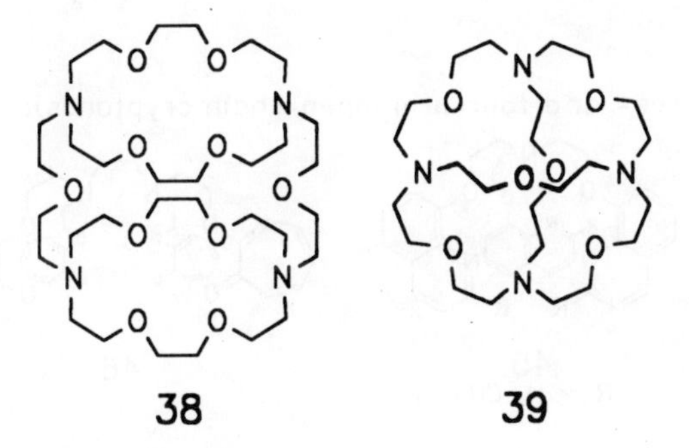

Figure 9. Cryptands

2.2.1.2.3 Podands

In contrast to most of what has been described so far, podands are distinguished by their notable lack of ring and bridge systems (Figure 10). In principle, this type of compound had been known for a long time, e.g. in the guise of polyethylene glycols [109] [cf. pentaethylene glycol (**40**)]. However, it was the crown ether chemistry that led to the synthesis of specific podands, especially those with terminal functionalities (**41**–**43**) [60] and pronounced lipophilicity (**44**) [110].

Linear (open-chain crown ethers/coronands):

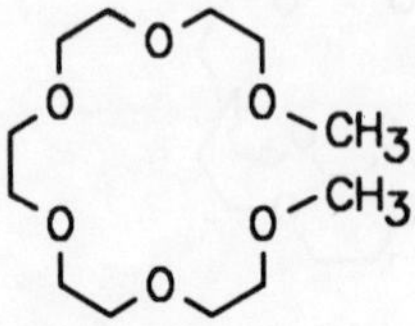

40
Pentaglyme
(Glyme–6)

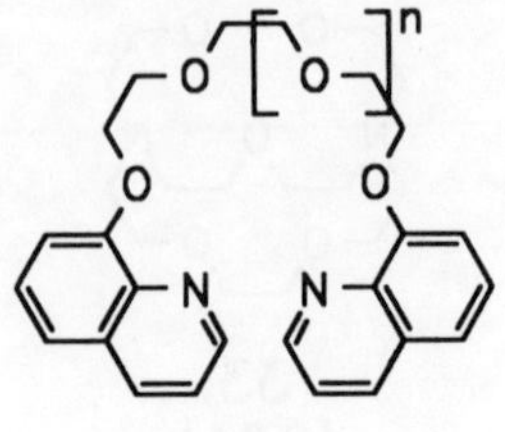

41
n = 0–5
n = 2 : Kryptofix–5

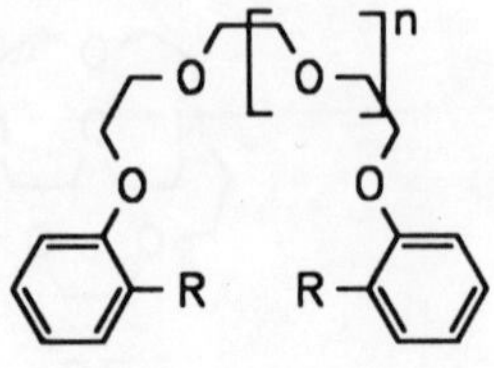

42
n = 0–2
R = H, OCH_3, NO_2, COOH, COOEt, CONHR, NHCOR

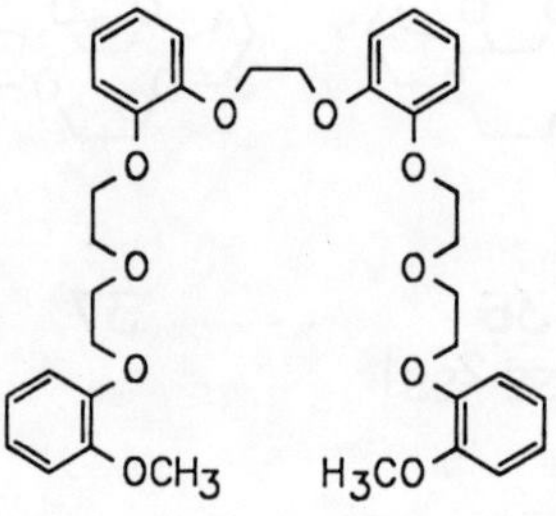

43

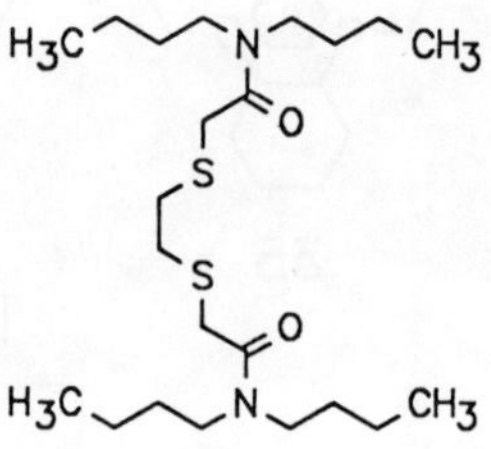

44

Three- and four-arm (open-chain cryptands):

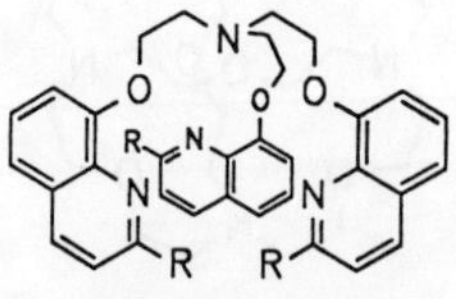

45
R = H, CH_3

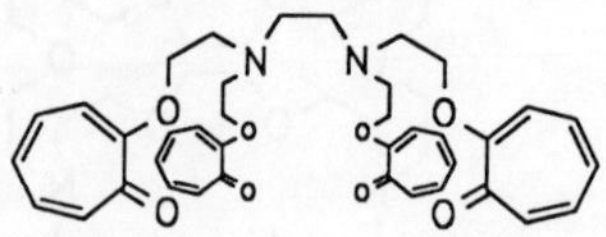

46

Figure 10. Some podands

Podand **41**, which ends with quinoline groups (n = 2, trade name Kryptofix-5®), is one of the most widely used crown compounds [111].

With podands too, topologically homologous series are possible, e.g. the so-called 'open-chain' cryptands with three or four arms (**45**, **46**) [70]. Because ligands with multiple arms, such as those shown in Figure 11 (**47** [112], **48** [113]), resemble squids, they are called 'octopus' molecules (and also 'tentacle' molecules) [60]. Topological hybrids, which contain a crown ether ring with a podand-arm attached (**49**–**51**), are casually called 'lariat ethers' ('lasso' ethers) [114].

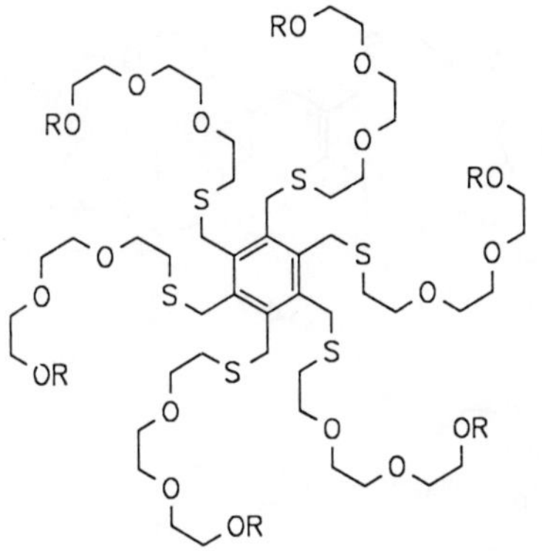

Figure 11. Octopus molecules and lariat ethers

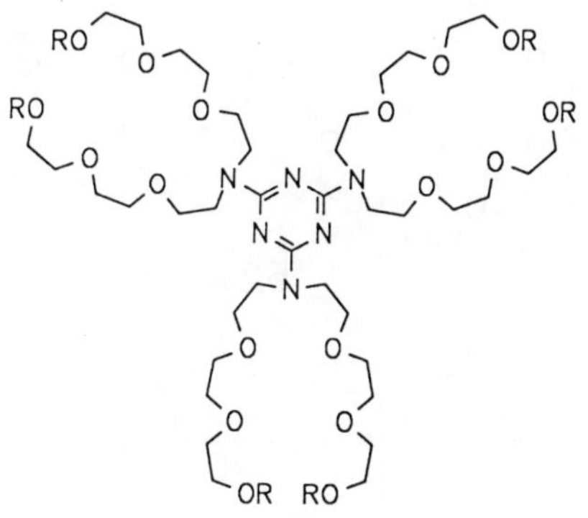

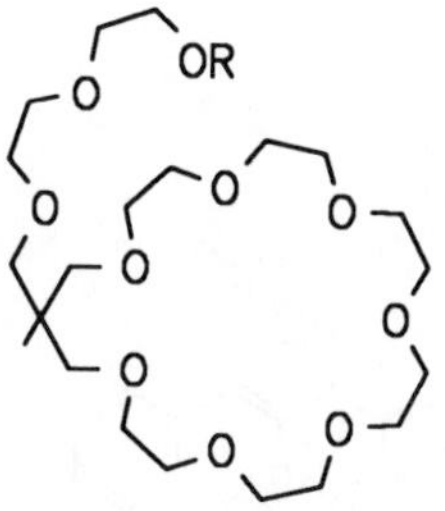

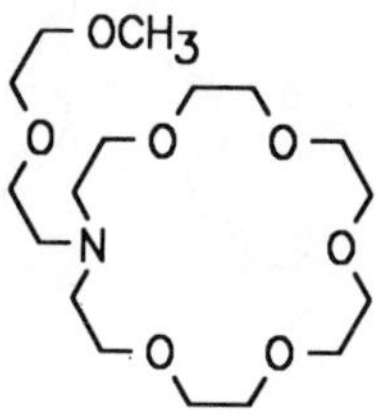

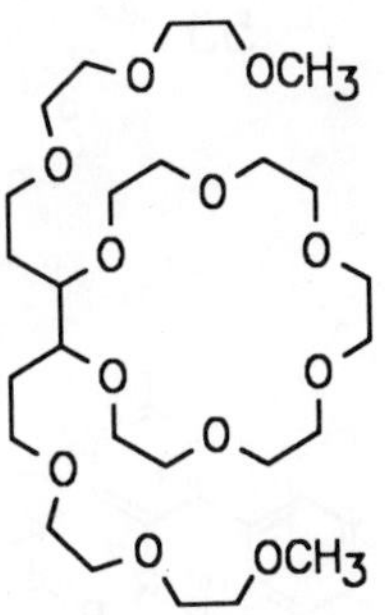

In Figure 12, multiple crowns (**52–54**) are shown, which consist of an assembly of the same (**53**) or different crown ethers or coronands (**52**, **54**) [115]. The spherical hexaether **53** [116] is aesthetically very attractive.

2.2.1.2.4 Spherands

Cram suggested that compounds such as **55** and **56** should be called spherands [117]: in these systems, the donor centres (OCH_3, OH, O^-) are parts of intraannular substituents which point into the interior of a rigid ring.

Polycyclic crown compounds and cryptands, as in Figures 12 and 9, respectively, require a great deal of synthetic effort. For many of the cyclic ring constitutions, well established procedures exist [118,119] and with the help of special effects, called 'template effects' [66,99], usually high yields are achieved [5]. The easiest of these syntheses are those of the podands, because there is no cyclization step involved. Preparative details may be found in many books [5,6] and reviews [24,41,62,66,75].

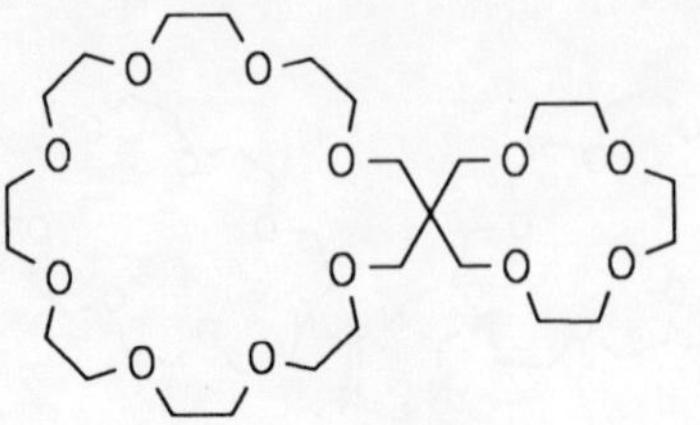

52

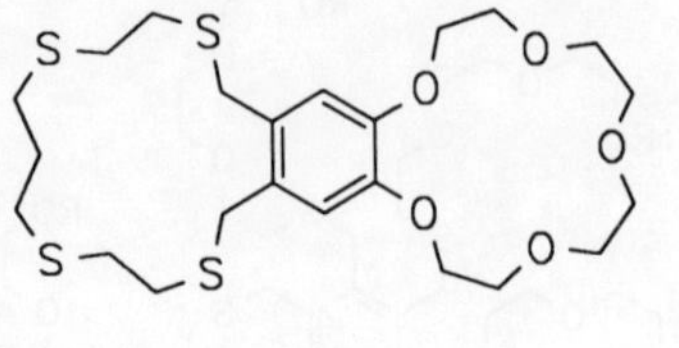

54

53

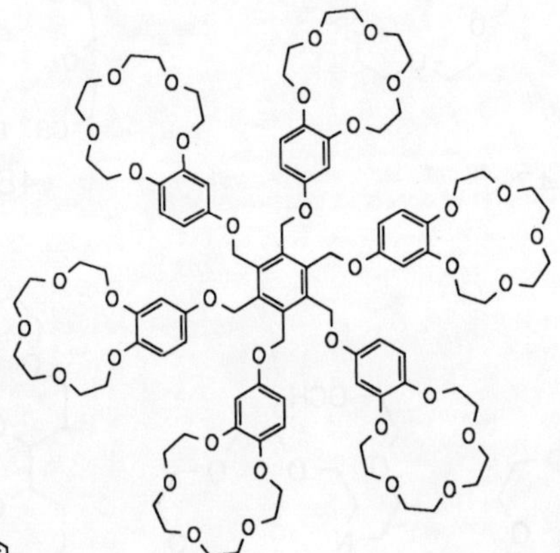

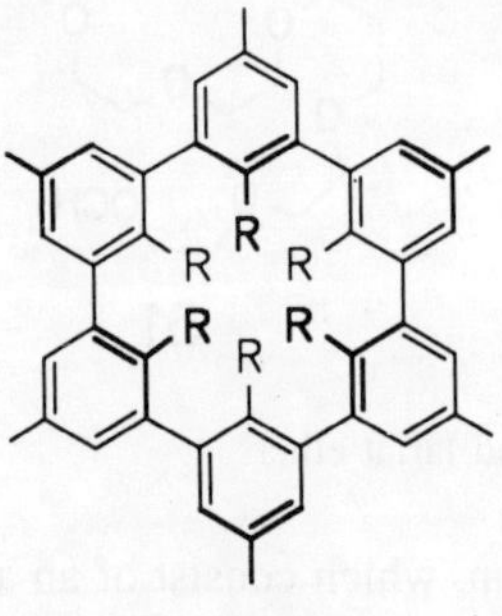

55

R = OCH_3; OH

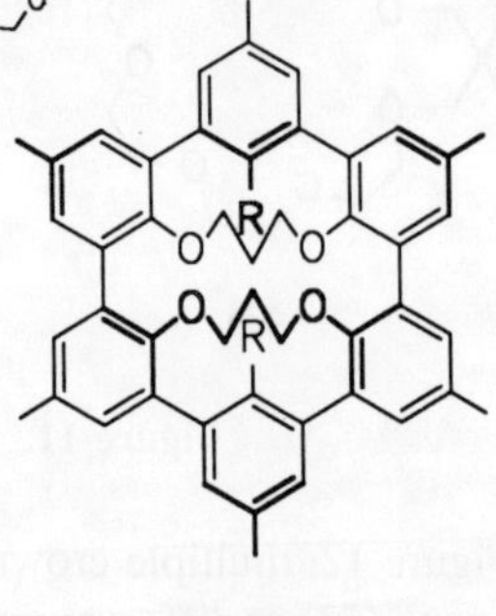

56

R = OCH_3

Figure 12. Multiple crowns and spherands

2.2.1.3 The properties of crown compounds

Polarity and solubility. Because for every hydrophilic oxygen atom there is also a lipophilic ethylene moiety, crown ethers show a remarkable balance with regard to binding symmetry and polarity. On the scale of lipophilicity according to Hansch [120], a value of exactly zero is found, i.e. a perfect balance between hydrophilicity and lipophilicity. This explains the universal solubility properties of crown ethers; many of them are readily soluble in hydrophilic media, such as water and alcohols,

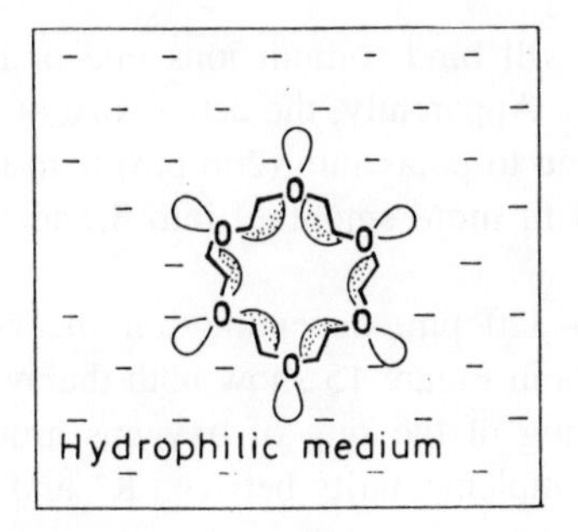

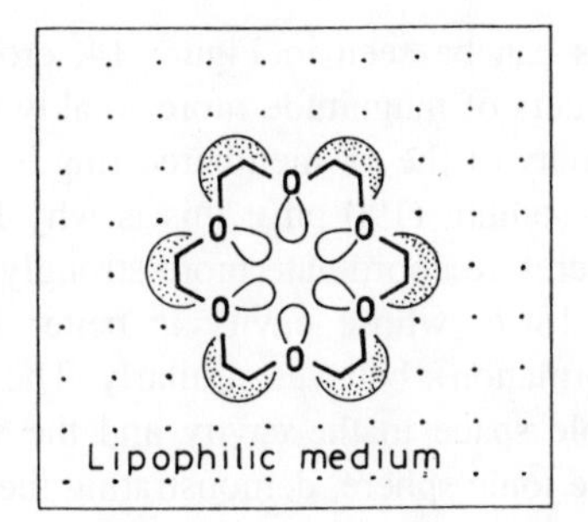

Figure 13. The solubility of [18]crown-6. Left, hydrophilic medium; right, lipophilic medium

and also in lipophilic solvents, e.g. benzene and chloroform. The type of solvent used, though, determines the conformation of the crown ether ring (Figure 13):

In a hydrophilic medium, the oxygen atoms will be pointing outwards, thus creating a lipophilic hydrocarbon core. The interior of the crown ether can, under these circumstances, be compared with a droplet of oil in water. In a lipophilic medium the polarities are inverted: conformatively, the oxygen atoms will be pressed inwards, with the hydrophobic CH_2 groups being turned outwards. The interior now behaves as a droplet of water in oil. In this case, a hydrophilic electron-rich cavity is formed, in which cations can now be incorporated, i.e. coordinated.

Influence of solvent, ring size and donor sites on the formation of complexes of crown ethers with cations [4,71,93]. The simple considerations above allow us to guess at the properties of crowns [36,99]. Moreover, it becomes apparent why the type of solvent will influence the crown's ability to form complexes. The same complex will be far more stable in a non-polar solvent than it will be in a polar one (cf. Figure 14).

DCH18C6 (**8**) binds potassium ions. In water, the formation constant (K_f) is approximately 10^2; in methanol, however, it is considerably larger, circa 10^6 [71].

8
DCH18C6
Isomer A
(*cis–syn–cis*–Isomer)

log K_f at 25 °C (potentiometrically)

	$Na^{\oplus}$	$K^{\oplus}$
H_2O	1.21	2.02
MeOH	4.08	6.01

Figure 14. The effect of the solvent on the complexation of crown ethers with cations

As can be seen in Figure 14, crown ether **8** will bind sodium ions one or two orders of magnitude more weakly than K^+ ions. Apparently, the actual size of the cavity of the 18-membered ring is better adapted to potassium (266 pm) than it is to sodium (190 pm); this is why K^+ ions will fit more smoothly into **8** and will therefore coordinate more strongly.

18C6, whose cavity is better known (260–320 pm, depending on the conformation), behaves similarly. The CPK models in Figure 15 show both the available space in the cavity and the seamless fitting of the ring of oxygens around the ionic sphere, demonstrating the optimum complementarity between K^+ and the interior of 18C6.

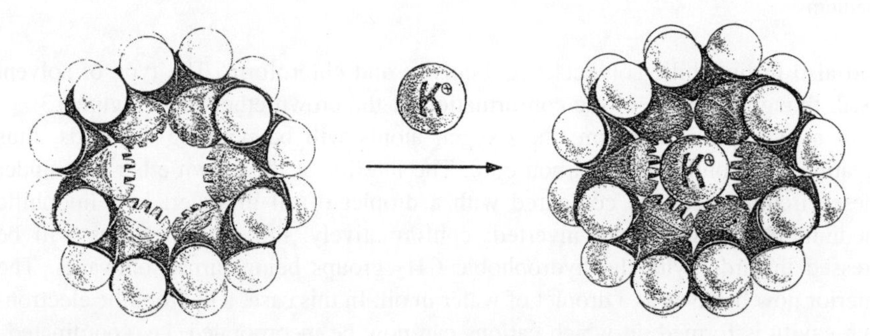

Figure 15. CPK model of [18]crown-6 (**7**) and its K^+ complex. (Uncomplexed **7** adopts other conformations, depending on solvent and state of matter.)

There are also crown ethers which fit the smaller and larger alkali metal ions (Table 1). Thus, the small Li^+ will fit into the tighter ring of 12C4, the larger Na^+ will go into the wider 15C5 (**10**), K^+ (as mentioned) likes 18C6 and, finally, Cs^+ fits inside 21C7 (**11**) [70].

Table 1. Comparison of the diameters of different alkali metal cations and crown ethers

Cation	Diameter (pm)	Crown	Diameter (pm)
Li^+	136	12C4 (9)	120–150
Na^+	190	15C5 (10)	170–220
K^+	266	18C6 (7)	260–320
Cs^+	338	21C7 (11)	340–430

In Figure 16, the logarithms of the complexation constants for different crown ethers with various ring sizes and number of oxygens are plotted against the ionic radii of Na^+, K^+, and Cs^+. As the ring size of the crown ethers increases, the larger cations will bind more strongly. The relative differences in stability for the different complexes determines the shape of the curve. This again provides information on the selectivity of the complexation [70], i.e. the favoured complexation of a given ion in the presence of others.

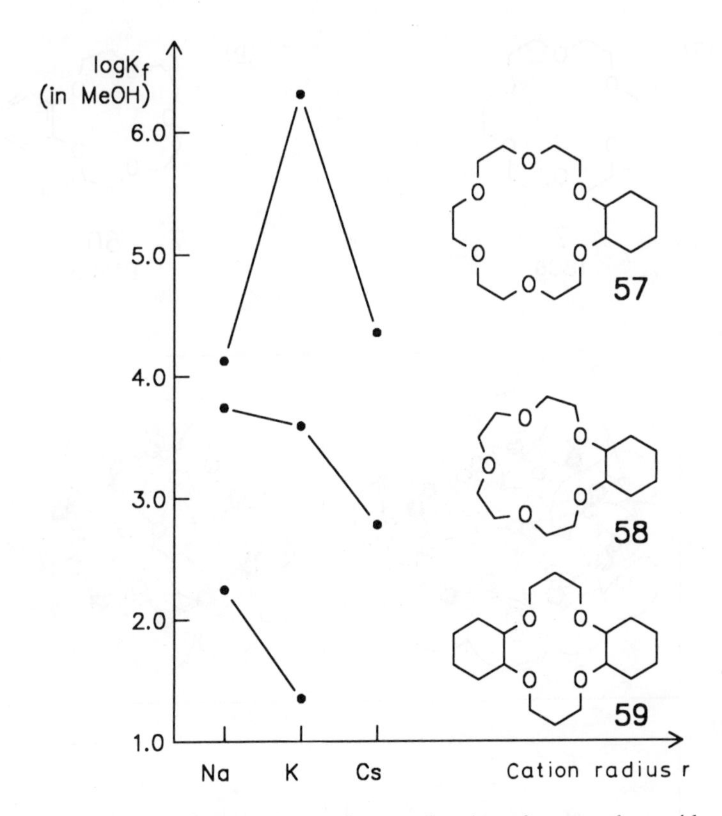

Figure 16. Influence of the ring size on the complexation of crown ethers with cations

In the case of crown ether **57**, one speaks of 'peak selectivity' for K^+, as Na^+, the next smallest ion, and also Cs^+, the next largest ion, will bind noticeably more weakly than potassium. When the curve is sloping to only one side, the term 'plateau selectivity' is used. This is roughly the case for the 15-membered crown **58**, where sodium binds most strongly, potassium only little less and then caesium clearly more weakly.

The relationship between ligand cavity and cation radius is reflected in the crystal structures of the complexes [42,65,70,121,122] (Figure 17). Thus, in the crystalline complex of potassium thiocyanate and 18C6, the potassium ion is found exactly at the centre of the ring (Figure 17a, cf. Figure 15). In the case of B15C5 (**60**), though, where the cavity is to small to accommodate K^+, a sandwich complex is formed (Figure 17b).

With larger ring sizes and increasing flexibility, the crown ligands tend to entangle cations of small size: the 1:1 complex of DB30C10 (**13**) with potassium iodide is one such example and its conformation looks like the seams of a tennis ball. On the other hand, a crown ether with a cavity of excessive size can accommodate two cations simultaneously, as exemplified by the disodium salt of the 30-membered crown **13**. In Figure 17c, these and some other structures of crown ether–cation complexes are represented schematically [123].

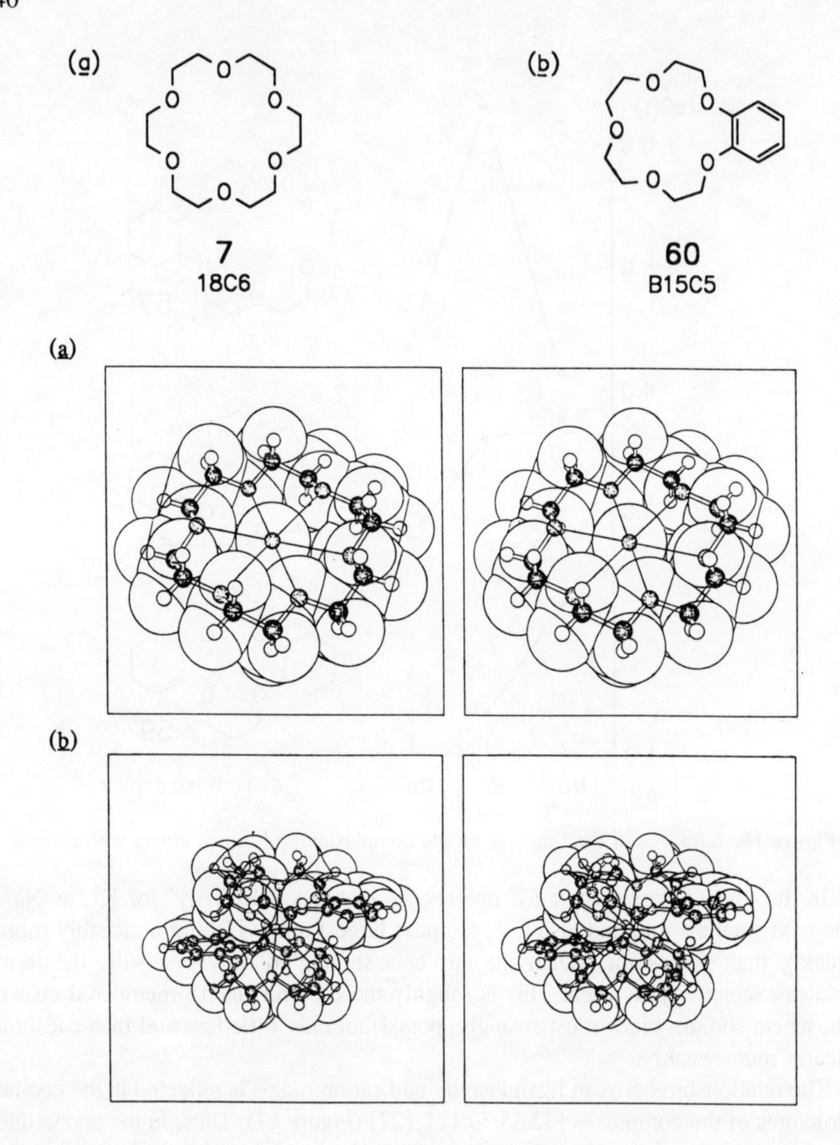

Figure 17. Crystal structures of the K^+ complexes of (*a*) 18C6 and (*b*) B15C5; (*c*) topologies of crown ether complexes (the ring size increases from top to bottom; see p. 41)

The ability to form complexes with different structures conflicts with a a good selectivity of complexation. This is true especially for crown ethers with small (sandwich complexes) and large (binuclear complexes) ring sizes. The spatial factors are, regardless of all else, of fundamental importance to the complexation process.

The exchange of donor sites in coronands, e.g. N or S for O (cf. Figure 7), will have even more drastic effects on the complexation behaviour (Figure 18):

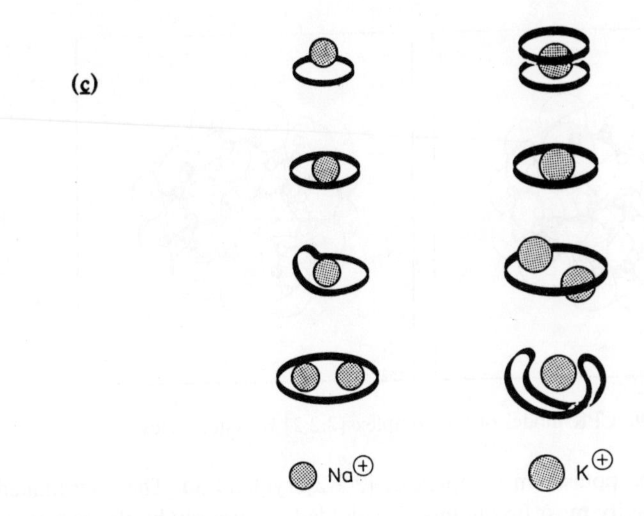

Figure 17. (*continued*)

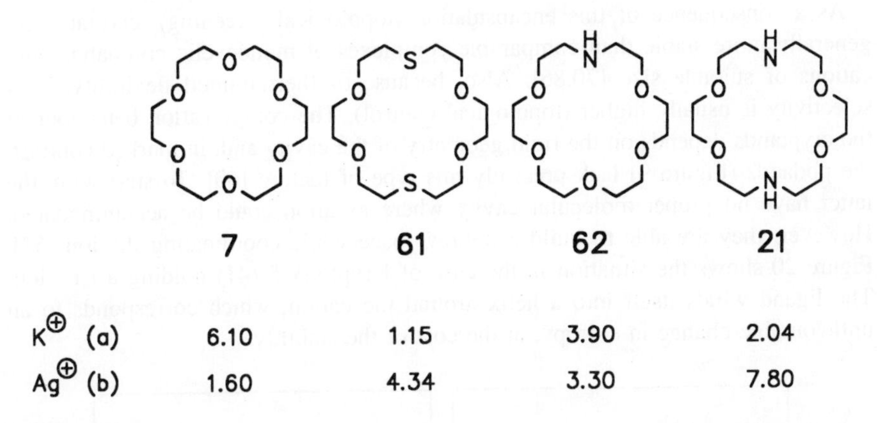

	7	61	62	21
$K^{\oplus}$ (a)	6.10	1.15	3.90	2.04
$Ag^{\oplus}$ (b)	1.60	4.34	3.30	7.80

(a): in methanol; (b): in water

Figure 18. Influence of the variation of the donor sites on the complexation behaviour (log K_f values)

whereas the complexation of alkali metal ions, e.g. K^+, falls dramatically when the ring oxygens are replaced with sulphur or nitrogen, it rises significantly in this order in the case of transition metal ions, such as Ag^+ [39,70]. This observation is in accordance with the HSAB principle (Hard and Soft Acids and Bases [124]), which predicts that energetically, hard oxygen atoms and hard alkali metal ions on the one hand, and soft sulphur and nitrogen atoms and soft transition metal ions on the other, will afford the better pairs.

Effects of ligand topology. The ligand topology is fundamentally responsible for the complexation properties of the cryptands (Figure 8) [22,47,66]. Figure 19 shows

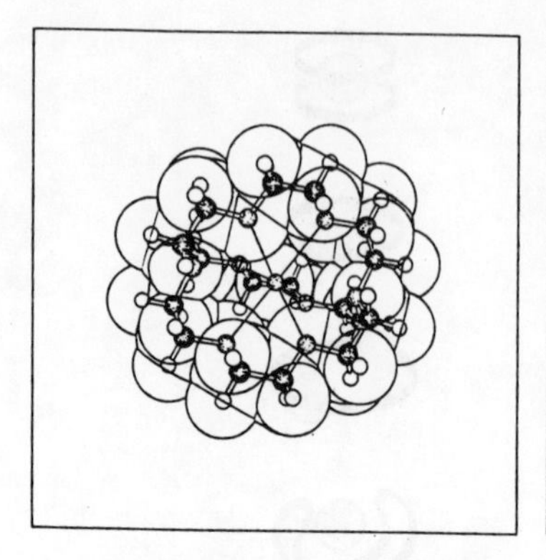
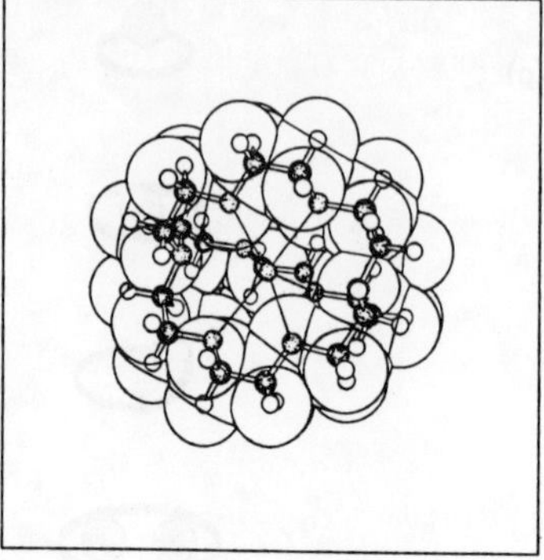

Figure 19. CPK model of the complex [2.2.2]·K^+ (stereoview)

the CPK model of the potassium complex of [2.2.2]cryptand **34**. The coordinated ion, which fills the cavity most favourably, is shielded so strongly by the polyether bridges that it is hardly visible. It is almost as if it was hidden in the guest molecule's cavity, hence the name cryptate for this type of complex.

As a consequence of this encapsulation (topological screening), cryptates are generally more stable than comparable complexes of monocyclic coronands with cations of suitable size [70,86]. Also, because of their limited flexibility, their selectivity is usually higher (topological control). The complexation behaviour of the cryptands depends on the rigid geometry of the cavity and, in marked contrast, the podands (Figure 9) lack precisely this type of feature [60]. To start with, the latter have no proper molecular cavity where a cation could be accommodated. However, they are able to build a hollow space while coordinating the ion [57]. Figure 20 shows the situation in the case of Kryptofix-5 (**41**) holding a Rb^+ ion. The ligand winds itself into a helix around the cation, which corresponds to an unfavourable change in entropy, at the cost of the stability.

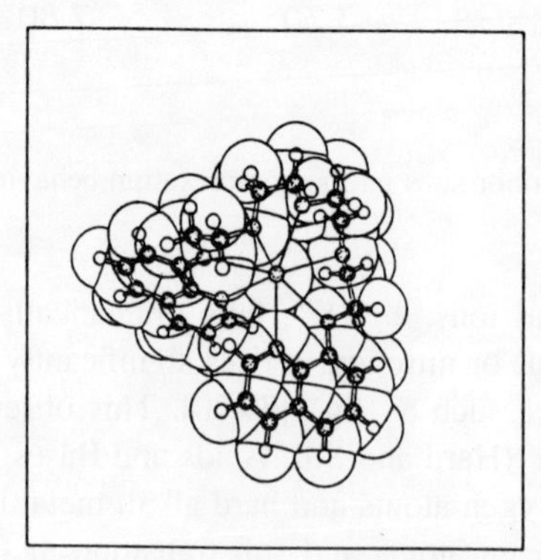
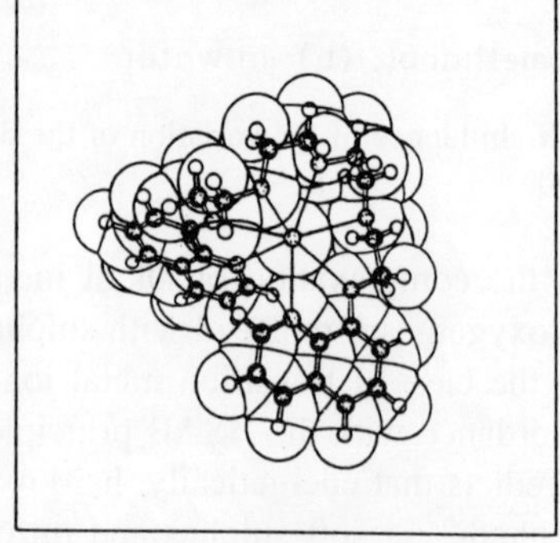

Figure 20. Crystal structure of the complex **41**·Rb^+ (stereoview)

Comparable complexes between podands and alkali and alkaline earth metal ions (*podates*) are more labile by orders of magnitude than crown ether and coronand complexes (*coronates*), and these in turn are weaker than cryptates by several powers of ten. Typical values for the formation constants (K_f), e.g. in methanol, are 10^2–10^4 for podates, 10^4–10^6 for coronates and 10^6–10^8 for cryptates (Table 2) [70]. Therefore, with each additional dimension in the ligand molecule, going from a thread-like to a cyclic and then a spherical structure, the stability of the corresponding complex increases by two to three orders of magnitude. This can also be shown with the help of some special effects (Table 2) [47]. However, with these only the thermodynamics of the complex formation are being described ('static complex stability'). There are also differences in their kinetic behaviour during both the formation and the decomposition of their complexes ('dynamic complex stability') [37,70,93].

Table 2. Relationship between type of complexation, complex formation constant and stabilization effect

Type of complex	K_f	Effect
Podate	10^2–10^4	Chelate
Coronate	10^4–10^6	Macrocyclic
Cryptate	10^6–10^8	Macrobicyclic

The conformatively rigid cryptands are slow in forming their complexes, whereas the flexible podands are very fast, and the coronands are somewhere in between. We may differentiate between 'cation receptors' and 'cation carriers,' depending on the properties shown: high stability with slow complexation/decomplexation and low stability with rapid complexation/decomplexation, respectively. It follows that, depending on the type of ligand, the formation of a complex can be considered from very different points of view.

Complexation of molecular cations, chiroselective complex formation. We face a more tricky situation in the case of the complexation of molecular cations, e.g. organic ammonium ions, as the correct geometrical arrangement of the binding sites, i.e. the geometrical complementarity or 'steric fit', is of even greater importance [32,46,47,53,69].

Figure 21a shows the 'trigonal recognition and binding' of a primary ammonium salt; the azacoronand is able to produce three hydrogen bonds to the guest molecule, i.e. a so-called 'three-point-interaction' [125]. The 'tetragonal binding' of an unsubstituted ammonium ion by a spherical (tricyclic) cryptand (**39**; cf. Figure 9) is depicted in Figure 21b [47,48]. In Figure 21c, the perfect complexation of an α,ω-alkyl diammonium salt in the cylindrical cavity of a cryptand (dihapto complex) is shown [86,99].

A step further is taken in the complex in Figure 21d. Here, both the central ring and the lateral arms (Figure 10) take part in the recognition of the bifunctional ammonium ion. The type of molecular guest ion used, a nicotinamide, illustrates enzyme-type binding. The ligand depicted in Figure 21d, a derivative of tryptophan

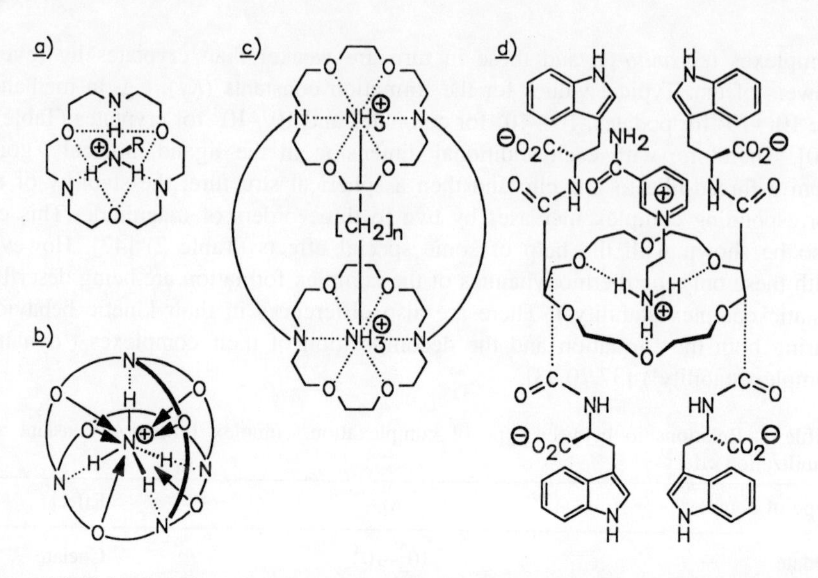

Figure 21. Selective binding of ammonium ions: trigonal, tetragonal, dihapto and lateral recognition

and tartaric acid, contains eight asymmetric centres and it should, therefore, be suitable for the 'chiroselective recognition' of enantiomeric ammonium guest-molecules. This is indeed the case [48], but a ligand of a different type (**63**, Figure 22) proved to be more efficient [29,32,46]. The main features of **63** are its C_2 symmetry (dissymmetry) and the presence of binaphtyl 'hinges'. Because of the steric hindrance of the *peri* hydrogen atoms, a twist of the molecule results and an inactive *meso* form (*R,S*) and two enantiomers (*R,R* and *S,S*; shown edge-on in Figure 23) are obtained.

Cram investigated the chiroselective complexation of optically active ammonium salts by enantiomerically pure **63** and found that indeed a good resolution of racemates could be achieved [32]. This is explained in Figure 23: when optically active **63** (e.g. *S,S*) is mixed with racemic phenylglycine esters [(*R*)-**64** and (*S*)-**64**] in their protonated form, two cases are possible [29,32,46]. The combinations of ligand (*S,S*)-**63** with the guest compounds (*R*)-**64** and (*S*)-**64** are shown as viewed through the axis of the C—N bond. Differences in energy result from the different geometrical arrangements: the complex at the bottom of the figure is held by four binding sites, one of which is to the ester function, while the complex above it has only three sites. The former is therefore energetically more favourable; the complex is more stable and is formed preferentially. Thus, a chromatographic column prepared with optically active **63** is able to resolve a racemate of **64** [29,32].

Complexation of anions and uncharged organic guest-molecules. There are also ligands for the selective complexation of anions (Figure 24) [86]. For this, the ligands need to have comparatively large cavities (Cl^- = 360 pm compared with Na^+ = 190 pm), which have so far proved difficult to synthesize. In addition, as the charge density of anions is low, the electrostatic forces with them are weaker than

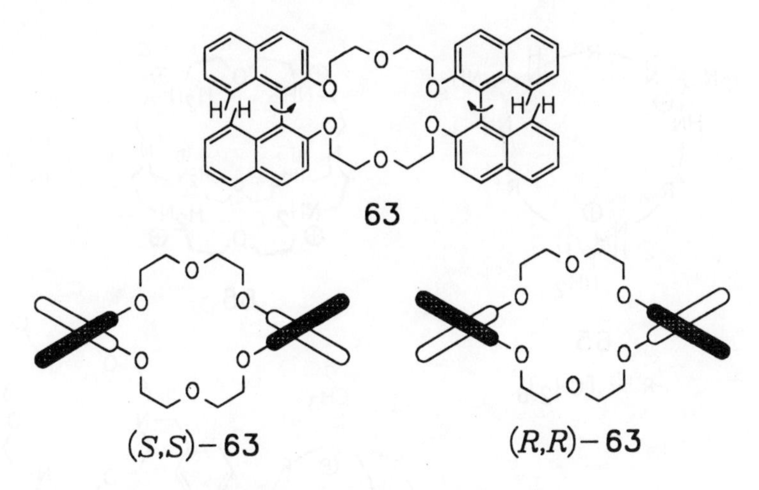

Figure 22. Chiral binaphthyl-crowns

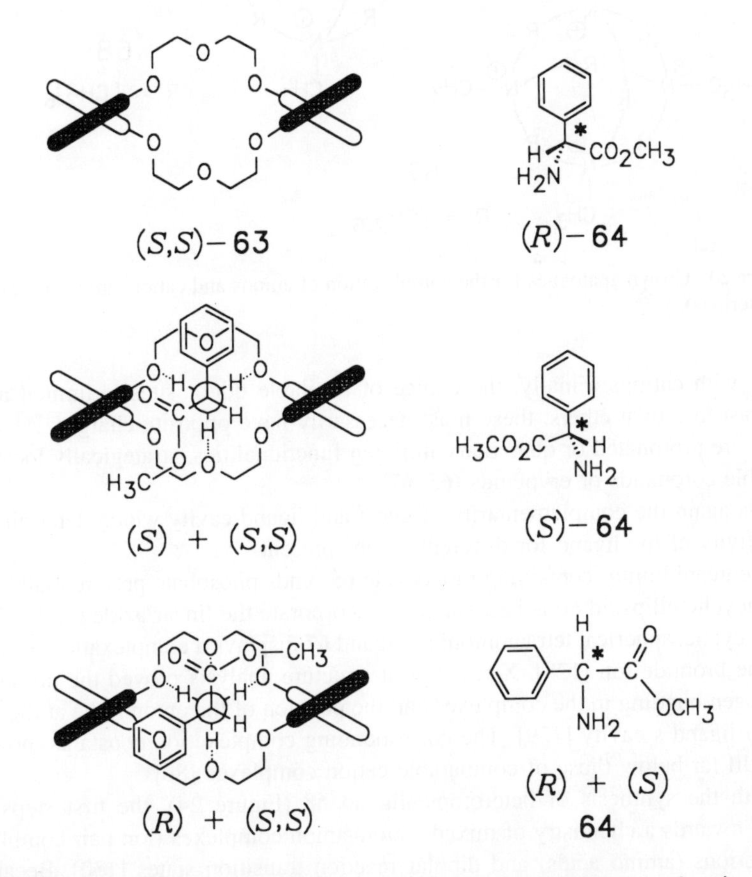

Figure 23. Chiroselective recognition and enantio-differentiating complexation

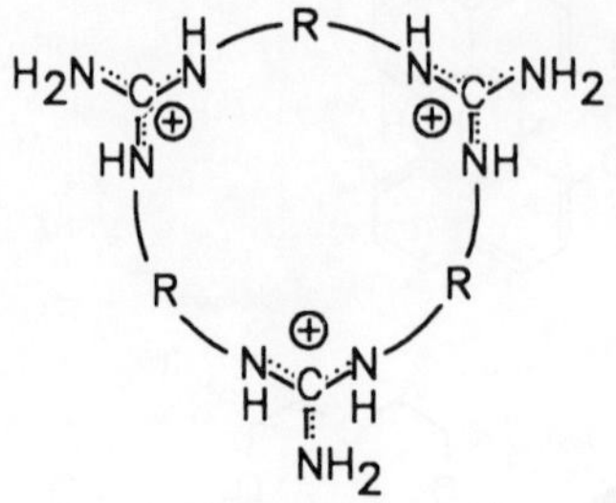

Figure 24. Crown analogues for the complexation of anions and cation–anion combinations (zwitterions)

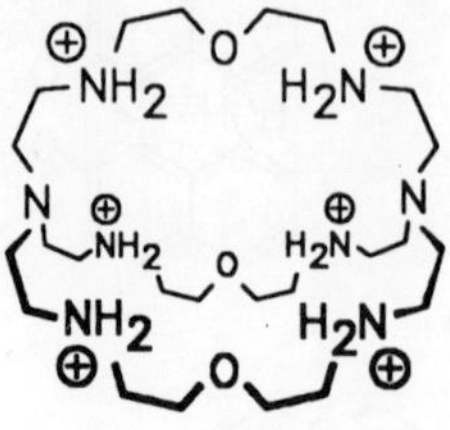

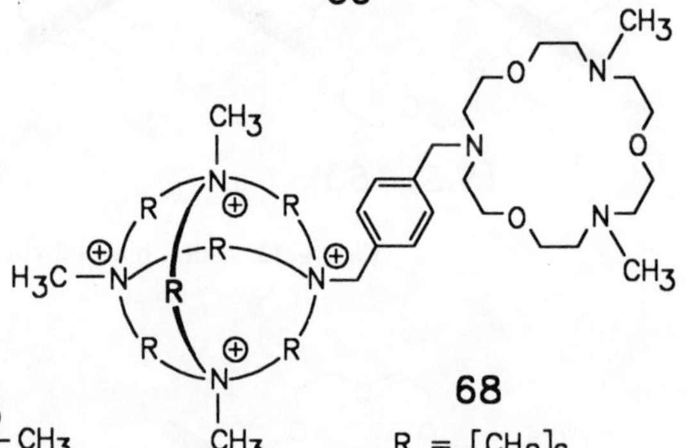

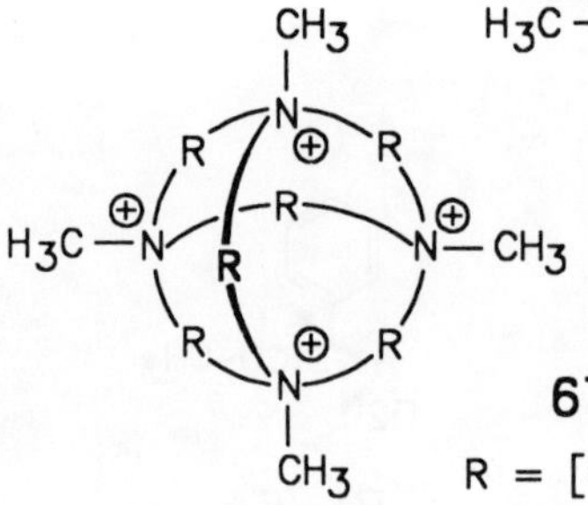

those with cations. Finally, the choice of available donor sites is limited and, in contrast to crown ethers, these must necessarily have opposite charges. As a rule, these are protonated or quaternary nitrogen functionalities, strategically located on suitable coronands or cryptands (**65–67**).

It is again the complementarity of guest and ligand cavity which determines the selectivity of the ligand for different anions present.

The guanidinium-containing monocycle **65** binds phosphate preferentially [126], the bicyclic ellipsoid **66** is best suited to incorporate the linear azide ion [127], and the tricyclic, spherical tetraammonium ligand **67** displays a complexation maximum for the bromide ion [128]. X-ray crystal structure analysis proved the presence of hydrogen bonding in the complexes and the position of the anion to be at the centre of the ligand's cavity [129]. The corresponding complexation constants, however, are still far below those of comparable cation complexes [86].

With the synthesis of heterotopic ligand **68** (Figure 24), the first steps were taken towards a chemistry of mixed cation–anion complexes (ion-pair complexes), zwitterions (amino acids) and dipolar reaction transition states [130]. Because at first crown compounds were regarded as ligands for cations only, their ability to form complexes with neutral organic molecules went unnoticed for a long time [131,132]. Subsequently, interest in this new aspect of crown compounds has increased considerably [88,91,133]. In Figure 25 two typical examples of such complexes are depicted [the guest molecules being dimethyl sulphate and methanol]. The host–guest binding is usually due to hydrogen bonding, for which weakly C—H acidic methyl groups (Figure 25a) [134] or CH_2 groups are

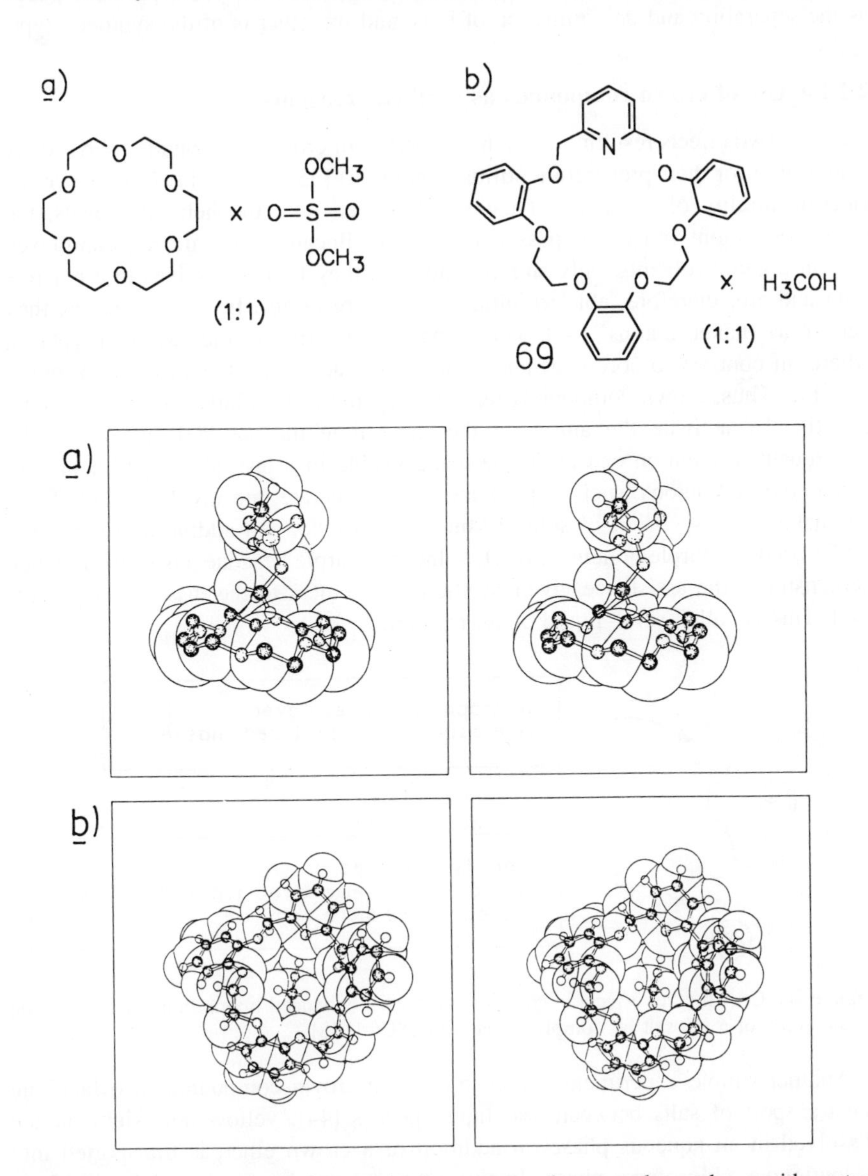

Figure 25. Crystal structures of representative crown and coronand complexes with neutral guest molecules (stereoviews). (a) Dimethyl sulphate complex with 18C6; (b) methanol complex with a tribenzopyridino crown (see formulae)

also used. Their mutual repulsion is reduced by bridging with oxygen or nitrogen. Beside that, sterically shielding groups, as in **69**, and the conformational rigidity of the ligand are also of importance (Figure 25b) [135].

Complexes of anions and neutral molecules [136–139] could also become important as applications of crown compounds [140]. At present, two main applications of crowns stand out: one is concerned with analytical problems, such as the separation and determination of ions, and the other is of the synthetic type.

2.2.1.4 Use of crown compounds as synthetic reagents

Basics. Two effects resulting from the properties of crown compounds are basically responsible for their preparative utility (Figure 26) [35,44,99,141]. To preserve the electroneutrality of solutions of cations dissolved with the help of crowns, the associated anions must also pass into solution. Because they are not complexed and, in organic solvents, only slightly solvated, they find themselves in a reactive state and are, therefore, able to initiate unusual reactions. In extreme cases, they behave as 'naked anions' [44]. This is particularly true in the case of cryptates, where, in contrast to coronates and podates, contact with the cation is no longer possible. Thus, crown compounds render inorganic salts soluble in organic media and thereby activate the anion (at the same time the cation is masked). The consequent dissolution can easily be made visible in a test-tube experiment, e.g. when coloured anions, such as permanganate or picrate, are involved [36]. Thus, on shaking a suspension of solid $KMnO_4$ in benzene, and adding a crown ether (18C6), a deep purple organic $KMnO_4$ solution ('purple benzene') is formed, which demonstrates the assistance given by the crown ethers to the dissolution process [44]. This is called solid–liquid phase transfer.

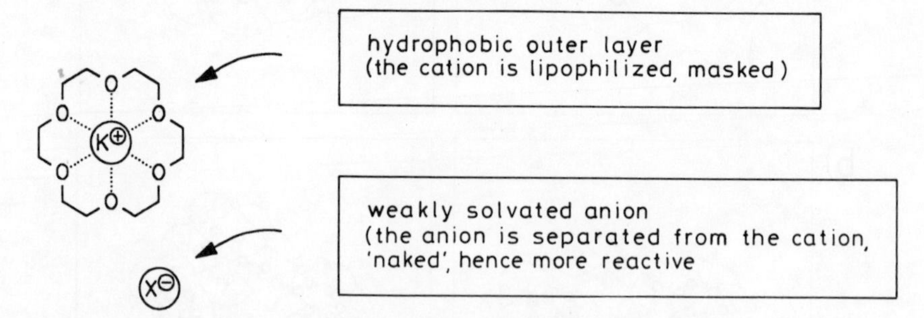

Figure 26. Crown ether effects in chemical synthesis: enhanced lipophilicity and activation of the anion of a K^+ salt by complexation with 18C6 [148]

Another simple experiment demonstrates that crown compounds also facilitate the transport of salts between two liquid phases [44]: yellow potassium picrate dissolved in an aqueous phase on addition of a crown ether, is transported into a contiguous chloroform phase. In this case, we speak of a liquid–liquid phase transfer.

The preparative utility is obvious: components of a reaction, which on account of their different polarities cannot be brought together in a common solvent, can now be made to react in a homogeneous phase. This opens up new uses for salt-like or only slightly soluble reagents. Numerous reactions, which previously were either impossible or gave only poor yields, now easily succeed in the often only catalytic presence of crown compounds (see below).

Synthetic applications of crown ethers. In principle, any chemical reaction which involves ions or ionic intermediates of any form can be either modified or improved by the use of crown compounds. This is true for nucleophilic substitution (e.g. halogen and pseudohalogen exchange, O, N and S alkylation), reactions with carbanions, C—C bond formation (e.g. Darzens or Knoevenagel condensation), additions, eliminations, the generation of carbenes, extrusion of gases (N_2, CO_2), oxidations, reductions ('dissolving metal,' 'electrides' [66], metal hydrides), rearrangements (Favorskii, Cope), isomerizations and polymerizations. Further applications of crown compounds lie in the fields of macrolide synthesis, reactions of organometallics and protected groups and in questions concerned with the chemistry of phosphorus and silicon. It is impossible to give a comprehensive compilation of all the applications of crown compounds in organic synthesis (see books [142–145] and reviews [3–35,45,78,84,141,146–151]). Nevertheless, a few fundamental preparative uses of crown compounds will be outlined below.

In Figure 27, the use of 'purple benzene' (permanganate dissolved in benzene) for oxidation purposes is demonstrated [152,153]. The yields for the formulated reactions are 100% throughout.

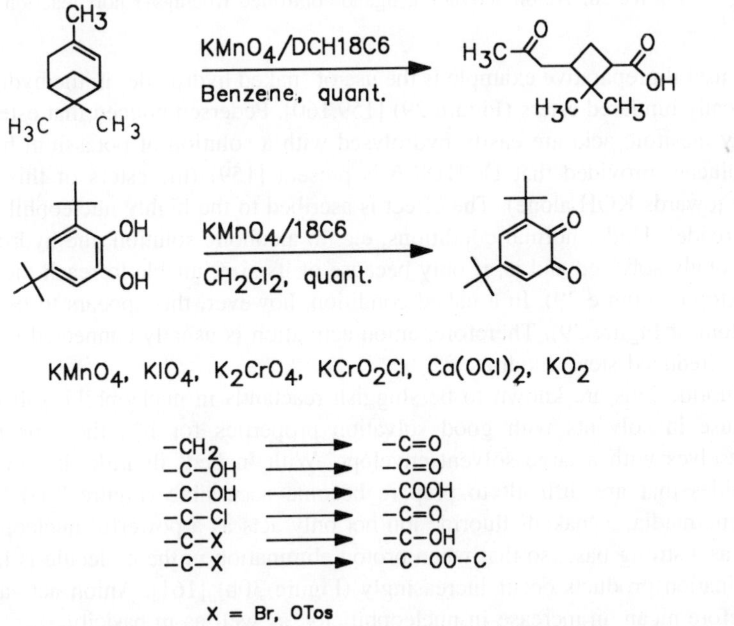

$KMnO_4$/DCH18C6, Benzene, quant.

$KMnO_4$/18C6, CH_2Cl_2, quant.

$KMnO_4$, KIO_4, K_2CrO_4, $KCrO_2Cl$, $Ca(OCl)_2$, KO_2

$-CH_2 \rightarrow -C{=}O$
$-C-OH \rightarrow -C{=}O$
$-C-OH \rightarrow -COOH$
$-C-Cl \rightarrow -C{=}O$
$-C-X \rightarrow -C-OH$
$-C-X \rightarrow -C-OO-C$

X = Br, OTos

Figure 27. 'Naked' permanganate and other crown-assisted oxidation reagents

Other inorganic ionic oxidizing agents (Figure 27) can also be used under similar conditions in organic media, e.g. for the transformations outlined at the bottom of Figure 27 [152–157]. Whereas these cases mostly depend on solubility being provided for the salts, other reactions give a measure for the activation of anions [44]. The Koenigs–Knorr-reaction is a good example, i.e. the reaction of acetobromoglucose with alcohols in the presence of silver nitrate (Figure 28) [158]. In the absence of ligands, the β-glycoside is formed almost exclusively. On addition of crowns, the nitrate ion is more or less activated and competitive, so that some amounts of β-nitric acid esters are formed. The bicyclic cryptand forms the most stable complexes and thus activates most strongly (Figure 28).

Catalyst	Relative yield (%)		
DB18C6	100	:	0
[2.2]	95	:	5
[2.2.2]	64	:	36

Figure 28. Anion activation: ligand-controlled Koenigs–Knorr reaction

A useful preparative example is the use of 'naked hydroxide' in the hydrolysis of sterically hindered esters (Figure 29) [159,160]. Pedersen noticed that esters of the bulky mesitoic acid are easily hydrolysed with a solution of potassium hydroxide in toluene, provided that DCH18C6 is present [159] (the esters of this acid are inert towards KOH alone). The effect is ascribed to the highly nucleophilic 'naked hydroxide': Under normal conditions, e.g. in alcoholic solution, the hydroxide ion is strongly solvated and is, if only because of its size, unable to reach the reacting site (top of Figure 29). In a naked condition, however, this appears to be possible (bottom of Figure 29). Therefore, anion activation is usually connected with steric relief (reduced steric hindrance).

Fluoride ions are known to be sluggish reactands in nucleophilic substitutions, because in solvents with good solvation properties for F^-, the ions surround themselves with a large solvent envelope. With 'naked' fluoride, however, alkyl fluorides that are difficult to prepare become accessible (Figure 30a) [161]. In organic media, a 'naked' fluoride ion not only acts as a powerful nucleophile, but also as a strong base, so that when proton elimination in the molecule is favoured, elimination products occur increasingly (Figure 30b) [161]. Anion activation can therefore mean an increase in nucleophilicity, as well as in basicity.

In Figure 31 more anions have been compiled which, by the action of crown

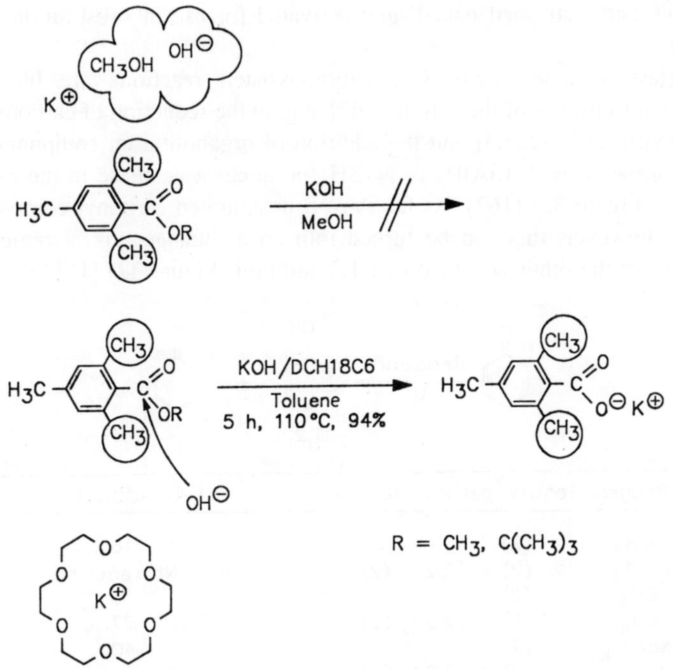

KOH / MeOH (no reaction)

KOH/DCH18C6, Toluene, 5 h, 110 °C, 94%

R = CH_3, $C(CH_3)_3$

Figure 29. Hydrolysis of sterically hindered mesitoic acid esters with 'naked hydroxide'

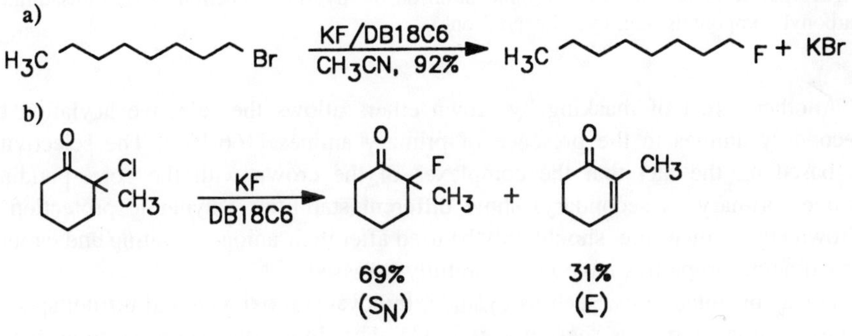

a) KF/DB18C6, CH_3CN, 92%: H_3C–(CH₂)ₙ–Br → H_3C–(CH₂)ₙ–F + KBr

b) KF, DB18C6: 69% (S_N) + 31% (E)

Figure 30. 'Naked fluoride' as nucleophile and base

'Naked anions' $F^{\ominus}$, $Cl^{\ominus}$, $Br^{\ominus}$, $I^{\ominus}$, $CN^{\ominus}$, $SCN^{\ominus}$, $OCN^{\ominus}$, $NO_2^{\ominus}$, $RO^{\ominus}$, $RCOO^{\ominus}$, $OH^{\ominus}$

Figure 31. Common 'naked anions'

ethers, have been 'stripped naked' and activated for use in substitution reactions [45,78,84].

In contrast to anion activation, cation-assisted reactions are hindered by complexation (masking of the cations) [67], e.g. in the reduction of carbonyl groups by metal hydrides [162,163] and the addition of organolithium compounds [164]. Cryptate formation with $LiAlH_4$ or $NaBH_4$ produces a decrease in the rate of the reaction (cf. Figure 32) [162]. As far as α,β-unsaturated carbonyl compounds are concerned, however, this can be turned into an advantage, viz. a regioselective 1,4-addition vs the otherwise favoured 1,2-addition (Figure 32) [165].

Reagent / DME → 1,2 + 1,4

Reagent (equiv. per mol ketone)		1,4-Adduct (%)
$LiAlH_4$	(2)	18
$LiAlH_4$	(2) + [2.2.1] (2)	No reaction
$LiBH_4$	(2)	35
$LiBH_4$	(2) + [2.2.1] (2)	77
$NaAlH_4$	(2)	40
$NaAlH_4$	(2) + [2.2.1] (2)	No reaction
$NaBH_4$	(2)	45
$NaBH_4$	(2) + [2.2.1] (2)	77

Figure 32. Influence of cation complexation on the hydride reduction of α,β-unsaturated carbonyl compounds (e.g. cyclohexen-2-one)

Another effect of masking by crown ethers allows the selective acylation of secondary amines in the presence of primary amines [166,167]. The selectivity is based on the fact that the complexes of the crown with the corresponding amine (primary vs secondary) show different stabilities ('dynamic protection'). Crown ethers, therefore, should only be used after their anion-activating and cation-deactivating properties have been carefully assessed.

For ambidentate ions, such as cyanide, thiocyanate, oximate and nitrite, special considerations apply, as seen in Figure 33a. The fluorenone oximate anion, being ambidentate, contains a hard oxygen and a soft nitrogen atom. This leads, when treated with iodomethane, to the formation of both alkylation products. In the presence of 18C6, the ratio of *O*- to *N*-alkylation is shifted in favour of *O*-alkylation, as more free oximate is available [168].

The addition of crown ethers also influences the prototropic equilibrium of ethyl acetate, which in the presence of a ligand is shifted to the left, and provokes the disruption of the ion-association (Figure 33b) [169]. The effects of ion association also determine the mechanism, and thus the product composition, of a β-elimination, as seen in Figure 34 [170], where X corresponds to the leaving

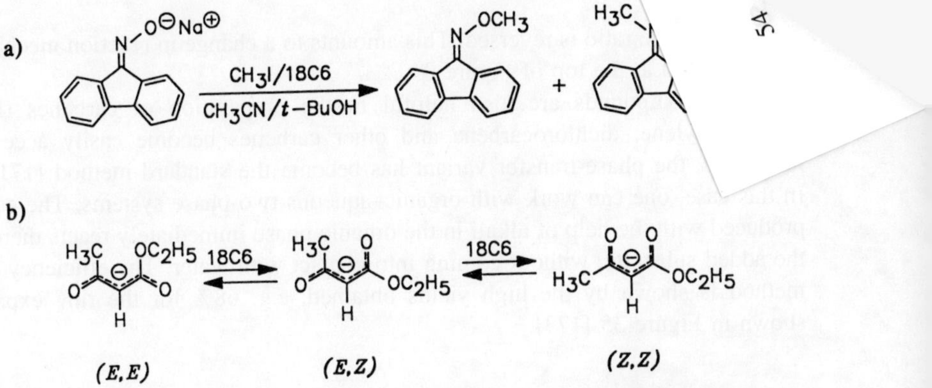

a) CH_3I/18C6, CH_3CN/t-BuOH

b) 18C6 ⇌ 18C6

(E,E) (E,Z) (Z,Z)

Figure 33. Complexation with crown ethers of ambidentate ions and configurationally flexible ions

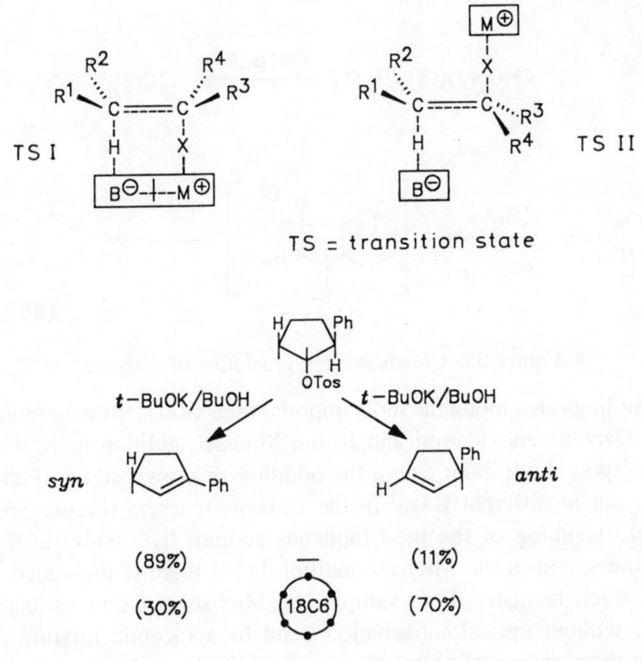

TS I TS II

TS = transition state

t-BuOK/BuOH t-BuOK/BuOH

syn anti

(89%) (11%)

(30%) 18C6 (70%)

Figure 34. Influence of 18C6 on the mechanism of α,β-elimination

group, B to the base and M to the counter ion. TS I represents an arrangement, where an associated base molecule is involved; TS II, however, is based on a dissociated base molecule.

In the model reaction at the bottom of Figure 34, 1-phenylcyclopentene is formed by syn elimination via TS I and 3-phenylcyclopentene by *anti* elimination via TS II. Under ordinary conditions, a *syn*/*anti* ratio of 89:11 is found, because potassium *tert*-butoxide is forming aggregates. In the presence of 18C6, the aggregates are reduced and the ratio is reversed. This amounts to a change in reaction mechanism, as formulated at the top of Figure 34.

Crown compounds are also helpful in the preparation of carbenes (Figure 35). Methylene, dichlorocarbene and other carbenes become easily accessible; nowadays, the phase-transfer variant has become the standard method [171,172]. In this case, one can work with organic–aqueous two-phase systems. The carbene produced with the help of alkali in the organic phase immediately reacts there with the added substrate, without coming into contact with water. The efficiency of the method is shown by the high yields obtained, e.g. 68% for the ring expansion shown in Figure 35 [173].

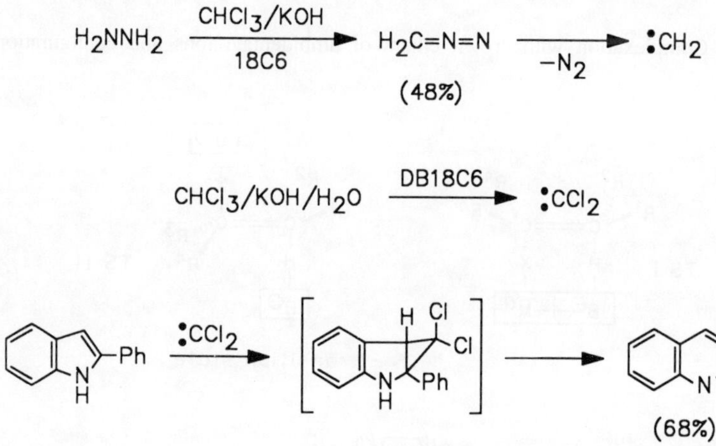

$$H_2NNH_2 \xrightarrow[18C6]{CHCl_3/KOH} H_2C{=}N{=}N \; (48\%) \xrightarrow{-N_2} :CH_2$$

$$CHCl_3/KOH/H_2O \xrightarrow{DB18C6} :CCl_2$$

Figure 35. Crown-assisted generation of carbenes

C—C couplings are among the most important reaction steps in organic synthesis, e.g. in the Darzens epoxidation and in the Michael addition to α,β-unsaturated carbonyl compounds. In both cases, the addition of crown ethers (Figure 36) has advantages, but in different ways: in the Darzens reaction (Figure 36a) [172], it is the simple handling of the base (aqueous sodium hydroxide) in the presence of crown ethers, and in the Michael addition [174] there is the added bonus of a favourable stereochemistry. For example, the Michael reaction outlined in Figure 36b would, without optical information, lead to a racemic mixture of addition products. In the presence of chiral crown ethers, however, an enantiomeric excess of almost 100% can be achieved (e.g. **63**, Figure 22) [175].

The mechanism of the stereodifferentiation is illustrated in Figure 37. Starting with the two enantiomerically pure crown ethers [(*S,S*) on the left, (*R,R*) on the right], it seems plausible that the reaction will go through the two corresponding transition states (TS I and TS II) shown underneath. There, the carbanion, with its fixed geometry, is bound to the complexed potassium ion and hence, its flat side is embedded in the crown ether's chiral niche. In this arrangement, the alkene can approach the reactive site of the Michael substrate only from the sterically open side; the formation of the product is enantioselective.

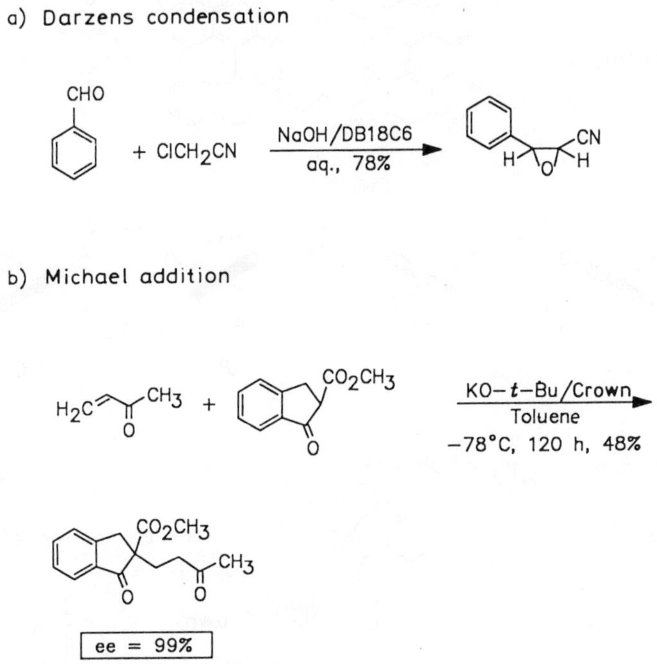

a) Darzens condensation

b) Michael addition

ee = 99%

Figure 36. Crown ether-assisted C–C coupling

From an economical point of view, it is important to know that many crown ether-assisted reactions can be run with only catalytic amounts of ligand [142–145]. However, in these cases, certain conditions must be met. Another economical procedure is based on the use of easily regenerated polymers or crown compounds bound to some polymeric support (see below) [78,150,176].

2.2.1.5 Use of crown compounds for chemical analysis

Whereas the chemical synthetic uses of crown ethers were first turned to the anionic part of the corresponding complexes, chemical analysis puts the complexed cation in the foreground. Depending on the constitution of the crown compound, cations are bound with varying strength (see above). Therefore, from a mixture of cations, those are chosen which form the most stable complexes with the crown compound concerned. This can be used to enrich, separate, mask and determine the concentration of ions [50,92,177–179].

Selective ion extractions [6,92,180,181]. Here, apart from the selective complexation of cations, the phase transfer of crown ether salt complexes from an aqueous into an adjacent organic phase is used (see below).

Which combination of ions, from a given mixture of salts, is extracted out of the aqueous phase by an organic crown ether solution depends on the ratio of the

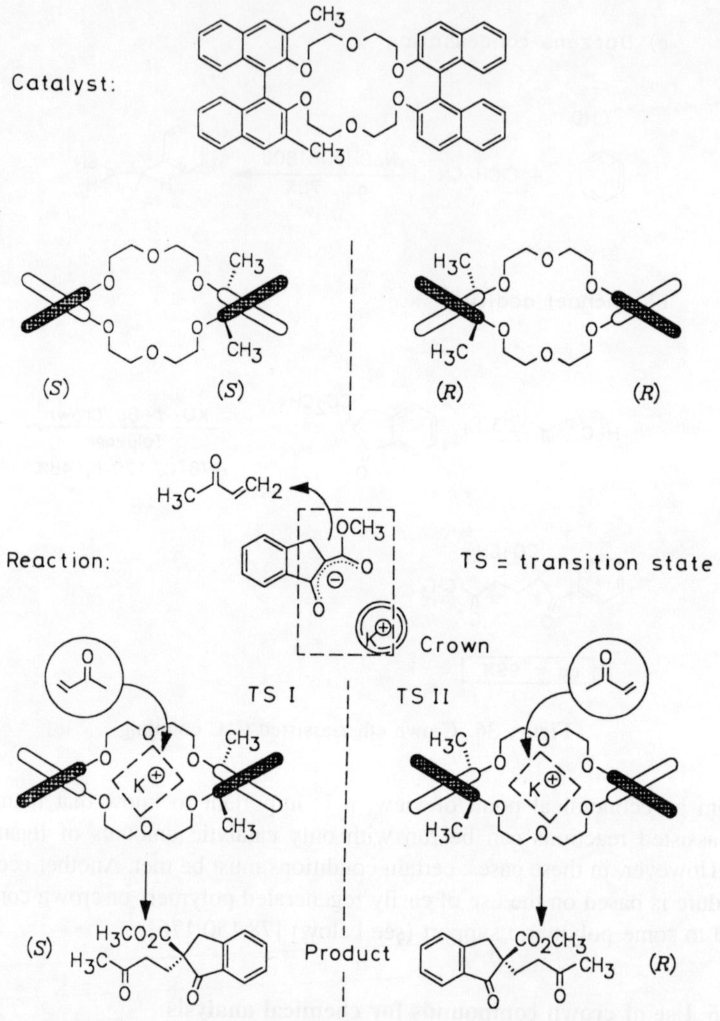

Catalyst:

Reaction:

TS = transition state

Crown

TS I

TS II

Product

Figure 37. Enantioselective Michael addition with chiral crowns as catalyst

extraction coefficients (K_{ex}) for the individual ions (Figure 38). This is roughly a function of the stability constants (K_s) and the partition coefficients (D) between the organic and aqueous phases of the complexes formed. Favourable conditions for salt transfer are, therefore, high stability and strong lipophilicity of the crown ether complexes formed [181]. This is represented in Figure 38 by the use of different symbols and their relative sizes (M_1^+ is transfered easily, M_3^+ only slightly and M_2^+ not at all).

For DB18C6 (**4**) and alkali metal ions, for instance, the extraction sequence in Figure 39 is obtained (phase-transfer system: water–nitrobenzene) [182]. K^+, because of the high stability of its complex with **4**, is extracted most efficiently. For the other cations, their lipophilicity is decisive.

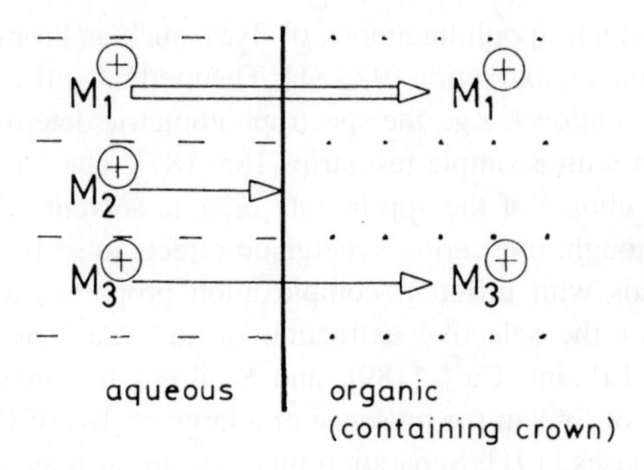

Figure 38. Ion extraction: selective ion transfer into an organic phase

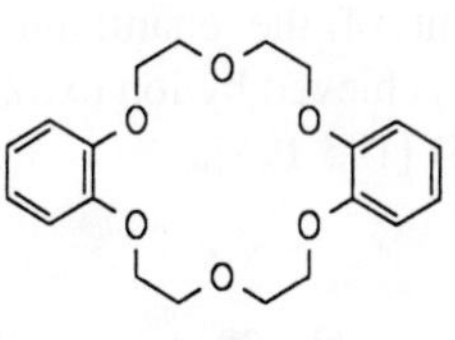

4

$K^{\oplus}$ (framed) $> Rb^{\oplus} > Cs^{\oplus} > Na^{\oplus} > Li^{\oplus}$ (H_2O/Nitrobenzene)

Picrate$^{\ominus}$ (framed) $> SCN^{\ominus} \approx ClO_4^{\ominus} > I^{\ominus} > NO_3^{\ominus} > Br^{\ominus} > OH^{\ominus} > Cl^{\ominus} > F^{\ominus}$ (H_2O/CH_2Cl_2)

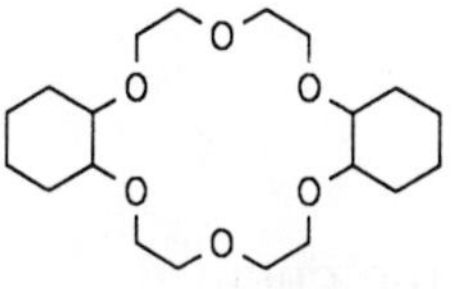

8

$Sr^{2\oplus}$ (framed) ($H_2O/CHCl_3$)

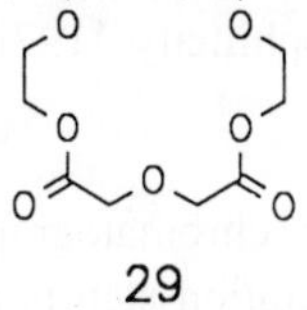

29

$Li^{\oplus}, Ca^{2\oplus}$ (framed) ($H_2O/CHCl_3$)

Figure 39. Extractive enrichment and separation of ions (the ions preferentially transported are framed)

The influence of lipophilicity on ion extraction is also apparent in the sequence obtained for the anions (Figure 39) [183,184]. With an equimolar mixture of all ions investigated, extraction with **4** would effect an enrichment of potassium picrate in the organic phase. Thus, the method can be used analytically in a second way, viz. the selective enrichment and separation of cation–anion pairs (co-extraction) [92]. This is the basis for spectrometric and fluorimetric determinations of ion concentrations, in which lipophilic anions of dyes, such as bromocresol green and eosin, are used for the co-extraction [92,181]. The method still works selectively in the μg/ml region and allows, e.g., the spectrophotometric determination of K^+ and Na^+ in blood serum with a simple test strip [185–187]. The ion selectivity can be fine-tuned with the choice of the appropriate organic solvent [181,182,188]. With mixtures of ions, though, interfering synergistic effects must be expected [181].

Crown compounds with different complexation properties to **4**, such as **8** or **29**, can be used for the selective extraction of Li^+, Ca^{2+} or Sr^{2+} (Figure 39); **29** is selective for Li^+ and Ca^{2+} [189], and **8** allows the analytically important separation of traces of Sr^{2+} in the presence of a large excess of Ca^{2+} [190], e.g. for radioanalytical purposes [191]. Separation methods are also available for mixtures of radioisotopic ions [94]. Thus, the separation of $^{40}Ca^{2+}$–$^{44}Ca^{2+}$ ($^{44}Ca^{2+}$ is used clinically as a non-radiating tracer) with DCH18C6 (**8**) (Figure 40) is more efficient by approximately one order magnitude in its ε-value, than conventional separation methods (ion exchange) [192]. Finally, the enrichment of the enantiomers of racemic ammonium salts on a preparative scale has been achieved by ion extraction with the help of chiral crown compounds (cf. Figure 23) [193,194].

$$^{40}Ca^{2\oplus}_{aq.} + [\,^{44}Ca^{2\oplus}\,Lig]_{CHCl_3} \overset{K_C}{\rightleftharpoons} \,^{44}Ca^{2\oplus}_{aq.} + [\,^{40}Ca^{2\oplus}\,Lig]_{CHCl_3}$$

$$Lig = DCH18C6 \quad (\mathbf{8}) \qquad \varepsilon = 4 \cdot 10^{-3}$$

$$K_C \equiv \alpha = \frac{(^{44}Ca/^{40}Ca)_{aq.}}{(^{44}Ca/^{40}Ca)_{org.}} = 1 + \varepsilon = 1.004$$

Figure 40. Isotopic enrichment by extraction (H_2O-$CHCl_3$)

Figure 41 shows some new examples of crown compounds with specific selectivity properties for the extraction of ions, e.g. **69** for Na^+ [195] and **70** for UO_2^{2+} [196]. Compound **71** is a ligand with a high lipophilicity [197], which guarantees that it remains in the organic phase.

Ion-exchange chromatography [50,92,179]. Ion-exchange chromatography has advantages over extractive separation methods: (a) the separation step is multiple and (b) there is no need to separate the salt from any crown compounds. In this context, high molecular weight supports are used in which crown compounds (cryptands) are used as anchoring groups; examples are complexation resins of both the polymerization or condensation type (72–74; Figure 41) [198]. The separation is based on the fact that on chromatographic columns prepared with such complexation resins, the cations will be eluted at different rates, according to the stability of the complexes formed; cations of labile complexes will move quickly and those of stable complexes slowly. Figure 42 shows a separation diagram obtained from a mixture of salts (mixture of cations) separated on P-DB21C7 (**73**, n = 2), a crown ether resin [198].

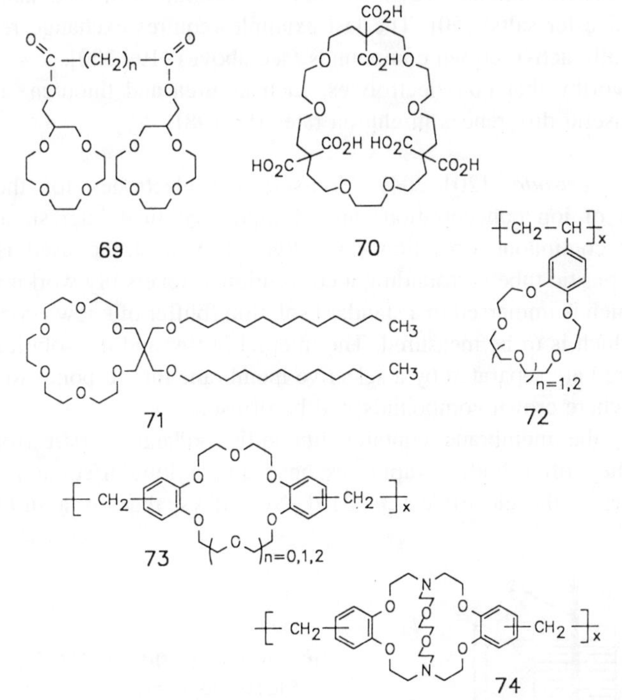

Figure 41. Carrier for ion extraction and ion-exchange resins with crown ethers or cryptands as anchoring groups

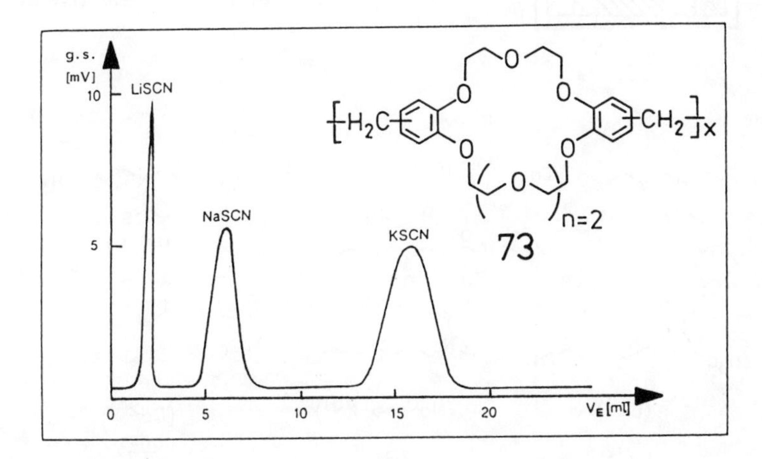

Figure 42. Separation diagram for a mixture of salts (LiSCN, NaSCN, KSCN; elution chromatography with water)

Apart from the separation of different cations with all the same anion, others are also possible: the separation of anions with the same cation, salt inversion with different cations and anions (see above) and the resolution of racemates of amino acids (as their ester salts) [50]. The last example requires exchange resins which contain optically active crown compounds (see above) [199,200].

It is noteworthy that non-electrolytes, such as urea and thiourea, also exhibit analytically useful differences in elution rates [50,198].

Ion-selective electrodes [201–204]. Ion-selective electrodes for the selective measurement of ion concentrations are of high analytical interest, as they are designed for continuous operation [205–208]. The electrode used (Figure 43) consists of a plastic tube surrounding a conventional, reversibly working reference electrode, which is immersed in a standard solution (buffer of known concentration) of the ion, which is to be measured. The internal buffer and the solution, which is to be monitored are separated by a selective membrane that responds to the chosen ion. This is where crown compounds will be of use.

As a rule, the membrane contains lipophilic podands. Apart from the ion selectivity, they offer both a rapid response and a long lifetime to this most important part of the electrode [209,210]. Nowadays, there is a suitable sensor

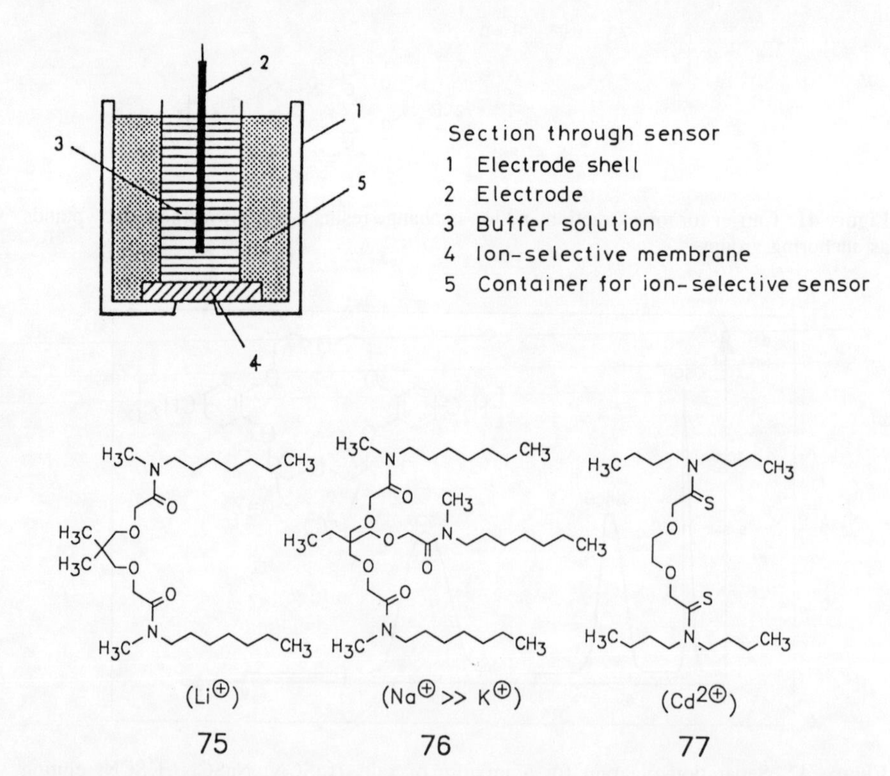

Figure 43. Ion-selective electrode technology: principle of an ion-selective liquid membrane electrode (top) and examples of sensor compounds (bottom)

(ionophore, crown compound) available for almost any cation of interest, e.g. Li^+, Na^+, K^+, Cs^+, Mg^{2+}, Ca^{2+}, Sr^{2+}, Ba^{2+}, Cd^{2+}, UO_4^{2+}, NH_4^+ [50,92,110]. Some (**75–77**) are shown in Figure 43, with the selectivities given.

On the basis of chiral crown compounds (see above), it has also been possible to produce enantiomer-selective electrode membranes, which allow the direct potentiometric determination of enantiomeric excesses of ammonium compounds (e.g. ephedrinium salts) [211,212].

Because of their easy handling and reliability, ion-selective electrodes are widely used in clinical and biological analysis, e.g. for the determination of Ca^{2+} in cytoplasm, blood serum or soil extracts, of Mg^{2+} in living cells and of Na^+ and K^+ in blood and urine [92,203,208,213,214]. For surgery and in intensive care units, it is important, with the help of continuous-flow multi-electrode systems, to be able to monitor continuously and simultaneously several ion concentrations (and activities) in the living body [215,216].

Chromoionophores and fluoroionophores. Chromoionophores (dyestuff crown compounds) are based on the idea, that the selective complexation of a cation by a crown compound can be made visible by a colour effect initiated within the same molecule [219]. This requires a chromophore attached next to the ligand moiety and an electronic coupling between the two parts of the molecule (Figure 44). [Whereas ionic acidic (ionizable by protons) chromoionophores have been known for some time, the neutral dyestuff-crowns are fundamentally new species.] The equivalent is true for fluoroionophores (crown compounds with fluorescent groups). Figure 44 depicts examples of molecules (**78–81**), which show this type of structure [92,217,218]. A chloroform–pyridine solution of the (acidic) dinitrophenylazo-phenyl crown ether **78** (n = 1) produces on addition of Li^+ ions a characteristic change of colour from yellow to red, whereas alkali metal ions with larger ionic radii and lower charge density remain inactive under analogous conditions [220]. These findings were the basis of a method for the determination of Li^+ in drugs [221]. This is the most sensitive colorimetric method for Li^+, and Na^+ does not interfere.

Crown ether **79** (n = 1), a type of phenol blue, also displays extreme bathochromic effects (and at the same time strong hyperchromicity, Figure 45) with ions of high charge density (Li^+ and divalent alkaline earth metal ions), which can be used spectrophotometrically [222]. The azulenyl crown ether **80** (n = 2) is particularly sensitive and selective for Ba^{2+} ions, which produce a colour change from yellowish orange to bluish violet [223]. Ligands of this type have already been incorporated in test strips [224].

Phenolic, i.e. dissociable, acidic chromoionophores, such as **78**, in contrast to neutral crown ether dyes, are well suited for extractive methods [217]. This is also true for the umbelliferone-containing fluoroionophore **81**, which in extraction experiments displays strong fluorescence and high selectivity for Li^+ and K^+, depending on its ring size (n = 1 and n = 2, respectively) [225]. Because of its high sensitivity, fluorimetry is becoming increasingly important in chemical trace analysis [226]. A useful development consists in the ‘visualization’ of the

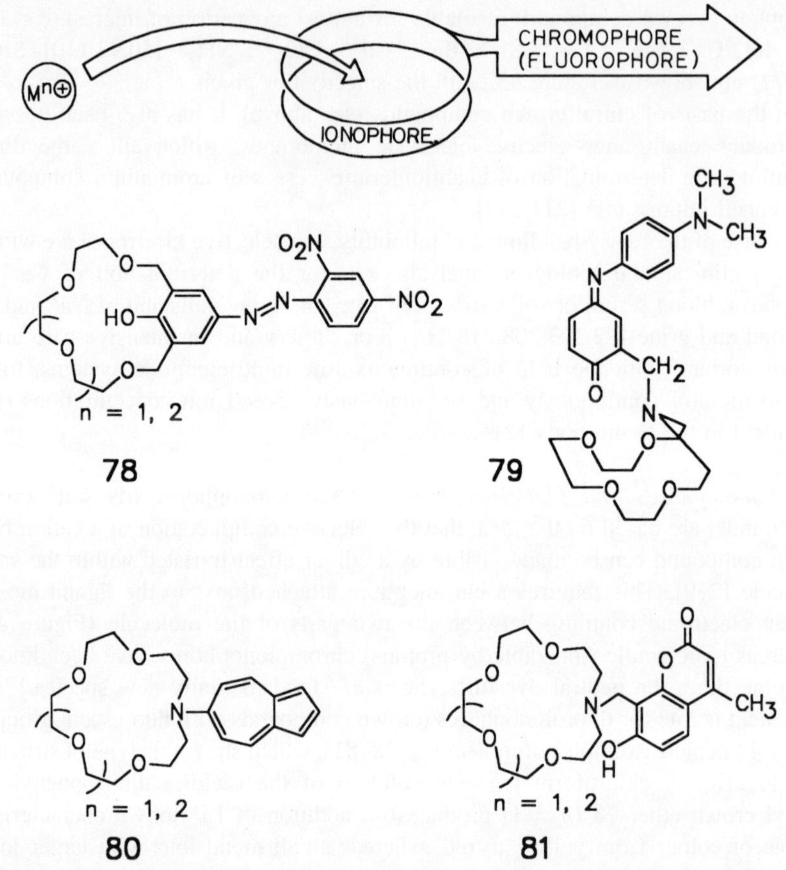

Figure 44. Chromoionophores and fluoroionophores. Molecular set-up and examples

enantiomeric purity of optically active compounds by chiroselective colour effects [218].

Transport through membranes [89,92,227–230]. As with extractive methods (see above), transport through membranes can be used to separate and enrich ions. However, in this case, there is no necessity for the separation of ligand and salt. In principle, ions are moved selectively from one aqueous phase (D = donor phase) through a lipophilic layer (M = liquid membrane) into a second aqueous phase (A = acceptor phase) (Figure 46). The selective transport of ions is effected by an (ionophoric) crown compound embedded in the organic layer [92,228].

The crown compound acts as a carrier by taking over the ion (ion pair) at the D/M interface, carrying it through the membrane as a complex and discharging it at the other interface M/A ('symport'). The carrier then shuttles back, and the process is repeated until an equilibrium concentration of the transported ions in both aqueous phases is reached (diffusion process).

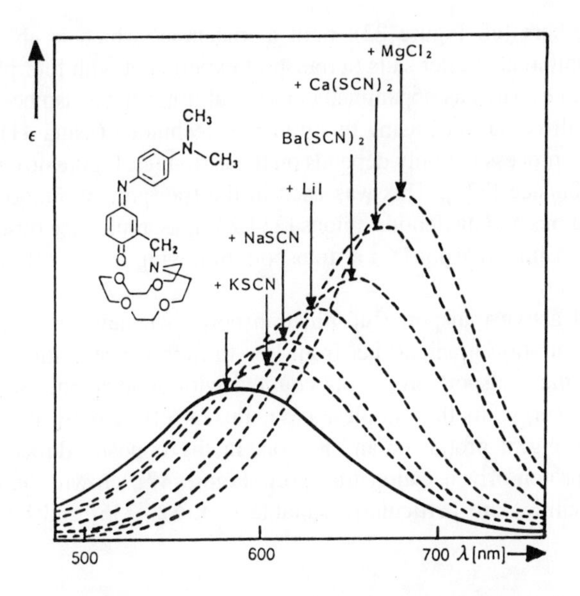

Figure 45. Selective detection of ions via cation-selective light absorption [bathochromic shift on addition of indicated salts (cations) to the solution of chromoionophore **79** in acetonitrile]

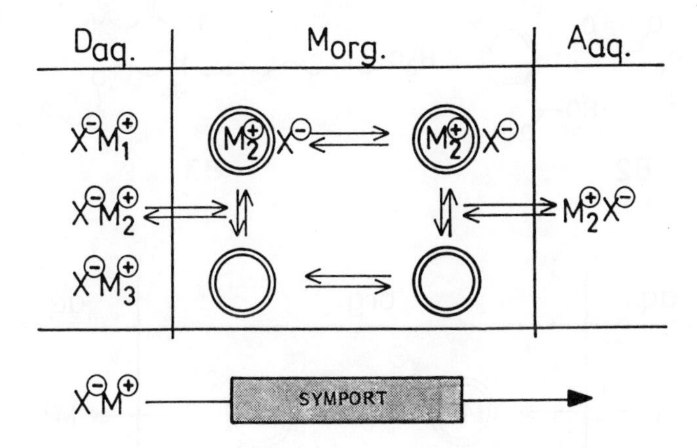

Figure 46. Selective carrier-induced transport of ions through a membrane (symport); crown carriers are shown as rings

To achieve the complete transport of a given ion, its equilibrium concentration in the membrane should remain low. Therefore, the crown compound should not form too strong a complex with the transported ion (suitable examples are podands, e.g. **41**) [229]. Thus, transport through membranes and solvent extraction produce different requirements for the crown to be used. If the transported ions in the acceptor phase are continually removed, the complete and selective transport of metal ions is then possible [92]. On this basis, Cram and co-workers using chiral crown ethers (cf. Figure 22), built a 'machine' which is able to separate enantiomeric amino acid ester salts (a transport experiment with four phases) [231].

Biogenic amines, such as dopamine, and several drugs have also been selectively freed from alkali metal ions, using this transport technique (using **41**) [232].

The transport process not only depends on the cation (cf. Figure 46), but the anion also has an influence [233]. This was used in the transport of amino acid anions, oligopeptide anions and nucleotide anions [234,235]; as a rule, one observes passive transport ('down hill transport'), i.e. transport following a concentration gradient [92].

We speak of active transport ('up hill transport') whenever ions are 'pumped' against a concentration gradient, i.e. from low to high concentration. This is only possible when the transport process is coupled with another one, which supplies the necessary energy. In the simplest case, this is effected by the transport of a different ion, e.g. a proton or an electron, in the opposite direction (antiport) [92]. For the proton-driven cation transport (Figure 47), crown compounds with acidic functionalities are particularly suitable (e.g. **82** [236] and **83** [237]). The

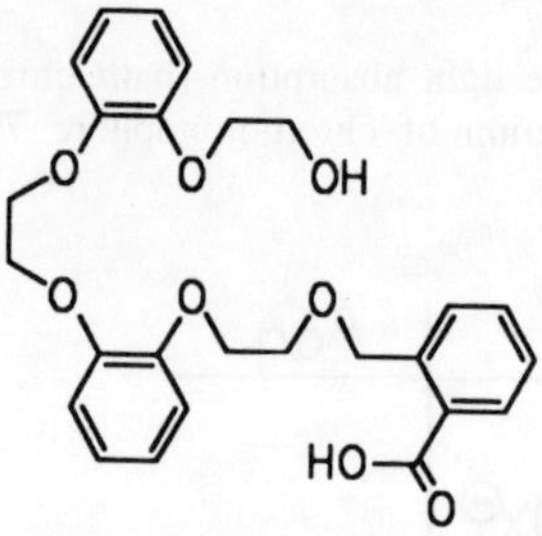

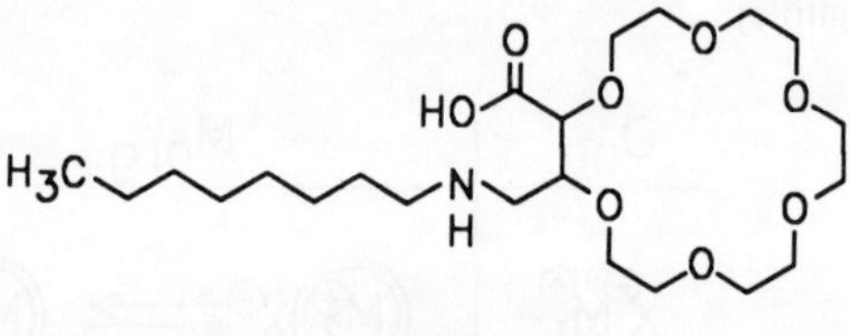

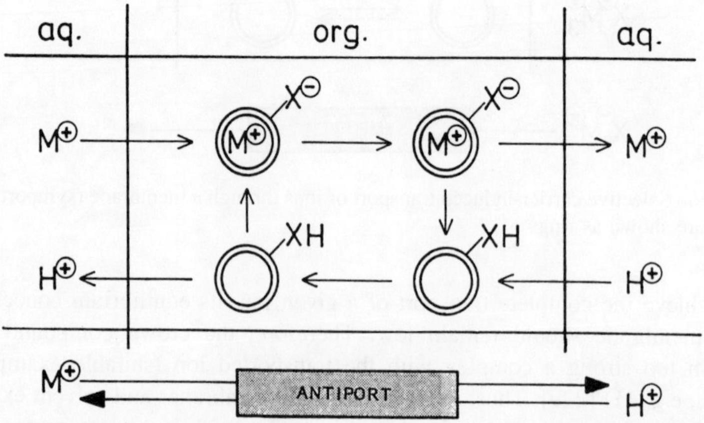

Figure 47. Carriers and transport system for the proton-driven cation transfer through a membrane (antiport; carriers shown as rings)

carrier shuttles back and forth between the cation-rich and the proton-rich aqueous phases, with the proton return transport overcompensating for the cation forward transport. Under such conditions, the total transport of a cation is possible, even without external help (e.g. removal of the transported cation from the acceptor phase) [238].

Even more interesting for practical applications and theoretical investigations are coupled ion-transport processes, which can be turned on and off with light or electrical energy (Figure 48) [239]. Crown compounds able to do this contain a molecular switch, which can be activated either photochemically or electrochemically and thus induce a change in the ligand's conformation (cf. the azo-group in **84** [240,241]) or configuration (cf. the disulphide bridge in **85** [242]). By triggering the switch, e.g. at an interface, the ion transport can be increased, changed to another ion or interrupted [239]. Chemical switches on a molecular level are deemed to be of the highest importance for technology in the years to come (cf. Chapters 7 and 10) [243].

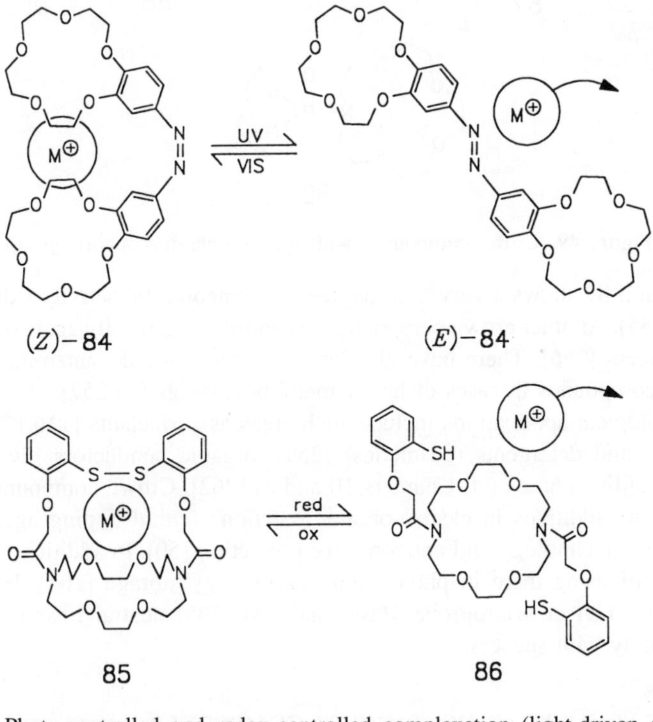

(*Z*)−84 UV/VIS (*E*)−84

85 red/ox 86

Figure 48. Photo-controlled and redox-controlled complexation (light-driven and redox-driven cation transport, respectively)

2.2.1.6 Biological [54,61] and other applications [51]

Alkali metal ions are of importance in a number of biological processes, e.g. the propagation of stimuli along nerves, the nervous control of secretion functions and muscle functions, protein synthesis and enzymatic regulation of metabolisms [51]. Crown compounds, as complexation agents, are able to intervene in such processes (cf. ionophores, Figure 2). These topics are covered in the literature [244–249].

Numerous bacteria, viruses (e.g. influenza and rhino viruses) and coccinic pathogenic agents can have their growth noticeably restrained by crown compounds, for which, it is believed, the metal complexing properties of the liglands are responsible [250–252]. This is also true for pharmacological crown compounds [61] built from moieties of classical drugs and crowns (Figure 49), e.g. isoprenaline as a pharmacophoric group in **87** [253] and eupaverin in **88** [254], both attached to B15C5.

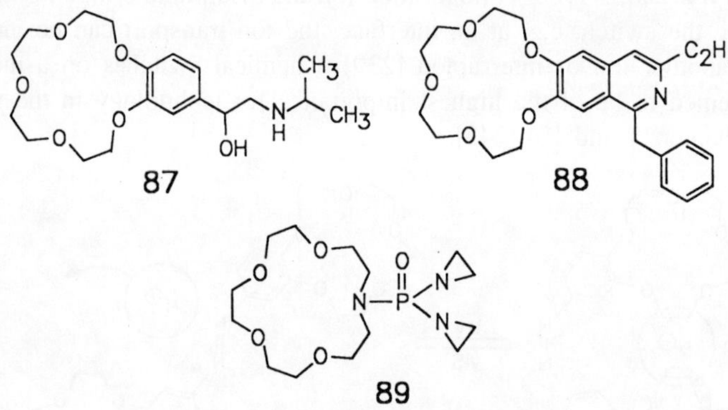

Figure 49. Crown compounds with attached pharmacophoric groups

Compound **89** shows a very high degree of antineoplastic activity ('drug for the future') [255]. Another crown compound acts antiulcerogenically against histamine-induced ulcers [256]. There have also been reports of the decontaminating action of crown compounds in cases of heavy metal poisonings [54,257].

Non-biological applications include such areas as surfactants [116,197,258] (cf. Chapter 9) and detergents (cosmetics) [259], organic conductors [260,261] and liquid-crystalline phases (cf. Chapters 10 and 8) [262]. Crown compounds are also of interest as additives in electro-organic reactions (chiral doping agents) [263], galvanization technology and anticorrosive protection [50]. In addition, there is the possibility of using them in photography and energy storage [264]. In LAMMA spectrometry (*LA*ser *M*icroprobe *M*ass *A*nalysis) [265] neutral ligands have been used routinely with success.

2.2.2 MULTI-NUCLEAR HOST–GUEST COMPLEXES

Beyond the monotopic 'molecular receptors' discussed above, which selectively recognize topologically defined guest-molecules, bind and transport them, or else submit them to catalysed reactions, macropolycyclic host-structures can be used as 'polytopic receptors' [266–270]: depending on their symmetry and their binding sites, several substrates (guests) of either the same or different species can be bound simultaneously by them. The properties of the host-compound are thereby enriched with cooperative, allosteric and regulatory effects and therefore go far beyond the discussed properties of monotopic receptors.

The binding properties of macropolycyclic cryptands towards metal cations, where extra coordination sites could intercalate another substrate with donor properties are of particular interest. Similarly, the complexation of organic and biologically relevant molecular cations and anions, in addition to neutral compounds, would also be worth investigating. Problems such as those described below can be studied with multi-nuclear guest compounds:

— selective binding and transport of one or several substrates;
— catalysis of multiple-centre, multi-electron processes;
— activation of bonded substrates;
— thermally and photochemically induced condensations and cross-reactions between different substrates, which are held close together by the host compound.

From a structural point of view, macro-oligocyclic cryptands and their guests have the possibilities shown schematically in Figure 50.

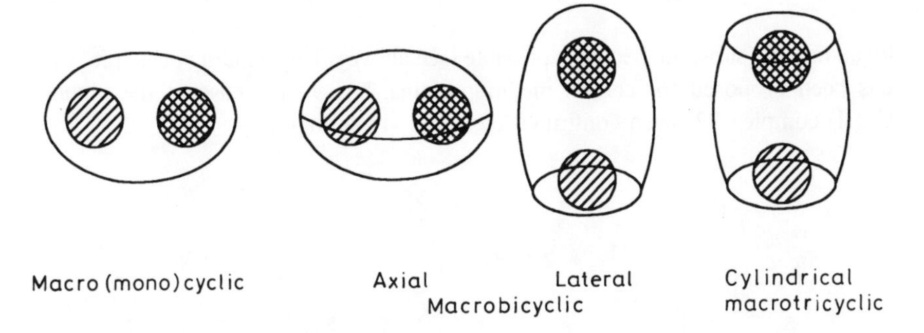

Figure 50. Various structures of oligocyclic cryptands (the circles within the oligocycles represent guest particles)

2.2.2.1 Binuclear macromonocycles

Following the HSAB principle, alkali, alkaline earth and also transition metals can be bound by crown compounds containing O, N or S donor atoms and thus be incorporated into a cyclic host-compound. In addition to a one-fold, a two-fold metal cation complexation is also possible, after which a complementary inclusion of an organic substrate may follow (Figure 51).

In 1970 several groups described the first binuclear metal complexes containing the components **90** and **91** (see below). A little later, Lehn [266–270] successfully synthesized hexaaza-monocyclic chelands, e.g. **92–94**. All three ligands form binuclear Cu(II) complexes, in which the distances between the metal ions are 598 and 479 pm for **92** and **93**, respectively. A special feature of such ligands is their ability to embed the metal ions in such a way that an extra bridging

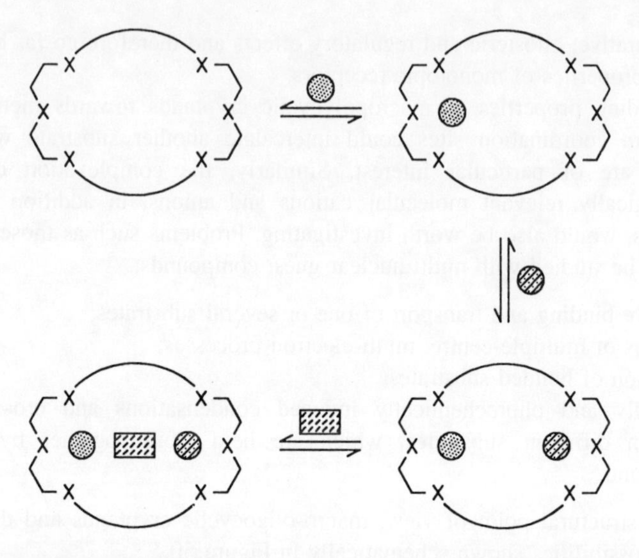

Figure 51. Multiple complexation of metal cations (circles) and organic substrates (rectangles) (schematic)

by an organic substrate becomes possible (see above). This bonding principle, which has been proposed for certain metalloproteins, has, in the case of the binuclear Cu(II) complex **92**, been confirmed by X-ray crystallography.

90 + 2 Cu^{II} **91** + 2 Ni^{II}

92 **93** **94**

As shown in Figure 52 and regarding the use of such compounds for highly specific reactions, very different functions can be assigned to the coordination sites of the metal ion used. For the binding of the metal ions to the macrocyclic cheland, two or three coordination sites are usually sufficient. The remaining coordination sites can then be used for the binding of modulators (e.g. Ph_3P) or substrates, which will change the cation's properties and thus allow reactions with enzymatic character. An example is carboxypeptidase A, which contains in its reactive centre a zinc ion, which is bonded to the polar groups of the protein backbone, but offers a coordination site to the incoming substrate.

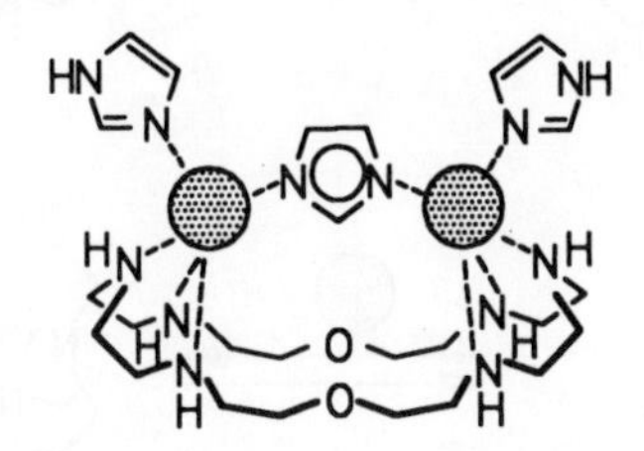

Figure 52. Structure of the imidazole-bridged binuclear cryptate **92**

In addition to the presented symmetrical ligands, unsymmetrical ones, built from different crown moieties would be of special interest. They would be able to bind two different metal ions simultaneously, e.g. Cu(II) and Zn(II), and could then be used in redox reactions.

2.2.2.2 Macrobicyclic binuclear cryptates

The tremendous complexing power of the ligand 'tren' [$N(CH_2CH_2NH_2)_3$] has been known for a long time. By incorporating two tren units in one macrobicyclic system, cryptand **95** was obtained. It has an ellipsoidal cavity and forms binuclear complexes with Ag^+, Zn^{2+}, Cu^{2+} or Co^{2+}, in which the average intermetallic distance is 450 ± 50 pm. They are formed in two successive steps, via the unsymmetrical mononuclear species. The stability constants are 16.6 for the mononuclear and 11.8 for the binuclear complex. Investigations using ESR spectroscopy suggest weak interactions between the two copper ions. Oxidation of the binuclear Co(II) complex leads to the formation of a μ-peroxo–μ-hydroxo species, as was shown by spectroscopic and titrimetric analysis. By comparison with the analogous bis(tren) complex, whose geometry is known, structure **96** was ascribed to the macrobicyclic compounds. Remarkably, the O_2 is not bonded as strongly in the bicycle as it is in the corresponding bis(tren) complex.

A similar structural complexation geometry is found in a Cu complex recently described by Karlin, with which the O_2-bonding behaviour of porphyrins was to be investigated and imitated on a low molecular level.

The existence of Cu-containing enzymes, such as haemocyanine, which transports O_2, and the monooxygenases tyrosinase and dopamine-β-hydroxylase, which incorporate oxygen atoms from O_2 into organic substrates, prompted the development of workable synthetic systems for the reversible binding of oxygen. X-ray crystal structure, chemical and spectroscopic analysis in particular have, over the last few years, demonstrated that two Cu(I) ions are surrounded by three histidine ligands, and when oxygen is absorbed a peroxo bridge is formed between the two metal centres. A similar complexation geometry was shown to exist in the low-molecular-weight model compound **97**.

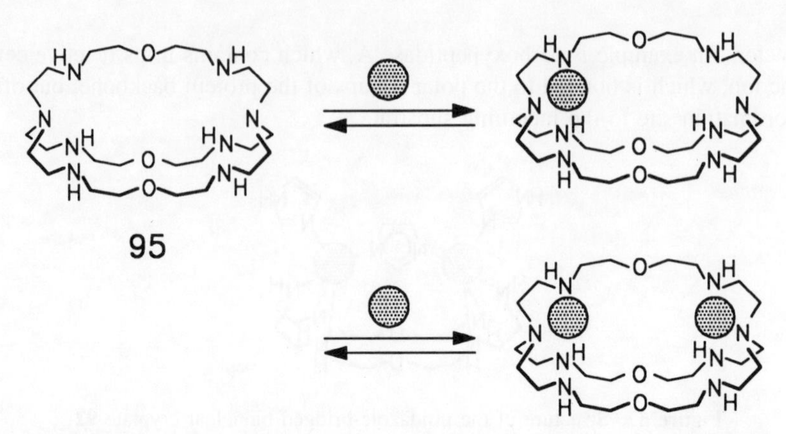

In dichloromethane solution and at low temperature, one equivalent of O_2 is absorbed, forming a purple two-centre complex with a trigonal bipyramidal structure and a peroxo bridge between the two metal centres. The binding of oxygen is linked with the reversible oxidation of Cu(I) to Cu(II). Under vacuum, the peroxo complex can be converted back reversibly to the Cu(I) complex. The whole process can be repeated several times, during which, however, increasing decomposition of the ligand is observed.

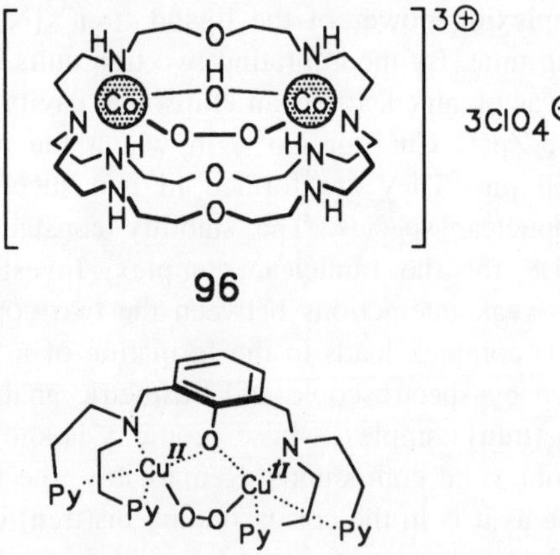

$3\oplus$ $3ClO_4^{\ominus}$

96

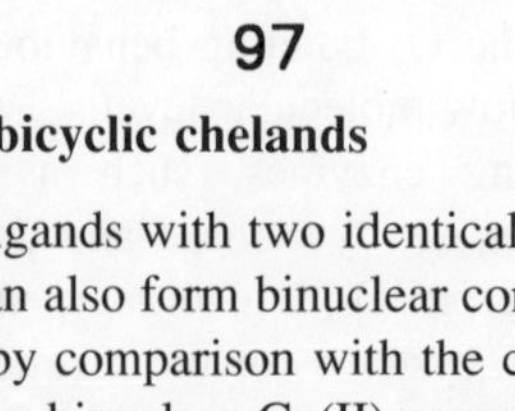

97

2.2.2.3 Lateral macrobicyclic chelands

Lateral macrobicyclic ligands with two identical bridges and a third, different one, as for example in **98**, can also form binuclear complexes, which exhibit one special feature: as was shown by comparison with the corresponding Cu(II) complexes of the partial structures, the binuclear Cu(II) complex has two redox potentials. This behaviour is explained by the dissimilar complexation geometries of the two Cu ions. Thus, a potential of +550 mV is ascribed to Cu ion I and one of +70 mV to Cu ion II. A similar behaviour is expected for compounds in which a porphyrin backbone is bridged with chelating chains.

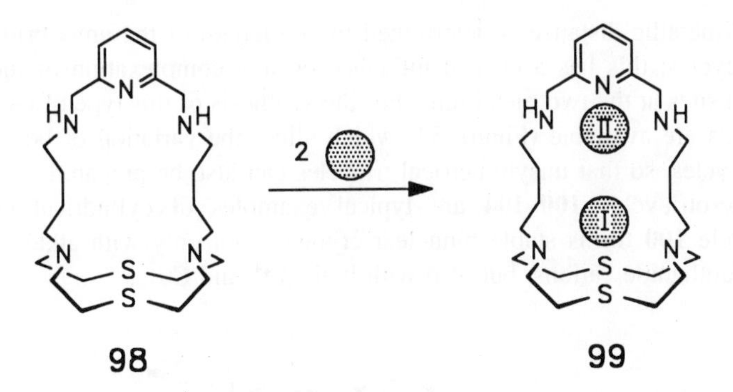

2.2.2.4 Binuclear cryptates of cylindrical macrotricyclic ligands

Macrotricyclic molecules of the cylindrical type (Figure 53) have three cavities: two lateral circular ones (drawn in boldface) and a central one (Figure 53a). In the case of binuclear complexes, the metal ions will be bonded by the lateral donor centres in different ways, depending on the ring sizes:

a. When the ring sizes are too small, e.g. in a tetracycle, the metal ion will be bonded on the surface of the cycle, i.e. it cannot dip into the ring. With this arrangement, the donor sites inevitably become aligned in one direction, so that in an octahedral complex a *cis*-octahedral arrangement is found [Figure 53a (a)].
b. With sufficiently large ring sizes, however, the metal ion can dip into the cycle, and thus a *trans*-octahedral complexation geometry becomes possible [Figure 53a (c)].

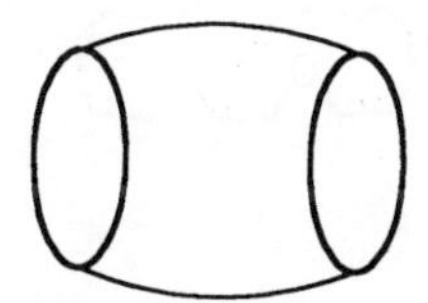

Figure 53. Cylindrical macrotricycle (schematic)

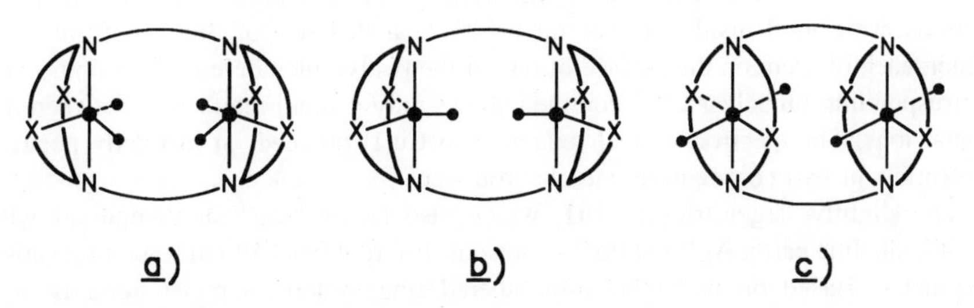

Figure 53a. Geometries of complexes of macrotricyclic ligands: (*a*) *cis*-octahedral; (*b*) *cis*-pentagonal-bipyramidal; (*c*) *trans*-octahedral

The intermetallic distance is determined by the length of the units bridging the two monocycles; this has a critical influence on any complexation of the as yet unoccupied sites at the two metal ions. For the synthesis of this type of compound, two methods are available (Figure 54) which allow the variation of both bridges and monocycles, so that unsymmetrical tricycles can also be prepared.

The macrotricycles **100–104** are typical examples of cylindrical chelands. Macrotricycle **100** forms stable binuclear cryptates not only with alkali, alkaline earth and lanthanide cations, but also with both Ag^{+} and Cu^{2+}.

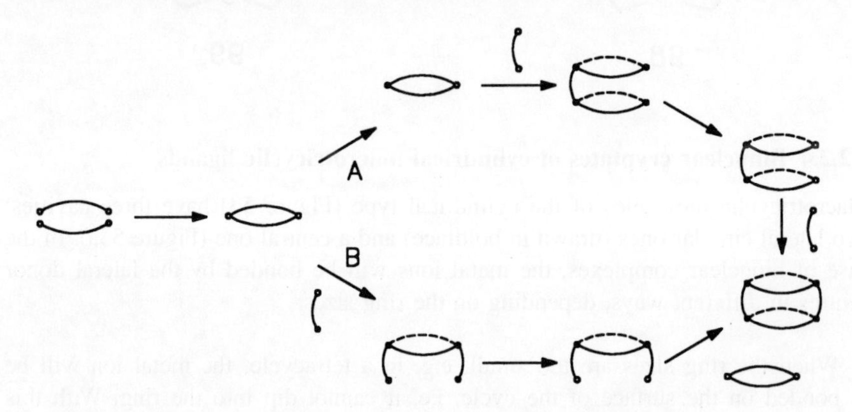

Figure 54. Strategies for the synthesis of cylindrical macrotricycles (schematic)

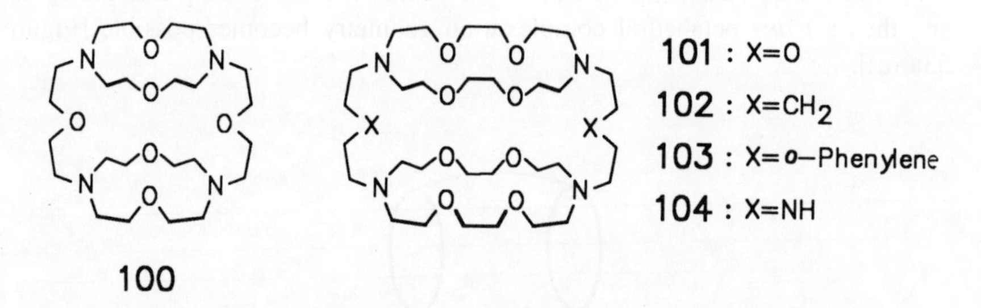

An analysis of the structure of the silver complex shows that the two metal cations are bound inside the central cavity at a distance of 388 pm from each other, each of them on the surface of one of the twelve-membered crown units. The corresponding binuclear Cu^{2+} complex shows only a small interaction between the metal ions. The electrochemical reduction to Cu(I) proceeds at distinctly positive potentials in two consecutive one-electron steps.

The slightly larger tricycle **101**, which also forms binuclear complexes with alkali, alkaline earth, Ag^{+} and Pb^{2+} cations, is different from **100** in its complexation geometry. Based on its eighteen-membered ring system, a metal bond is now possible, the metal ions dipping into the rings. Thus it is understandable that in the binuclear Na^{+} complex the intermetallic distance is 640 pm, despite the fact that in both **100** and **101**, the two crown subunits are at the same distance from each other.

Apart from the symmetrical homo-binuclear metal complexes, corresponding unsymmetrical mononuclear, and also heteronuclear complexes have been observed. Cryptand **105** has an interesting chiral structure, in which the two crown subunits are linked on one side with a binaphthol bridge. In addition to the expected binding of metal ions, it is also able to bind molecular cations and anions with limited chiral discrimination. The fixation of a chiral, negatively charged molecule by the chiral host is done by binding the anion to a previously complexed cation.

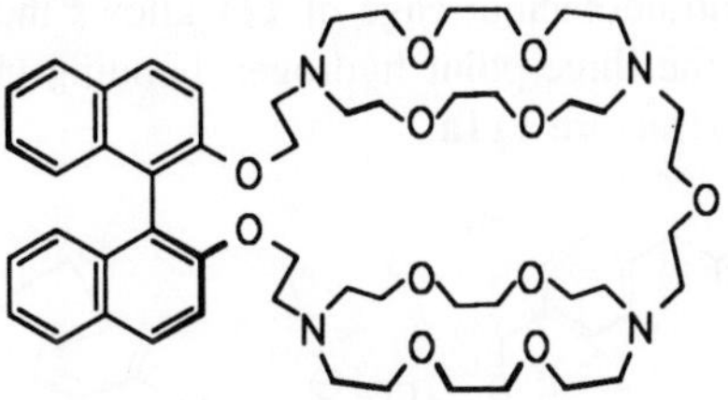

105

If the oxygen atoms in the crown ring are replaced with sulphur atoms (**106–108**), then the complexation of transition metal cations also becomes possible. Thus, **106** and two Cu(II) cations form a complex with structure **109**, in which the metal–metal distance is 562 pm. The complexation of the Cu ions proceeds in two steps with stability constants for the mononuclear and the binuclear complexes of 10.9 and 9.0, respectively (in aqueous solution). Electrochemical investigations of this complex showed a reversible two-electron step with a potential of +445 mV. Although ESR spectroscopy did not show any significant couplings between the metal

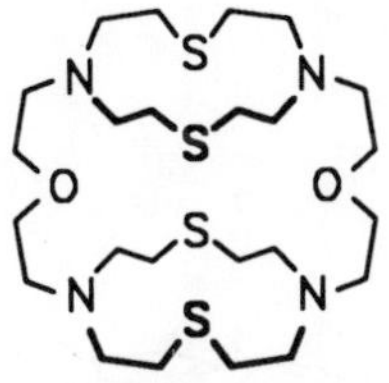

106 ● = $Cu^{2\oplus}$ 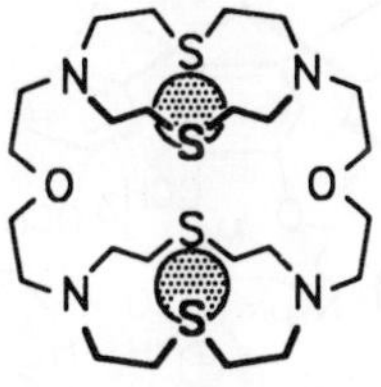 **109**

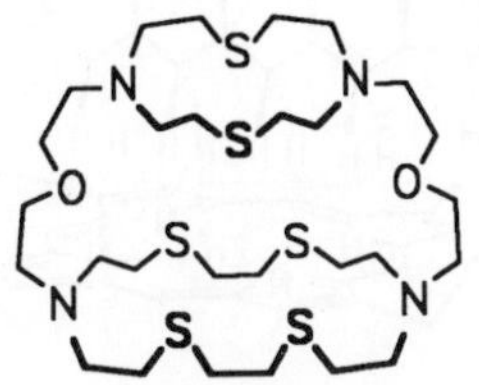

107 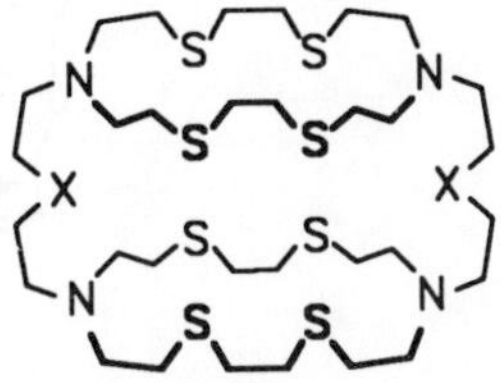 **108**

ions; measurements of the magnetic susceptibility revealed an antiferromagnetic coupling.

2.2.2.5 Speleands

Combinations of polar subunits and rigid non-polar backbone components afford macropolycyclic 'co-receptors' of the cryptand type, called *speleands*. In addition to speleand **110**, which binds molecular cations intramolecularly with electrostatic and hydrophobic interactions (primary, secondary, tertiary and quaternary ammonium compounds, such as acetylcholine), molecular structures like **111** also belong to the speleands. The rigid intramolecular cage of **111** allows the insertion of $CH_3NH_3^+$ ions with the help of the three-point hydrogen bonding characteristic of crowns, the complex adopting structure **111a**.

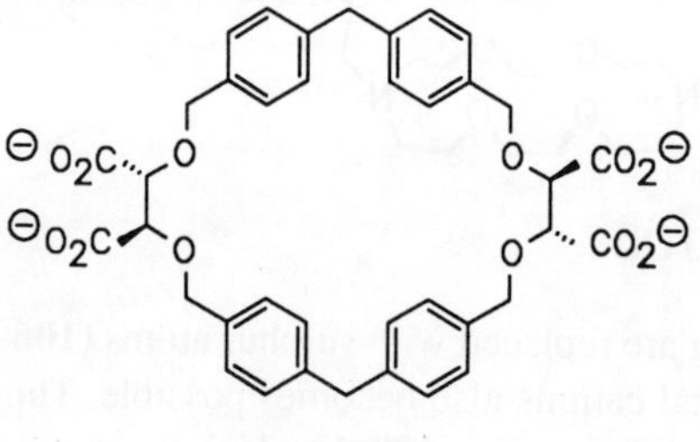

110

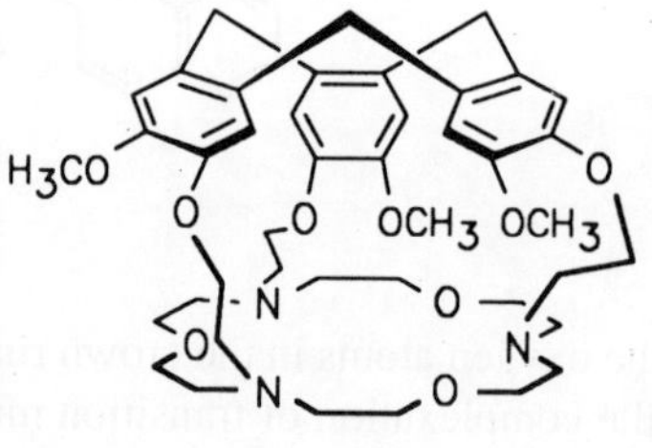

111

Porphyrin-containing macrocycles bridged with crowns, such as **112**, can also function as co-receptors. In addition to the complexation of Zn^{2+} ions into the tetrapyrrole ring system, an additional incorporation of organic diammonium ions is possible, as shown by complex **112**.

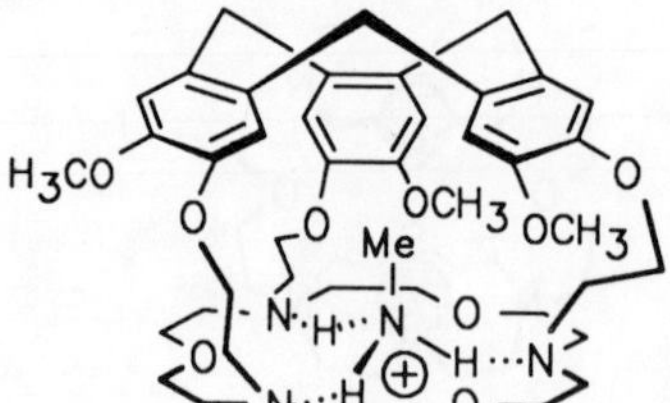

111a

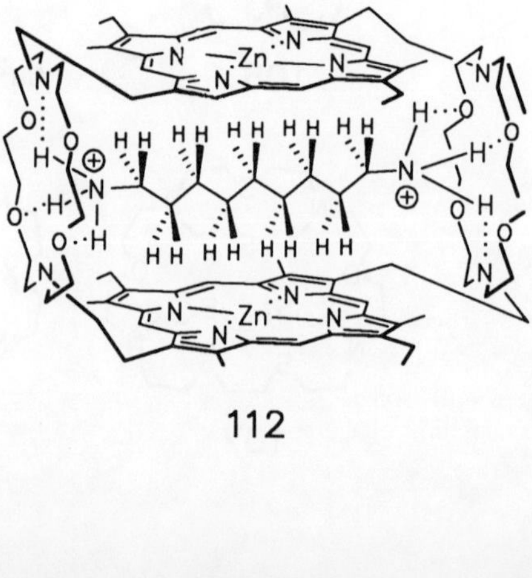

112

The simultaneous complexation of organic and inorganic substrates offers the possibility of physical or chemical interactions and reactions between a metal and an organic substrate.

2.2.2.6 Photoactive cryptands

In 1984, Lehn described *photoactive cryptands* [271]. Here, the donor centres are bipyridine and phenanthroline moieties. In 1987, the energy-transfer luminescence of europium(III) and terbium(III) cryptates was reported [272]. Eu(III) and Tb(III) cryptates with ligands **113** and **114** and Eu(III) cryptates with **115** can be obtained by reaction of the corresponding Na^{+} cryptates with Eu(III) and Tb(III) salts. Stable, kinetically inert complexes are formed, in which the metal ion is enclosed in the ligand's cavity, where interactions with the solvent and dissolved molecules are not possible [273]. The particularity of these Eu cryptates consists, in contrast to simple aquo complexes, in their strong luminescence in aqueous solution at room temperature, and a marked difference in redox potential compared with the aquo complexes.

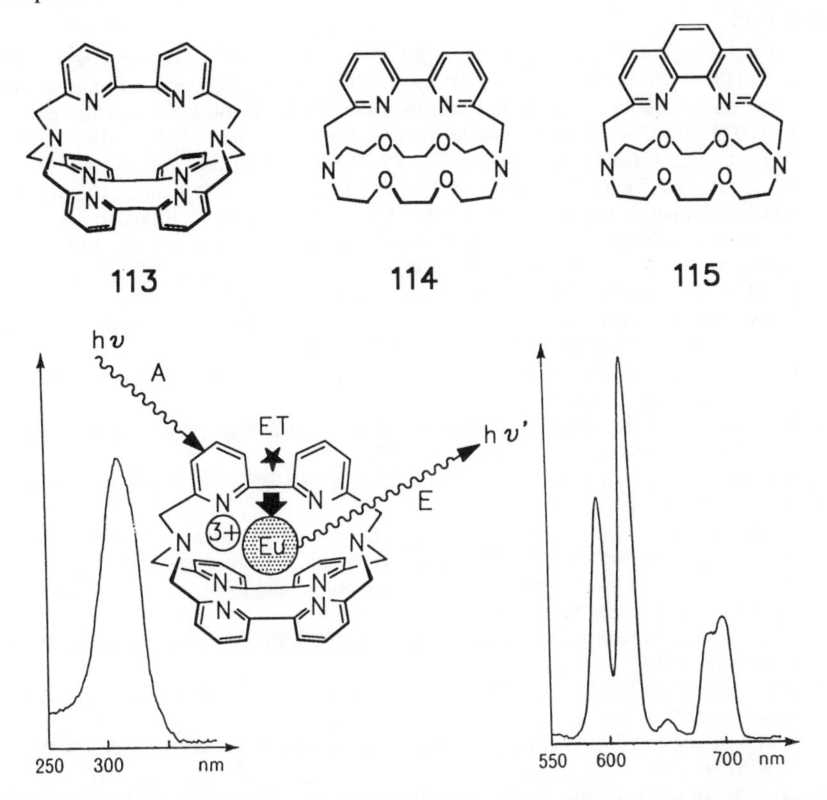

Figure 55. Scheme for the light conversion process, absorption–energy transfer–emission (A–ET–E) light conversion process, exemplified with [Eu^{3+}·**113**] (centre). Left, excitation spectrum (emission at 700 nm); right, emission spectrum (excitation at 320 nm); 10^{-6} M aqueous solution of the nitrate at 20 °C

While the excitation spectra of the complexes relate to those of the free ligands, the red and green emissions display spectra characteristic of Eu(III) or Tb(III) luminescence. According to these findings, the UV light absorbed by the ligand is used via intramolecular energy transfer to excite Eu and Tb states, which then emit from $^5D_0$ and $^5D_4$ states, respectively. The absorption–emission process in such light converters is shown schematically in Figure 55.

Electrochemical investigations have shown the reduction of the Eu cryptate of **113** to be reversible at a potential of −555 mV and the Eu(II) state to be stabilized. The duration of the emission is also longer in these complexes than in the corresponding aquo complexes. This is due, at least partially, to the better screening of the metal ion by the macrobicyclic ligand.

In addition to the interest in the purely photophysical and photochemical aspects [274] of this type of complexes (cf. Chapter 12), there are also a number of potential applications, e.g. in the development of luminescent materials and markers for biological uses.

References

1. (a) C.J.Pedersen, *J. Am. Chem. Soc.*, **89**, 2495, 7017 (1967); *Angew. Chem.*, **100**, 1053 (1988); (b) recent reviews: E.Weber, in *Phase Transfer Catalysts, Properties and Applications*, Merck-Schuchardt, Darmstadt, 1987; H.J.Schneider, *Chem. Unserer Zeit*, **6**, A60 (1987); E.Weber and F.Vögtle, *Nachr. Chem. Tech. Lab.*, **35**, 1149 (1987); (c) J.-M.Lehn, *Angew. Chem., Int. Ed. Engl.*, **27**, 89 (1988) (Nobel lecture); D.J.Cram, *Angew. Chem., Int. Ed. Engl.*, **27**, 1009 (1988) (Nobel lecture); H.M.Colquhoun, J.F.Stoddart and D.J.Williams, *Angew. Chem., Int. Ed. Engl.*, **25**, 487 (1986); E.Weber, Molecular Inclusion and Molecular Recognition—Clathrates II. *Top. Curr. Chem.*, **149** (1988), Springer, Berlin; (d) J.F.Stoddart, *Ann. Rep. Prog. Chem. Sect. B*, **85**, 353 (1988–9); (e) D.H.Busch and N.A.Stephenson, *Coord. Chem. Rev.*, **100**, 119 (1990).
2. *Alkali Metal Complexes with Organic Ligands, Structure and Bonding*, Vol.**16**, Springer, Berlin, Heidelberg, New York, 1973.
3. G.A.Melson (Ed.), *Coordination Chemistry of Macrocyclic Compounds*, Plenum Press, New York, London, 1979.
4. F.de Jong and D.N.Reinhoudt (Eds.), *Stability and Reactivity of Crown Ether Complexes*, Academic Press, New York, 1981.
5. G.W.Gokel and S.J.Korzeniowski (Eds.), *Macrocyclic Polyether Synthesis*, Springer, Berlin, Heidelberg, New York, 1982.
6. M.Hiraoka (Ed.), *Crown Compounds, Their Characteristics and Applications. Studies in Organic Chemistry*, Vol. 12, Elsevier, Amsterdam, Oxford, New York, 1982.
7. R.M.Izatt and J.J.Christensen (Eds.), *Synthetic Multidentate Macrocyclic Compounds*, Academic Press, New York, San Francisco, London 1978.
8. R.M.Izatt and J.J.Christensen (Eds.), *Progress in Macrocyclic Chemistry*, Vol.1. Wiley, New York, 1979.
9. R.M.Izatt and J.J.Chritsensen (Eds.), *Progress in Macrocyclic Chemistry*, Vol.2, Wiley, New York, 1981.
10. R.M.Izatt and J.J.Christensen (Eds.), *Progress in Macrocyclic Chemistry*, Vol.3. Wiley, New York, 1986.
11. (a) E.Weber and F.Vögtle, *Nachr.Chem.Tech.Lab.*, **35**, 1149 (1987); (b) F.Vögtle (Ed.), *Host Guest Complex Chemistry, I, Topics in Current Chemistry*, Vol.**98**, Springer, Berlin, Heidelberg, New York, 1981.
12. F.Vögtle (Ed.), *Host Guest Complex Chemistry II, Topics in Current Chemistry*, Vol.**101**, Springer, Berlin, Heidelberg, New York, 1982.

13. F.Vögtle and E.Weber (Eds.), *Host Guest Complex Chemistry, III, Topics in Current Chemistry*, Vol.**121**, Springer, Berline, Heidelberg, New York, 1984.
14. F.Vögtle and E.Weber (Eds.), *Host Guest Complex Chemistry—Macrocycles—Synthesis, Structures, Applications*, Springer, Berlin, Heidelbeg, New York, Tokyo, 1985.
15. F.Vögtle and E.Weber (Eds.), *Biomimetic and Bioorganic Chemistry*, Topics in Current Chemistry, Vol.**128**, Springer, Berlin, Heidelberg, New York, 1985.
16. C.J.Pedersen, *Aldrichim*. Acta, **4**, 1 (1971).
17. M.R.Truter and C.J.Pedersen, *Endeavour*, 142 (1971).
18. J.J.Christensen, J.O.Hill and R.M.Izatt, *Science*, **174**, 459 (1971).
19. H.Klamberg, *Chem. Lab. Betr*., **29**, 97 (1971).
20. D.St.C.Black and A.J.Hartshorn, *Coord. Chem. Rev*., **9**, 219 (1972).
21. C.J.Pedersen and H.K.Frensdorff, *Angew. Chem., Int. Ed. Engl*., **11**, 16 (1972).
22. B.Dietrich, J.-M.Lehn and J.P.Sauvage, *Chem. Unserer Zeit*, **7**, 120 (1973).
23. F.Vögtle and P.Neumann, *Chem. -Ztg*., **97**, 600 (1973).
24. J.S.Bradshaw and J.Y.K.Hui, *J.Heterocycl. Chem*., **11**, 649 (1974).
25. J.J.Christensen, D.J.Eatough and R.M.Izatt, *Chem. Rev*., **74** 351 (1974).
26. D.J.Cram and J.M.Cram, *Science*, **183** 803 (1974).
27. P.N.Kapoor and R.C.Mehrotra, *Coord. Chem. Rev*., **74**, 1 (1974).
28. C.Kappenstein, *Bull. Soc. Chim. Fr*., 189, (1974).
29. D.J.Cram, R.C.Helgeson, L.R.Sousa, J.M.Timko, M.Newcomb, P.Moreau, F.deJong, G.W.Gokel, D.H.Hoffman, L.A.Domeier, S.C.Peacock, K.Madan and L.Kaplan, *Pure Appl. Chem*., **43**, 327 (1975).
30. L.F.Lindoy, *Chem. Soc. Rev*., **4**, 421 (1975).
31. J.Lipkowski, *Wiad. Chem*., **29**, 435 (1975).
32. D.J.Cram, in *Application of Biochemical Synthesis in Organic Systems, Part II* (J.B.Jones, C.J.Sih and D.Perlmann, (Eds.), *Techniques of Chemistry*, Vol. X, Wiley, New York, 1976, p.815.
33. G.W.Gokel and H.D.Durst, *Aldrichim. Acta*, **9**, 3 (1976).
34. G.W.Gokel and H.D.Durst, *Synthesis*, 168 (1976).
35. A.C.Knipe, *J. Chem. Educ*., **53**, 618 (1976).
36. E.Weber and F.Vögtle, *Chem. Exp. Didakt*. **2**, 115 (1976).
37. W.Burgermeister and R.Winkler-Oswatitsch, *Top. Curr. Chem*., **69**, 91 (1977).
38. R.C.Hayward, *Chem. Tech. Lab*., **25**, 15 (1977).
39. R.M.Izatt, L.D.Hansen, D.J.Eatough, J.S.Bradshaw and J.J.Christensen in *Metal–Ligand Interactions in Organic Chemistry and Biochemistry, Part 1* (B.Pullman and N.Goldblum, Eds.), Reidel, Dordrecht, 1977 p.337.
40. J.-M.Lehn, *Pure Appl. Chem*., **49**, 857 (1977).
41. G.R.Newkome, J.D.Sauer, J.M.Roper and D.C.Hager, *Chem. Rev*., **77**, 513 (1977).
42. M.R.Truter, in *Metal–Ligand Interactions in Organic Chemistry and Biochemistry, Part 1* (B.Pullman and N.Goldblum, Eds.), Reidel, Dordrecht, 1977, p.317.
42. (a) For X-ray structure of 18C6–potassium complexes, see P.Seiler, M.Dobler and J.D.Dunitz, *Acta Crystallogr., Sect.B*, 30, 2744, 2733 (1974).
42. (b) For X-ray structure of potassium cryptates, see D.Moras, B.Metz and R.Weiss, *Acta Crystallogr., Sect.B*, **29**, 383 (1973).
43. F.Vögtle and E.Weber, *Kontakte (Darmstadt)*, No.1, 11 (1977).
44. F.Vögtle and E.Weber, *Kontakte (Darmstadt)*, No.2, 16 (1977).
45. E.Weber and F.Vögtle *Kontakte (Darmstadt)*, No.3, 36 (1977).
46. D.J.Cram and J.M.Cram, *Acc. Chem. Res*., **11**, 8 (1978).
47. J.-M.Lehn, *Acc. Chem. Res*., **11**, 49 (1978).
48. J.-M.Lehn, *Pure Appl. Chem*., **50**, 871 (1978).
49. V.Prelog, *Pure Appl. Chem*. **50**, 893 (1978).
50. E.Weber and F.Vögtle, *Kontakte (Darmstadt)* No.2, 16 (1978).
51. F.Vögtle, E.Weber and U.Elben, *Kontakte (Darmstadt)*, No.3, 32 (1978).

52. J.S.Bradshaw, G.E.Maas, R.M.Izatt and J.J.Christensen, *Chem. Rev.*, **79**, 37 (1979).
53. A.C.Coxon, W.D.Curtis, D.A.Laidler and J.F.Stoddart, *J.Carbohydr. Nucleosides Nucleotides*, **6**, 167 (1979).
54. R.M.Izatt, J.D.Lamb, D.E.Eatough, J.J.Christensen and J.H.Rytting, *Drug Design*, Vol.**3**, 356 (1979).
55. J.-M.Lehn, *Pure Appl. Chem.*, **51**, 979 (1979).
56. N.S.Poonia and A.V.Bajaj, *Chem. Rev.*, **79**, 389 (1979).
57. W.Saenger, I.H.Suh and G.Weber, *Isr. J. Chem.*, **18**, 253 (1979).
58. J.F.Stoddart, *Chem. Soc. Rev.*, **8**, 85 (1979).
59. F.Vögtle, *Chimia*, **33**, 239 (1979).
60. F.Vögtle and E.Weber, *Angew. Chem., Int. Ed. Engl.*, **18**, 753 (1979).
61. F.Vögtle, E.Weber and U.Elben, *Kontakte (Darmstadt)*, No.1, 3 (1979).
62. J.S.Bradshaw and P.E.Stott, *Tetrahedron*, **30**, 461 (1980).
63. J.Dale, *Isr. J. Chem.*, **20**, 3 (1980).
64. F.de Jong and D.N.Reinhoudt, *Adv. Phys. Org. Chem.*, **17**, 279 (1980).
65. I.Goldberg in *The Chemistry of the Ether Linkage, Suppl. E., Part 1* (S.Patai, Ed.), Wiley, Chichester, 1980, p.175.
66. J.L.Dye, 'Elektride,' *Spektrum der Wissenschaft*, 50 (1987).
67. J.-M.Lehn, *Pure Appl. Chem.*, **52**, 2303 (1980).
68. J.-M.Lehn, *Pure Appl. Chem.*, **52** 2441 (1980).
69. J.F.Stoddart, *Lect. Heterocycl. Chem.*, **5**, 47 (1980).
70. F.Vögtle and E.Weber, in *The Chemistry of the Ether Linkage, Suppl. E, Part 1* (S.Patai, Ed.), Wiley, Chichester, 1981, p.59. MM-calculations of metal ion recognition: R.D.Hancock, *Acc. Chem. Res.*, **23**, 253 (1990); cf. also P.A.Kollman and K.M.Merz, Jr., *Acc. Chem. Res.*, **23**, 246 (1990).
71. F.Vögtle, E.Weber and U.Elben, *Kontakte (Darmstadt)*, No.2, 36 (1980).
72. R.G.H.Kirstetter, *Math. Naturwiss. Unterricht* **34**, 78 (1981).
73. J.-M.Lehn, *Recherche*, **127**, 1213 (1981).
74. E.Weber and F.Vögtle, *Kontakte (Darmstadt)*, No.1, 24 (1981).
75. G.W.Gokel, D.M.Dishong, R.A.Schultz and V.J.Gatto, *Synthesis*, 997 (1982).
76. S.T.Jolley, J.S.Bradshaw ans R.M.Izatt, *J. Heterocycl. Chem.*, **19**, 3 (1982).
77. V.K.Majestic and G.R.Newkome, *Top. Curr. Chem.*, **106**, 79 (1982).
78. E.Weber, *Kontakte (Darmstadt)*, No.1, 24 (1982).
79. R.C.Hayward, *Chem. Soc. Rev.*, **12**, 285 (1983).
80. J.-M.Lehn, in *Proceedings of the 2nd International Kyoto Conference on New Aspects of Organic Chemistry*, Kodansha, Tokyo, 1983.
81. J.-M. Lehn, in *Physical Chemistry of Transmembrane Ion Motions* (G.Spach, Ed.), Elsevier, Amsterdam, 1983, p.181.
82. D.Parker, *Adv. Inorg. Chem. Radiochem.*, **27**, 1 (1983).
83. J.F.Stoddart, *Annual Reports B*, Royal Society of Chemistry, London, 1983, p.353.
84. E.Weber, *Kontakte (Darmstadt)* No.1, 38 (1983).
85. J.C.G.Bünzli and D.Wessner, *Coord. Chem. Rev.*, **60**, 191 (1984).
86. (a) B.Dietrich, in *Inclusion Compounds*, Vol.2, (J.L.Atwood, J.E.Davies and D.D.MacNicol, Eds.), Academic Press, London, 1984, p.337; (b) for binding of ATP, see M.W.Hosseini, A.J.Blacker and J.-M.Lehn, *J. Chem. Soc., Chem. Commun.*, 596 (1988); (c) for anion binding with oxygen-free macrotricycles, see F.P.Schmidtchen and G.Müller, *J. Chem. Soc., Chem. Commun.*, 1115 (1984); (d) for anion receptors of the guanidinium type, see J.-M.Lehn *et al.*, *Helv. Chim. Acta*, **71**, 685 (1988); G.Müller, J.Riede and F.P.Schmidchen, *Angew. Chem., Int. Ed. Engl.*, **27**, 1516 (1988); (e) Fluoride as guest in 'expanded-porphyrin' ligands: J.L.Sessler, M.J.Cyr and V.Lynch, *J. Am. Chem. Soc.*, **112**, 2810 (1990).
87. M.Dobler, *Chimia*, **38**, 415 (1984).
88. I.Goldberg in *Inclusion Compounds*, Vol.2, (J.L.Atwood, J.E.D.Davies and D.D.MacNicol, Eds.), Academic Press, London, 1984, p.261.

89. D.W.McBride, R.M.Izatt, J.D.Lamb and J.J.Christensen in *Inclusion Compounds*, Vol.3, Academic Press, London, 1984 p.261.
90. I.O.Sutherland, *Heterocycles*, **21**, 235 (1984).
91. F.Vögtle, W.M.Müller and W.H.Watson, *Top. Curr. Chem*., **125**, 131 (1984).
92. E.Weber, *Kontakte (Darmstadt)* No.1, 26 (1984).
93. R.M.Izatt, J.S.Bradshaw, S.A.Nielsen, J.D.Lamb, J.J.Christensen and D.Sen, *Chem. Rev*., **85**, 271 (1985).
94. K.G.Heumann, *Top. Curr. Chem*., **127**, 77 (1985).
95. J.Jurczak and M.Pietraszkiewicz, *Top. Curr. Chem*., **130**, 183 (1986).
96. B.C.Pressman, H.J.Harris, W.S.Jagger and J.H.Jonson, *Proc. Natl. Acad. Sci. USA*, **58**, 1948 (1967).
97. Übersicht: U.Gräfe, R.Schlegel and M.Bergholz, *Pharmazie*, **39**, 661 (1984).
98. E.Weber and F.Vögtle, *Inorg. Chim. Acta*, **45**, L65 (1980); see also E.Weber and H.-P.Josel, *J. Inclus. Phenom*., **1**, 79 (1983).
99. For reviews, see E.Weber and F.Vögtle, in ref.11, p.1; ref.14, p.1. Template effect at the formation of C=N double bonds: L.-Y.Chung, E.C.Constable, A.R.Dale, M.S.Khan, M.C.Liprot, J.Lewis and P.R.J.Raithby, *J. Chem. Soc., Dalton Trans*., 1397 (1990).
100. E.Weber, *Chem. Ber*., **118**, 4439 (1985).
101. C.J.Pedersen, *J. Am. Chem. Soc*., **92**, 391 (1970).
102. For review, see D.N.Reinhoudt and F.de Jong in ref.8, p.157.
103. E.Weber and F.Vögtle, *Chem. Ber*., **109**, 1803 (1976).
104. For a review, see E.Kimura, in ref.15, p.113.
105. T.Kauffmann and J.Ennen, *Chem. Ber*., **118**, 2714 (1985).
106. For a review, see J.-M.Lehn, in ref.2, p.1.
107. A.C.Coxon and J.F.Stoddart, *J. Chem. Soc., Perkin Trans. 1*, 767 (1977).
108. D.G.Parson, *J. Chem. Soc., Perkin Trans. 1*, 451 (1978).
109. For a review, see J.Smid, *Angew. Chem*., **84**, 127 (1972); *Angew. Chem., Int. Ed. Engl*., **11**, 112 (1972).
110. For a review, see W.E.Morf, D.Ammann, R.Bissig, E.Pretsch and W.Simon, in ref.8, p.1.
111. E.Weber and F.Vögtle, *Tetrahedron Lett*. 2415 (1975).
112. F.Vögtle and E.Weber, *Angew. Chem., Int. Ed. Engl*., **13**, 814 (1974).
113. R.Fornasier, F.Montanari, G.Podda and P.Tundo, *Tetrahedron Lett*. 1381 (1976).
114. G.W.Gokel, D.M.Dishong and C.J.Diamond, *J. Chem. Soc., Chem. Commun*., 1053 (1980).
115. E.Weber, *Angew. Chem., Int. Ed. Engl*., **18**, 219 (1979).
116. E.Weber, *Angew. Chem., Int. Ed. Engl*., **22**, 16 (1983).
117. For reviews, see D.J.Cram and K.N.Trueblook, in ref.11, P.43; ref.14, p.125; D.J.Cram, *Angew. Chem*. **98**, 1041 (1986).
118. C.J.Pedersen, *Org. Synth*., **52**, 66 (1972).
119. G.W.Gokel, D.J.Cram, C.L.Liotta, H.P.Harris and F.L.Cook, *Org. Synth*., 57, 30 (1977).
120. For a review, see A.Leo, C.Hansch and D.Elkins, *Chem. Rev*., **71**, 525 (1971).
121. M.R.Truter, in ref.2, p.71.
122. R.Hilgenfeld and W.Saenger, in ref.12, p.1; ref.14, p.43.
123. N.S.Poonia, in ref.8, p.115.
124. T.-L.Ho (Ed.), *Hard and Soft Acids and Bases Principle in Organic Chemistry*, Academic Press, New York, 1977.
125. J.-M.Lehn and P.Vierling, *Tetrahedron Lett*. 317 (1977).
126. B.Dietrich, T.M.Fyles, J.-M.Lehn, G.Pease and D.L.Fyles, *J. Chem. Soc., Chem. Commun*., 934 (1978).
127. J.-M.Lehn, E.Sonveaux and A.K.Willard, *J. Am. Chem. Soc*., **100**, 4914 (1978).
128. F.P.Schmidtchen, *Angew. Chem., Int. Ed. Engl*., **16**, 720 (1977).
129. B.Metz, J.M.Rosalky and R.Weiss, *J. Chem. Soc., Chem. Commun*., 533 (1976).

130. F.P.Schmidtchen, *Tetrahedron Lett*., 4361 (1984).
131. C.J.Pedersen, *J. Org. Chem*., **36**, 1690 (1971).
132. G.W.Gokel, D.J.Cram, C.L.Liotta, H.P.Harris and F.L.Cook, *J. Org. Chem*., **39**, 2445 (1974).
133. For reviews, see F.Vögtle, H.Sieger and W.M.Müller, in ref.11, p.107; ref.14, p.319.
134. G.Weber, *J. Mol. Struct*., **98**, 333 (1983).
135. G.Weber and P.G.Jones, *Acta Crystallogr*., Sect. C, **39**, 1577 (1983).
136. D.J.Cram, *Science*, **219**, 1177 (1983).
137. F.Diederich, *Chem. Unserer Zeit*, **17**, 105 (1983).
138. Y.Murakami, *Top. Curr. Chem*., **115**, 107 (1983).
139. F.Vögtle and W.M.Müller, *J. Inclus. Phenom*., **2**, 369 (1984).
140. F.Vögtle and W.M.Müller, *Naturwissenschaften*, **67**, 255 (1980).
141. G.W.Gokel and G.W.Weber, *J. Chem. Educ*., **55**, 350, 429 (1978).
142. W.P.Weber and G.W.Gokel, *Phase Transfer Catalysis in Organic Synthesis. Reactivity and Structure Concepts in Organic Chemistry*, Vol.4, Springer, Berlin, Heidelberg, New York, 1977.
143. C.M.Starks and C.L.Liotta (Eds.), *Phase Transfer Catalysis, Principles and Techniques*. Academic Press, New York, San Francisco, London 1978.
144. W.E.Keller (Ed.), *Compendium of Phase Transfer Reactions and Related Synthetic Methods*, Fluka Buchs, 1979.
145. E.V.Dehmlow and S.S.Dehmlow (Eds.), *Phase Transfer Catalysis. Monographs in Modern Chemistry*, Vol.117, Verlag Chemie, Weinheim, 1980, p.1983.
146. K.Koga, *Yuki Gosei Kagaku Kyokai Shi*, **33**, 163 (1975).
147. E.V.Dehmlow, *Angew. Chem., Int. Ed. Engl*., **16**, 493 (1977).
148. C.L.Liotta, in ref. 7, p.111.
149. C.L.Liotta, in *The Chemistry of Functional Groups, Suppl. E, Part 1* (S.Patai, Ed.), Wiley, Chichester, 1980, p.157.
150. F.Montanari, D.Landini and F.Rolla, in ref.12, p.147.
151. P.Cocagne, R.Gallo and J.Elguero, *Heterocycles*, **20**, 1379 (1983).
152. D.J.Sam and H.E.Simmons, *J. Am. Chem. Soc*., **94**, 4024 (1972).
153. M.Mack and H.D.Durst, unpublished information see ref.44.
154. K.Ganboa, J.M.Aizpurua and C.Palomo, *J. Chem. Res.(S)*, 92 (1984).
155. O.Cardillo, M.Orena and S.Sandri, *J. Chem, Soc., Chem. Commun*., 190 (1976).
156. J.San Filippo, Jr, C.I.Chern and J.S.Valentine, *J. Org. Chem*., **40**, 1678 (1975).
157. R.A.Johnson and E.G.Nidy, *J. Org. Chem*., **40**, 1680 (1975).
158. A.Knöchel and G.Rudolph, *Tetrahedron Lett*., 3739 (1974).
159. C.J.Pedersen, *U.S.Pat*., 3 686 225 (1972); *Br. Pat*., 1 149 229 (1969) [*Chem. Abstr*., 71, 60685m (1969)].
160. B.Dietrich and J.-M.Lehn, *Tetrahedron Lett*., 1225 (1973).
161. C.L.Liotta and H.P.Harris, *J. Am. Chem. Soc*., **96**, 2250 (1974).
162. H.Handel and J.L.Pierre, *Tetrahedron Lett*., 741 (1976).
163. A.Loupy, J.Seyden-Penne and P.Tchoubar, *Tetrahedron Lett*., 1677 (1976).
164. J.L.Pierre, H.Handel and R.Perraud, *Tetrahedron Lett*., 2013 (1977).
165. H.Handel and J.L.Pierre, *Tetrahedron*, **31**, 2799 (1975).
166. A.G.M.Barrett and J.C.A.Lana, *J. Chem. Soc., Chem. Commun*., 471 (1978).
167. A.G.M.Barrett, J.C.A.Lana and S.Tograie, *J. Chem. Soc., Chem. Commun*., 300 (1980).
168. S.G.Smith and M.P.Hanson, *J. Org. Chem*., **36**, 1931 (1971).
169. C.Cambillau, G.Braun, J.Corset, C.Richa and C.Pascard-Billy, *Tetrahedron*, **34**, 2675 (1978).
170. For a review, see R.A.Bartsch, *Acc. Chem. Res*., **8**, 239 (1975).
171. D.T.Sepp, K.V.Scherer and W.P.Weber, *Tetrahedron Lett*., 2983 (1974).
172. M.Makosza and M.Ludwikow, *Angew. Chem*., **86**, 744 (1974); *Angew. Chem., Int. Ed. Engl*., **13**, 665 (1974); cf. also the catalysis of MeSiCN-addition: S.Kim, R.Bishop, D.C.Craig, I.G.Dance and M.L.Scudder, *J. Org. Chem*., **55**, 355 (1990).

173. S.Kwon, Y.Nishimura, M.Ikeda and Y.Tamura, *Synthesis*, 249 (1976).
174. I.Belsky, *J. Chem. Soc., Chem. Commun*., 237 (1977).
175. D.J.Cram and G.D.Y.Sogah, *J. Chem. Soc., Chem. Commun*., 625 (1981).
176. For a review, see S.L.Regen, *Angew. Chem., Int. Ed. Engl*., **18**, 421 (1979).
177. H.M.N.Irving, *Pure Appl. Chem*., **50**, 1129 (1978).
178. I.M.Kolthoff, *Anal. Chem*., **51**, 1R (1979).
179. E.Blasius and K.P.Janzen, in ref.11, p.163; ref.14, p.189.
180. T.Sekine and Y.Hasegawa, *Kagakuno Ryoiki*, **33**, 464 (1979).
181. Y.Takeda, in ref.13, p.1.
182. P.R.Danesi, H.Heider-Gorican, R.Chiarizia and G.Scibona, *J. Inorg. Nucl. Chem*., **37**, 1479 (1975).
183. V.V.Y.Yakshin, V.M.Abashkin, N.G.Zhukova, N.A.Tsarenko and B.N.Laskorin, *Dokl. Akad. Nauk SSSR*, **252**, 373 (1980).
184. M.Jawaid and F.Ingman, *Talanta*, **25**, 91 (1978).
185. H.Sumiyoshi, K.Nakahara and K.Ueno, *Talanta*, **24**, 763 (1977).
186. M.Takagi, H.Nakamura, Y.Sanui and K.Ueno, *Anal. Chim. Acta*., **126**, 185 (1981).
187. A.Sanz-Medel, D.B.Gomis and J.R.G.Alvarez, *Talanta*, **28**, 425 (1981).
188. T.Iwachido, M.Miniami, H.Naito and K.Toei, *Bull. Chem. Soc. Jpn*., **55**, (1982).
189. A.V.Bogatskii, N.G.Luk'yanenko, M.U.Mamina, V.A.Shapkin and D.Taubert, *Dokl. Chem*., **250**, 82 (1980).
190. T.Kimura, T.Maeda and T.Shono, *Talanta*, **26**, 945 (1979).
191. K.Kimura, K.Iwashima, T.Ishimori and H.Hamaguchi, *Chem. Lett*., 563 (1977).
192. B.E.Jepson and R.DeWitt, *J. Inorg. Nucl. Chem*., **38**, 1175 (1976).
193. L.R.Sousa, G.D.Y.Sogah, D.H.Hoffman and D.J.Cram, *J. Am. Chem. Soc*., **100**, 4569 (1978).
194. V.Prelog, *Chimia*, **37**, 12 (1983).
195. T.Maeda, M.Ouchi, K.Kimura and T.Shono, *Chem. Lett*., 1573 (1981).
196. I.Tabushi, Y.Kobuka, K.Ando, M.Kishimoto and E.Ohara, *J. Am. Chem. Soc*., **102**, 5947 (1980).
197. E.Weber, *Liebigs Ann. Chem*., 770 (1983).
198. E.Blasius, K.P.Janzen, W.Adrian, G.Klautke, R.Lorschneider, P.G.Maurer, B.V.Nguyen Tien, G.Scholten and J.Stockemer, *Fresenius' Z. Anal. Chem*., **284**, 337 (1977).
199. E.Blasius, K.P.Janzen and G.Klautke, *Fresenius' Z. Anal. Chem*., **227**, 374 (1975).
200. G.Dotsevi, G.D.Y.Sogah and D.J.Cram, *J. Am. Chem. Soc*., **98**, 3028 (1976).
201. G.J.Moody and J.D.R.Thomas (Eds.), *Selective Ion Sensitive Electrodes*., Merrow, Watford, 1971.
202. K.Cammann (Ed.), *Das Arbeiten mit ionenselektiven Elektroden*, Springer, Berlin, Heidelberg, New York, 1973.
203. M.Kessler, K.C.Clark, Jr, D.W.Lübbers, I.A.Silver and W.Simon (Eds.), *Ion and Enzyme Electrodes in Biology and Medicine*, Urban and Schwarzenberg, Munich, Berlin, Vienna, 1976.
204. J.Koryta (Ed.), *Ion-selective Electrodes*, Wiley, Chichester, New York,m Brisbane, Toronto, 1980.
205. D.C.Cornish, *Chimia*, **29**, 398 (1975).
206. E.Pretsch, R.Büchi, D.Ammann and W.Simon, in *Analytical Chemistry, Essays in Memory of Anders Ringbohm* (E.Wänninin, Ed.). Pergamon Press, Oxford, New York, 1977, p.321.
207. W.E.Morf and W.Simon, in *Ion-selective Electrodes in Analytical Chemistry*, Vol.1 (H.Freiser, Ed.), Plenum Press, New York, London, 1978, p.211.
208. D.Ammann, F.Lauter, *et al*., in *Ion-selective Mircoelectrodes and Their Use in Excitable Tissues* (E.Sykova, P.Hnik and L.Vyk-licky, Eds.), Plenum Press, New York, London, 1981, p.13.
209. E.Pretsch, D.Ammann and W.Simon, *Res. Dev*., **25**, 20 (1974).

210. R.B.Fischer, *J. Chem. Educ*., **51**, 387 (1974).
211. A.P.Thoma, Z.Cimerman, U.Fiedler, D.Bedekovic, M.Güggi, P.Jordan, K.May, E.Pretsch, V.Prelog and W.Simon, *Chimia*, **29**, 344 (1975).
212. W.Bussmann, J.-M.Lehn, U.Oesch, P.Plumeré and W.Simon, *Helv. Chim. Acta*, **64**, 657 (1981).
213. P.C.Meier, D.Ammann, H.F.Osswald and W.Simon, *Med. Prog. Technol*., **5**, 1 (1977).
214. P.C.Meier, D.Ammann, W.E.Morf and W.Simon, in *Medical and Biological Applications of Electrochemical Devices* (J.Koryta, Ed.), Wiley, Chichester, New York, Brisbane, Toronto, 1980, p.13.
215. J.G.Schindler, R.Dennhardt and W.Simon, *Chimia*, **31**, 404 (1977).
216. J.G.Schindler and M.v.Gülich, *J. Clin. Chem. Biochem*., **19**, 49 (1981).
217. M.Takagi and K.Ueno, in ref.13, p.39; ref.14, p.217.
218. For a review, see H.-G.Löhr and F.Vögtle, *Acc. Chem. Res*., **18**, 65 (1985); see also M.E.Huston, C.Engleman and A.W.Czarnil, *J. Am. Chem. Soc*., **112**, 7054 (1990); A.Prasana de Silva and K.R.A.Saman Kumara Sandanayake, *Angew. Chem., Int. Ed. Engl*., **29**, 1173 (1990).
219. F.Vögtle, *Pure Appl. Chem*., **52**, 2405 (1980).
220. T.Kaneda, K.Sugihara, H.Kamiya and S.Misumi, *Tetrahedron Lett*., **22**, 4407 (1981).
221. K.Nakashima, S.Nakatsuji, S.Akiyama, T.Kaneda and S.Misumi, *Chem. Lett*., 1781 (1982).
222. J.P.Dix and F.Vögtle, *Chem. Ber*., **114**, 638 (1981).
223. H.-G.Löhr and F.Vögtle, *Chem. Ber*., **118**, 905 (1985).
224. Merck, *Eur. Pat. Appl*., 83 100 281 (1983).
225. H.Nishida, Y.Jatayama, H.Katsuki, H.Nakamura, M.Takagi and K.Ueno, *Chem. Lett*., 1853 (1982).
226. T.J.Rink, *Pure Appl. Chem*., **55**, 1977 (1983); S.Fery-Forgues, M.T.LeBris, J.P.Guetté and B.Valeur, *J. Chem. Soc., Chem. Commun*., 384 (1988).
227. S.Lindenbaum, J.H.Rytting and L.A.Sternson, in ref.8, p.225.
228. J.D.Lamb, J.J.Christensen and R.M.Izatt, *J. Chem. Educ*., **57**, 227 (1980).
229. J.D.Lamb and R.M.Izatt, in ref.9, p.41.
230. G.R.Painter and B.C.Pressman, in ref.12, p.83.
231. M.Newcomb, J.L.Toner, R.C.Helgeson and D.J.Cram, *J. Am. Chem. Soc*., **101**, 4941 (1979).
232. H.Tsukube, *Tetrahedron Lett*., **23**, 2109 (1982).
233. J.J.Christensen, J.D.Lamb, S.R.Izatt, S.E.Starr, G.C.Weed, M.S.Astin, B.D.Stitt and R.M.Izatt, *J. Am. Chem. Soc*., **100**, 3219 (1978).
234. H.Tsukube, *J. Chem. Soc., Perkin Trans. 1*, 2359 (1982).
235. H.Tsukube, *J. Chem. Soc., Perkin Trans. 1*, 29 (1983).
236. N.Yamazaki, S.Nakahama, A.Hirao and S.Negi, *Tetrahedron Lett*., 2429 (1978).
237. T.M.Fyles, V.A.Malik-Diemer and D.M.Whitfield, *Can. J. Chem*., **59**, 1734 (1981).
238. T.M.Fyles, V.A.Malik-Diemer, C.A.McGavin and D.M.Whitfield, *Can. J. Chem*., **60**, 2259 (1982).
239. For reviews, see S.Shinkai and O.Manabe, in ref.13, p.67; ref.14, p.245.
240. S.Shinkai, T.Ogawa, Y.Kusano and O.Manabe, *Chem. Lett*., 283 (1980).
241. S.Shinkai, T.Ogawa, Y.Kusano, O.Manabe, K.Kikukawa, T.Goto and T.Matsuda, *J. Am. Chem. Soc*., **104**, 1960 (1982).
242. S.Shinkai, K.Inuzuka and O.Manabe, *Chem. Lett*., 747 (1983).
243. F.Riedlberger and T.Näbauer, *Bild Wiss* No.3, 118 (1985).
244. R.Günther, O.Hauswirth and R.Ziskoven, *Naunyn-Schmiedeberg's Arch. Pharmakol*., **310**, 79 (1979).
245. C.Achenbach, O.Hauswirth, J.Kossmann and R.Ziskoven, *Physiol. Chem. Phys*., 12, 277 (1980).
246. R.C.Kolbeck, L.B.Hendry, E.D.Bransome and W.A.Speir, *Experientia*, **40**, 727 (1984).

247. E.J.Harris, B.Zaba, M.R.Truter, D.G.Parsons and J.N.Wingfield, *Arch. Biochem. Biophys*., **82**, 311 (1977).
248. P.Georgiou, C.H.Richardson, K.Simmons, M.R.Truter and J.N.Wingfield, *Inogr. Chim. Acta*, **66**, 1 (1982).
249. P.Georgiou, K.Simmons, A.Sharp, M.R.Truter and J.N.Wingfield, *Inorg. Chim. Acta*, **69**, 89 (1983).
250. V.A.Popova, J.V.Podgornaya, I.Y.Postovskii and N.N.Frovola, *Khim. Farm. Zh*., **10**, 66 (1976).
251. C.J.Kauer, *U.S. Pat*., 3 997 565; Appl. 615 185 (1975) [*Chem. Abstr*., **86**, 121388s].
252. G.R.Brown and A.J.Foubister, *J. Med. Chem*., **26**, 590 (1983).
253. F.Vögtle and B.Jansen, *Tetrahedron Lett*., 4895 (1976).
254. U.Elben, B.Fuchs, K.Frensch and F.Vögtle, *Liebigs Ann. Chem*., 1102 (1979).
255. G.Sosnovsky and J.Lukszo, University of Wisconsin, Mulwaukee, Wisconsin, U.S.A., and National Foundation for Cancer Research (U.S.A.).
256. M.G.Voronkov, I.G.Kuznetsov, S.K.Suslova, G.M.Tizenberg, V.I.Knutov and M.K.Butin, *Khim. Farm. Zh*., **19**, 819 (1985).
257. J.F.Pilichowski, J.Michelot, M.Borel and G.Meyniel, *Naturwissenschaften*, **70**, 201 (1983).
258. J.Le Moigne and J.Simon, *J.Phys.Chem*., **84**, 170 (1980).
259. L'Oreal SA, 17.03.78-Fr.Pat.-007914 (31.10.1979).
260. M.Fujimoto, T.Nogami and H.Mikawa, *Chem. Lett*., 547 (1982).
261. K.Matsuoka, T.Nogami, T.Matsumoto, H.Tanaka and H.Mikawa, *Bull. Chem. Soc. Jpn*., **55**, 2015 (1982).
262. I.Haller, W.R.Young, G.L.Gladstone and D.T.Teaney, *Mol. Cryst. Liq. Cryst*., **24**, 249 (1973).
263. L.Horner and W.Brich, *Chem. Ber*., **111**, 574 (1978).
264. Übersicht: N.J.Turro, M.Grätzel and A.M.Braun, *Angew. Chem., Int. Ed. Engl*., **19**, 675 (1980).
265. R.Kaufmann, F.Hillenkamp and P.Wechsung, *Eur. Spectros. News*, **20**, 3 (1978).
266. J.-M.Lehn, *Angew. Chem., Int. Ed. Engl*., **27**, 89 (1988).
267. J.-M.Lehn, *Pure Appl. Chem*., **52**, 2441 (1980).
268. J.-M.Lehn, *Science*, **227**, 849 (1985).
269. E.Weber and F.Vögtle, *Nachr. Chem. Tech. Lab*., **35**, 1149 (1987).
270. D.Karlin, *J. Am. Chem. Soc*., **109**, 2668 (1987); see also J.B.Vincent, J.C.Huffmann, G.Christou, Q.Li, M.A.Nanny, D.N.Hendrickson, R.H.Fong and R.H.Fisch, *J. Am. Chem. Soc*., **110**, 6898 (1988); P.P.Paul, Z.Tyeklar, A.Farooq, K.D.Karlin, S.Liu and J.Zubieta, *J. Am. Chem. Soc*., **112**, 2430 (1990); H.Adams, N.A.Bailey, W.D.Carlisle, D.E.Fenton and G.Rossi, *J. Chem. Soc., Dalton Trans*, 1271 (1990).
271. J.-M.Lehn, *Helv. Chim. Acta*, **67**, 2264 (1984).
272. J.-M.Lehn, *Angew. Chem., Int. Ed. Engl*., **26**, 266 (1987).
273. B.Alpha, V.Balzani, J.-M.Lehn, S.Perathoner and N.Sabbatini, *Angew. Chem., Int. Ed. Engl*., **26**, 1266 (1987).
274. V.Balzani, N.Sabbatini and F.Scandola, *Chem. Rev*., **86**, 319 (1986); for the photophysics and photochemistry of luminescent ruthenium complexes, see L.De Cola, F.Barigelletti, V.Balazani, P.Belser, A.von Zelewsky, F.Vögtle, F.Ebmeyer and S.Grammenudi, *J. Am. Chem. Soc*., **100**, 7210 (1988); P.Belser, *Chimia*, **44**, 226 (1990); V.Balzani, F.Barigelletti and L.De Cola, *Top. Curr. Chem*., **158**, 31 (1990); B.M.Krasovitskii and B.M.Bolotin, *Organic Luminescent Materials*, VCH-Verlagsgesellschaft, Weinheim, 1988.

2.3 THE SIDEROPHORES

2.3.1 INTRODUCTION: ENTEROBACTIN AND THE NATURAL COMPLEXATION OF IRON

In 1911, scientists discovered that mycobacteria needed an 'essential substance', vital to their growth [1]; but it was only 40 years later that this growth factor was isolated by crystallization of the corresponding aluminium complex [2]. The metal-free growth factor was obtained by a special purification method [3] and its main chemical properties were then investigated [4]. This compound was given the name *mycobactin*. After 30 more years, it became obvious that the mycobactins belong to the *siderophores*, an important class of substances promoting microbiological growth. Siderophores are iron-binding, low-molecular-mass compounds (molecular mass 500–1000 dalton) with remarkable chemical properties.

The siderophore enterobactin (**1**) was first described in 1970 by both Neilands [5] and Gibson [6]. It was isolated in small amounts from cultures of *Aerobacter aerogenes*, *Escherichia coli* and *Salmonella typhimurium*. To these bacteria, it serves as a ligand for the complexation and the transportation of iron, and plays a role in their growth.

Enterobactin is the cyclic triester of 2,3-dihydroxybenzoyl-L-serine. It binds Fe^{3+}-ions so strongly that at pH 7 no competitive reaction with EDTA is observed. Its complex formation constant, a measure of the stability of a complex, is $K_f = 10^{52}$, the highest ever reported for the iron complex of a natural compound.

It is only recently that specifically engineered macrobicyclic iron ligands have been prepared, whose complex formation constants surpassed enterobactin's exceptionally high value. Thus, a synthetic macrobicyclic ligand with a complex formation constant of $K_f \approx 10^{59}$ has been reported; this is seven orders of magnitude higher than that for enterobactin, and the highest ever for any iron complex (see below and Chapter 3).

Siderophores (Greek for iron bearer; the expression was coined in 1973 by Lankford [1]), such as enterobactin, are ligands for iron ions, produced by microorganisms in cases of iron deficiency. They generally take over the important role of solvating and transporting iron(III) ions in organisms.

On exposure to circulating blood, microorganisms produce siderophores. In the organism of an adult person there are 4–5 g of iron, of which approximately 65% is bound by the oxygen-transporting protein haemoglobin and a further 30% is found in the iron-storing proteins ferritin and haemosiderin. The protein transferrin (synonym siderophilin) promotes the transport of iron and maintains a small concentration of free iron.

In spite of the human organism's ability to regulate the level of iron, larger amounts of this element are toxic. The misapplication of iron-rich vitamin concentrates has led to acute iron poisoning, especially in young children. After a blood transfusion, a chronic iron excess is observed as a side effect in patients who suffer from Cooley's anaemia (approximately 3 million world-wide). This is a genetic disease, where the β-chains of haemoglobin are not formed to the normal extent. Continual blood transfusions lead to a constant enrichment of iron in the organism, as the human body does not have a specific physiological mechanism for the elimination of iron; a healthy human can eliminate a maximum of 1 mg of iron per day. There is particular damage to heart, liver and pancreas. Drugs therefore need to be found that will complex the iron and convert it into a form that the body can eliminate. This can be achieved with the drug desferrioxamine (Desferal, Ciba). The compound complexes iron and the complex can then be excreted with the urine. However, there is much room for further improvement, and specific iron-binding molecules are therefore still intensely sought. With new drugs, the necessary daily intake could be reduced and the side-effects diminished. The biological necessity for the production of siderophores by microorganisms derives from the irreplaceable role of iron in almost every oxidation and reduction process in the cell, combined with the near-insolubility of iron hydroxide at physiological pH. At pH 7 the equilibrium concentration of iron ions (as free Fe^{3+}) is nearly 10^{-18} mol/l. Microorganisms, such as the intestinal bacteria, need, for optimum growth an end-concentration of at least 5×10^{-7} mol/l. The available concentration is therefore too low by several orders of magnitude. Only good chelands, such as the siderophores, can mobilize the iron in this environment and transport it into the microbial cell.

This scavenging of iron by siderophores requires a highly specific recognition at the cell surface. In the case of intestinal bacteria, these receptors consist of either one or several proteins with high molecular masses. To study the interactions between receptor and substrate, it is essential to isolate the proteins involved. Important aspects of the action of siderophores have been investigated with radioactively labelled synthetic iron ligands of related structure.

Plants also need a continuous import of metal salts to maintain a healthy growth. Iron is one of the elements needed by plants for the biosynthesis of chlorophyll. Although green plants contain substantial amounts of iron, it is not transferred from old to new leaves, which is why the plant needs an uninterrupted supply of iron. An insufficient supply of iron leads to 'iron chlorosis' (green sickness), in the course of which the young leaves turn yellow or white. As a consequence, the plants stop growing and eventually perish.

Usually there is enough iron in the soil. However, the available amount of Fe^{3+} in calcareous soils is insufficient for plant growth and iron chlorosis develops. The solubility of Fe(III) at pH 4 and higher decreases by one thousandth per additional pH unit. Nevertheless, some plants thrive in alkaline soils. This indicates that such plants display a biochemical function, which changes insoluble iron in soils into useful Fe(III) ions. From this, we can distinguish between iron-efficient and iron-inefficient plants. In relation to the availability of iron, there seem to be four essential biochemical factors for the intake and the transport of iron in iron-efficient plants:

a. release of protons by the roots;

b. release of reduced compounds by the roots;

c. reduction of iron(III) to iron(II) in the roots;

d. increase in organic acids (especially citrate) in the roots.

It is hardly surprising that a number of iron-chelating plant siderophores ('phytosiderophores') have been isolated from roots. In the last 20 years, scores of siderophores have been isolated (mostly from bacteria) which contain catechol units (as in enterobactin) or hydroxamate groups as chelating functionalities (**1–11**).

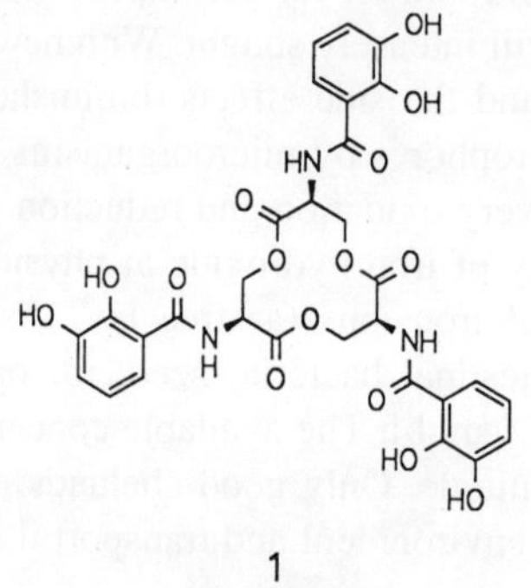

1
Enterobactin

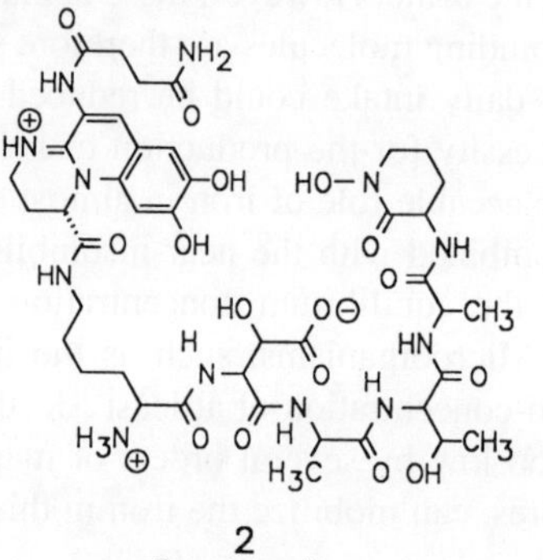

2
Pseudobactin

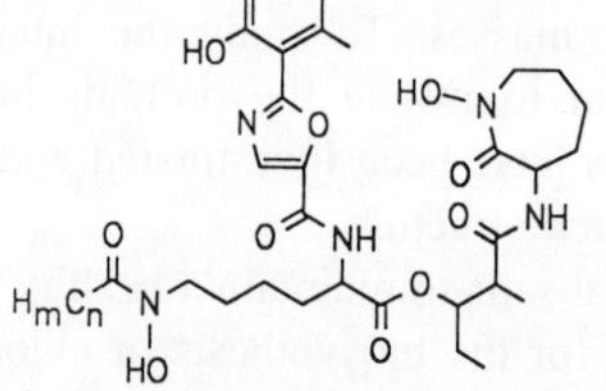

3
Mycobactin
n > 12

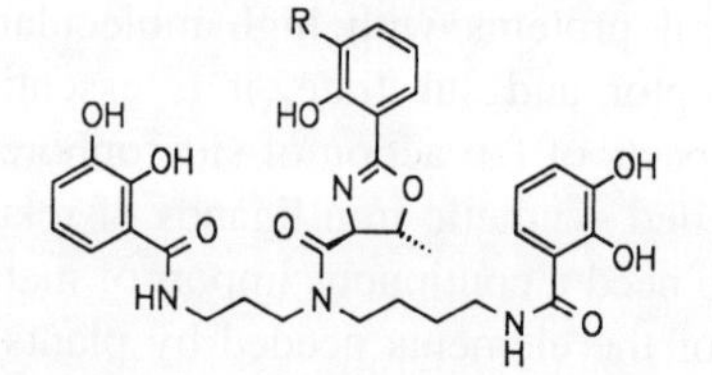

4, 5
Agrobactin (R = OH)
Parabactin (R = H)

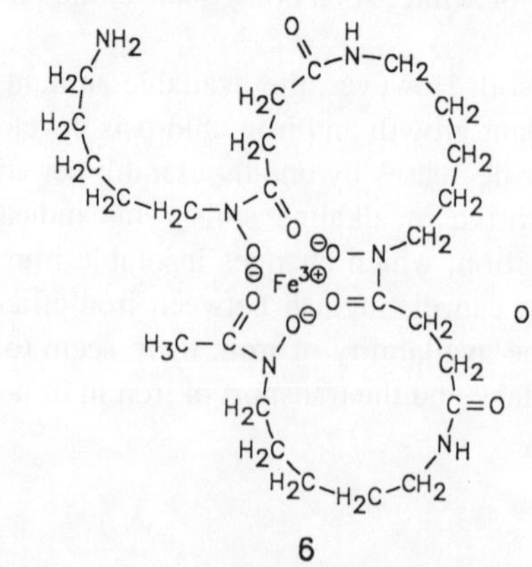

6
Ferrioxamine B

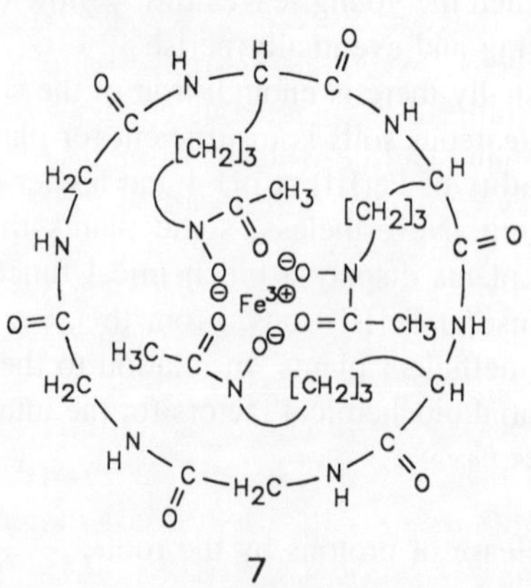

7
Ferrichrome

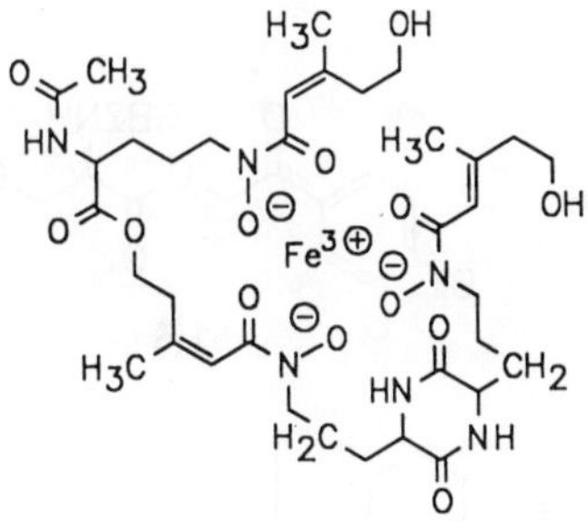

8 Coprogen

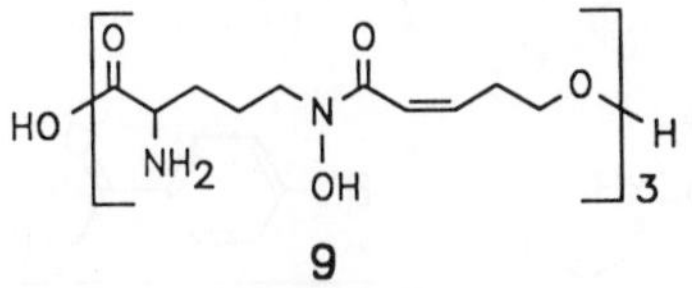

9 Fusarinin C

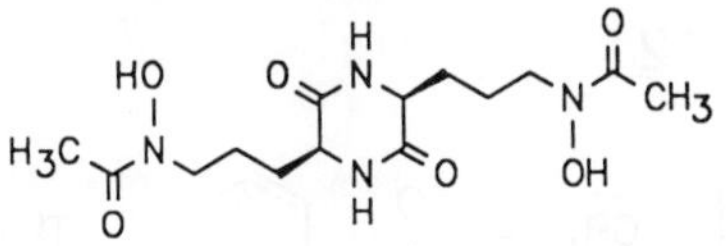

10 Rhodotorulcic acid

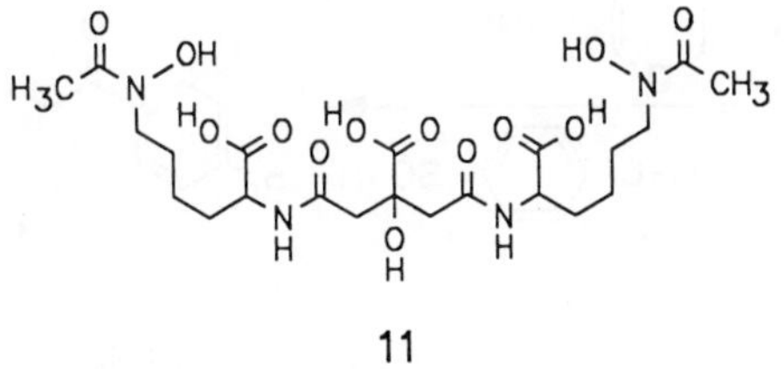

11 Aerobactin

2.3.2 ISOLATION OF ENTEROBACTIN; SYNTHESIS AND BIOSYNTHESIS

Enterobactin is the prototype to the catechol siderophores. It was isolated in 1970 independently by Pollack and Neilands from *Salmonella typhimurium* [5] and by O'Brien and Gibson from *Escherichia coli* [6].

Iron-poor nutrients promote the growth of *Escherichia coli* in fermentation vats [6]. Sterile solutions of L-methionine, thiamine and glucose are added to the nutrients. An optimum rate of growth is reached at 37 °C and the process is finished after 6 h. The cells are centrifuged at 4 °C and the residue is acidified with concentrated hydrochloric acid and extracted with ethyl acetate. Further purification is effected by column chromatography. The crude enterobactin thus obtained is recrystallized from aqueous ethanol. With this elaborate procedure, 15 mg of pure compound can be isolated from 1 l of nutrient solution. With regard to the small amounts obtained, this isolation from nutrients is too complicated, and synthesis consequently became a challenge for preparative organic chemists.

In 1977, Corey and Bhattacharyya published the synthesis of enterobactin and its non-natural mirror isomer enantioenterobactin [7]. They first built the corresponding trilactones from L- and D-serine, then subjected those to reaction with 2,3-dihydroxybenzoyl chloride (Scheme 1).

Rastetter *et al*., in 1981 reported a further synthesis of enterobactin and its antipode [8]. In growth experiments with bacteria, the non-natural enantio-enterobactin not only proved to be totally inactive, but indeed to inhibit growth by keeping the iron in the nutrient.

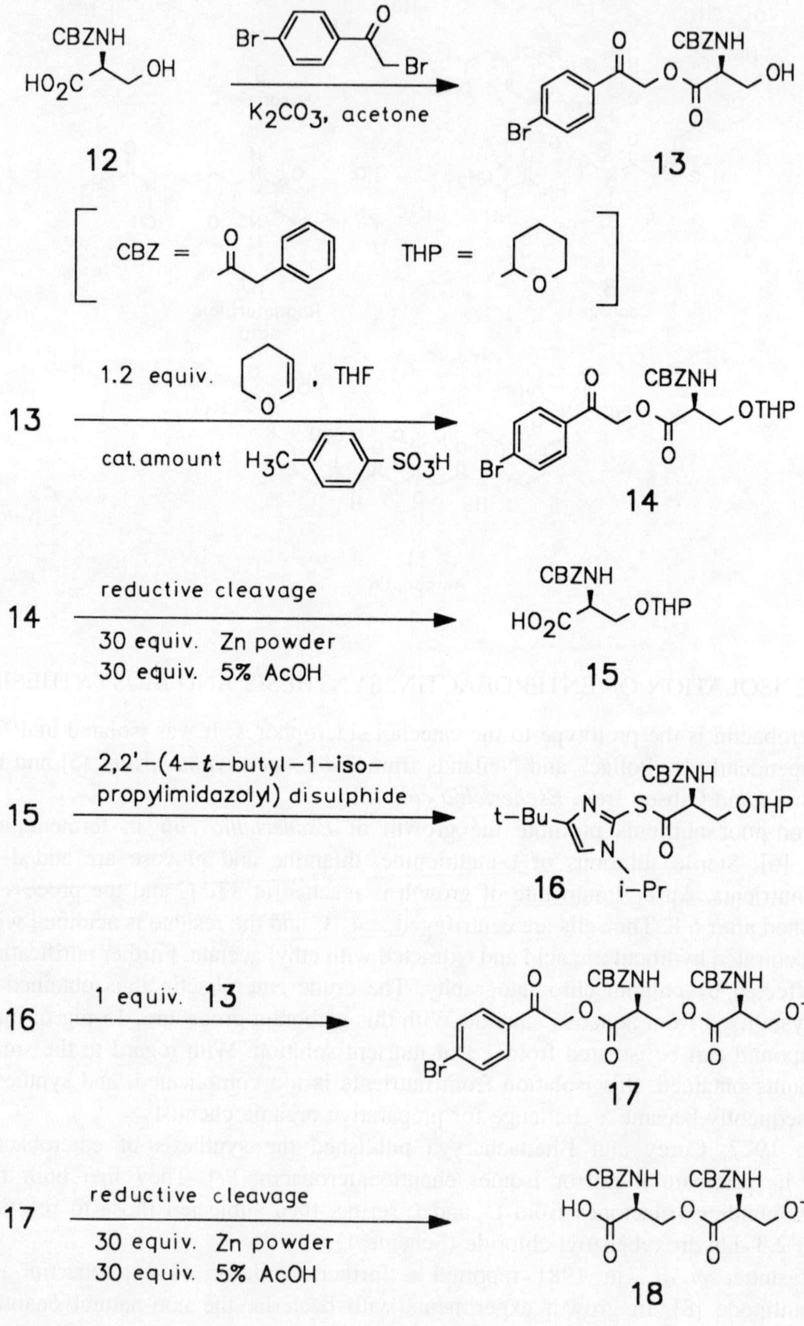

CBZNH
HO_2C OH
12

Br–C₆H₄–CO–CH₂–Br
K_2CO_3, acetone

13

[CBZ = PhCH₂CO– THP =]

13
1.2 equiv. (dihydropyran), THF
cat. amount H_3C–C₆H₄–SO_3H
14

14
reductive cleavage
30 equiv. Zn powder
30 equiv. 5% AcOH
15

15
2,2'-(4-*t*-butyl-1-isopropylimidazolyl) disulphide
16
t-Bu, i-Pr, OTHP

16
1 equiv. **13**
17

17
reductive cleavage
30 equiv. Zn powder
30 equiv. 5% AcOH
18

Scheme 1

18 —— 1 : activation, in analogy with 16 / 2 : 1 equiv. 13 ——→ 19

19 —— AcOH, CH_3OH, THF ——→ 20

20 —— reductive cleavage / 30 equiv. Zn powder / 30 equiv. 5% AcOH ——→ 21

21 —— cyclization of the 2,2'-(4-*t*-butyl-1-isopropyl-imidazolyl) thiol ester, formed in situ from 1 equiv. 21, 1.2 equiv. 2,2'-(4-*t*-butyl-1-isopropylimidazolyl) disulphide and 1.3 equiv. PPh_3 in CH_2Cl_2 ——→ 22

22 —— H_2/Pd–C, THF ——→ 23

23 —— 2,3-dihydroxybenzoyl chloride / NEt_3, THF ——→ 1

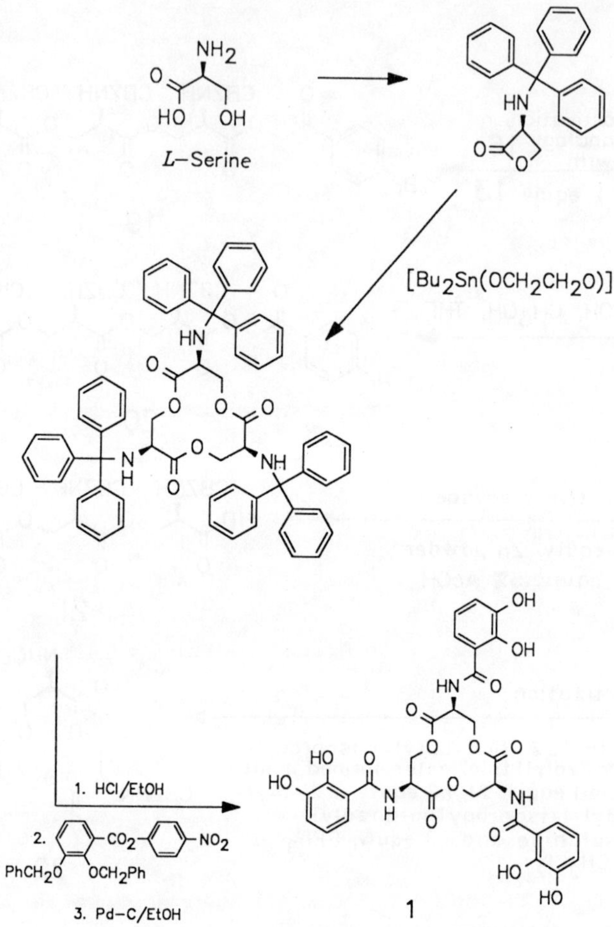

Scheme 2

The latest and simplest synthesis is due to Shanzer *et al*., and takes advantage of the template effect [9] (Scheme 2).

Biosynthesis of enterobactin. As was demonstrated, iron ions inhibit the biosynthesis of most siderophores. The mechanism by which this control is effected is as yet unknown.

Cell-free extracts capable of the biosynthesis of enterobactin (and also parabactin) need only ATP as cofactor; the presence of coenzyme A is not required. For the biosynthesis of enterobactin from dihydroxybenzoic acid and L-serine, at least four proteins are required (Scheme 3) [10].

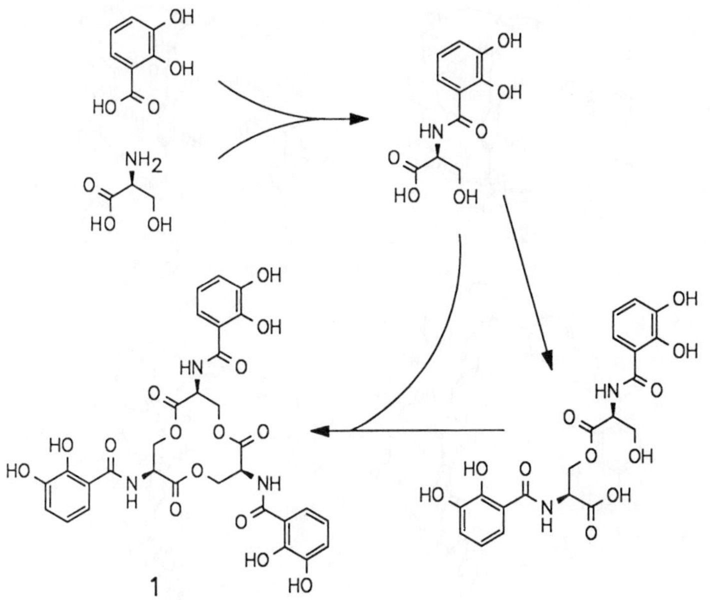

Scheme 3

2.3.3 SYNTHETIC SIDEROPHORE ANALOGUES

In order to obtain a better understanding of the extraordinary properties of enterobactin, a series of comparative ligands (**24–28**) have been investigated [31a–c,e]. Like enterobactin, they contain three catechol units, which are attached to an 'anchoring frame' having a threefold (axial) symmetry. Open-chain anchor groups have also been used (**29–33**). Molecular modelling studies showed the rigid mesitylene systems of **30** and **31** to be able to mimic the triester ring of enterobactin. The complex formation constant of Fe^{3+} and MECAM is correspondingly high: $K_f = 10^{46}$. However, this is still six orders of magnitude lower than that of enterobactin itself. As the binding sites (three catechol units) are the same, it can be surmised that stereochemical factors must be the basis for enterobactin's special position. However, $[Fe(MECAM)]^{3-}$ is biologically similar to $[Fe(ent)]^{3+}$ (Figure 1): it uses the enterobactin receptor system of *E.coli* and promotes its growth! For the experimenter, MECAM even has the decisive advantage of lacking enterobactin's hydrolysis-prone triester ring.

2.3.4 PROPERTIES

Enterobactin is not a very stable compound: the triester ring is prone to hydrolyse and the catechol units are sensitive to oxidation. Enterobactin ('ent') binds Fe^{3+} ions so strongly that at pH 7 no competitive reaction with EDTA occurs. Only at pH 5 is there a photometrically useful distribution of Fe^{3+} between the two ligands.

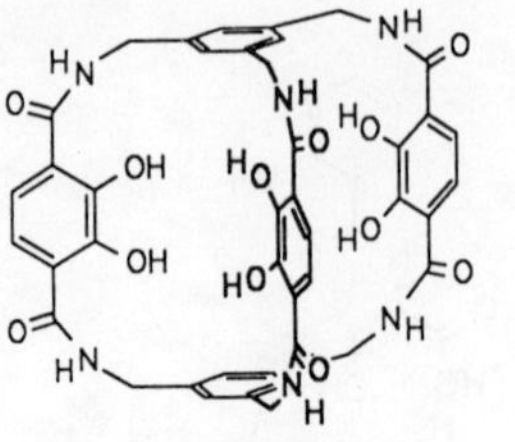

The complex formation constant is approximately 10^{20} times larger than those of ferrichrome and ferrioxamine B:

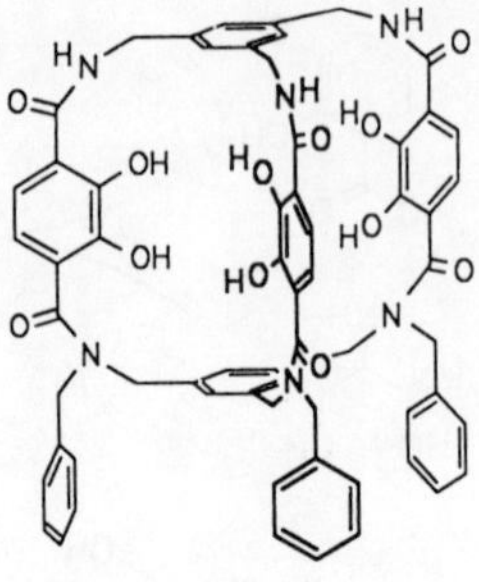

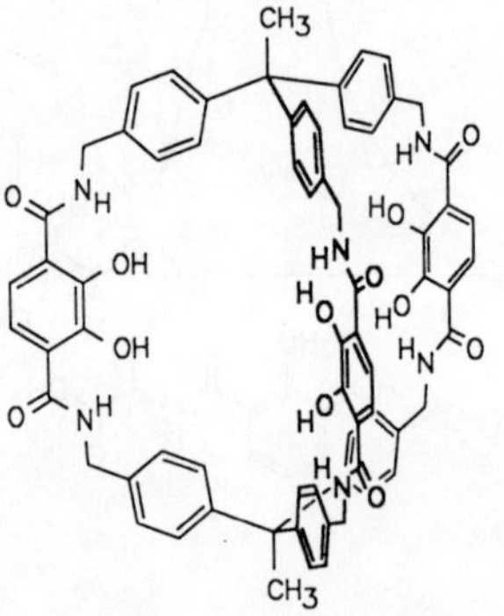

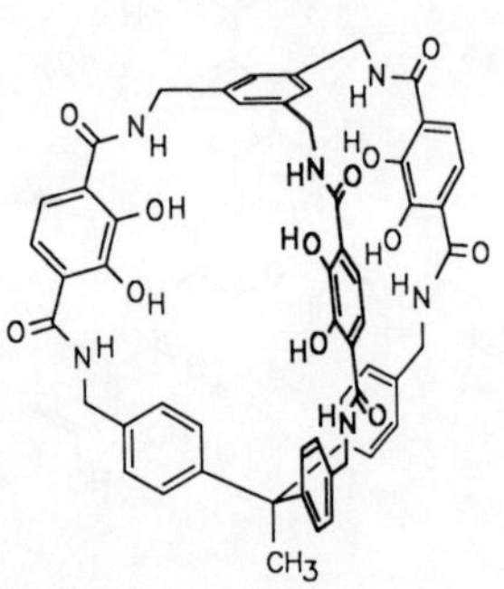

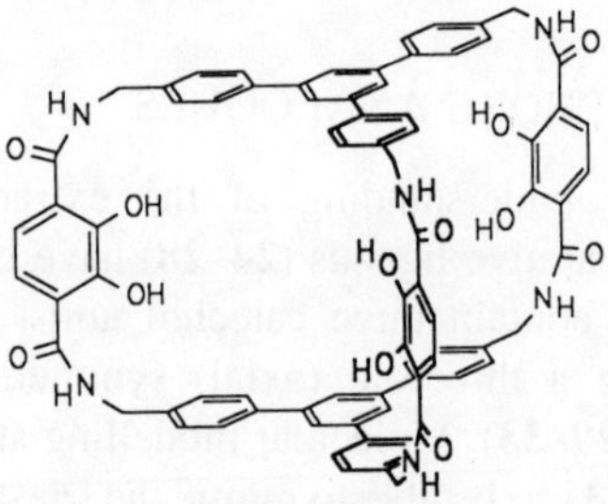

$$K_f = \frac{[\text{Fe(ent)}]^{3-}}{[\text{Fe}^{3+}][\text{ent}^{6-}]} = 10^{52}$$

(the higher the complex formation constant K_f or its negative logarithm, the higher is the stability of the corresponding complex).

Figure 2 gives an idea of the (slight) pH dependence of the visible spectrum of Fe^{3+}•enterobactin.

Catechol is a weak acid; the pK_a values are between 9.2 and 13. This is why protonation reactions compete with the formation of the complex. As a result of the pH dependence of the complexation constants, comparisons of the iron-binding properties of the ligands are inadequate. For this reason, the pM value has been introduced:

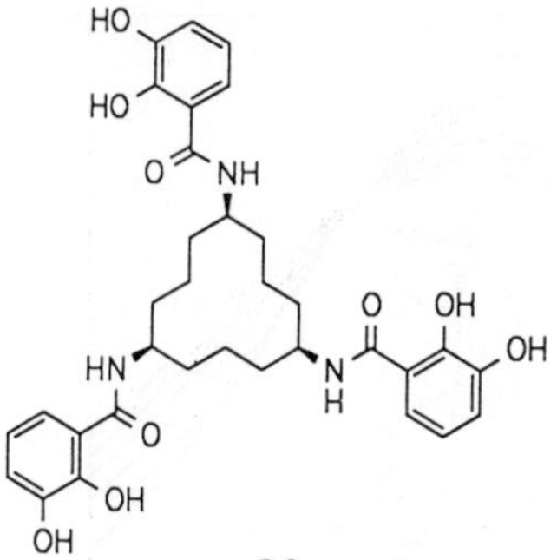

29

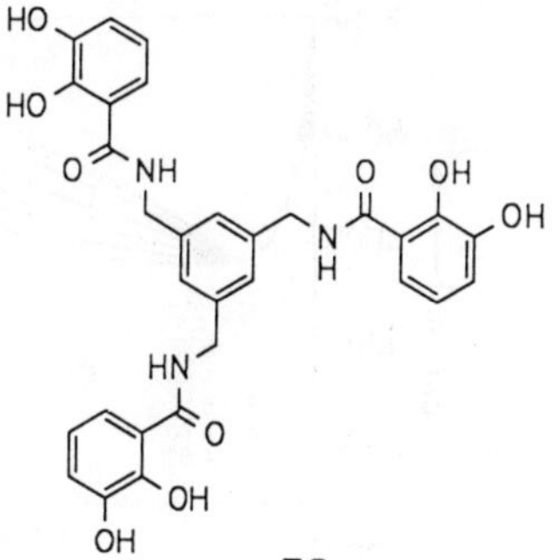

30

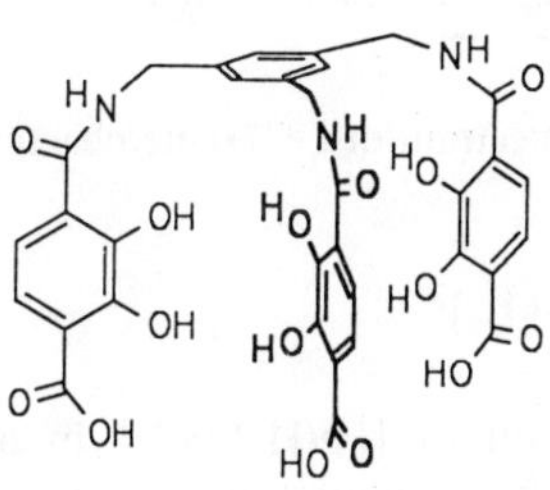

31

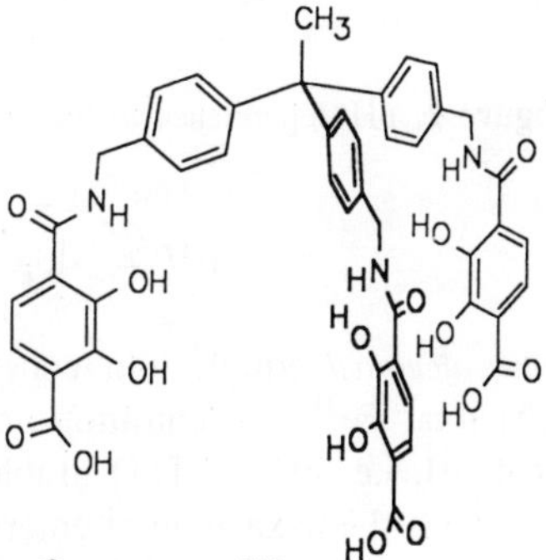

32

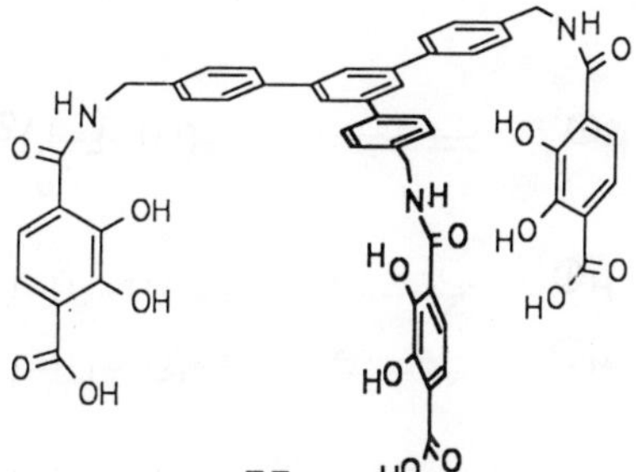

33

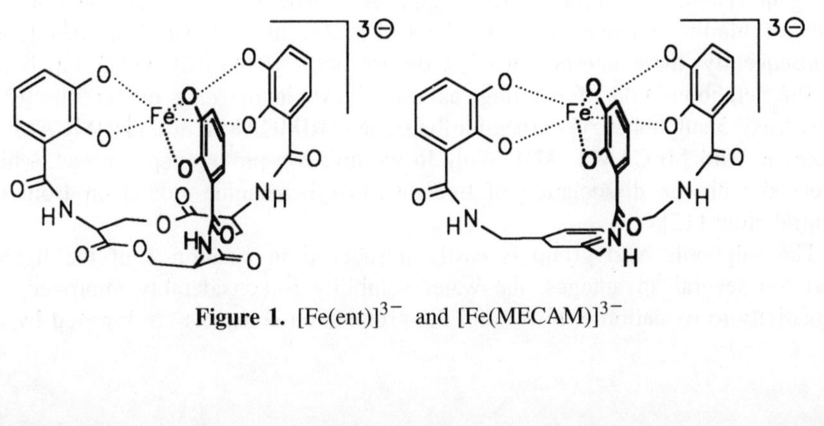

Figure 1. $[Fe(ent)]^{3-}$ and $[Fe(MECAM)]^{3-}$

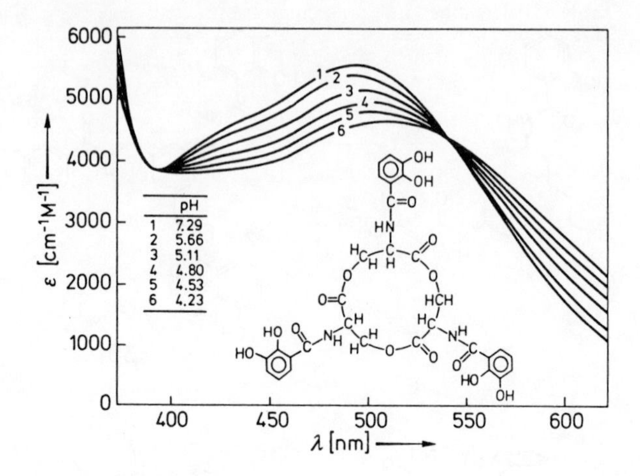

Figure 2. pH dependence of the visible spectrum of Fe^{3+}•enterobactin

$$pM = -\log[Fe(H_2O)_6]^{3+}$$

It gives the *calculated* equilibrium concentration of $[Fe(H_2O)_6]^{3+}$ in a solution with a 1 μM total Fe^{3+} concentration and a 10 μM total ligand concentration at physiological pH, i.e. pH 7.4 [11] (Table 1). The protonation of $[Fe(L\text{–}L\text{–}L)]^{3-}$ [where (L–L–L) is the hexaanion of enterobactin or MECAM] proceeds one proton at a time:

$$Fe(L\text{–}L\text{–}L)^{3\ominus} + H^{\oplus} \rightleftharpoons Fe(LH\text{–}L\text{–}L)^{2\ominus}$$

$$Fe(LH\text{–}L\text{–}L)^{2\ominus} + H^{\oplus} \rightleftharpoons Fe(LH\text{–}LH\text{–}L)^{\ominus}$$

$$Fe(LH\text{–}LH\text{–}L)^{\ominus} + H^{\oplus} \rightleftharpoons Fe(LH\text{–}LH\text{–}LH)$$

This behaviour becomes apparent in the visible spectrum during a titration. These findings are supported by the IR spectrum of $Fe(H_3MECAM)$. Here, the C=O stretching of $[Fe(MECAM)]^{3-}$ at 1610 cm^{-1} is missing, but an OH deformation is again visible [14]. From this, the authors concluded that on protonation, the type of binding changes from catechol-like (**34**) to salicylate-like (**35**) [14a]. Consequently, these one-proton steps do not occur if the carbonyl group is not in the neighbourhood of the ring, as was shown in the case of 1,3,5-tris[(2,3-dihydroxy-5-sulphobenzyl)carbamido]benzene (TRIMCAMS, **36**). [TRIMCAMS is isomeric with MECAMS (**37**)]. With **36** we find two-proton steps instead, which proceed with the dissociation of one dihydroxybenzamide side-chain from the central atom [12].

The sulphonic acid group is easily introduced in position 5 of the ligands and has several advantages: the water solubility is considerably improved, the sensitivity to oxidation is lower and the protonation constants are lowered by the

Table 1. p*M* values for some enterobactin analogues [12,13]

Ligand	p*M* value	Log K_f
Enterobactin	35.5	52
HBED	31.0	
MECAMS	29.4	41
MECAM	29.1	46
LICAMS	28.5	41
MeMECAMS	26.9	40.6
Ferrioxamine B	26.6	30.5
EHPG	26.4	
Ferrichrome A	25.2	32
TRICAMS	25.1	
NacMECAMS	25.0	
CYCAMS	24.9	
Diethylenepentaacetic acid	24.7	
Transferrin	23.6	
EDTA	12.2	
1,2-Dihydroxy-3,5-disulphobenzene	19.5	
N,N-Dimethyl-2,3-dihydroxy-5-sulphobenzamide	19.2	

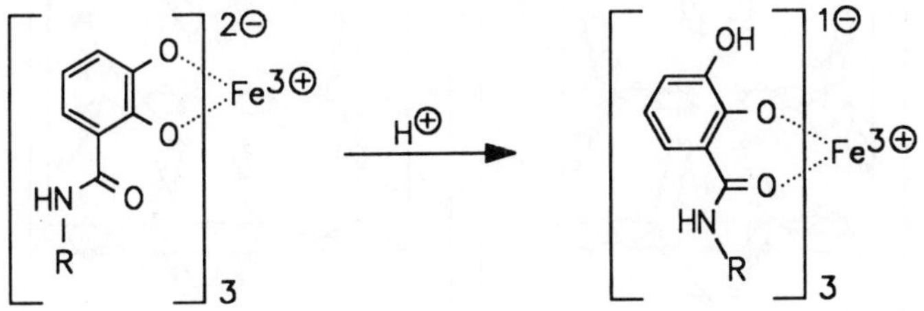

34

35

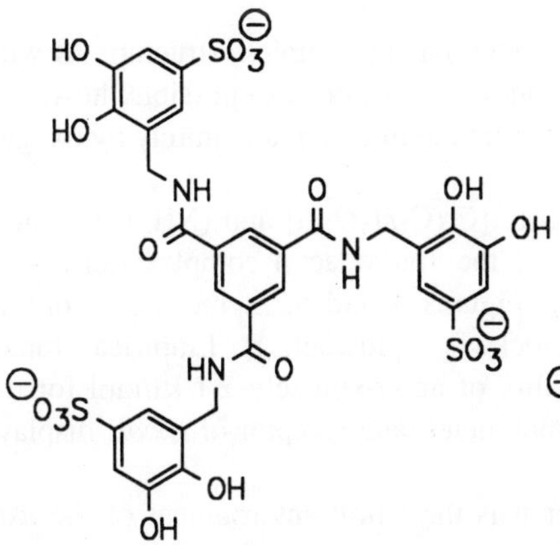

36

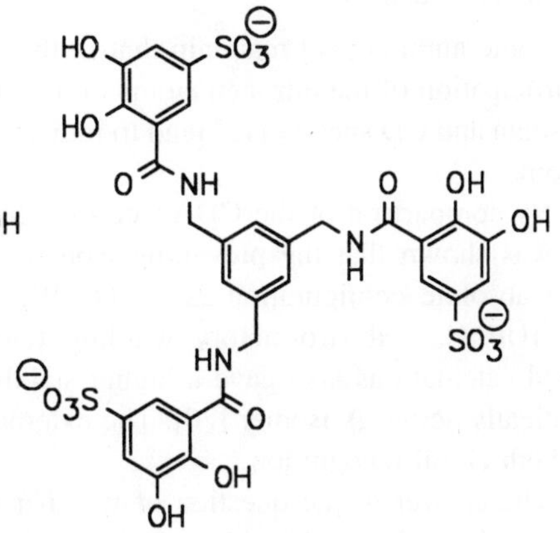

37

electron-withdrawing effect. In 1976, Raymond *et al*. [15] reliably established the coordination number of iron(III) complexes with catechol ligands by using X-ray structure analysis. The iron atom in the iron complex with unsubstituted catechol is surrounded by six catechol oxygens in a distorted octahedron, as shown in Figure 3.

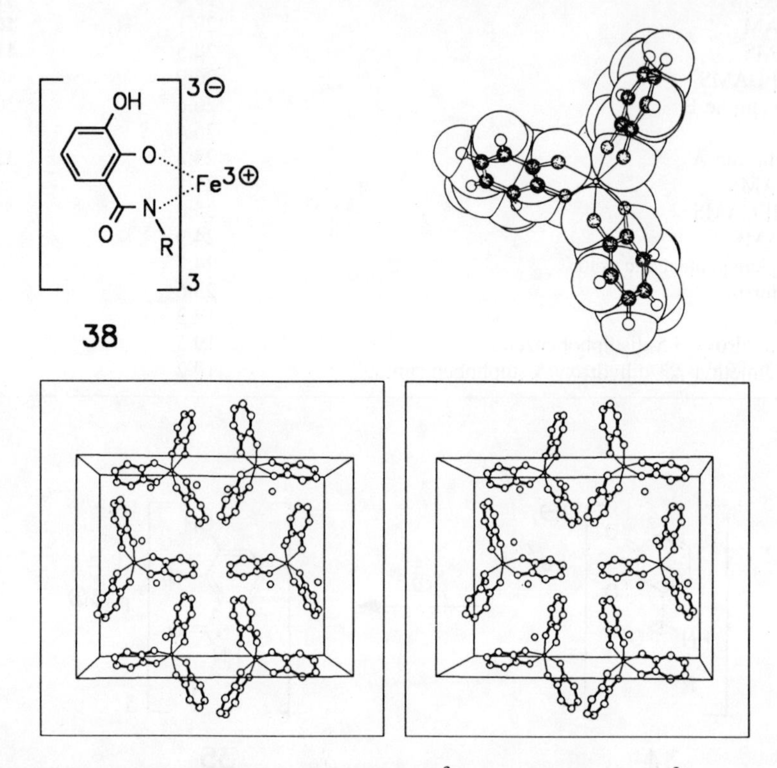

Figure 3. Perspective drawing of $[Fe(C_6H_6O_2)_3]^{3-}$, the complex of Fe^{3+} with the unsubstituted catechol (top right) and packing diagram of $K_3[Fe(C_6H_6O_2)_3]\cdot 1.5H_2O$ [15] (stereoview, bottom)

Some authors [16] maintain that in the enterobactin complex, structure **38** with participation of the nitrogen atoms in the binding of iron is also possible; however, Raman and CD spectra [17] tend to indicate that the iron is bound purely by oxygen atoms.

By comparison of the CD spectra of Δ-$K_3[Cr(C_2H_4O_2)_3]$ and $(NH_4)_3[Cr(ent)]$, it was shown that the prevailing isomer of the enterobactin complex must have the absolute configuration Δ-*cis* [18,19] [Figures 4 and 5; Δ means, according to IUPAC, a dextrorotatory winding (clockwise, plus-helix)]. Empirical force-field calculations also gave a higher stability of approximately 2.1 kJ/mol for the optically active Δ-isomer [20]. The external membrane receptor of *E.coli* displays a high chiral recognition.

The answer to the question of whether it is the chiral environment of the iron atom (Λ or Δ) or the chiral template (from L or D-serine) which is playing the greater role is probably that with a size of 1.4 nm the complexation centre is of

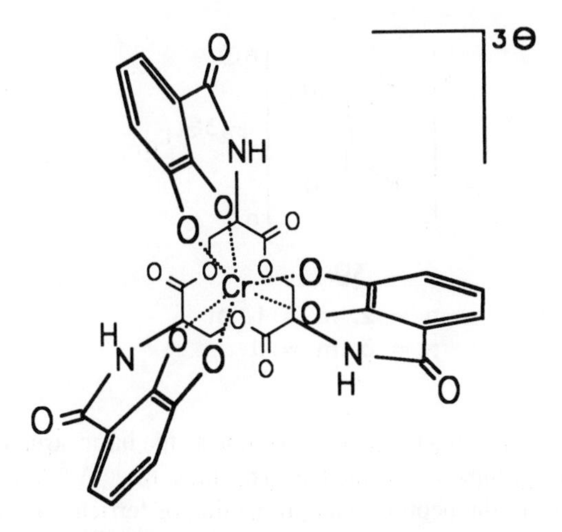

Figure 4. Δ-*cis*-$[Cr(ent)]^{3-}$ [18,19]

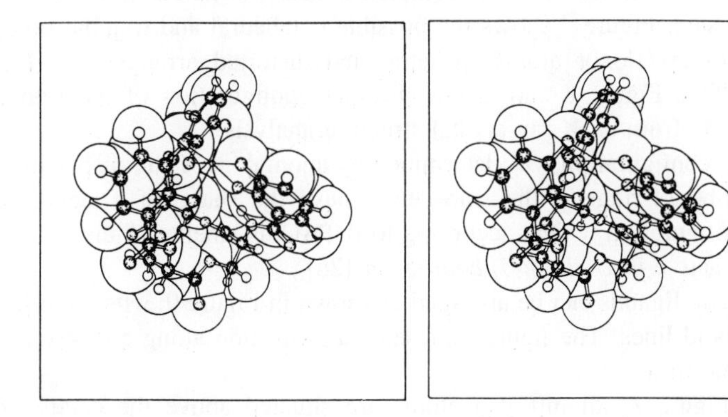

Figure 5. Stereoview of the $[Fe(ent)]^{3-}$ complex in its Λ-configuration (minus-helix; empirical force-field calculation [20])

greater importance. These findings are supported by the fact that the biological activity of a racemate (Λ and Δ) of the carbocyclic enterobactin analogue **29** is only half as large [21].

2.3.5 FURTHER LIGANDS OF THE SIDEROPHORE TYPE

2.3.5.1 Ligands with hydroxamic acid donors

Hydroxamic acids form with Fe(III) salts intensely red complexes ($\lambda_{max} \approx 430$ nm). Generally, they can be formulated as in **39**.

Numerous microorganisms produce such hydroxamate containing compounds, which belong to the class of the siderophores, also called siderochromes [22,23]. There are two types of siderochromes containing hydroxamate groups, the ferrichromes and the ferrioxamines.

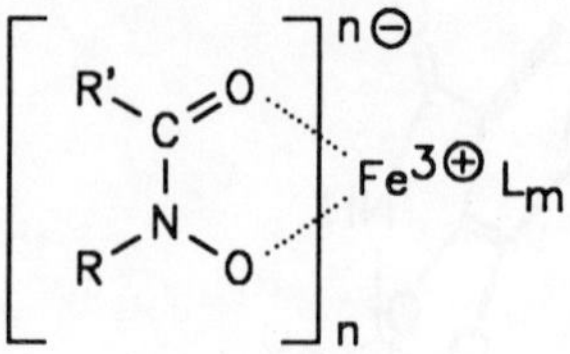

39

n = 2, m = 1 or
n = 3, m = 0

The ferrichromes contain a cyclic hexapeptide as the basic structure element. The three hydroxamate groups are formed by attaching three N^{δ}-acyl-N^{δ}-hydroxy-L-ornithine molecules to the peptide. The production of ferrichrome, e.g. by *Ustilago sphaerogena*, is highest, when the microorganisms grow in iron-poor nutrients [24]. As can be seen in Figure 6, the complex is nearly octahedral, not ideally so, but slightly twisted. Figure 7 shows the possible octahedral and trigonal-prismatic arrangements of six donor atoms. A tetragonal distorted arrangement (D_{4h}) is also possible [27]. Figure 8 shows the absolute configuration of the iron atom in ferrichrome A (from an X-ray crystal structure analysis).

The absolute configuration was determined by anomalous X-ray dispersion. The three five-membered rings at the iron centre have the shape of a laevorotatory propeller, a configuration which, according to IUPAC, is labelled Λ-cis, and which is biologically active in *Ustilago sphaerogena* [28].

Three bidentate ligands can be arranged as shown in Figure 9, where the ligands are shown as bold lines. The figure represents a projection along a C_3-axis into a plane orthogonal to it.

If, e.g., in Figure 8, all nitrogen atoms are situated above the centre of the octahedron, the configuration is *cis*, but if carbon and nitrogen atoms alternate, then these are *trans* configurations. All these diastereoisomers can also be resolved into enantiomers, as in the case of ferrichrome, where optically active amino acids form the backbone.

The other group consists of the linear ferrioxamines (sideramines). Ferrioxamine B, for example, is produced by *Streptomyces pilosus*. It was isolated by Prelog in 1960. The linear trihydroxamic acid is built from alternating 1-amino-5-(hydroxyamino)pentane and succinic acid units. It also serves as a growth factor, e.g. for *Microbacterium lacticum*. Figure 10 shows the molecular structures of some ferrioxamines [29].

Schwarzenbach *et al*. demonstrated that desferrioxamine B has a remarkable affinity for Fe(III) but a low affinity for other ions of different charges and sizes, and almost none for Fe(II) [30]. Here, five enantiomeric pairs are possible, one *cis* and four *trans*. The chelate ring closest to the free amine end is designated as 'ring 1' (Figure 11). If the nitrogen in ring 1 is situated beneath the carbon, then the configuration is named N. By analogy, the reverse also applies. Rings 2 and 3 are labelled *cis* or *trans* by comparison with ring 1.

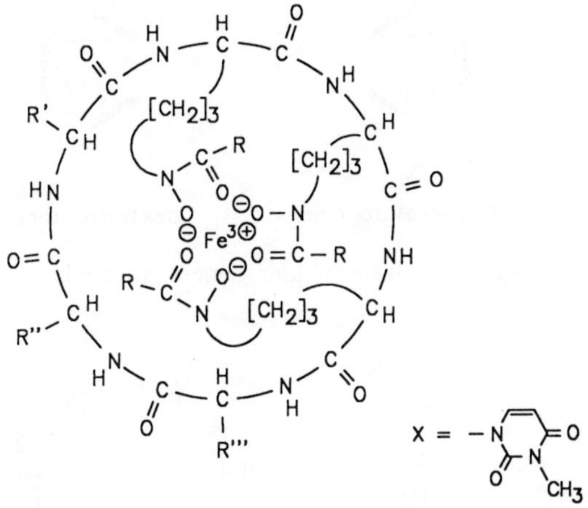

Ferrichrome	R	R'	R''	R'''
Ferrichrome	H	H	H	CH_3
Ferrichrysin	CH_2OH	CH_2OH	H	CH_3
Ferricrocin	H	CH_2OH	H	CH_3
Ferrichrome C	H	CH_3	H	CH_3
Ferrichrome A	CH_2OH	CH_2OH	H	$-CH{=}C(CH_3)-CH_2CO_2H$ (*E*)
Ferrirhodin	CH_2OH	CH_2OH	H	$-CH{=}C(CH_3)-CH_2CH_2OH$ (*Z*)
Ferrirubin	CH_2OH	CH_2OH	H	$-CH{=}C(CH_3)-CH_2CH_2OH$ (*E*)
Albomycin δ_1	CH_2OSO_2X	CH_2OH	CH_2OH	CH_3

Figure 6. Molecular structure of ferrichrome [25,26]

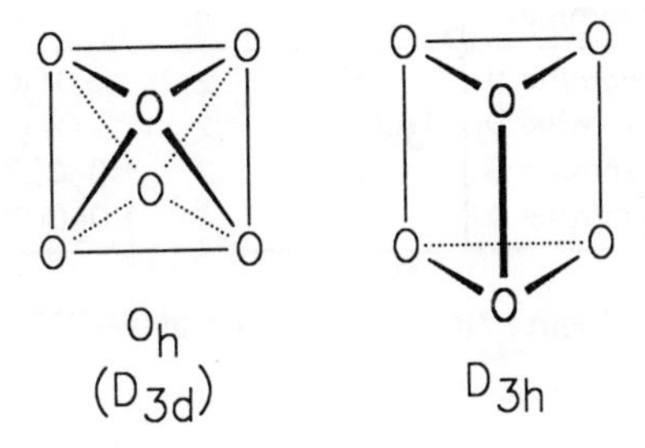

Figure 7. Arrangement of the ligand oxygens for a coordination number of 6

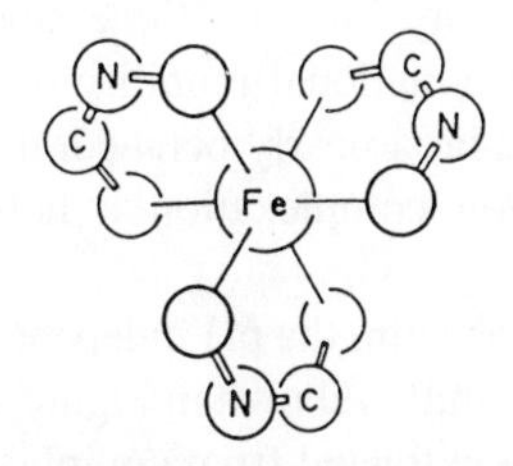

Figure 8. Absolute configuration at the iron atom in ferrichrome A

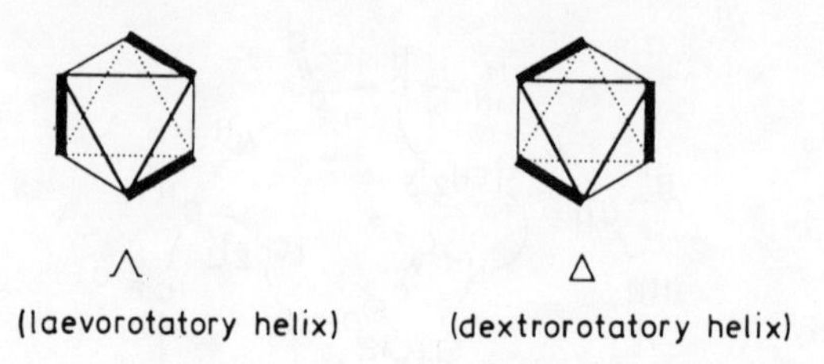

Figure 9. Octahedral arrangement of three bidentate ligands

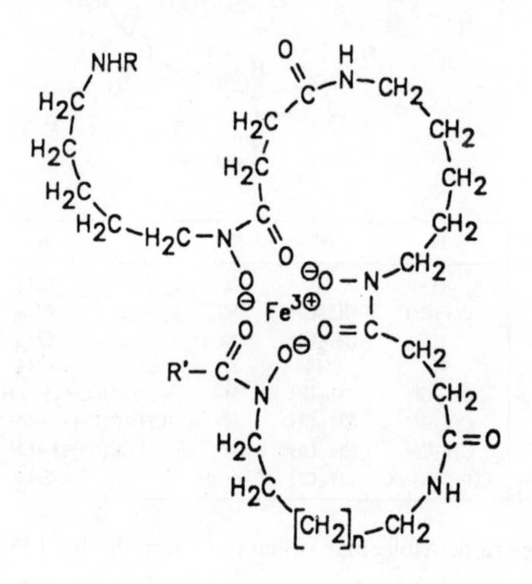

Ferrioxamine	R	n	R'
Ferrioxamine B	H	2	H_3C-
Ferrioxamine D_1	$H_3C(C{=}O)-$	2	H_3C-
Ferrioxamine G	H	2	$HO_2C(CH_2)_2-$
Ferrioxamine A_1	H	1	$HO_2C(CH_2)_2-$

Figure 10. Various ferrioxamines [29]

2.3.5.2 Macrobicyclic 'siderands'

Kiggen and Vögtle synthesized the novel complexing ligand **24a** [31], a cage-like macrobicycle containing a cavity lined on three sides with functional groups for complexation. Here, the conformational mobility of the catechol units is severely restricted. In the hexaanion **24a**, a nearly octahedral donor geometry for the metal cation has been prepared; on complexation a helically chiral configuration is adopted.

For comparison with enterobactin, the pH independence of the Fe^{3+} complexation with this macrobicyclic 'siderand' is shown in Figure 12. The binding of Fe^{3+} ions is so strong, that the cations are extracted from complexes with open-chain siderands. This is shown in Figure 13.

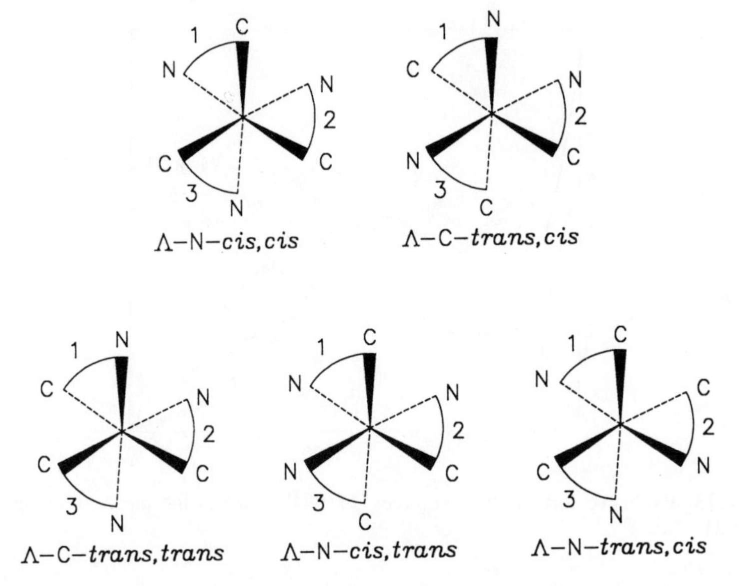

Figure 11. Configurations at the iron atom in the desferrioxamine B complex [29]

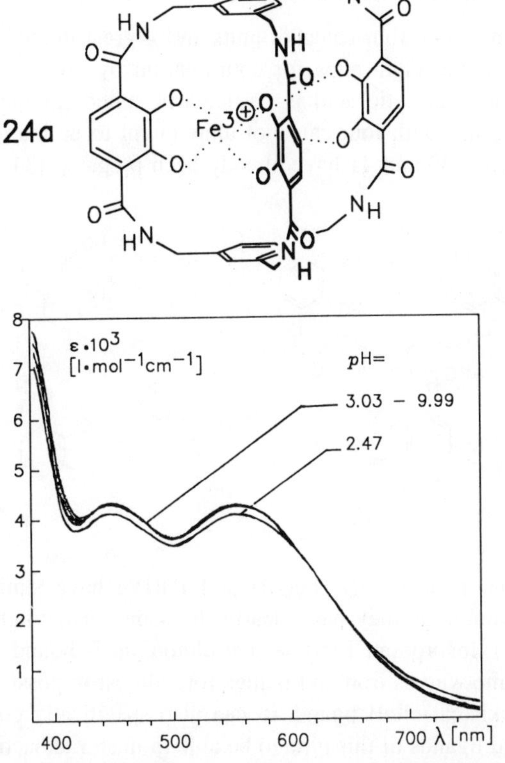

Figure 12. pH independence of **24a**, the complex of Fe^{3+} with **24** (UV–visible spectra)

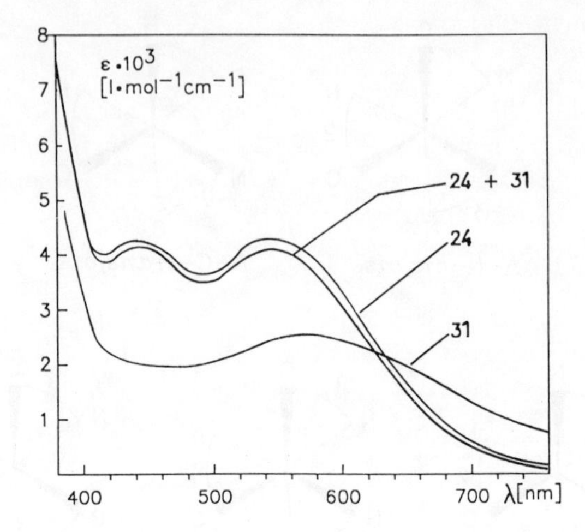

Figure 13. Exclusive formation of complex **24**•Fe^{3+} (**24a**) in the presence of open-chain ligand **31**

2.3.5.3 Octadentate ligands with four catechol donors

Siderophore ligands with four catechol units are of great interest. Large ions often offer not only six, but eight sites for coordination by donor atoms; this is why diethylenetriaminepentaacetic acid is particularly suited for the complexation of Pu^{4+} [32,33]. Ligands with four catechol units ought to be interesting octadentate ligands. Compounds **40** and **41** have already been prepared [34–38].

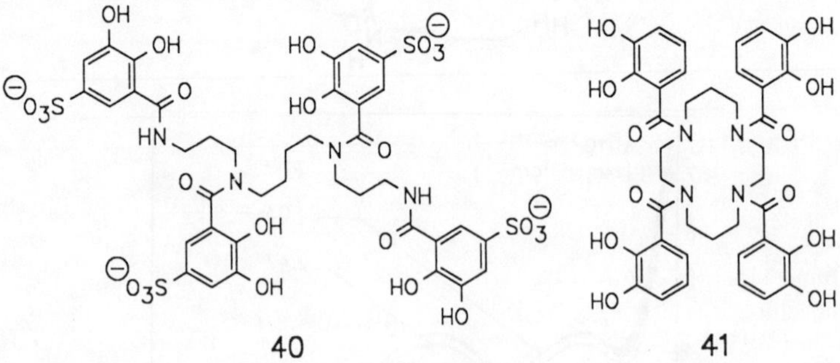

40 41

Apart from the radioactivity, Fe(III) and Pu(IV) have similar chemical and biological behaviours, as they have nearly the same charge/radius ratio (4.6 and 4.2, respectively). Incorporated radioactive plutonium is bound by transferrin and ferritin in the same way as iron and is therefore almost impossible to excrete. The octadentate tetrakis(catechol) ligand **40** can also specifically complex Pu(IV). *In vivo* tests showed ligands of this type to be able to flush radioactive actinides, such as Pu(IV), out of the organism ('sequestering agents').

2.3.5.4 The tunichromes

Nakanishi *et al*. have been engaged in the study of a group of sea organisms, the tunicates, which have the ability to accumulate metals such as vanadium, iron, molybdenum and niobium [39]. *Ascidia nigra*, for instance, is able to absorb pentavalent vanadium, to concentrate it 10^6-fold and to store it at physiological pH (approximately 7.2) as the reduced form V(III), or as V(V). A complexing agent is probably used for that purpose, which at the same time acts as a strong reducing agent. For this, the tunichromes would be suitable; they are the bright yellow sanguineous dyes of *A.nigra* and several other tunicates. Because the tunichromes are labile and occur as complex mixtures of closely related compounds, they evaded isolation and structure determination for some time.

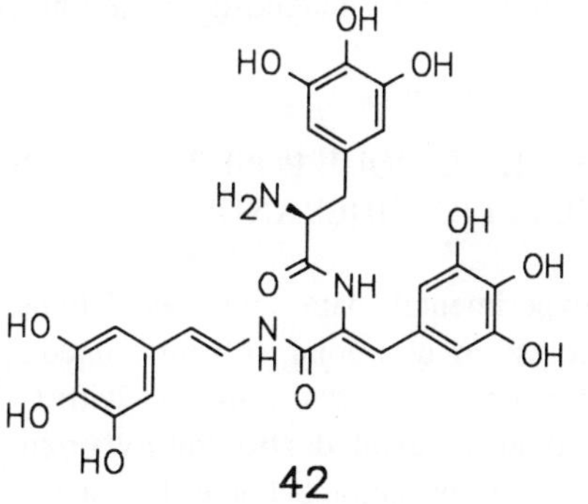

42

The isolation and characterization of tunichrome B-1 (**42**), the first member of this group of new biological reducing agents, was successful. The reducing moieties of tunichrome B-1 are the pyrogallol units, and it is known that the molecule is able to reduce V(V) to V(IV) and bind it in a complex.

2.3.6 APPLICATIONS OF CATECHOL-TYPE LIGANDS

In analytical chemistry, catechol-3,5-disulphonic acid (**43**) is of some importance. It is used as an indicator in complexometric titrations [40], and for the photometric determination of titanium and iron, even in each other's presence [41]. This is also where the name Tiron comes from.

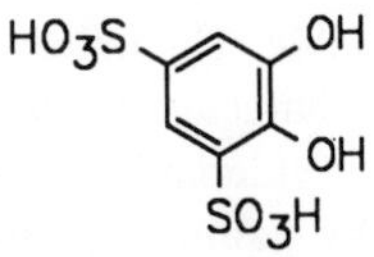

43

At the end of the 1920s, the sodium salt of the Sb^{3+} complex of catechol-3,5-disulphonic acid, known as stibophen (Fuadin), gained some importance as a drug against bilharziosis. In this context, the chelate complexes of alizarin should also be mentioned.

Polydentate complexing ligands are used for the selective extraction of metal ions from solutions of various types and origins. In the preparatory enrichment of ion solutions and the performance of analytical methods, they have been playing a classical role for a long time. Metal complexes of various types have catalytic properties. Also, there are both existing and potential uses in medicine: in the case of heavy metal poisoning, decontamination of the organism can be achieved with the help of complexing ligands. A pathological accumulation of metals leads to damage to heart, lungs and pancreas. As even a healthy human can eliminate only 1 mg of iron per day, iron must in any case be eliminated with desferrioxamine. With the help of complexing ligands, radioisotopes can be fixed onto antibodies and used for tumour diagnosis (early diagnosis of metastasis) and tumour therapy.

2.3.7 CONCLUSIONS: THE UNIQUE POSITION OF ENTEROBACTIN AND THE MACROBICYCLIC 'SIDERANDS'

Using all available experimental data and calculations on enterobactin, the following picture emerges: in non-polar solvents, uncomplexed enterobactin is overwhelmingly found in one stable conformation with C_3 symmetry, where the side-chains are arranged in an axial dextrorotatory orientation. With increasing polarity of the solvent this conformation becomes less stable. The axial arrangement of the side-chains is stabilized by intramolecular hydrogen-bonding between the amide protons of the side-chains and the oxygen atoms in the lactone units of the ring. The bonding of the amide proton to the lactone oxygen of the same asymmetric unit (i.e. L-lysine) is favoured over the bonding to the lactone oxygen of the neighbouring unit; this produces the predominance of the Δ-conformation of free enterobactin over its diastereoisomeric Λ-conformer.

On complexation of Fe^{3+} ions the side-chains rotate inwards and the hydrogen bonds of the amide to the ring are broken, while those to the catechol oxygens are shortened. The enterobactin molecule incurs only very little strain when forming the complex $[Fe{\cdot}enterobactin]^{3-}$.

An additional stabilization is observed in the best of the catechol ligands, BEL (**24**), because an octahedral donor geometry is already available, and the cation becomes enclosed all around within the rigid macrobicyclic cavity. The conformational mobility of the catechol units is largely reduced and the iron(III) complex is particularly stable. The extremely high complexation constant, $K_f \approx 10^{59}$, is the outstanding characteristic of this new macrobicyclic ligand, which belongs to the group of the 'siderands' (synthetic siderophores). Moreover, macrobicycle **24** is distinguished by stability towards oxidation and hydrolysis, and therefore meets all the requirements expected of a tailored ligand.

Recently, strong Fe^{2+} complexing agents of the siderand type (**44**–**47**) have been prepared, which contain bipyridine building blocks instead of catechol units [31e] (cf. Section 2.1). Here too, the size of the cavity was widely varied using spacer

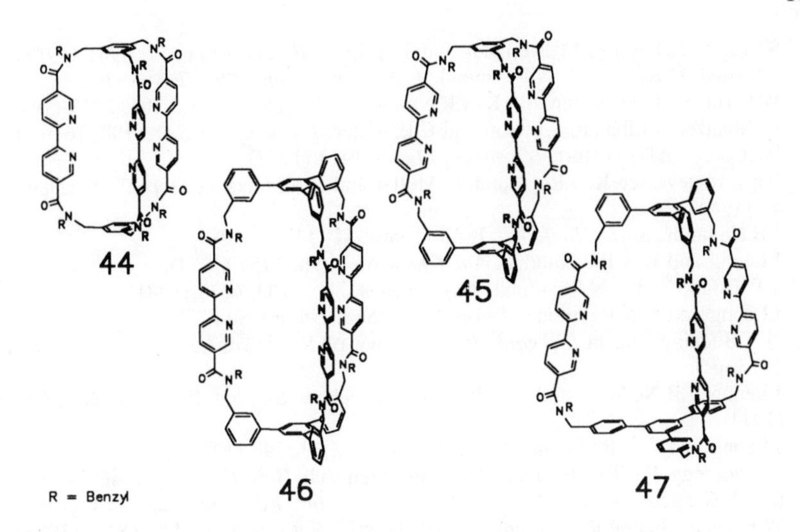

units (system of building blocks), making 'receptors' available for small ions, in addition to organic molecules (e.g. phloroglucinol as guest in **46**).

The search for high-performance ligands, not only for the complexation of iron(III) but also for gallium(III) and indium(III), will remain of some importance in the future [42].

References

1. C.E.Lankford, *Crit. Rev. Microbiol*. **2**, 273 (1973); for review, see G.Winkelmann, D.van der Helm and J.B.Neilands (Eds.), *Iron Transport in Microbes, Plants and Animals*, VCH, Weinheim, 1987.
2. J.B.Nielands, *Annu. Rev. Biochem*., **50**, 715 (1981).
3. J.Francis, J.Mandinaveitia, H.M.Macturk and G.Snow, *Nature (London)*, **163**, 365 (1949).
4. A.G.Lockhead, M.O.Burton and R.H.Thexton, *Nature (London)*, **170**, 282 (1952).
5. J.R.pollack and J.B.Neilands, *Biochim. Biophys. Res. Commun*., **38**, 989 (1970).
6. I.G.O'Brien and F.Gibson, *Biochim. Biophys. Acta*, **21**, 393 (1970).
7. E.J.Corey and S.Bhattacharyya, *Tetrahedron Lett*., 3919 (1977).
8. W.H.Rastetter, T.J.Erichson and M.C.Venuti, *J. Org. Chem*., **46**, 3579 (1981).
9. A.Shanzer and J.Libmann, *J. Chem. Soc., Chem. Commun*., 846 (1983); see also Y.Tor, J.Libman, A.Shanzer and S.Lifson, *J. Am. Chem. Soc*., **109**, 6517 (1987); as to the 'template effect' generally, cf. D.H.Busch and N.A.Stephenson, *Coord. Chem. Rev*., **100**, 119 (1990).
10. K.T.Greenwood and K.R.Luke, *Biochim. Biophys. Acta*, **614**, 183 (1980).
11. F.L.Weitl, W.R.Harris and K.N.Raymond, *J. Med. Chem*., **22**, 1281 (1979).
12. W.R.Harris, K.N.Raymond and F.L.Weitl, *J. Am. Chem. Soc*., **103**, 2667 (1981).
13. V.L.Pecoraro, F.L.Weitl and K.N.Raymond, *J. Am. Chem. Soc*., **103**, 5133 (1981).
14. W.R.Harris and K.N.Raymond, *J. Am. Chem. Soc*., **101**, 6534 (1979).

14a. See also R.C.Hider, D.Bickar, I.E.G.Morrison and J.Silver, *J. Am. Chem. Soc*., **106**, 6983 (1984).

15. K.N.Raymond, St.S.Isied, L.D.Brown, F.R.Fronczek and J.H.Nibert, *J. Am. Chem. Soc*., **98**, 1767 (1976).
16. B.F.Anderson, D.A.Buckingham, G.B.Robertson, J.Webb, K.S.Murray and P.E.Clark, *Nature (London)*, **262**, 722 (1976).

17. S.Salana, J.D.Stong, J.B.Neilands and T.G.Spiro, *Biochemistry*, **17**, 3781 (1978).
18. S.S.Isied, G.Kao and K.N.Raymond, *J. Am. Chem. Soc*., **98**, 1763 (1976).
19. W.R.Harris, C.J.Carrano and K.N.Raymond, *J. Am. Chem. Soc*., **101**, 2213 (1979).
20. A. Shanzer, J.Libman, S.Lifson and C.E.Felder, *J. Am. Chem. Soc*., **108**, 7619 (1986).
21. E.J.Corey and S.D.Hurt, *Tetrahedron Lett*., 3923 (1977).
22. For a review, see K.N.Raymond, G.Müller and B.F.Matzanke, *Top. Curr. Chem*., **123** 49 (1984).
23. J.B.Neilands, *Struct. Bonding*, **1**, 59 (1966); **11**, 145 (1972).
24. J.Leong and K.N.Raymond, *J. Am. Chem. Soc*., **96**, 1757 (1974).
25. T.P.Tufano and K.N.Raymond, *J. Am. Chem. Soc*., **103**, 6617 (1981).
26. J.Leong and K.N.Raymond, *J. Am. Chem. Soc*., **96**, 6628 (1974).
27. D.A.Buckingham, in *Inorganic Biochemistry* (G.Eichhorn, Ed.), Elsevier, New York, 1973, p.3.
28. J.Leong, J.B.Neilands and K.N.Raymons, *Biochem. Biophys. Res. Commun*., **60**, 1066 (1974).
29. J.Leong and K.N.Raymond, *J. Am. Chem. Soc*., **97**, 293 (1975).
30. G.Anderegg, F.L'Eplattenier and G.Schwarzenbach, *Helv. Chim. Acta*, **46**, 1409 (1963).
31. (a) W.Kiggen and F.Vögtle, *Angew. Chem., Int. Ed. Engl*., **23**, 714 (1984); (b) W.Kiggen, F.Vögtle, S.Franken and H.Puff, *Tetrahedron*, **42**, 1859 (1986); (c) P.Stutte, W.Kiggen and F.Vögtle, *Tetrahedron*, **43**, 2065 (1987); (d) T.J.McMurry, M.W.Hosseini, T.M.Garrett, F.E.Hahn, Z.E.Reyes and K.N.Raymond, *J. Am Chem. Soc*., **109**, 7196 (1987); (e) for analogous bipyridine 'siderands,' see S.Grammenudi and F.Vögtle, *Angew. Chem*., **98**, 1119 (1986); *Angew. Chem., Int. Ed. Engl*., **25**, 1122 (1986); F.Ebmeyer and F.Vögtle, *Angew. Chem*., **101**, 95 (1989); *Int. Ed. Engl*., **28**, 79 (1989); F.Ebmeyer and F.Vögtle in H.Du-gas (Ed.), *Bioorganic Chemistry Frontiers 1*, Springer, Berlin, p.143; (f) T.J.McMurry, S.J.Rodgers and K.N.Raymond, *J. Am. Chem. Soc*., **109**, 3451 (1987); T.J.McMurry, M.W.Hosseini, T.M.Garrett, F.Hahn, Z.E.Reyes and K.N.Raymond, *J. Am. Chem. Soc*., **109**, 7196 (1987).
32. G.Anderegg, P.Nägeli, F.Müller and G.Schwarzenbach, *Helv. Chim. Acta*, **42**, 827 (1952).
33. T.J.Wenzel, M.E.Ashley and R.E.Sievers, *Anal. Chem*., **54**, 615 (1982).
34. F.R.Weitl, K.N.Raymons and P.W.Durbin, *J. Med. Chem*., **24**, 203 (1981).
35. K.N.Raymond and W.L.Smith, *Struct. Bonding*, **43**, 159 (1981).
36. S.P.Sinha, *Struct. Bonding*, **25**, 69 (1976).
37. P.W.Durbin, E.S.Jones, K.N.Raymond and F.L.Weitl, *Radiat. Res*., **81**, 170 (1980).
38. F.L.Weitl, K.N.Raymond, W.L.Smith and T.R.Howard, *J. Am. Chem. Soc*., **100**, 1170 (1978).
39. R.C.Bruening, E.M.Oltz, J.Farukawa and K.Nakanishi, *J. Am. Chem. Soc*., **107**, 5298 (1985).
40. Brochure: *Komplexometrische Bestimmungsmethoden mit Titriplex*, Third Edn, E.Merck, Darmstadt.
41. J.Fries and H.Getrost, *Organische Reagenzien für die Spurenanalyse*, E.Merck, Darmstadt, 1975.
42. See G.Winkelmann (Ed.), *Biology of Metals*, Springer, Berlin, 1988; T.M.Loehr (Ed.), Iron Carriers and Iron Proteins, in *Physical Bioinorganic Chemistry*, Vol. 5, VCH, Weinheim, 1989.

2.4 THE π-SPHERANDS

2.4.1 THE CYCLOALKENE COMPLEXES OF SILVER

McMurry used the synthetic method named after him to prepare a sixteen-membered hydrocarbon macrocycle from cyclohexane-1,4-dione (**1**; four cyclohexadiylidene units; pentacyclo[$12.2.2.2^{2,5}.2^{6,9}.2^{10,13}$]-1,5,9,13-tetracosatetraene) [1]. It was not cyclohexanedione, however, that was subjected to the McMurry reaction, but the open-chain tetrameric diketone **7**, which gave **1** (m.p. >300 °C) in an excellent yield of 90%. Remarkably, **7** was not prepared by the McMurry procedure, but by the Barton method, via a spirothiadiazoline. The crucial McMurry cyclization **7**→**1** was carried out by slow dropwise addition, over a 48-hour period, of diketone **7** to a refluxing mixture of $TiCl_3$–Zn–Cu in dimethoxyethane.

An X-ray crystal structure analysis confirmed the constitution of hydrocarbon **1** and showed the distance between two facing double bonds in the ring to be 511 pm.

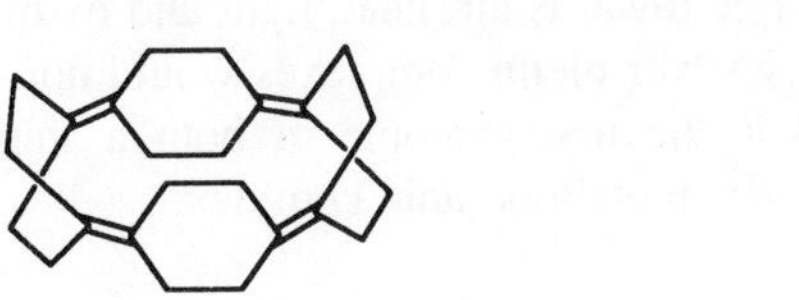

1

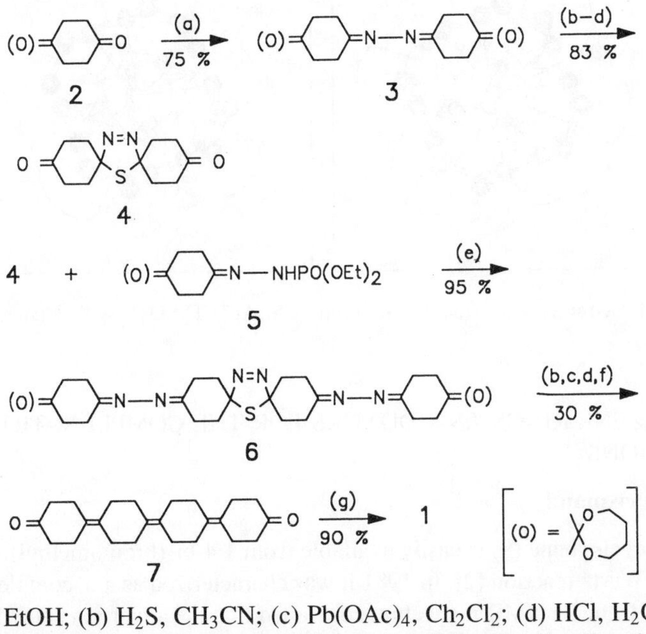

(O)= ⟨ ⟩ =O —(a), 75 %→ (O)= ⟨ ⟩ =N—N= ⟨ ⟩ =(O) —(b–d), 83 %→

2 3

O= ⟨ N=N, S ⟩ =O

4

4 + (O)= ⟨ ⟩ =N—NHPO(OEt)$_2$ —(e), 95 %→

5

(O)= ⟨ ⟩ =N—N= ⟨ N=N, S ⟩ =N—N= ⟨ ⟩ =(O) —(b,c,d,f), 30 %→

6

O= ⟨ ⟩ =O —(g), 90 %→ 1 [(O) = ketal]

7

(a) N_2H_4; EtOH; (b) H_2S, CH_3CN; (c) $Pb(OAc)_4$, Ch_2Cl_2; (d) HCl, H_2O, CH_2Cl_2; (e) NaH, THF; (f) toluene, Δ; then $P(OEt)_3$, Δ; (g) $TiCl_3$–Zn–Cu, dimethoxyethane, 48 h addition, Δ

The size of the cavity requires a metal-to-carbon bonding distance of approximately 250 pm. The cation would have to be stripped of all ligands, including solvent molecules, in order to enter the cavity. If tetraene **1** is to act as an eight-electron donor and square-planar ligand, then this would only be possible with d^{10} metals, as d^8 metals do not adopt a square-planar geometry. Based on these arguments, Ag(I) was chosen as the cation, because it is known that bond lengths in silver–olefin complexes are of the order of 240–260 pm and that its weakly bonded trifluoromethanesulphonate (triflate salt) is readily available. Indeed, after silver triflate has been added, the relatively insoluble tetra-alkene **1** dissolves in THF within a few minutes to form a homogeneous solution, from which a colourless solid then slowly precipitates. After recrystallization from THF, a 66% yield of a crystalline silver–alkene complex is obtained, the ^{1}H NMR and ^{13}C NMR spectra of which suggest a highly symmetrical structure. Moreover, the ^{13}C NMR spectrum shows carbon–silver couplings which point to a stable, static complex.

The structure of the CF_3SO_3Ag complex **8** was determined by X-ray structure analysis: the silver ion is located at the centre of the cavity (Figure 1). This silver complex is stable towards air, heat, light and hydroxylic solvents under conditions where ordinary silver-olefin complexes would normally be destroyed immediately. Compound **8** is the first example of both a static silver–olefin complex and a square-planar d^{10} metalloorganic complex.

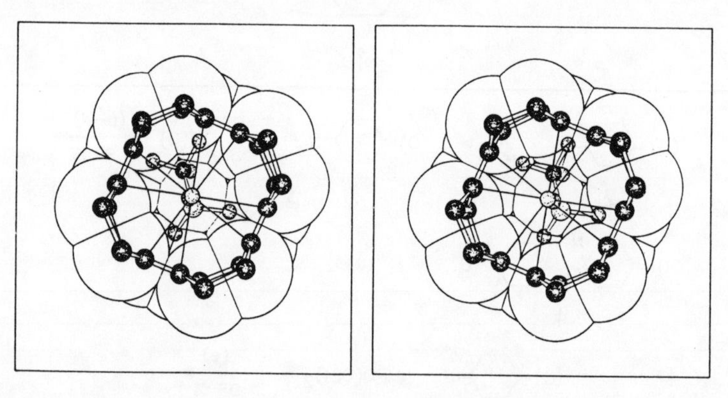

Figure 1. X-ray crystal structure of complex **8**, $Ag^+CF_3SO_3^-$ with **1** (stereoview)

2.4.2 BENZENE RINGS AS π-DONORS FOR THE COMPLEXATION OF CATIONS

2.4.2.1 π-Prismand

[2.2.2]Paracyclophane (**9**) is easily available from 1,4-bis(bromomethyl)benzene by a modified Wurtz reaction [2]. In 1981 it was characterized as a π-complex-forming compound: from ^{1}H NMR it was first concluded that the silver ion should be sitting in the middle, simultaneously complexed by all three benzene rings, which have to stand perpendicular to the plane of the large ring (18-membered).

3 BrH_2C–C₆H₄–CH_2Br —(Alkali metal; $Ph_2C{=}CPh_2$)→ **9** + H_3C–C₆H₄–$[CH_2]_2$–C₆H₄–$[CH_2]_2$–C₆H₄–CH_3 **10**

The geometry of 'π-prismand' **9** itself was proved by X-ray crystal structure analysis (Figure 2) [3]. The benzene rings are parallel to the threefold molecular axis which runs perpendicularly through the centre of the 18-membered macrocycle. The above-mentioned 1 : 1 silver trifluoromethanesulphonate complex of **9**, triflate complex **11**, is formed as a stable crystalline powder on mixing equimolar amounts of **9** and silver triflate in dry THF at room temperature. The melting point of the recrystallized complex (**11**) is 192 °C and sharp, 24 °C higher than that for the free host compound (**9**). The field desorption mass spectrum shows ions which correspond to $[\mathbf{9}{\cdot}Ag]^+$ and $[\mathbf{9}]^+$.

Whereas the solubility of CF_3SO_3Ag in chloroform is hardly changed by the addition of reference substance **10**, a marked increase in solubility is observed with the addition of **9**.

The complexation properties of **9** emerge clearly from ^1H NMR studies in $CDCl_3$ or CD_3OD solutions: the ^1H NMR spectrum of reference compound **10** remains unchanged in the presence of one equivalent of silver triflate, whereas the proton resonances of **9** are strongly shifted. The aromatic protons of **9** are shifted to lower field, which is attributed to a diminution of the π electron density in the complexing benzene rings.

If a 2 : 1 mixture of **9** and CF_3SO_3Ag in $CDCl_3$ is heated in a sealed tube, broadening of the proton signals of both the uncomplexed and complexed species is observed. A coalescence is observed at 52 °C for the aromatic and at 31 °C for the benzylic protons. This corresponds to exchange rates of $k(\text{arene}) = 77\ s^{-1}$ and $k(\text{methylene}) = 27\ s^{-1}$. At a coalescence temperature of 52 °C the free enthalpy of activation according to the Eyring equation is $\Delta G^{\neq} = 67$ kJ/mol.

The stability constant of the complex was determined by ^1H NMR using the Benesi–Hildebrand equation (concentration dependence of chemical shifts) and was found to be 195 ± 10 l/mol at 24 °C. For comparison, the known values for ordinary interactions between aromatics and silver salts in methanol are of the order of 2–3 l/mol, i.e. much lower.

Whereas the ^1H NMR spectrum of a 2 : 1 mixture of **9** and CF_3SO_3Ag is temperature dependent, the spectrum of the 1 : 1 complex does not change with temperature. A comparison with **2** also seemed to support the hypothesis that the

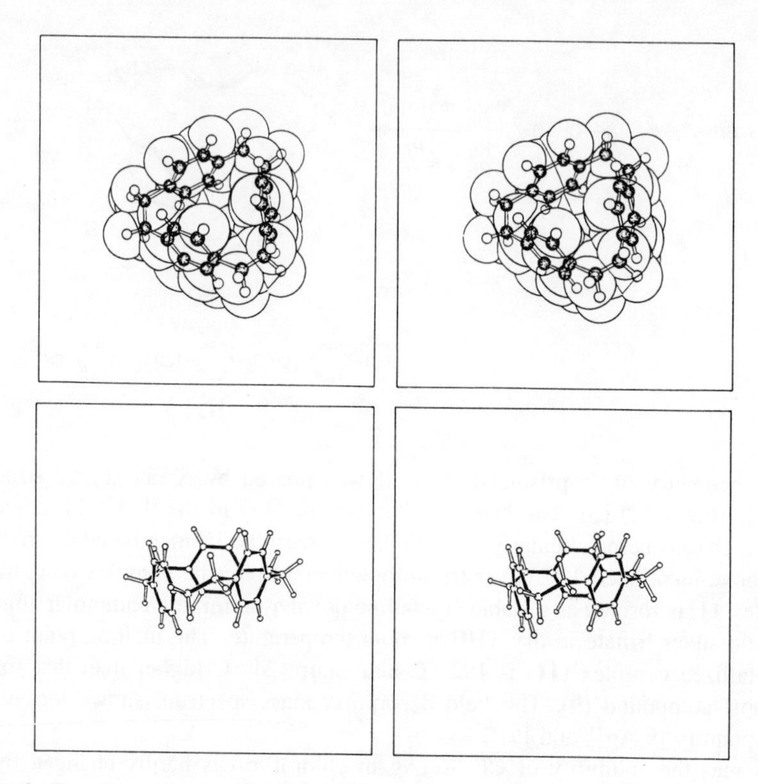

Figure 2. X-ray crystal structure of [2.2.2]paracyclophane (**9**): (*a*) view along the threefold molecular axis; (*b*) view sideways (stereoviews)

silver ion should be enclosed within the cavity of the [2.2.2]paracyclophane. This π-cryptate effect seems to enhance the complex's stability 100-fold, even though only π-binding sites are involved (benzene rings as π-donors).

Later dynamic ^{1}H-NMR investigations of silver triflate complex **11** by Boekelheide *et al*.[4] indicated that in this case the silver ion should be situated outside the cavity; for **11**, as for deltaphane (see below), a rapid exchange in solution is also observed. These findings were corroborated by X-ray crystal structure analysis of the silver triflate complex (**11**) and the silver perchlorate complex (**12**) of **9** (Figures 3 and 4).

Both structure analyses show that the Ag^{+} ion is located on the (pseudo)-threefold axis of the original hydrocarbon (**9**), outside the cavity, approximately 20 pm above the plane defined by the centres of the three benzene rings.

2.4.2.2 Deltaphane

Boekelheide *et al*. were able to synthesize $[2_6]$(1,2,4,5)cyclophane (**19**), which he named deltaphane, and to prepare its Ag^{+} complex [4]. On gas-phase pyrolysis of benzo[1,2;4,5]dicyclobutene (**13**) in a stream of nitrogen at 425 °C, **19** is formed, in addition to $[2_4]$(1,2,4,5)cyclophane (**16**) and methyl-substituted cyclophanes (**20**).

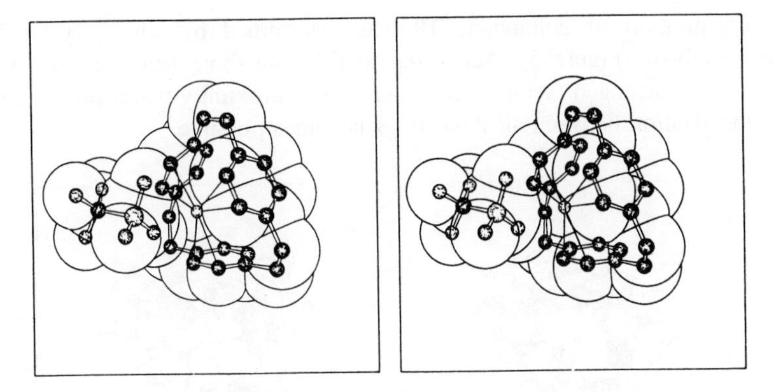

Figure 3. X-ray crystal structure of complex **11** of silver triflate with [2.2.2](1,4)cyclophane (**9**) [4]. The drawn Ag-to-C distances are 240–260 pm (stereoview)

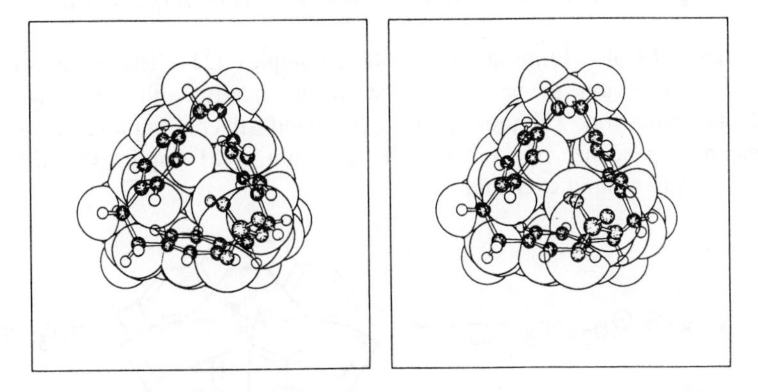

Figure 4. X-ray crystal structure of complex **12** of silver perchlorate with [2.2.2](1,4)cyclophane (**9**) [3]. The (Ag-to-C distances are 255–267 pm; stereoview)

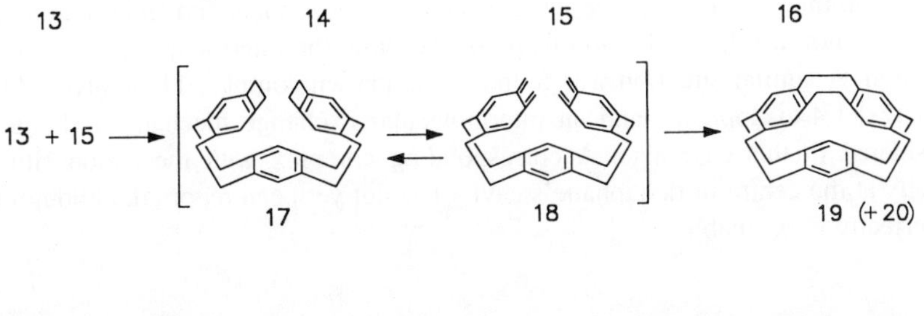

N_2, 425°C

13   14   15   16

13 + 15

17   18   19 (+20)

The constitution of deltaphane **19** was determined by single-crystal X-ray structure analysis (Figure 5). According to this, the three benzene rings in this rigid frame are arranged face-to-face in such close proximity that a certain amount of π-delocalization between all three rings becomes possible.

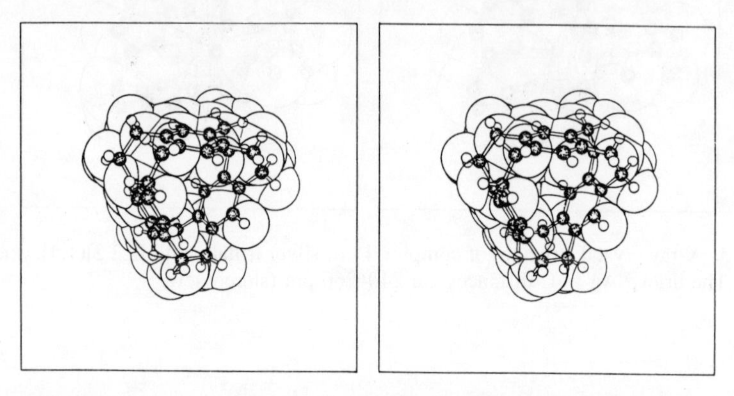

Figure 5. X-ray crystal structure of deltaphane (**19**) [4] (stereoview)

Deltaphane **19** also forms a silver triflate complex (**21**). The X-ray structure shows the silver ion to be complexed symmetrically, although outside the deltaphane framework (Figure 6; cf. π-prismand–silver complex, above) The temperature dependence of the ^{1}H NMR spectrum of **21** implies that the silver ion is subject to a rapid exchange.

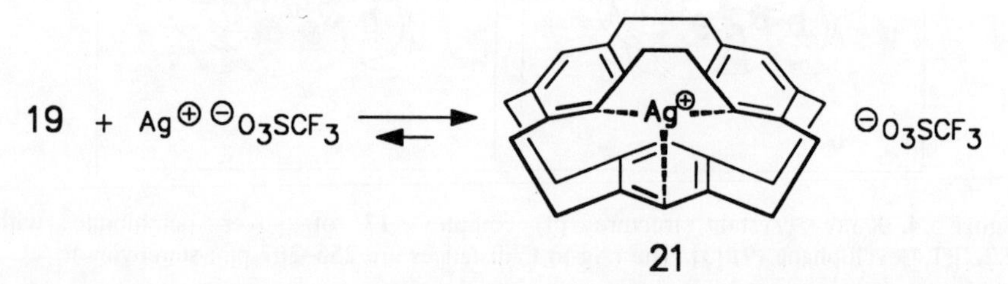

In solutions of the deltaphane–silver triflate complex (**21**), a dynamic equilibrium of the type **21a** $\rightleftharpoons$ **21b** is probable. This process could happen either by an external intermolecular exchange or by an interesting intramolecular tunnelling of the silver through the cavity of the deltaphane. The existence of an intermolecular exchange of the silver ion was demonstrated when it was found that increasing the concentration of **21** by 30% reduced the coalescence temperature by 10 °C; because an intermolecular process is concentration dependent and an intramolecular process is not, it follows that in this particular case, the intermolecular exchange is favoured. A similar situation was found in the known complex **11** of silver triflate with [2_3](1,4)cyclophane, and an intermolecular exchange mechanism should be considered in this case also. A corresponding complex with the cation situated exactly at the centre of deltaphane's cavity has not yet been reported, although this is perfectly conceivable.

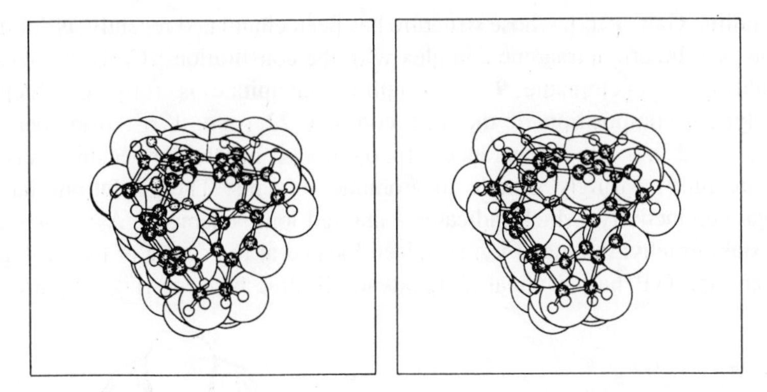

Figure 6. X-ray crystal structure of complex **21** of silver triflate with deltaphane (**19**) [4]. The Ag-to-C distances are 241–248 pm (stereoview)

$F_3CSO_3^{\ominus}\ {}^{\oplus}Ag$ ⇌ $Ag^{\oplus}\ {}^{\ominus}O_3SCF_3$

21a **21b**

2.4.2.3 The Ga$^+$ complex of [2.2.2]paracyclophane

A macrocyclic tris-arene complex of Ga(I) with η^{18}-coordination was prepared for the first time in 1987 by Schmidbaur *et al.* [5]. Whereas silver does not enter the cavity of [2.2.2](1,4)cyclophane (**9**), it was demonstrated that Ga(I) is located at the centre of this same π-ligand.

ns^2-Configured metals of the Main Groups of the Periodic Table form centric η^6-complexes with aromatic hydrocarbons, but the long distances between the metal and the centre of the arene indicate weak interactions. For Ga(I), In(I) and Tl(I) 1:1 complexes exist, and 1:2 complexes, as well, where the rings are inclined towards each other. When the arene ligands are linked together, a large stabilization in solution, due to the entropy effect, is to be expected for this type of bonding. An optimum situation, and therefore a high tendency to form the complex, should be found in cyclophanes, particularly in [2.2.2](1,4)cyclophane (**9**). However, attempts to incorporate Ga$^+$, In$^+$ and Tl$^+$ between the benzene rings of [2.2]- and [3.3]paracyclophanes and related naphthalene systems have failed to date: in the adducts obtained, the metal atoms were coordinated from outside to the benzene rings of the cyclophane. Based on its prefered metal-to-arene distances and its modes of coordination, it appeared possible to put Ga$^+$ at the centre of [2.2.2]paracyclophane's cavity (**9**). Experimentally, this new type of coordination succeeded surprisingly easily.

Crystalline Ga[$GaBr_4$], whose structure has been elucidated recently, is soluble in benzene as a dimeric tetraarene complex with the constitution [(C_6H_6)2Ga·$GaBr_4$]. On addition of cyclophane **9** a crystalline precipitate is formed, which has the analytical composition of the 1:1 complex **22** (m.p. 176°, from benzene). Compound **22** has the appearance of colourless, hardly air-sensitive crystals, which are only sparingly soluble in aromatic solvents. Their solutions have no electrical conductivity, which indicates a marked ionic character. As shown by X-ray crystallography, the metal ion is indeed located at the centre of the cyclophane cavity and the Ga^+ has identical η^6-bonds to all three benzene rings (Figure 7).

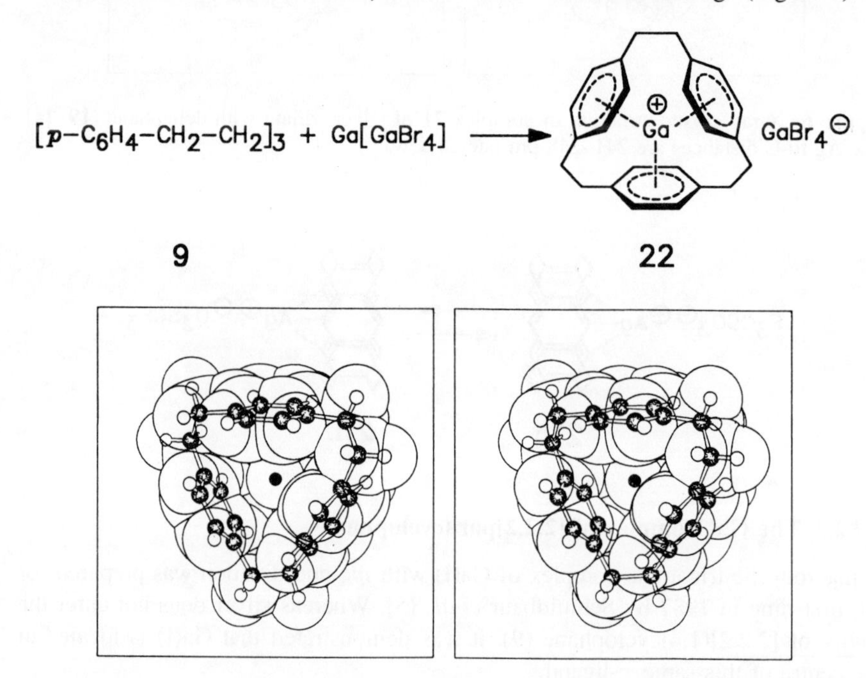

Figure 7. X-ray crystal structure of complex **22** of Ga^+ with **9** (stereoview)

The gallium ion is nearly at the same distance from all the 18 carbon atoms of the arenes. The metal ion, however, is not exactly in the plane defined by the three centres of the arene rings, but is located 43 pm above it. Considering the position of the anion, it becomes obvious that this displacement is due to the approach of a bromine atom, which induces a deformation of the coordination geometry towards a tetrahedron (Figure 7). In solution, the ion pairs should exist in a similar spatial arrangement. CPK models reveal the perfect spatial fitting of cyclophane **9** and the metal atom.

Compound **22** is the first metal complex with three centrically bound neutral arene ligands. The central atom therefore has no less than 19 atoms in the first sphere of coordination. This high 'hapticity' distinguishes it from its analogous silver complex [$C_6H_4CH_2CH_2$]$_3AgClO_4$. In the latter case, the metal remains outside the cavity and only forms three η^2-bonds to the arene rings [3,4]. The coordination with the arenes is obviously so efficient that all eight Ga^+—bromine bonds found in the $Ga[GaBr_4]$ crystal, except one, are broken. In all other structurally investigated cases, several bonds to the anions are retained.

References

1. J.E.McMurry, G.J.Harley, J.R.Matz, J.C.Clardy and J.Mitchell, *J. Am. Chem. Soc.*, **108**, 515 (1986).
2. F.Vögtle and W.Kissener, *Chem. Ber*., **117**, 2538 (1984).
3. C.Cohen-Addad, P.Baret, P.Chautemps and J.-L.Pierre, *Acta Crystallogr*., Sect.C **39**, 1346 (1983).
4. H.C.Kang, A.W.Hanson, B.Eaton and V.Boekelheide, *J. Am. Chem. Soc.*, **107**, 1979 (1985).
5. H.Schmidbauer, R.Hager, B.Huber and G.Müller, *Angew. Chem*., **99**, 354 (1987); see also C.Elschenbroich, J.Schneider, M.Wünsch, J.L.Pierre, P.Baret and P.Chautemps, *Chem. Ber*., **121**, 177 (1988).

2.5 CATENANES, CATENANDS AND CATENATES

2.5.1 TOPOLOGICAL STEREOCHEMISTRY AND TOPOLOGICAL CHIRALITY

The description of most chemical structures is possible by considering their molecular composition (topology), on the one hand, and their Euclidian geometry (stereochemistry), on the other. There are a few important structures with stereochemical differences based on topological factors. Such chemical structures recently ceased to be of purely theoretical interest and became the subject of organic chemical syntheses.

The concept of topology comes from mathematics. Topology is a branch of geometry, where geometrical structures are considered from a general point of view. In doing this, quantities such as lengths, areas and angles are ignored. Whereas in Euclidian geometry equivalence or similarity is produced by identical quantities and ratios, topological equivalence is given by type of connection within a structure.

The three structures a, b and c are distinguished by having identical connectivities, i.e. by being homeomorphic. In addition, they can be interconverted by deformation and are therefore topologically equivalent (isotopic). The three drawings represent the geometric structure of the closed curve.

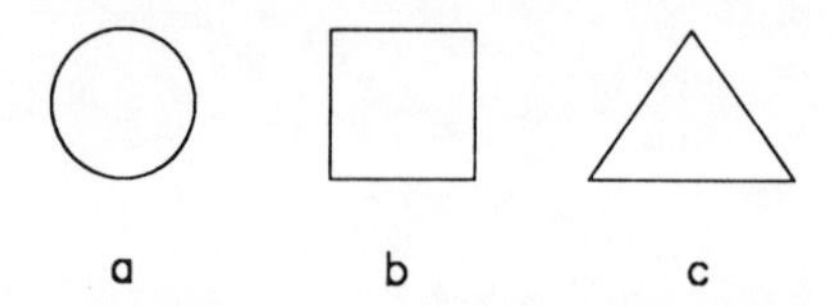

Although topological properties are correctly described only by abstract mathematics, simple relationships can be represented by 'graphs'. A graph is a two-dimensional representation made of points and lines. A point is called a vertex, and a line linking two vertices an edge; occasionally, a vertex with more than two edges is termed a branched vertex. A graph specifies the neighbouring relationships within an object and, therefore, corresponds to the representation of a molecular structure, a molecular graph. Here, vertices are atoms and edges chemical bonds.

Molecular graphs are often simplified to the extent of showing only the topologically relevant structural elements (I and II).

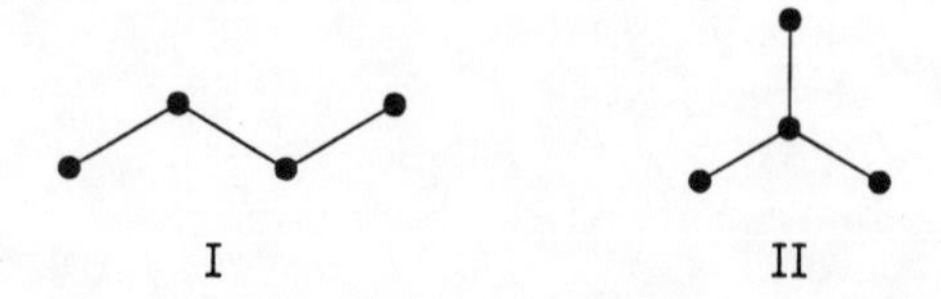

Because constitutional isomers always have different connectivities, they are neither homeomorphic nor isotopic. The geometric differentiation between the configurationally isomeric alkenes (*Z*)- and (*E*)-but-2-ene (III and IV) is based on

their rigidity. As they can be interconverted by deformation, both configurational and conformational isomers are isotopic and therefore also homeomorphic.

Apart from constitutional isomers and stereoisomers, there are also isomers which do not fit into either of the two categories. The three drawings d, e and f are homeomorphic, but not isotopic. If these are homeomorphic cycloalkanes, we call them topological stereoisomers. Structures d and e (f) are topological diastereoisomers, and e and f are topological enantiomers. This is a rare case of topological chirality. Topological chirality is found when:

1. the structure is a molecular knot;
2. the connections of chains are oriented;
3. the rings are connected in a different chiral way;
4. there is no achiral graph for a given topological structure.

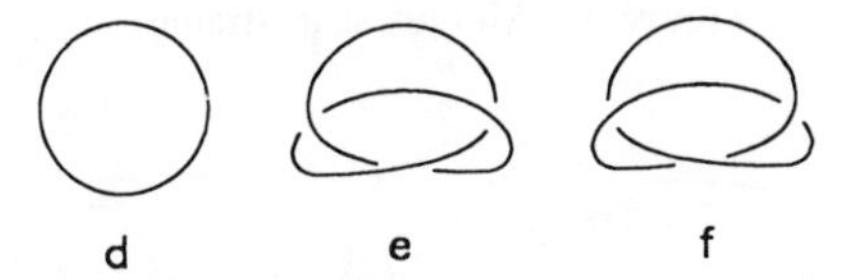

After these general considerations on topology, the question arises of how such topologically chiral structures might be prepared. Possible strategies were proposed by Wasserman as early as 1961 [1]. The strategy known as *threading* is shown by molecular models to be conceivable with alkyl chains of at least 50 carbon atoms: a loop is formed, one end is pulled through and then both ends are connected (Figure 1). The probability of producing a detectable molecular knot by this method is remote, however. This strategy of threading can also be considered for the synthesis of catenanes. An alkyl chain is pulled through a preformed ring, after which its loose ends are tied together (Figure 2). Catenane syntheses based on similar methods have already been successfully used [2].

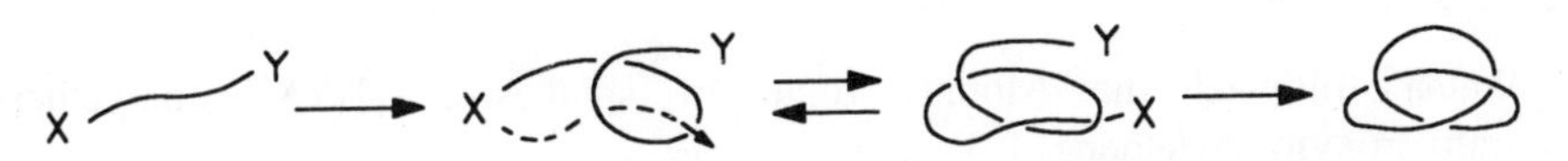

Figure 1. Threading to form a molecular knot

X
Y
+
X
Y

Figure 2. Threading to form a catenane

The second method is called the 'strategy of the Möbius strip.' A Möbius strip is formed by twisting the strip 180° along its axis and then connecting the ends, thus producing a structure with only one surface. A molecular ladder, twisted into a strip of this type, could have the rungs cut out to yield a single macrocycle (Figure 3). The corresponding structures, twisted once or one and a half times, would, on breaking the rungs, produce catenanes and knots, respectively (Figure 4).

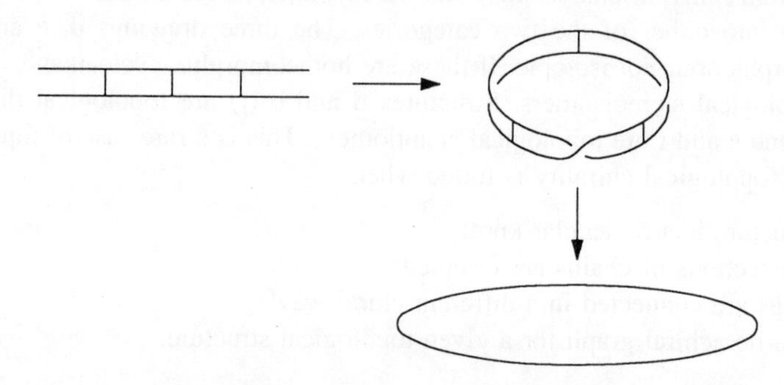

Figure 3. 'Möbius strip strategy'

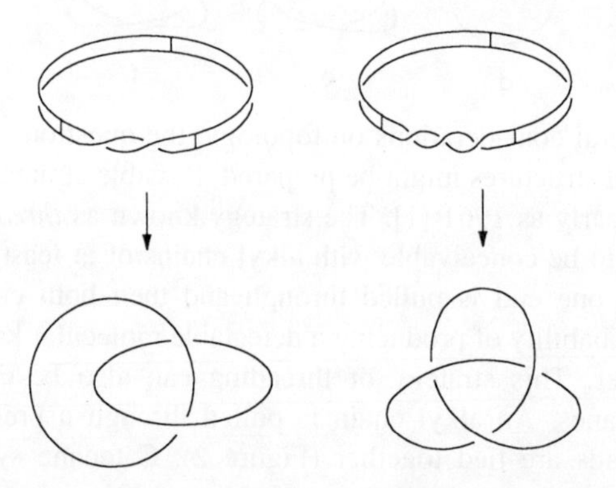

Figure 4. Synthesis of catenanes and molecular knots according to the 'Möbius strip strategy'

Walba followed the Möbius route to synthesize THYME polyethers (**tetrahy**droxy**m**ethyl**e**thene) [3].

Whereas the double head-to-tail connection of diolditosylate V (n = 1) only yields a cylindrical macrocycle of type VI [4], the 'three-runged' diolditosylate V (n = 2) reacts to produce a separable mixture of VI and VII (n = 2), VII being a racemate [5]. The cutting of the rungs is easily achieved by ozonolysis of the double bonds. The isolation of more highly twisted molecules has not yet been successful. On cyclization of the 'four-runged' diolditosylate V (n = 3) most products observed are of types VI and VII [6].

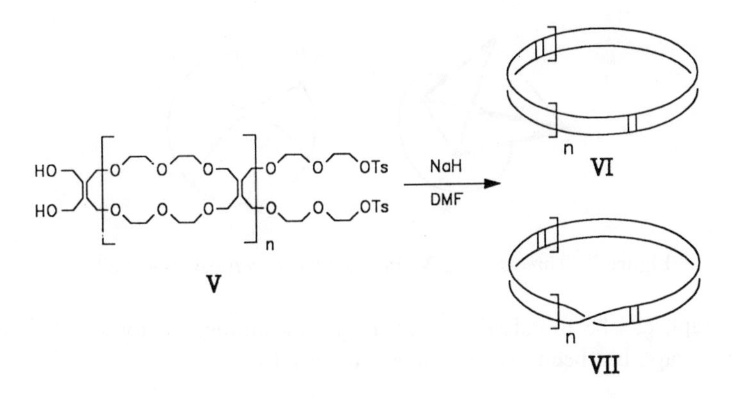

Möbius strip-like molecules, such as VII, are chiral. In order to decide whether they are also topologically chiral, they must be treated as Möbius ladders. Following the fourth condition for topological chirality (see above), each structure is topologically deformed and an achiral graph is sought, if possible. A planar graph is a figure in which vertices are connected without any edges crossing each other. Because a planar graph has a mirror plane, its corresponding structure is topologically achiral. If a planar graph is not found, the most symmetrical representation possible may help. Thus, a 'two-runged' Möbius ladder is achiral. Molecules with this sort of topology have already been prepared (e.g. VIII) (Figure 5) [7].

The topological achirality of the hypothetical 'two-runged' THYME polyether is easily demonstrated if one realizes that rotating one enantiomer around the double bond produces the other. In the case of the 'three-runged' polyether this is impossible. If all edges of the 'three-runged' Möbius ladder are considered to be equal, then the structure is topologically achiral (Figure 6), but all edges of the THYME polyether are not identical and bearing this in mind no topologically

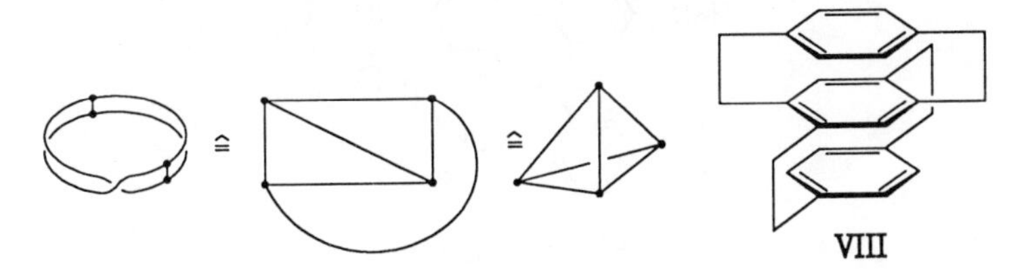

Figure 5. Two-runged Möbius ladder: 4 vertices, 6 edges

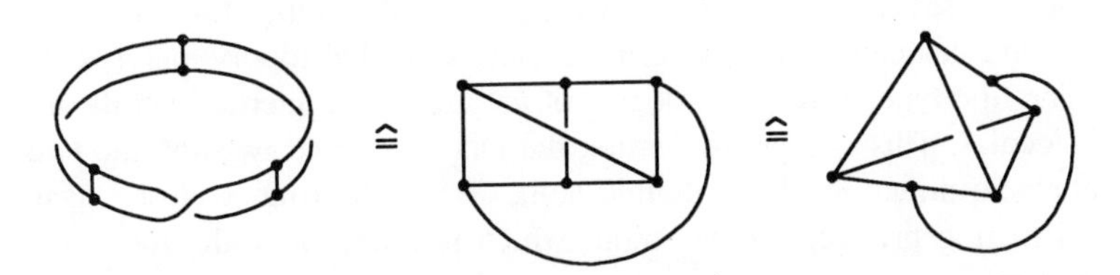

Figure 6. Three-runged Möbius ladder: 6 vertices, 9 edges

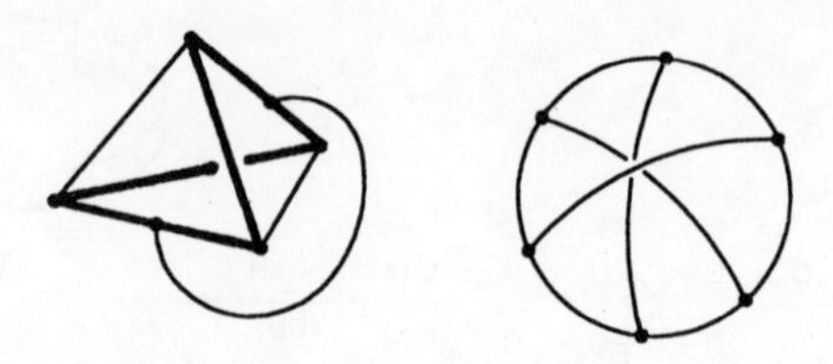

Figure 7. Three-runged Möbius ladder: 6 vertices, 6 + 3 edges

achiral graph can be found (Figure 7). In the meantime, the topological chirality of such a graph has been proved mathematically [8].

2.5.2 CATENANES AND ROTAXANES

Catenanes [9] are made up of two separate subunits; as the two rings of the catenane molecule are not held together by classical chemical bonding, their cohesion is often called mechanical bonding or, according to Wasserman, topological bonding [10].

At first sight, catenane-type molecules seem interesting only from a theoretical point of view. However, such linked molecules occur naturally, e.g. chained DNA dimers in human leucocytes with leukaemia. Their DNA contains 26% of cyclic dimeric DNA with a 3% portion of linked dimer. Catena-DNA molecules form intercalation complexes with aminoacridines or ethidium bromide, which can be isolated by sedimentation in a centrifuge. Catena-DNA has also been observed directly using an electron microscope. DNA-catenanes have been found in the mitochondria of cells of mice, after certain viruses had transformed them, as well as in the kinetoplastic DNA of *Trypanosoma cruzi*.

H_2N — — NH_2 $=N^{\oplus}$ $Br^{\ominus}$ Et

Ethidium bromide

The existence of such chained rings raises the question of their formation in Nature. In fact, several groups have been able to isolate enzymes, which produce supercoils and catenanes of genetic molecules, so-called topoisomerases (types I and II). The first type reduces the degree of netting of the interlaced coils of cyclic DNA molecules. This is done by cutting the DNA ring and twisting the two ends in the opposite direction of the double helix. When the ring is closed again, the decreased netting produces some strain, which is relieved by the formation of a superstructure.

The chaining of double-stranded DNA is effected by a type II topoisomerase

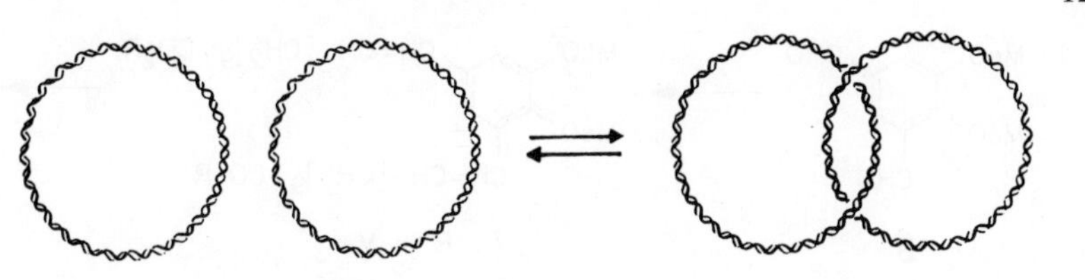

Figure 8. Two double-stranded DNA rings form a catenane

(Figure 8). This enzyme was first found in *Escherichia coli* and was called DNA-gyrase.

The first synthesis of organic catenanes (**4**) succeeded by statistical means. The strategy for this 'statistical catenane synthesis' is the following [10]: diethyl tetratriacontanedicarboxylate (**1**) was cyclized to the acyloin **2** with sodium and then reduced according to the Clemmensen method with deuterium chloride, to afford cyclic hydrocarbon **3** (which contains five deuterium atoms). Diester **1** was then cyclized in the presence of the deuterated macrocycle in xylene, using a 100-fold excess of **3**. Unchanged deuterated macrocycle was separated by chromatography and the simple acyloin ring oxidized with H_2O_2 in alkaline medium and separated as the diacid **5**. The success of this synthesis, though, is still in doubt, as no mass spectrum was recorded. The yield of **4** was given as 0.0001%.

$EtO_2C{-}[CH_2]_{32}{-}CO_2Et$ (**1**) → 1) Na, Xylene, 140° C; 2) HOAc → $[CH_2]_{32}$ acyloin ring (=O, OH) (**2**) → Zn/DCl → $C_{34}H_{63}D_5$ (**3**) (50% yield)

2 → $HO_2C{-}[CH_2]_{32}{-}CO_2H$ (**5**)

$D_5H_{63}C_{34}$ interlocked with $[CH_2]_{32}$ acyloin ring (O, OH) (**4**)

The first directed synthesis of a catenane was achieved by Schill and Lüttringhaus in 1964 by breaking certain bonds in a precatenane prepared in a multi-step sequence of reactions [9]. The starting material for this more effective strategy was dialdehyde **6**, to which two unbranched 'alkyl ester arms' were attached by a Wittig reaction with 10-methoxycarbonyldecylidenetriphenylphosphorane. After hydrolysis to dicarboxylic acid **8**, the double bonds were reduced and the acids esterified to afford compound **9**, as shown in the reaction scheme.

Diester **9** was reduced to the diol **10** with $LiAlH_4$, converted to dibromide **11** and then dinitrile **12**, which on Ziegler cyclization formed the cyanoketimide. After hydrolysis and complementary methylation, ketone **13** was obtained. A Huang–

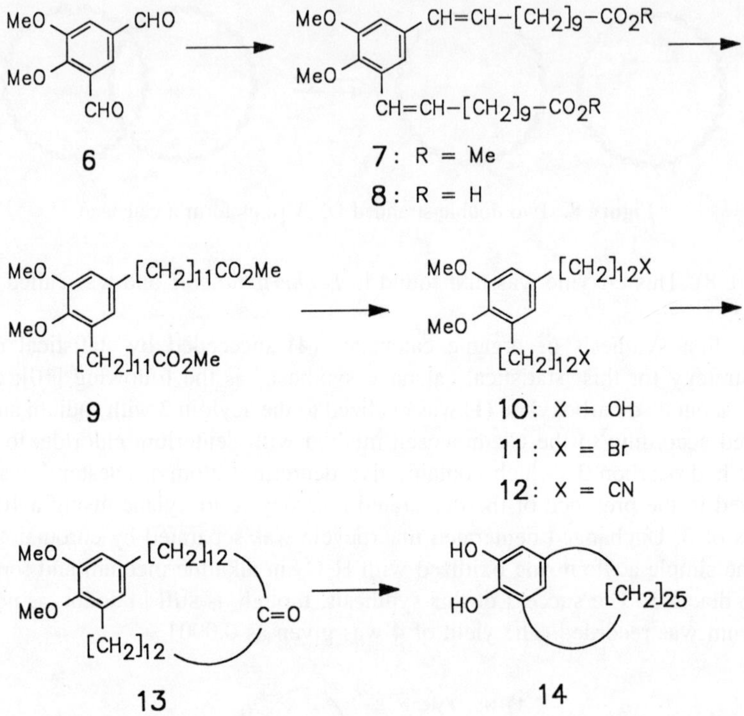

Minlon reduction, followed by an ether cleavage, yielded 3,5-pentacosamethylene-catechol **14**. This *ansa*-catechol was converted to the corresponding acetal with 1,25-dichloropentacosan-13-one, nitrated with copper nitrate in acetic anhydride and reduced to amine **17**. Cyclization with potassium carbonate and potassium iodide in pentan-1-ol afforded 2,2,*N*,*N*-bis(didecamethylene)pentacosamethylene-5-aminobenzene-4,6-diol (**18**) in 29% yield.

Acetal **18** was then cleaved with HBr, converted to diacetate **20** and oxidized to the amino-*o*-benzoquinone **21** with iron(III) sulphate. Acid hydrolysis afforded catenane **22** and its tautomer **23**. The *O,N*-diacetate **24** was obtained with acetic anhydride–sodium acetate; reductive acetylation yielded tetraacetate **25**, which in turn was hydrolysed under alkaline conditions to catenahydroxy-*p*-quinone **26**. The acetates were particularly suitable for the characterization of the catenane.

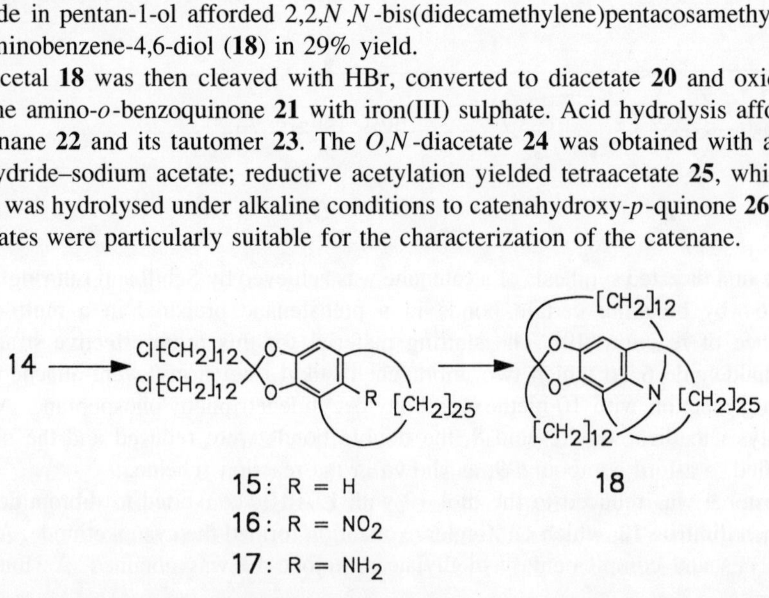

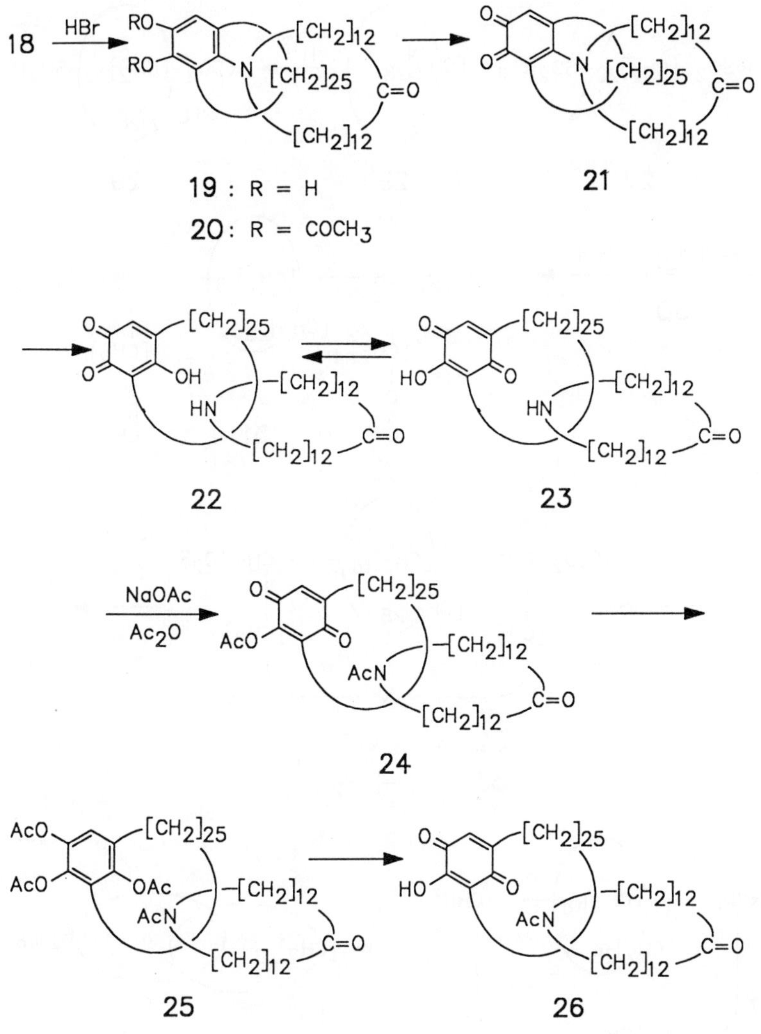

Schill *et al*. managed the synthesis of the first purely hydrocarbon catenane [11]: [2]cyclohexatetracontane ([cyclooctacosane]catenane, **35**). First, **27** was threaded through cyclooctacosane (**28**) with the help of toluenesulfonic acid to yield the rotaxane **29** (10% yield, an oil). After a twofold metallation with lithium diisopropylamide (LDA) in THF, alkylation with 13-bromotridec-1-yne (**30**) afforded **31**.

The rotaxane **31** was cyclized to the catenane **32** by a Glaser coupling with copper(II) acetate in diethyl ether–pyridine. Its reduction, and the cleavage of both triphenylmethyl groups, followed by its conversion to the dichloride **34**, its reduction with sodium amalgam and catalytic hydrogenation afforded the purely carbocyclic catenane **35** (colourless crystals, m.p. 58–59 °C).

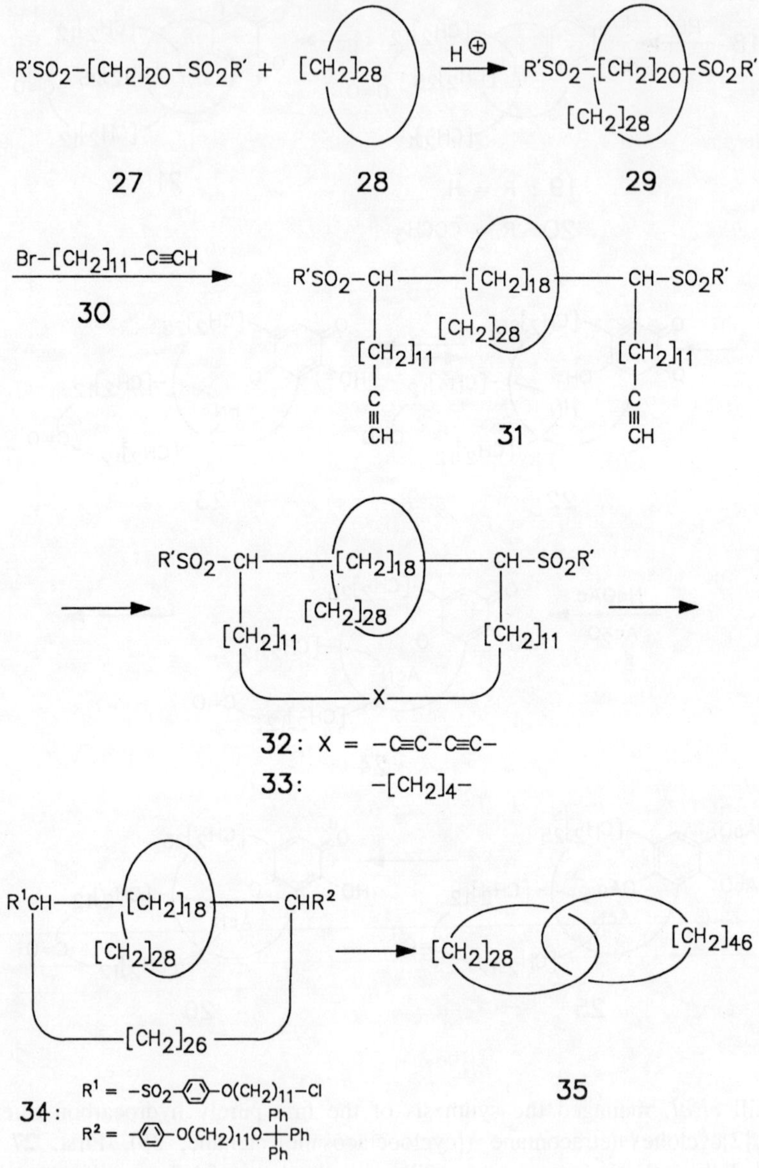

$R'SO_2-[CH_2]_{20}-SO_2R'$ + $[CH_2]_{28}$ $\xrightarrow{H^\oplus}$ $R'SO_2-[CH_2]_{20}-SO_2R'$ / $[CH_2]_{28}$

27 28 29

$\xrightarrow{Br-[CH_2]_{11}-C\equiv CH}$ **30**

$R'SO_2-CH-[CH_2]_{18}-CH-SO_2R'$, $[CH_2]_{28}$, $[CH_2]_{11}-C\equiv CH$ (×2) **31**

→ $R'SO_2-CH-[CH_2]_{18}-CH-SO_2R'$, $[CH_2]_{28}$, $[CH_2]_{11}$–X–$[CH_2]_{11}$ →

32: X = $-C\equiv C-C\equiv C-$

33: $-[CH_2]_4-$

$R^1CH-[CH_2]_{18}-CHR^2$, $[CH_2]_{28}$, $-[CH_2]_{26}-$ → $[CH_2]_{28}$ $[CH_2]_{46}$

34: $R^1 = -SO_2-C_6H_4-O(CH_2)_{11}-Cl$

$R^2 = -C_6H_4-O(CH_2)_{11}O-C(Ph)_3$

35

2.5.3. CATENANDS AND CATENATES

New, simpler and more productive syntheses of heterocyclic catenanes, although rigidified with aromatics, were found by Sauvage *et al*.[12]. The tedious preparation of precatenanes is replaced by the use of the 'template effect,' a well known tool in coordination chemistry. These template-directed syntheses are based on the idea that a large enough cyclic ligand is able to bind a metal ion, which is used to bind a second ligand with its free coordination sites.

As shown in Scheme 1, the complexing N atoms, because of the given arrangement of the rings, are forced away from the more favourable square-planar configuration into a tetrahedral configuration. The second ligand can then be cyclized to a ring, which inevitably will be chained to the first one. The template-directed synthesis of 'catenands' (i.e. catenanes with ligand properties) can even be

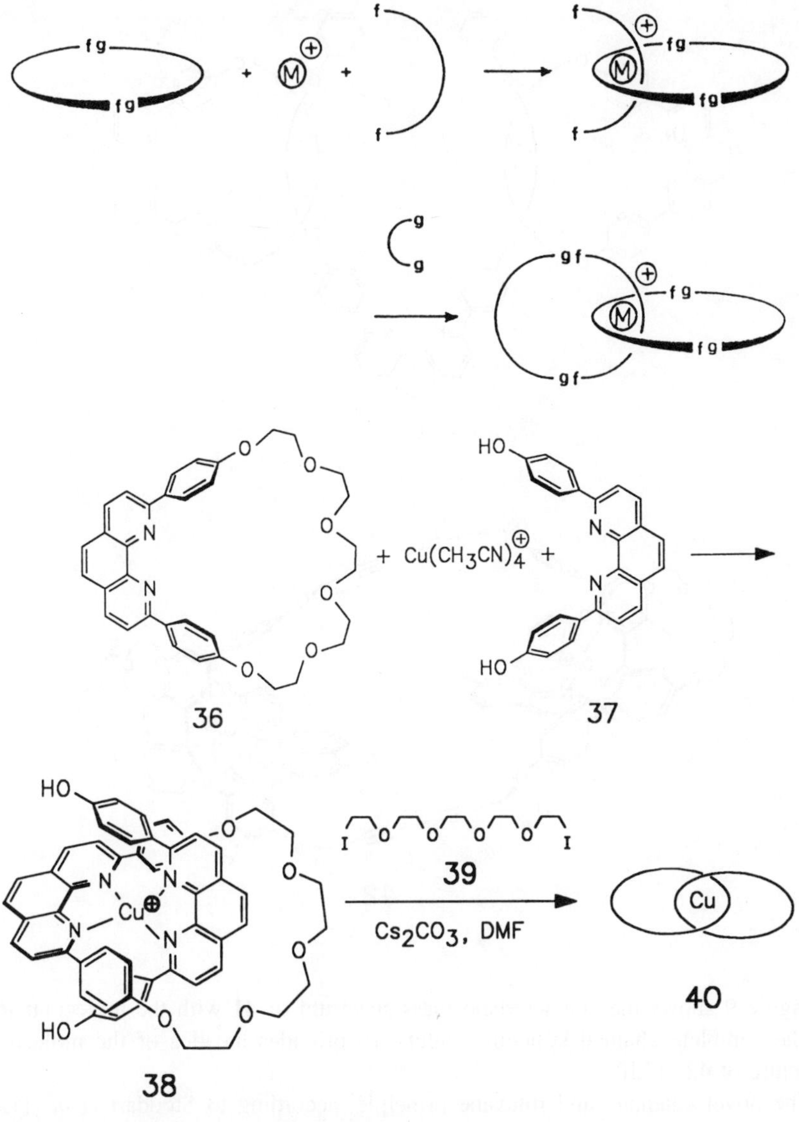

fg fg + M⊕ + f f → f ⊕ fg M fg f

g g → gf ⊕ fg M fg gf

O O O O O O N N + $Cu(CH_3CN)_4^{\oplus}$ + HO N N HO →

36 37

HO N N Cu⊕ N N O O O O O HO

I O O O O I
39
Cs_2CO_3, DMF

Cu

40

38

Scheme 1. Template strategy for catenane synthesis: (a) schematically (fg = functional group); (b) building blocks and reactions

done starting from the open-chain building blocks **37** and **39**, using copper ions and caesium carbonate (caesium effect). The use of this and other synthetic strategies also allowed the preparation of [3]catenanes, such as **41**, in acceptable yields.

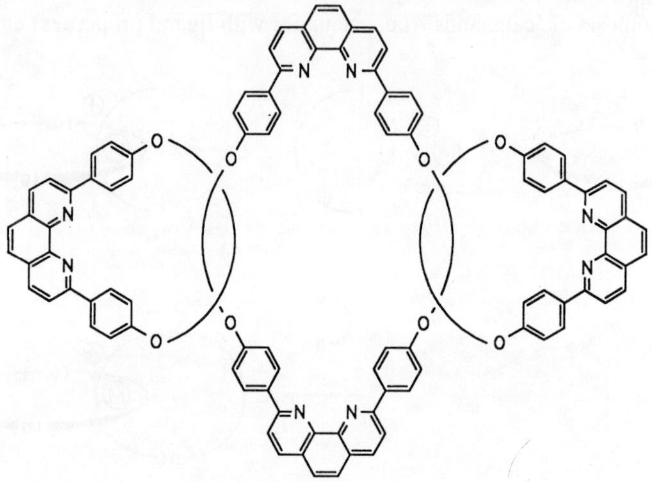

41

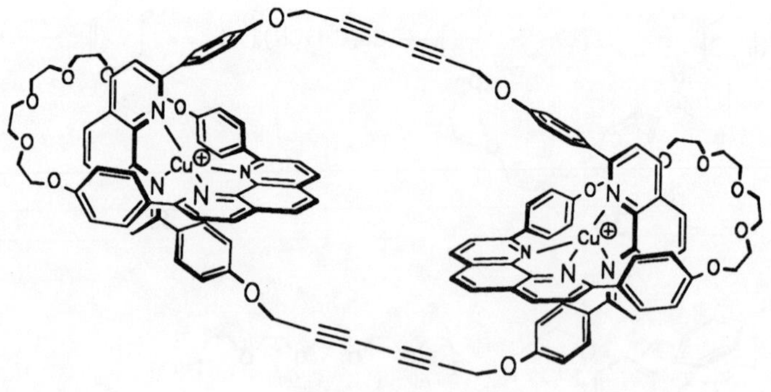

42

Figure 9 shows the characteristic mass spectrum of **41** with the molecular ion of the complete chained structure. Figure 10 provides an idea of the molecular structure of 42 [12d].

The novel catenane and rotaxane principle, according to Stoddart *et al*. [13], using a strong charge-transfer interaction between 4,4′-bipyridinium and crown ether building blocks, is shown in Figure 11.

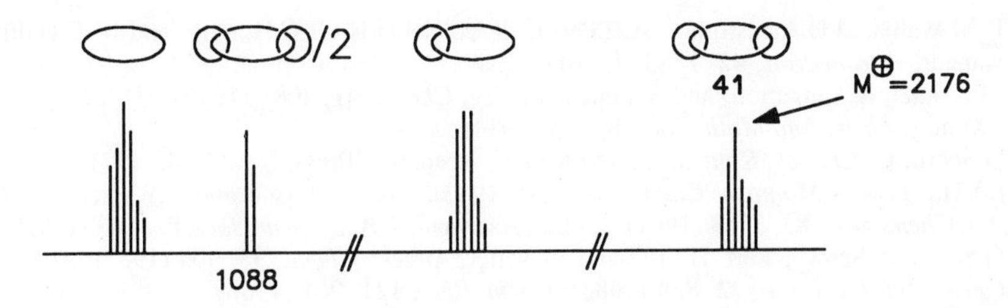

Figure 9. Mass spectrum of [3]catenand **41** (with fragment ions)

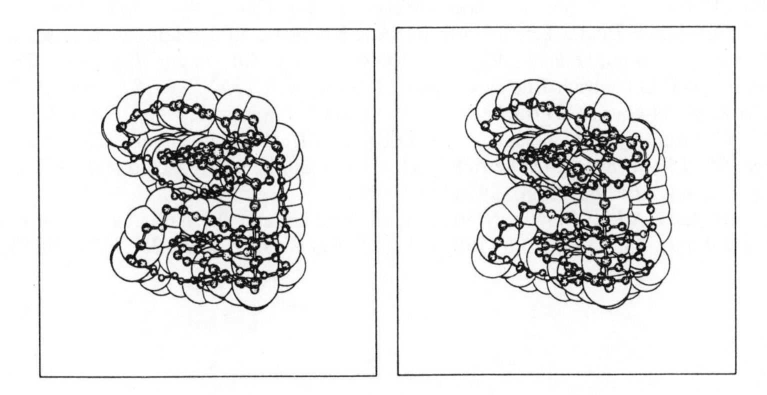

Figure 10. X-ray structure of binuclear copper(I) complex **42** ([3]catenate **42**) (stereoview)

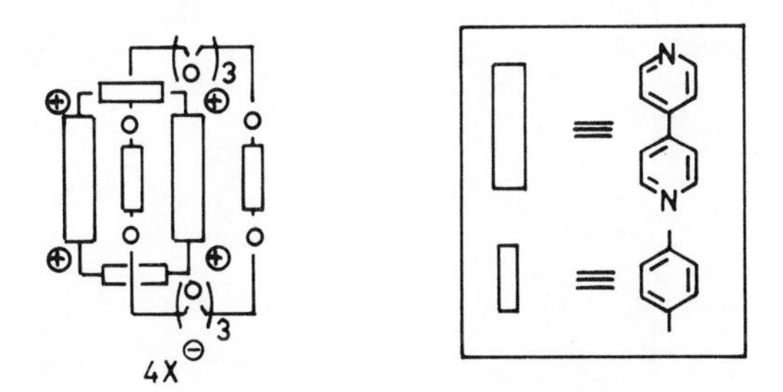

Figure 11. Scheme of catenanes according to Stoddart *et al.* [13]

References

1. H.L.Frisch and E.Wasserman, *J. Am. Chem. Soc.*, **83**, 3789 (1961).
2. (a) C.O.Dietrich-Buchecker, J.P.Sauvage and J.P.Kintzinger, *Tetrahedron Lett.*, 5095 (1983); (b) C.O.Dietrich-Buchecker, J.P.Sauvage and J.M.Kern, *J. Am. Chem. Soc.*, **106**, 3043 (1984).
3. D.M.Walba, Tetrahedron, 41, 3161 (1985); D.M.Walba, R.M.Richards, M.Hermsmeier and R.C.Haltiwanger, *J. Am. Chem. Soc.*, **109**, 7081 (1987).
4. D.M.Walba, R.M.Richards, S.P.Sherwood and R.C.Haltiwanger, *J. Am. Chem. Soc.*, **103**, 6213 (1981).
5. D.M.Walba, R.M.Richards and R.C.Haltiwanger, *J. Am. Chem. Soc.*, **104**, 3112 (1982).

6. D.M.Walba, J.D.Armstrong, A.E.Perry, R.M.Richards, T.C.Homan and R.C.Haltiwanger, *Tetrahedron*, **42**, 1883 (1986).
7. M.Nakazi, K.Yamamoto and S.Tanaka, *J. Org. Chem.*, **41**, 4081 (1976).
8. J.Simon, *Abstr. Am. Math. Soc.*, **5**, 185 (1984).
9. G.Schill, *Catenanes, Rotaxanes, and Knots.*, Academic Press, New York, 1971.
10. (a) H.L.Frisch, *Montash. Chem.*, **84**, 250 (1983); (b) H.L.Frisch and E.Wasserman, *J. Am. Chem. Soc.*, **83**, 3789 (1961); (c) E.Wasserman, *J. Am. Chem. Soc.*, **82**, 4433 (1960).
11. G.Schill, N.Schweickert, H.Fritz and W.Vetter, *Angew. Chem.*, **95**, 909 (1983); *Angew. Chem., Int. Ed. Engl.*, **22**, 889 (1983); *Chem. Ber.*, **121**, 961 (1988).
12. (a) C.O.Dietrich-Buchecker, J.P.Sauvage and J.M.Kern, *J. Am. Chem. Soc.*, **106**, 3043 (1984); (b) C.O.Dietrich-Buchecker, J.P.Sauvage and J.P.Kintzinger, *Tetrahedron Lett.*, 5095 (1983); (c) J.P.Sauvage and J.Weiss, *J. Am. Chem. Soc.*, **107**, 6108 (1985); (d) C.O.Dietrich-Buchecker, J.Guilhem, A.K.Khemiss, J.P.Kintzinger, C.Pascard and J.P.Sauvage, *Angew. Chem.*, **99**, 711 (1987); *Angew. Chem., Int. Ed. Engl.*, **28**, 661 (1987); (e) C.O.Dietrich-Buchecker and J.P.Sauvage, *Chem. Rev.*, **87**, 795 (1987); (f) A.-M.Albrecht-Gary, C.D.Buchecker, Z.Saad and J.-P.Sauvage, *J. Am. Chem. Soc.*, **110**, 1467 (1988); (g) J.-P.Sauvage and C.O.Dietrich-Buchecker, *Angew. Chem., Int. Ed. Engl.*, **28**, 189 (1989); C.O.Dietrich-Buchecker, J.Guilhem, C.Pascard and J.P.Sauvage, *Angew. Chem., Int. Ed. Engl.*, **29**, 1154 (1990).
13. J.Y.Ortholand, A.M.Z.Slawin, N.Spencer, J.F.Stoddart and D.J.Williams, *Angew. Chem., Int. Ed. Engl.*, **28**, 1394, 1396 (1989); *Angew. Chem., Adv. Mat.*, **28**, 1405 (1989).

3 Bioinorganic Model Compounds

In the following, host–guest complexations, which go beyond the already discussed crown ethers and cryptates, will be divided, admittedly in a slightly formalistic fashion, into two parts: one on more inorganic topics (cf. Section 2.2.2 on polynuclear cryptates) and the other on more organic examples. In the first part, metal cations play the role of guests, and in the second these are organic molecules.

Many bioinorganic model compounds have already been discussed in Chapter 2: mono- and polynuclear neutral or charged ligands and complexes. Here, we shall outline a few recent developments going beyond what was described previously.

In 1973, Collman *et al*. [1] published a model for haemoglobin which functions at 25 °C (**1**). The Fe^{2+} complex of 'picket fence porphyrin' **1** binds oxygen reversibly when substituted imidazoles, pyridine or THF are present. With a clever choice of imidazoles attached, which can be either sterically hindering or not, even the control mechanisms of the oxygen binding can be mimicked. Sterically demanding imidazoles occupying the fifth coordination site of the iron on the unsubstituted side of the porphyrin cannot be pulled by the metal ion towards the porphyrin plane. In this strained state (T state), the binding of oxygen is not possible. On the other hand, sterically less demanding imidazoles allow the uptake of oxygen (R state).

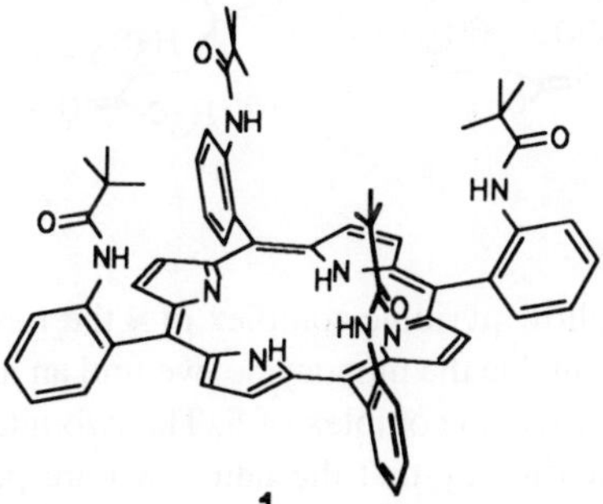

O NH NH HN HN O NH N HN N O O

1

In subsequent years, a number of capped and bridged porphyrins (e.g. **2** and **3**) were prepared, mainly because of frequently observed oxodimerization (the formation of binuclear complexes) and irreversible oxidation of Fe(II) to Fe(III) [2]. In this way, the central atom was effectively protected from the mentioned dimerization process. So far, it has not been possible to stop the oxidation of the metal completely [only Fe(II) is able to bind oxygen], as is achieved with proteins in natural models.

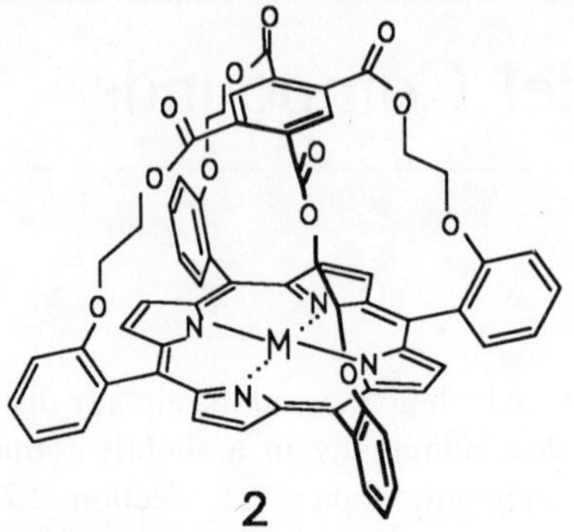

2

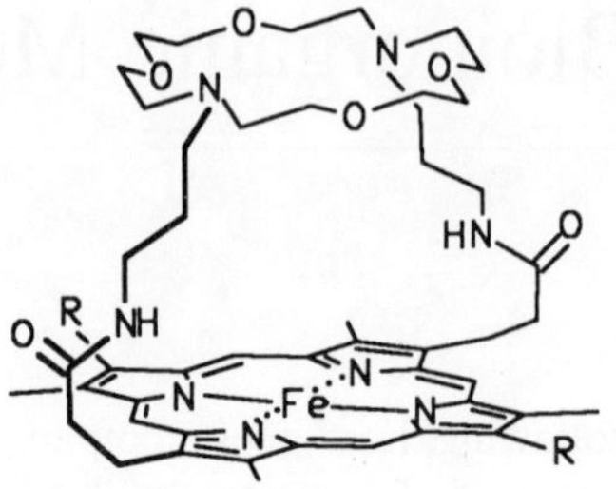

3 R=C_6H_{13}

In 1986, Maverick *et al*. were able to prepare the two binuclear model complexes **4** and **5** [3]. The two chelate complexes are green in chloroform solutions. The distances between the Cu(II) ions were measured to be 490.8 pm (**4**) and 734.9 pm (**5**) and the angles between the planes of the complexes and the aromatic units are 89.8° (**4**) and 82–83° (**5**). On addition of Lewis bases, colour changes are observed: to light yellow (**4** with pyridine) and turquoise [**5** with pyridine, diazabicyclooctane (dabco), pyrazine and quinuclidine]. Line broadening in ^{1}H NMR studies indicates the presence of hydrogen bonding between the ketone and the chloroform.

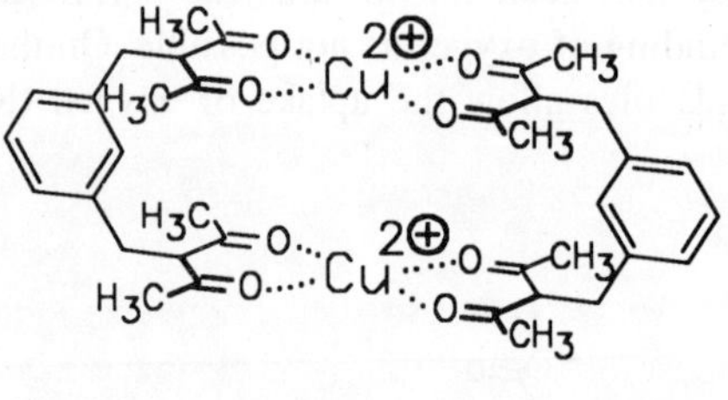

4

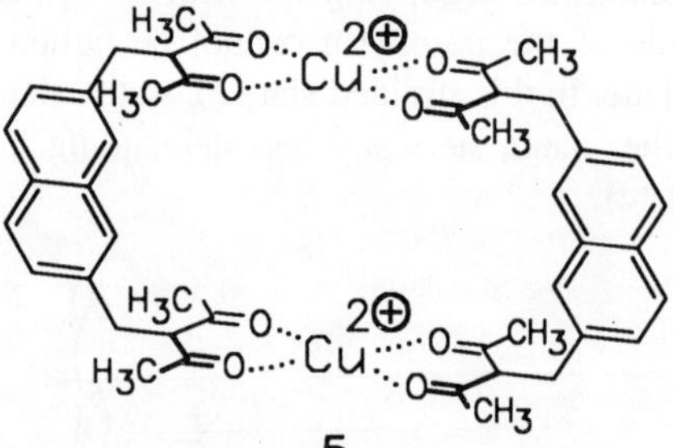

5

Whereas in the light yellow pyridine complex of **4** the Lewis base is coordinated to the Cu(II) ions axially outside the monocycle, we find an intramolecular inclusion in the case of the turquoise dabco complex of **5**. The two nitrogen atoms coordinate to the two Cu ions in such a way that the latter two are pulled inside by 14 pm, toward the guest and away from the plane formed by the four ketones. Thus, a double square-pyramidal complex is formed, where the Cu(II)–Cu(II) distance is now 740 pm. The intramolecular inclusion was investigated using X-ray crystal structure analysis. Figure 1 shows the ORTEP drawing of the complex. Even though this metal complex has no direct natural equivalent, it impressively illustrates the function of the metal ion. Only by this particular coordination is the desired cyclic arrangement produced. In addition, in the case when dabco is incorporated, the two copper ions act as binding sites for the substrate [4].

Not only in modern technology, e.g. selective metal ion extraction (Purex

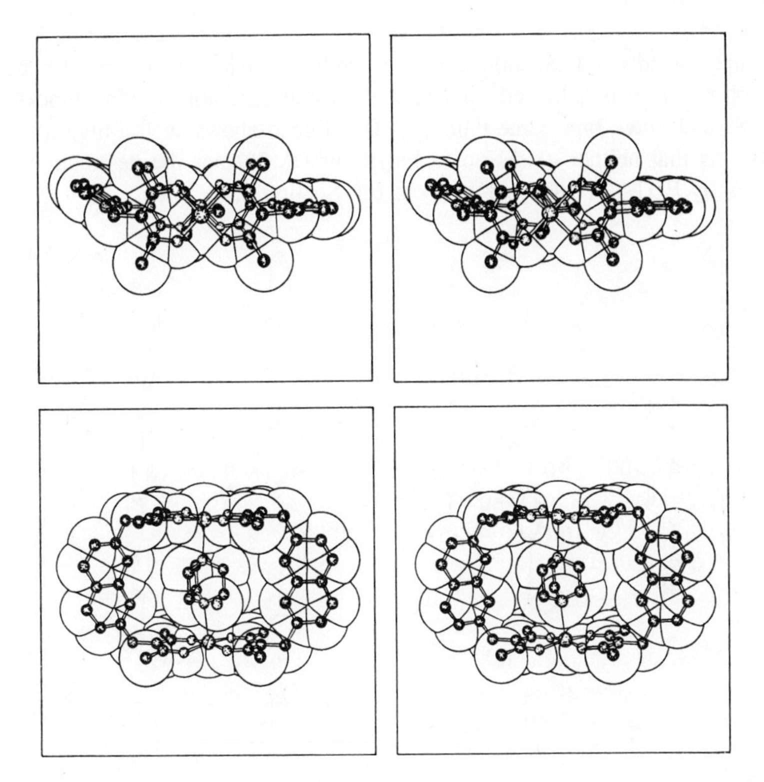

Figure 1. ORTEP drawing of the dabco complex of **5** [3] (stereoviews of two different views)

procedure) or metal-catalysed processes, but also in numerous other areas does the chemistry of organic, polydentate complexing ligands attract attention. Above all, the two classes of the 'siderophores' (cf. Section 2.3) and the 'ionophores' (cf. Section 2.2) are of particular interest, because of the stability of their complexes, the selectivity of the complexation and their physiological relevance.

Apart from ligands with phenolic, carboxylic acid, nitrogen or hydroxamic acid donors, those with catechol units have been studied in particular depth. Probably the best known catechol siderophore is enterobactin, which forms the Fe^{3+} complex depicted below (for details, see Section 2.3).

After the preparation of numerous synthetic siderophores, culminating with the

Fe^{3+} superligand 6 (cf. Section 2.3), the synthesis of a new bicyclic hexalactam siderophore (**7**) was achieved, in which the usual catechol building blocks were replaced with three bipyridine units [5]. This ligand shows such a high affinity to Fe(II) ions that neither oxidation to Fe(III) in DMSO nor the destruction of the complex by EDTA or HCl (pH 1) occur (cf. Section 2.1).

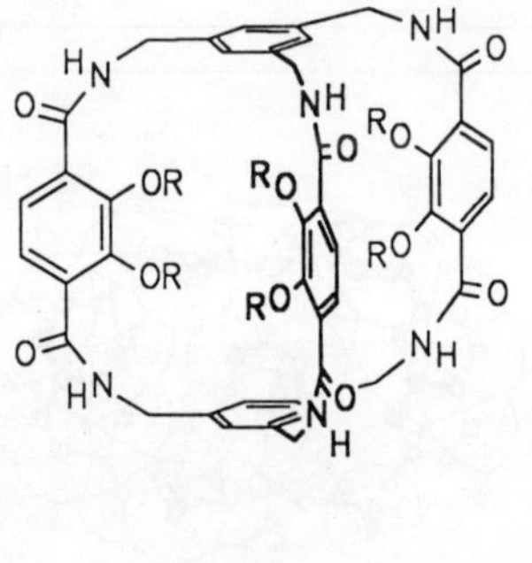

6

a: R = CH_3
b: R = H

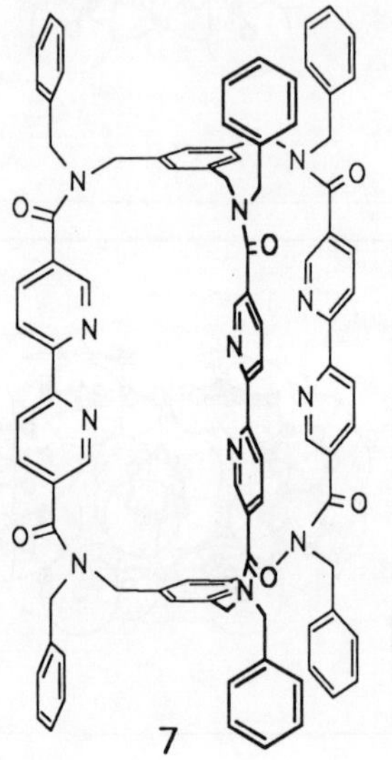

7

As can be shown with molecular models, in an octahedral coordination of the Fe(II) ions, the ligand adopts a helical conformation. The high stability of the complex indicates an intramolecular inclusion of the iron ion. The exolipophilic properties of the complex allow its extraction into chloroform. The complexation of precious metals, such as Rh and Ru, has already been reported [5]. Potential applications could include indirect electrochemical redox reactions and the photochemical cleavage of water (cf. Chapter 12).

A molecule with an interesting combination of donor sites, able to complex several metal ions concurrently but leaving plenty of room in the cavity, is due

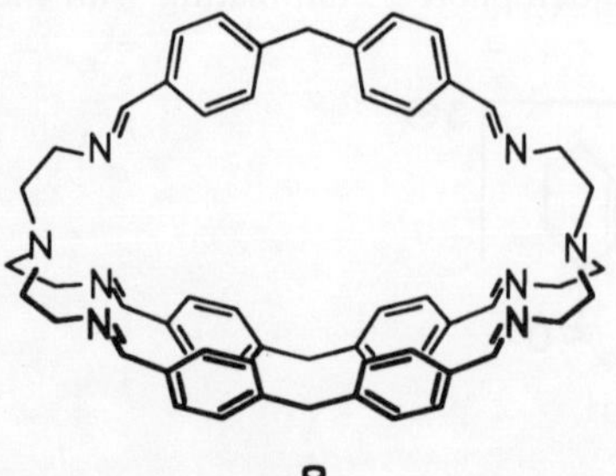

8 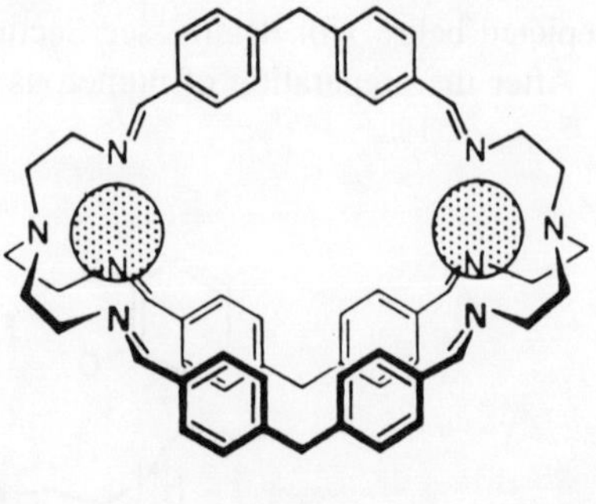 **9**

to Lehn[6]: by reacting 'tren' with 4,4′-diphenylmethane dialdehyde, the Schiff's base **8** was obtained, which incorporates two metal cations at a defined distance from each other (**9**; cf. Figure 2).

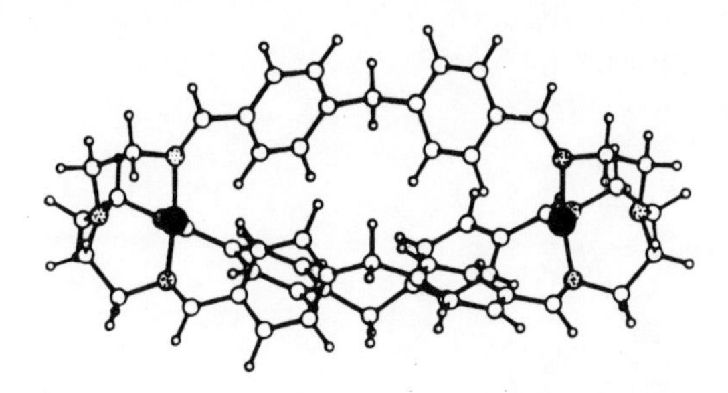

Figure 2. Simultaneous complexation of two metal cations by ligand **8** (complex **9**, X-ray crystal structure) [6]

Spatially ordered binuclear [7] and even tetranuclear macrocyclic ligands, which for instance bind four Ni(II) or Cu(II) ions, have been prepared in recent years [8].

Finally, we should mention the preparation of new oxygen-binding and activating complexes [9] and Fe and Mo cluster complexes [10] with a biomimetic background.

References

1. J.P.Collman, R.R.Gagne, T.R.Halbert, J.C.Harchon and C.A.Reed, *J. Am. Chem. Soc.*, **95**, 7868 (1973).
2. J.Almog, J.E.Baldwin, R.L.Dyer and M.Peters, *J. Am. Chem. Soc.*, **97**, 226 (1975); for a review, see J.E.Baldwin and P.Perlmutter, *Bridged, Capped and Fenced Porphyrins, Topics in Current Chemistry, Vol.* 121, (F.Vögtle and E.Weber Eds.), Springer, Berlin, Heidelberg, 1984, p.181.
3. A.W.Maverick, S.C.Buckingham, Q.Yao, J.R.Bradbury and G.G.Stanldy, *J. Am. Chem. Soc.*, **108**, 7430 (1986).
4. See also D.Ramprasad, W.K.Lin, K.A.Goldsby and D.H.Busch, *J. Am. Chem. Soc.*, **110**, 1408 (1988).
5. S.Grammenudi and F.Vögtle, *Angew. Chem., Int. Ed. Engl.*, **25**, 1122 (1986); S.Grammenudi, M.Franke, F.Vögtle and E.Steckhan, *J. Inclus. Phenom.*, **5**, 695 (1987); P.Belser, *Chimia*, **44**, 226 (1990); V.Balzani, F.Barigelletti and L.De Cola, *Top. Curr. Chem.*, **158**, 31 (1990).
6. J.Jazwinski, J.-M.Lehn, D.Lilienbaum, R.Ziessel, J.Guilhem and C.Pascard, *J. Chem. Soc., Chem. Commun.*, 1691 (1987); J.-M.Lehn, *Angew. Chem., Int. Ed. Engl.*, **27**, 89 (1988).
7. D.Beltrán *et al.*, *Inorg. Chem.*, **27**, 19 (1988); L.K.Thompson, F.L.Lee and E.J.Gabe, *Inorg. Chem.*, **27**, 39 (1988); R.T.Stibrany and S.M.Gorun, *Angew. Chem., Int. Ed. Engl.*, **29**, 1156 (1990); K.Wieghardt, *Angew. Chem., Int. Ed. Engl.*, **28**, 1153 (1989).
8. (a) M.Bell, A.J.Edwards, B.F.Hoskins, E.H.Kachab and R.Robson, *J. Chem. Soc., Chem. Commun.*, 1852 (1987); (B) V.McKee and S.S.Tandon *J. Chem. Soc., Chem. Commun.*, 385 (1988); K.P.McKillop, S.M.Nelson, J.Nelson and V.McKee, *J. Chem. Soc., Chem. Commun.*, 387 (1988).

9. K.D.Karlin and Y.Gultneg, *Prog. Inorg. Chem.*, **35**, 217 (1987).
10. T.D.P.Stack and R.H.Holm, *J. Am. Chem. Soc.*, **110**, 2484 (1988); T.Tanaka *et al.*, *Inorg. Chem.*, **27**, 137 (1988); E.Sappa *et al.*, *Prog. Inorg. Chem.*, **35**, 437 (1987).

4 Bioorganic Model Compounds

4.1 THE SELECTIVE COMPLEXATION OF TOPOLOGICALLY COMPLEMENTARY ORGANIC MOLECULES

The investigation and mimicking of biological processes is the main topic of an area of research which in recent years has continually gained in importance: *bioorganic chemistry*. One section of host–guest chemistry, which goes beyond crown ethers and cryptands and which follows Emil Fischer's key–lock principle, concerns itself with the properties of space-enclosing, concave, mostly macrocyclic *host molecules*. The basis for these investigations is the synthesis of tailored hosts, which in solution, like biological receptors or enzymes, are able to bind or incorporate selectively smaller, spatially complementary *guest molecules* [1]. This primarily leads to a more or less pronounced 'molecular encapsulation' and 'masking' of the guest compound (Figure 1).

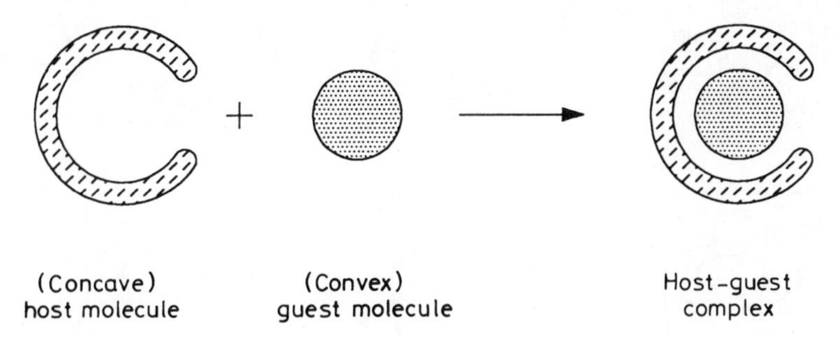

Figure 1. Complexation (encapsulation) of molecular guests by topologically complementary organic host molecules (schematic)

Apart from the studies on the inclusion behaviour and intermolecular interactions between guests and hosts, the synthesis of 'artificial enzymes' is of particular interest; with these, using their specific, catalytic properties, starting materials can be transformed quickly and in high yields into desired enantiomerically or diastereomerically pure products [2].

4.2 THE CYCLODEXTRINS

Since their discovery, the cyclodextrins have been prototypes for novel host compounds and catalysts. We shall describe them as an introduction to the topic.

The cyclodextrins were first isolated in 1891 by Villiers as degradation products of starch; in 1904 Schardinger characterized them as cyclic oligosaccharides; and in 1938 Freudenberg *et al*. [3] described them as being macrocyclic compounds built from glucopyranose units linked by α-(1,4)-glycosidic bonds.

Cyclodextrins can be obtained with the help of cyclodextrin glycosyltransferases by enzymatic degradation of starch. The latter is a polysaccharide consisting of α-(1,4)-linked glucose units and is found as a left-handed helix with six units per turn. In the process, compounds with six to twelve glucopyranose units per ring are produced. Depending on the enzyme and how the reaction is contolled, the main product is α, β or γ-cyclodextrin (6, 7 and 8 glucopyranose units, respectively). These are toxicologically harmless, crystalline compounds, which both in the solid state and in solution adopt a circular, conical conformation, where the height is about 800 pm and the cavity's inner diameter is between 500 and 800 pm wide (Figure 2). They have an endolipophilic cavity, which is made water-soluble by the many outward-pointing OH groups.

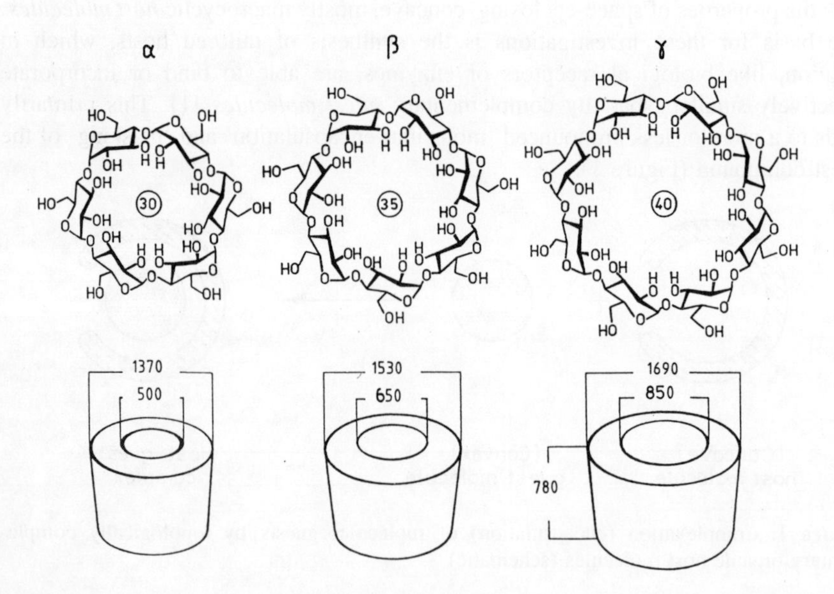

Figure 2. Cyclodextrins: degradation products of starch (measurements in pm; the corresponding ring sizes are given in the centre of the rings)

A fascinating property of the cyclodextrins is their ability to incorporate other organic compounds into their cavity, both in the solid state and in solution. Such host–guest complexes must clearly be separated from the clathrates (lattice inclusion compounds), which occur only in the solid state [4] (cf. Chapter 5).

While the cyclodextrins usually form 1 : 1 complexes with their guests or, in the case of the incorporation of iodine (I_3^-), stack in tubes (in analogy with the iodine–starch reaction), the guests in clathrates occupy intermediate sites in the lattice of the host crystal. The considerable power of complexation of the cyclodextrins was first recognized by Freudenberg and, later in the 1950s, substantiated by complexation studies by Cramer, Saenger [5] and others. Evidence that the incorporated guest is actually sitting at the centre of the cavity was obtained by the determination of the association constant for the solvated complex (Benesi–Hildebrand method) and confirmed by X-ray crystallography. Figure 3 shows a stereoview of α-cyclodextrin (host) containing an incorporated 3-nitroaniline molecule (guest).

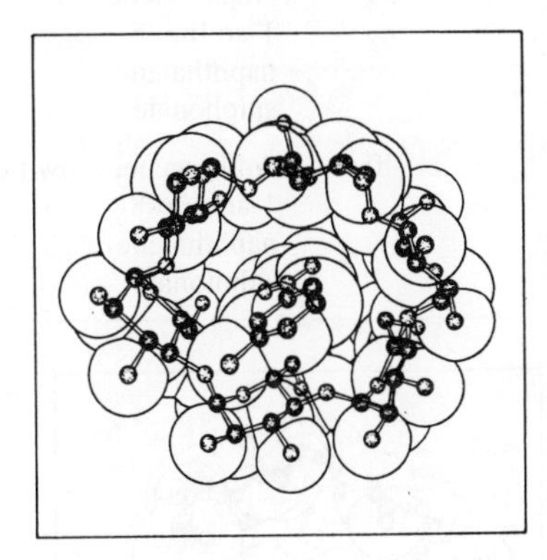
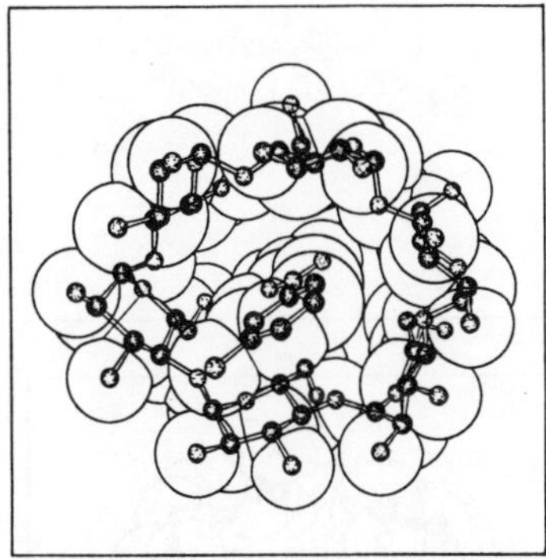

Figure 3. X-ray crystal structure of m-nitroaniline enclosed in α-cyclodextrin (stereoview) [6]

A primary criterion for the inclusion of guests within the host's cavity is obviously their size (size selectivity). This is illustrated in Table 1.

Without doubt, one of the nicest examples of a guest inclusion with optimum concave–convex complementarity seen lately is the inclusion of crown ethers within the well of γ-cyclodextrin with the additional incorporation of a metal ion into the cavity of the crown (like matrioshkas, the traditional Russian wooden dolls, encased one inside another) [7]. Figures 4 and 5 show the high symmetry of these complexes [8].

The complexation is based, depending on the solvent and the type of host and guest, on a combination of several intermolecular interactions (and possibly hydrophobic effects):

a. steric fit;
b. van der Waals interactions;
c. dispersive forces;
d. dipole–dipole interactions;
e. charge–transfer interactions;
f. electrostatic interactions;
g. hydrogen bonding.

Entropy effects must also be considered: on entering the cavity from an aqueous solution, the guest has to shed its hydrate envelope. The incorporation of a guest

Table 1. Correlation between the size of the cavity of (currently available) cyclodextrins and the size of the guests

Cyclodextrin	Number of glucose units	Internal diameter of cavity (centre) (pm)	Ring size	Incorporated guests
α-	6	500	30	Benzene, phenol
β-	7	620	35	Naphthalene, 1-anilino-8-naphthalene-sulphonate
γ-	8	800	40	Anthracene, crown ethers, 1-anilino-8-naphthalene-sulphonate

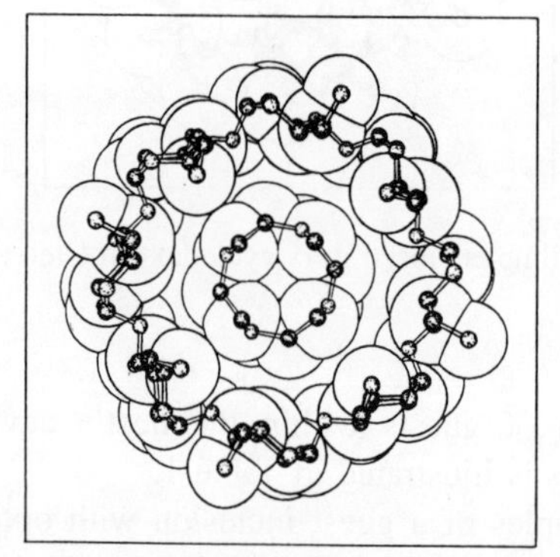
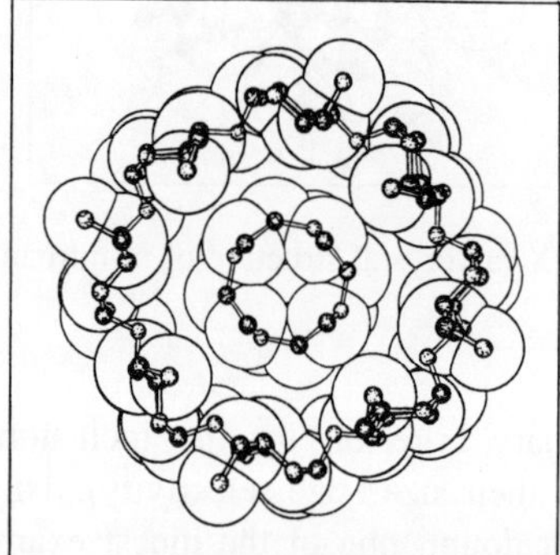

Figure 4. [12]Crown-4 as a guest in the cavity of γ-cyclodextrin (fourfold symmetry; X-ray structure; stereoview) [8]

leads through the decrease in the number of degrees of freedom of the guest to a decrease in entropy (increasing degree of order). This loss of entropy has to be compensated for by favourable interactions between host and guest. A stabilizing effect on the complex results from the increase in entropy during the complexation process, when the water molecules within the cavity are expelled from it [5c]. Changes in conformation of the cyclodextrin while exchanging guests also play a role ('induced fit').

Although the complexing properties of cyclodextrins are relatively unselective and the sugar units insufficiently resistant to harsh reaction conditions (e.g. acidic media), wide fields of research and applications have opened up. Thus, cyclodextrins are used to encapsulate light- and oxygen-sensitive substances, fragrances, drugs and toxic compounds. In addition, polymer-bound cyclodextrins can be used advantageously in gel-inclusion chromatography and the separation of enantiomers.

Vigorous developments were seen in the area of 'enzyme models' and 'artificial enzymes.' The aim is to mimick, and if possible to surpass, the ability of enzymes

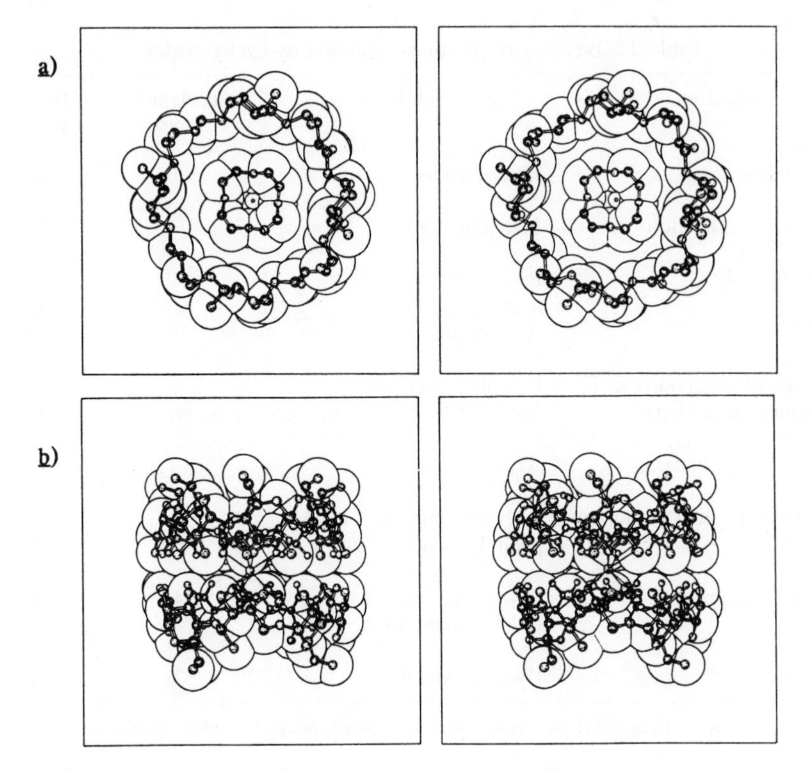

Figure 5. [12]Crown-4 complex of Li^+SCN^- in the cavity of γ-cyclodextrin (C_4-symmetry; X-ray structure [8]): (*a*) view into the cavity; (*b*) side view (stereoviews)

to bind certain substrates quickly and selectively, reversibly and non-covalently, and to catalyse possible reactions.

A number of properties seem to make cyclodextrins a reasonable choice as enzyme models, including:

a. Their water solubility, since under physiological conditions water is the most frequent solvent.
b. Guests are bound reversibly in the cavity; the release of the guest, however, is normally slower as with enzymes.
c. On incorporation of achiral molecules into the chiral cavity of cyclodextrins, the latter may exhibit an induced circular dichroism. Enantioselective complexation is also a possibility.
d. A number of chemical reactions may be catalysed by the addition of cyclodextrins. For instance, the rate of hydrolysis of *p*-nitrophenol esters is increased up to 750 000 times by the addition of β-cyclodextrin.

An overview of reactions catalysed by cyclodextrins is given in Table 2.

Table 2. Overview of reactions catalysed by cyclodextrins

Reaction	Substrate	Acceleration factor	Type of catalysis*
Ester hydrolysis	Phenyl esters	300	C
Asymmetric induction	Mandelic acid	1.4	U
Amide hydrolysis	Penicillin	89	C
	N-Acylimidazole	50	C
	Acetanilide	16	C
Cleavage of phosphoric and phosphonic acid esters	Pyrophosphate ester	200	C
	Methylphosphonic acid ester	66	C
Cleavage of carbonates	Aryl carbonate	19	N
Intramolecular acyl migration	2-Hydroxymethyl-4-nitrophenylpivalate	6	N
Decarboxylation	Glyoxylate ion	4	N
	Cyanoacetate ion	44	N
Oxidation	Hydroxy ketones	3	N

* C, intermediate is covalently bound; N, intermediate is not covalently bound; U, mechanism unknown.

How cyclodextrin might catalyse a reaction is shown schematically in Figure 6.

Finally, cyclodextrins have been functionalized with active groups of natural enzymes. In the case of the histamine residue, a marked increase in the rate of hydrolysis of *p*-nitrophenol acetate is observed.

4.3 BIOORGANIC MOLECULAR COMPLEXES

Model compounds are always a simplification of their natural analogues. Thus, in artificial 'receptors' and 'enzymes,' the amino acids of peptides and proteins are replaced with aliphatic or aromatic building blocks. Also, the bonds between these building blocks are only occasionally amide bonds. However, even natural enzymes normally do not contain macrocyclic peptides; in enzyme models, niches, clefts and cavities are often formed by macrobicyclic structures. For this reason, molecules with trigonal symmetry are often used in the synthesis of host compounds. They serve as rigid building blocks (spacers), stabilizing the cavity.

One of the first host compounds to incorporate, in organic media, uncharged organic molecules within its cavity was hexabenzyl[6.6.6]paracyclophane (**I**). This 30-membered macrocycle complexes small halogen hydrocarbons, such as chloroform and bromoform, in both the solid state (proved by X-ray crystal structure analysis [10]) and in solution [11,12].

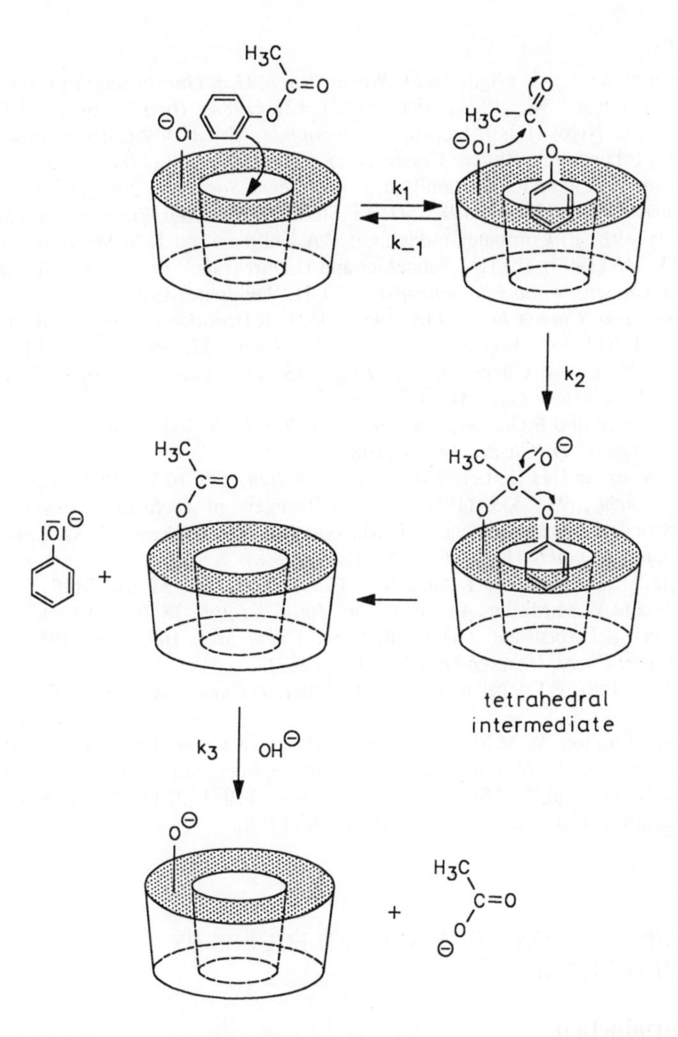

Figure 6. Mechanism of the cyclodextrin-catalysed hydrolysis of phenyl acetate (according to Diederich, modified) [9]. For the involvement of water molecules, see J.F.Kennedy and Ch.A.White, *Bioactive Carbohydrates*, Wiley, New York, 1983, p.134; for the role of ternary complexes see O.S.Tee and M.Bozzi, *J. Am. Chem. Soc.*, **112**, 7815 (1990)

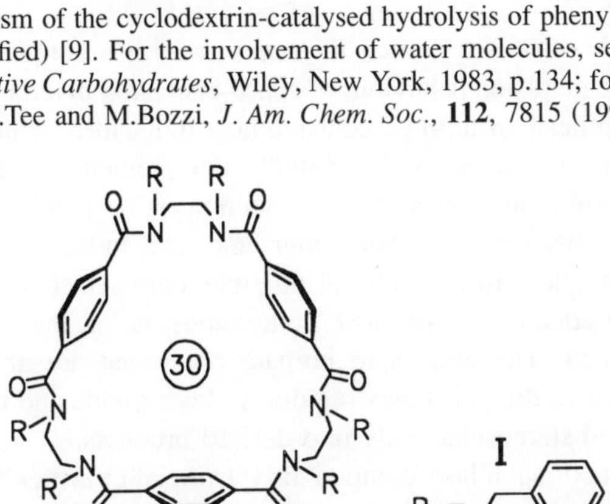

References

1. (a) For reviews, see F.Vögtle and E.Weber (Eds.), *Host Guest Complex Chemistry I–III, Top. Curr. Chem.*, **98**, (1981), **101**, (1982), **121** (1984); (b) J.L.Atwood, J.E.D.Davies and D.D.MacNicol (Eds.), *Inclusion Compounds*, Vols.1–3, Academic Press, London, 1984; (c) F.Diederich, *Angew. Chem., Int. Ed. Engl.*, **27**, 362 (1988); (d) for vancomycin model, see N.Pant and A.D.Hamilton, *J. Am. Chem. Soc.*, **110**, 2002 (1988); (e) T.Endo, *Top. Curr. Chem.*, **128**, 91 (1985); (f) J.F.Stoddart, *Ann. Rep. Progr. Chem., Sect. B*, **85**, 353 (1988–9); (g) Computer modelling: P.A.Kollman and K.M.Merz, Jr., *Acc. Chem. Res.*, **23**, 246 (1990); (h) H.-J.Schneider and H.Dürr (Eds.), *Frontiers in Supramolecular Organic Chemistry and Photochemistry*, VCH, Weinheim, 1990.
2. *I.Tabushi, Top. Curr. Chem.*, **113**, 145 (1983); R.Breslow, *Science*, **218**, 532 (1982); see also J.-M.Lehn, *Angew. Chem., Int. Ed. Engl.*, **27**, 89 (1988); H.D.Lutter and F.Diederich, *Angew. Chem., Int. Ed. Engl.*, **25**, 1125 (1986); S.Sasaki, Y.Takase and K.Koga, *Tetrahedron Lett.*, **31**, 6051 (1990).
3. K.Freudenberg and F.Cramer, *Z.Naturforsch., Teil B*, **3**, 464 (1948).
4. E.Weber, *Top. Curr. Chem.*, **140**, 1 (1987).
5. (a) F.Cramer and H.Hettler, *Naturwissenschaften*, **54**, 625 (1967); (b) W.Saenger, *Angew. Chem.*, **92**, 343 (1980); (c) W.Saenger, in *Inclusion Compounds*, Vol.2 (J.L.Atwood, J.E.D.Davies and D.D.MacNicol, Eds.) Academic Press, London, 1984; (d) J.Szejtli, *Kontakte (Darmstadt)* No.1, 31 (1988).
6. K.Harata, H.Uedaira and J.Tanaka, *Bull. Chem., Soc. Jpn.*, **51**, 1627 (1978).
7. F.Vögtle and W.M.Müller, *Angew. Chem., Int. Ed. Engl.*, **18**, 628 (1979).
8. S.Kamitori, K.Hirotsu and T.Higuchi, *J. Am. Chem. Soc.*, **109**, 2409 (1987).
9. F.Diederich, *Chem. Unserer Zeit*, **17**, 105 (1983).
10. F.Vögtle, H.Puff, E.Friedrichs and W.M.Müller, *J. Chem. Soc., Chem. Commun.*, 1398 (1982).
11. F.Behm, W.Simon, W.M.Müller and F.Vögtle, *Helv. Chim. Acta*, **68**, 940 (1985).
12. For selective binding of imidazoles or barbiturates in organic media, see (a) J.D.Kilburn, A.R.MacKenzie and W.C.Still, *J. Am. Chem. Soc.*, **110**, 1307 (1988); (b) S.K.Chang and A.D.Hamilton, *J. Am. Chem. Soc.*, **110**, 1318 (1988).

4.3.1 COMPLEXATION OF SMALL MOLECULES BY CRYPTOPHANES

4.3.1.1 Introduction

An area of organic chemistry which in recent years has gained in importance is concerned with the inclusion behaviour of space-enclosing macrocyclic compounds. The aim is the synthesis of tailored concave host molecules, which in solution are able to incorporate as guests small, spatially complementary, convex molecules (molecular encapsulation; masking, cf. Section 4.1, Figure 1). Apart from investigations on the inclusion behaviour and the intermolecular interactions between host and guest (molecular, if possible chiroselective recognition), the synthesis of catalytically active host compounds is at the centre of current research in this area. The idea is to prepare hosts that are able specifically to transform, as enzymes do, substrates rapidly, in high yields and chemoselectively, regioselectively and stereoselectively into desired products.

For the synthesis of such host compounds ('artificial enzymes'), building blocks with trigonal symmetry proved to be convenient. Amongst these, cyclotriveratrylene

(CTV) has established itself as a good molecular 'floor' or 'roof' in suitable host molecules:

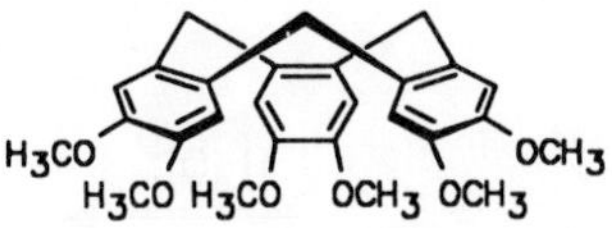

H_3CO H_3CO H_3CO OCH_3 OCH_3 OCH_3

1

Cyclotriveratrylene

4.3.1.2 Historical aspects

Stimulated by work done by Ewins [1] in 1909, where, after reacting homopiperonyl alcohol with PCl_5, a high-melting and almost insoluble solid with suspected constitution **2** was isolated. Robinson in 1915 [2] succeeded in a similar reaction with homoveratryl alcohol to isolate and identify 2,3,6,7-tetramethoxy-9,10-dihydroanthracene (**3**). With this, the constitution of **2** seemed to be confirmed. However, it was only in 1965 that it was recognized that the compound isolated by Robinson had a mass 1.5 times that of the molecule originally assumed [3]. 10,15-Dihydro-2,3,7,8,12,13-hexamethoxy-5*H*-tribenzo[*a,d,g*]cyclononene (cyclotriveratrylene, **1**) is a molecule with C_{3v} symmetry. The three benzene rings occupy the faces of a three-sided pyramid, while the CH_2 groups are placed on the edges facing the tip of the pyramid. For this molecule two conformations are possible: a 'saddle' conformation and a crown conformation (Figure 7). The latter, for steric reasons, is the more stable, as was shown by ^1H NMR studies.

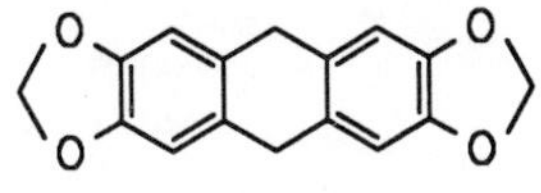

H_3CO OCH_3 H_3CO OCH_3

2 3

H_3CO H_3CO H_3CO OCH_3 OCH_3 OCH_3

H_3CO OCH_3 H_3CO H_3CO OCH_3 OCH_3

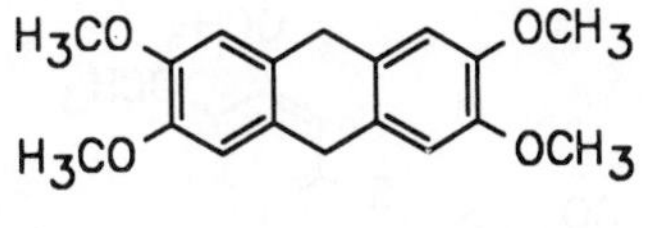

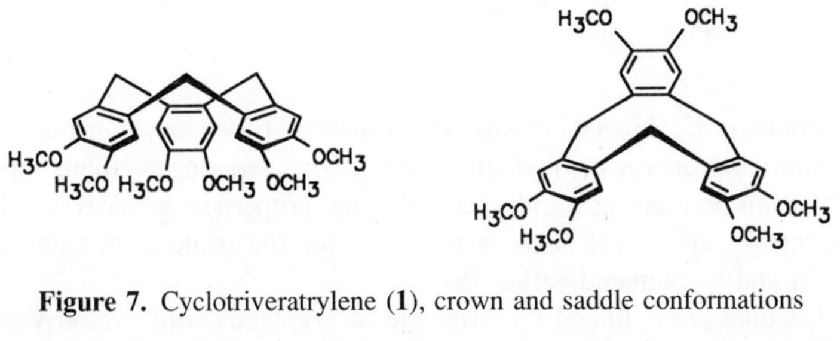

Figure 7. Cyclotriveratrylene (**1**), crown and saddle conformations

Cyclotribenzylene (**4**) and its derivatives can be obtained comparatively easily by any of the routes shown.

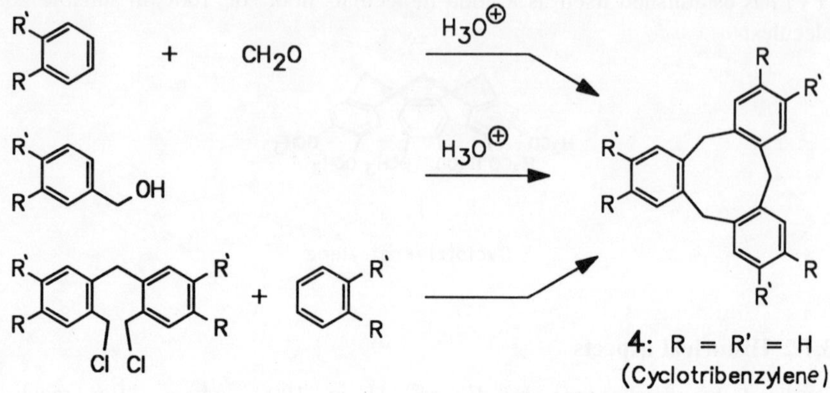

The reaction succeeds especially well if the residues R and R′ are electron donors (+M effect). In all cases where R = R′, the product is achiral. If, on the other hand, R ≠ R′, a racemate is formed, which, in the case of OH or ether substituents, can be separated by esterifying with chiral, enantiomerically pure acids and separating the diastereoisomers. Thus, in 1966 Lüttringhaus and Peters succeeded for the first time in separating chiral cyclotriveratrylene derivatives into their enantiomers [4].

In the following years, the many inclusion properties of cyclotriveratrylene became apparent in several areas of organic chemistry. Bhagwat *et al*. described clathrates of cyclotriveratrylene with benzene, chlorobenzene, toluene, chloroform, acetone, carbon disulphide, acetic acid, thiophene, decalin, butan-2-one and ethanol [5]. Analogues of CTV, such as trithiacyclotriveratrylene (**5**) or cyclotricatechylene (**6**), also show clathrate-forming properties.

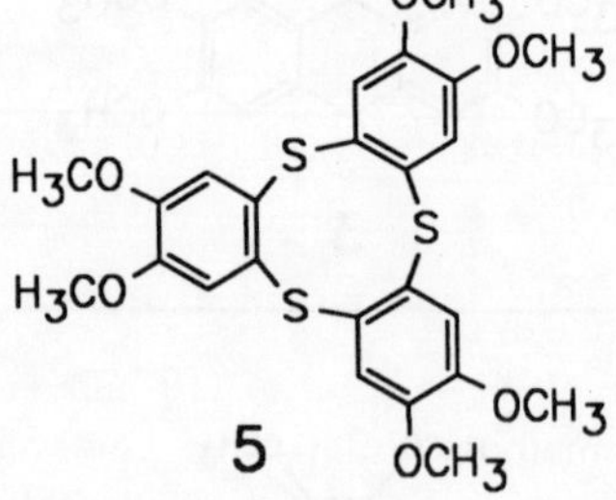

Zimmermann *et al*. showed hexaalkoxy-substituted tribenzocyclononenes to have liquid-crystalline properties [6]. Hyatt succeeded in preparing 'octopus molecules' with eight arms exhibiting similar complexing properties towards alkali metal cations as crown ethers [7]. This is also true for the triple crown ethers based on CTV (**7a** and **b**) prepared earlier [8].

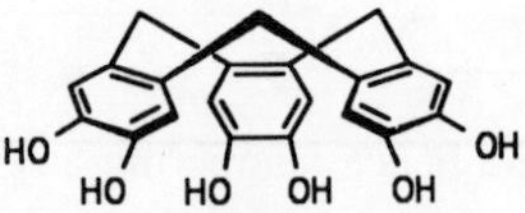

In 1982, Collet *et al*. linked the principle of cryptands with cyclotriveratrylene (cf. Section 2.2) [9]. The 'speleands' **8** thus obtained are able to form complexes with small ammonium salts in the same way as ordinary cryptands do. By linking

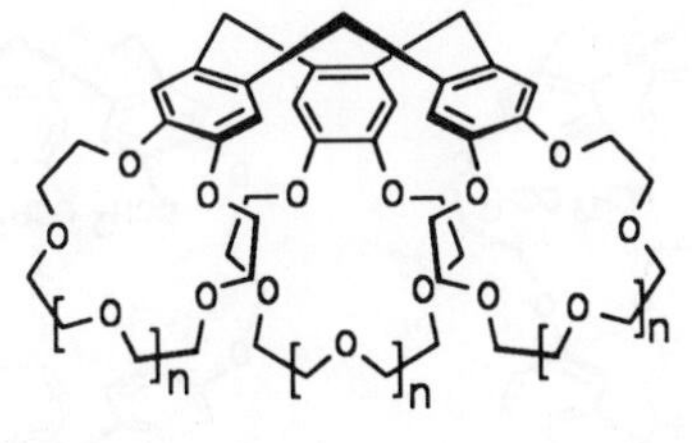

7a: n = 1
b: n = 2

two cyclotriveratrylene units with three alkane chains, Collet *et al*. obtained interesting 'cryptophanes.' Because of their novelty, their complexation behaviour with small guest molecules will be discussed in more detail.

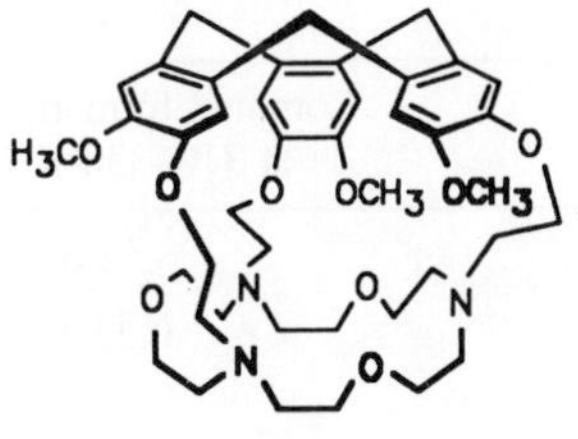

8

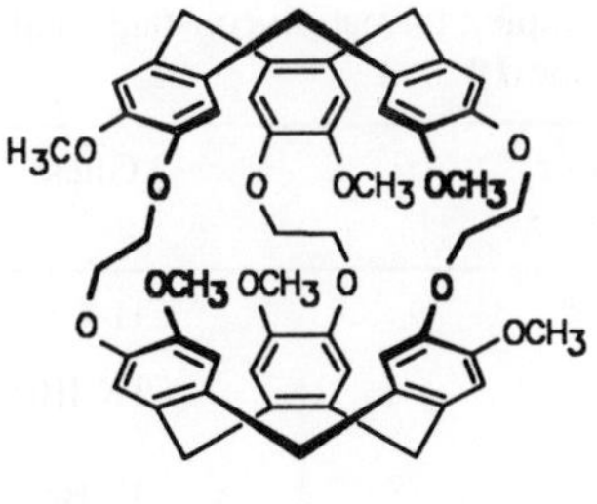

9

4.3.1.3 Host–Guest complexation

The first bicyclic hexaether prepared by Collet *et al*. was the chiral cryptophane A which has constitution **9** (see above) [10]. The crystalline, high-melting solid forms with chloroform a complex, which can also be crystallized. ^{1}H NMR studies, however, were not sufficient to prove unambiguously the inclusion of chloroform or dichloromethane. This is in agreement with observations made on models, which show the six methoxy groups in **9** to be obstructing the entrance to the molecular cavity and hence hampering any molecular inclusion.

On the basis of these considerations, cryptophanes C and D (**10** and **11**, both racemic) were prepared, in which three of the methoxy groups were replaced with H [11]. By enlarging the entrance to these host molecules, the inclusion of small guest molecules of the type CH_2XY were thought to have been made easier.

Complexation experiments on (±)-cryptophane C (**10**) were run, based on the weak complexation of chloroform (K_s = 0.1 l/mol at 310–330 K), in $CDCl_3$ solution. For the tested guest molecules of the types CH_2XY and CHXYZ, the complex formation constants and energy barriers listed in Table 3 were obtained.

The intramolecular inclusion of dichloromethane (DCM) as guest was observed in the ^{1}H NMR spectrum as a high-field shift of the methylene protons ($\Delta\delta$ = 4 ppm!) and a line broadening due to slow complexation–decomplexation. An X-ray crystal structure was also obtained.

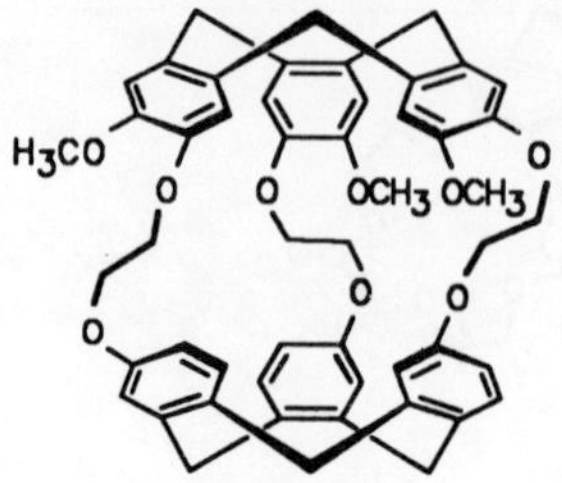

10

(–)-Cryptophane C

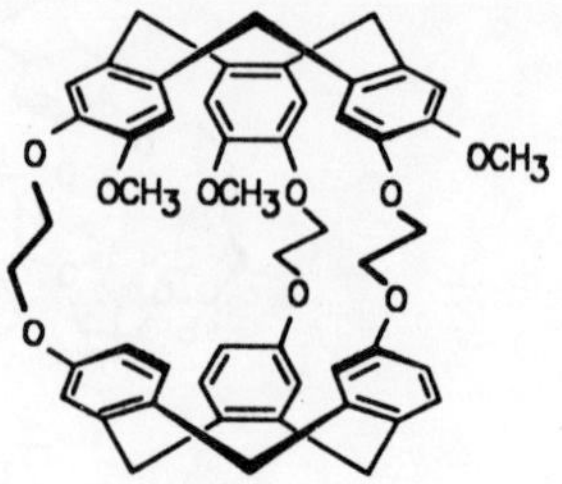

11

(+)-Cryptophane D

Table 3. Complex formation constants and energy barriers for the inclusion of halomethanes in cryptophane **10**

Energy barrier (kJ/mol)	Guest	Complex formation constant at 310–330 K (l/mol)
46	CH_2Cl_2	2.6
	CHClBrF	0.22–0.3
46	CH_2Br_2	0.7
63	$CHCl_3$	0.1

Analogous complexation studies to those made with cryptophane C (**10**) were then conducted with (+)-cryptophane D (**11**). For dichloromethane as a guest, similarly significant high-field shifts were measured ($\Delta\delta$ = 4.2 ppm). However, here the complex formation constants were lower by a factor of 10 (K_s = 0.28 l/mol for DCM) than for (–)-cryptophane C (10). Dibromomethane, which is a larger molecule than DCM, is complexed more strongly by (+)-cryptophane D (**11**); the difference in energy between the two complexes was determined as being 6 kJ/mol.

X-ray crystal structure analysis of the complex of cryptophane D with DCM showed that here, too, the guest is enclosed within the host's cavity. However, there is a fundamental difference between the crystal structures of the DCM complexes of the two cryptophanes (Figure 8). From a racemic mixture of (±)-cryptophane D (**11**) in DCM–acetone solution, only the (+)-enantiomer crystallizes as a complex with two to three molecules of DCM (separation of enantiomers by crystallization; formation of conglomerates). In contrast, the corresponding (±)-cryptophane C complex is made up of homochiral layers of (+)- and (–)-enantiomers with one guest molecule per host (racemate).

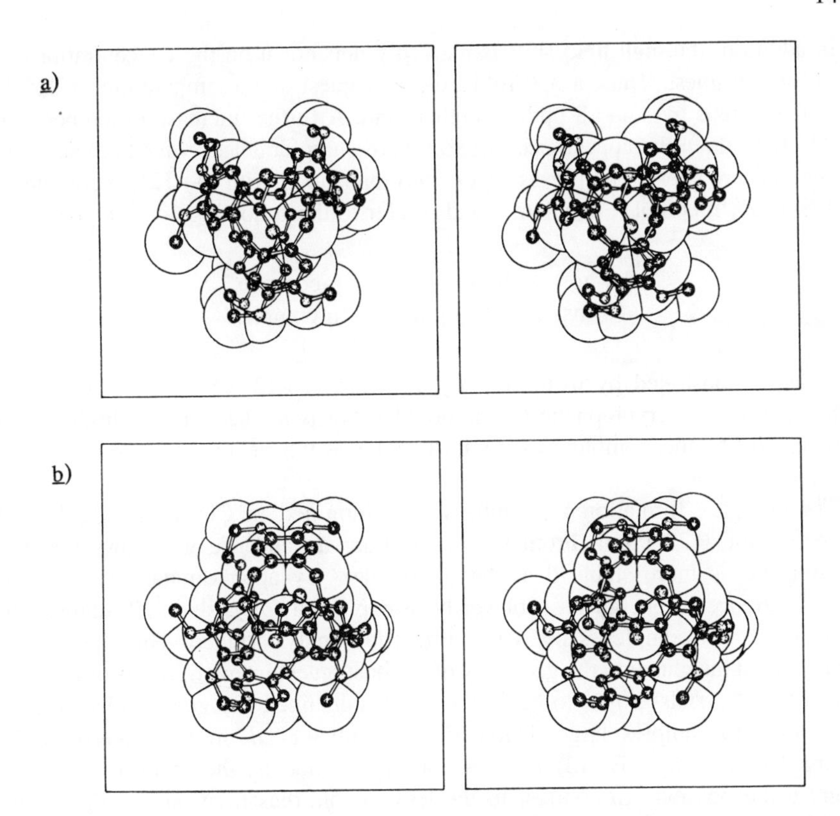

Figure 8. (a) (+)-Cryptophane D•CH_2Cl_2 conglomerate; (b) (+)/(−)-cryptophane C•CH_2Cl_2 inclusion (racemic). The crystal consisted of alternating, homochiral layers of (+)- and (−)-host (stereoviews)

4.3.1.4 Chiral recognition by complexation

In the complexation experiments discussed so far, the interactions of a chiral host with an achiral guest have been described. The complexation behaviour of (±)-cryptophane (**10**) with the likewise chiral halomethane guest CHBrClF [(±)-**12**] provided the possibility of studying chiroselective recognition phenomena between molecules. In this particular case even specific rotation and absolute configuration could be determined, a problem unsolved for a long time.

For the investigation of stereochemical and selectivity problems, all four stereoisomers of cryptophane C and D were prepared as easily separated pairs of diastereoisomers. Their differences are more in the shape rather than in the size of the cavity (size vs shape selectivity). ^{1}H NMR studies showed that (+)-cryptophane C and racemic (±)-**12** only complex on warming, as monitored by the high-field shift of the broadened signals of the guest. Whereas at 233 K the rate of complexation is slow and only a doublet (δ = 7.62 ppm, J_{FH} = 52 Hz) is observed in the ^{1}H NMR spectrum (cf. Figures 10 and 11), on raising the temperature to 273 K an increase in the rate of complexation is observed, coupled with a high-field shift and line broadening of the guest's NMR signals.

In addition, the high-field shift is markedly dependent on the concentrations of the host and guest. Thus, a 300-fold excess of guest and a temperature of 215 K shift the signals by 4.4–4.8 ppm to higher field. A further increase in temperature to 333 K leads to the appearance of two sets of sharp doublets, which are ascribed to the diastereomeric complexes of (+)-cryptophane C with (+)-**12** ('p-complex') and (−)-**12** ('n-complex'), where p and n refer to the following combinations:

(+)-host with (+)-guest → 'p-complex'
(+)-host with (−)-guest → 'n-complex'

As was demonstrated by using slightly enriched (±)-**12** (ee = 4.3 ± 1%) with either (+)- or (−)-cryptophane C, the doublet that is shifted most to higher field corresponds to the complex where host and guest have the same direction of rotation.

The complex formation constants were determined in $CDCl_3$ at 332 K and were 0.3 and 0.22 l/mol. From the temperature dependence it follows that the 'p-complex' is more stable than the 'n-complex' by approximately 1.1 kJ/mol. This difference in energy has not yet been explained fully, since all attempts to characterize the complexes by X-ray diffraction analysis failed. This would also have yielded insights into the geometry of the host–guest interaction. Unfortunately, none of the crystals grown so far in this series contained any guest molecules.

In order to complex larger halomethanes, Collet *et al*. in 1986 prepared the enlarged cryptophane E (**13**) [12]. According to models, the inclusion of CCl_4 ought to be possible. In contrast to the less soluble diastereoisomer cryptophane F (**14**), it was possible to grow crystals of cryptophane E (**13**) with dichloromethane, chloroform and bromoform. NMR spectroscopy showed cryptophane E to form a complex with two molecules of chloroform and to have a stability constant of 470 l/mol at 300 K (100 and 40 l/mol for DCM and bromoform, respectively). Calculations of the energy barriers for the consecutive inclusion–exclusion processes at the same temperature gave values of 56 and 72 kJ/mol. Figure 9 shows a ^1H NMR spectrum of cryptophane E in $(CDCl_2)_2$, where the proton signal of the complexed chloroform is shifted by 4.44 ppm (!) to higher field compared with uncomplexed $CHCl_3$.

Thermodynamic considerations demonstrated a strong dependence of the complexation on the enthalpy ($\Delta H = -29$ kJ/mol, $\Delta S = -46$ kJ/mol•K). Moreover, weaker complexations of acetone, benzene and ethanol have also been observed.

The syntheses of cryptophanes **9, 10, 11, 13** and **14**, described so far, are shown in Scheme 1.

In 1987, Collet *et al*. were able to prepare a new hydrophilic cryptophane, which is available from cryptophane A (**9**) by selective ether cleavage of the methoxy functions by lithium diphenylphosphide and subsequent alkylation with methyl α-bromoacetate [13]. By hydrolysis of the ester functions, the corresponding hexaacid **28** was obtained. This now water-soluble cryptophane for the first time allowed investigations on the role of the 'hydrophobic effect' in complexation processes in aqueous medium.

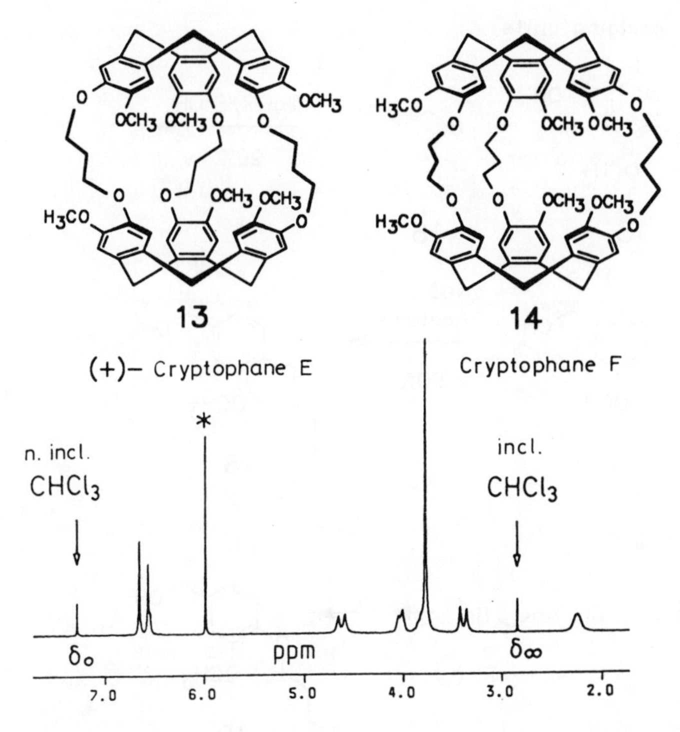

Figure 9. ^{1}H NMR spectrum of the complex of (+)-cryptophane E with $CHCl_3$ in $(CDCl_2)_2$. The considerable high-field shift is obvious. Asterisk, partially deuterated solvent; incl., included chloroform; n. incl., non-included chloroform

^{1}H NMR studies were conducted in D_2O/NaOD. Here, it is essential to keep the concentration of cryptophane below the critical micellar concentration (cmc), since otherwise aggregation (formation of micelles) of the host molecules is observed. It appears that under these conditions, chloroform and dichloromethane are incorporated inside the cavity of **28**. The high-field shifts of the proton NMR signals of both $CHCl_3$ and CH_2Cl_2 are again in the same order as seen earlier ($\Delta\delta$ = 4.5 ppm). Thermodynamic calculations and competition experiments produced a stability constant for the $CHCl_3$ complex which is 1 kJ higher compared to the one with CH_2Cl_2.

For comparison, measurements with the same guests and the lipophilic, water-insoluble cryptophane A were made [$(CDCl_2)_2$ as solvent]. In this case, the NMR high-field shift observed for chloroform is slightly lower ($\Delta\delta$ = 4.33 ppm), but still in the same range as the values obtained with **28**. The major difference, however, between cryptophane A and its water-soluble 'rival' lies in their reversed affinities: of the two guests, **28** complexes $CHCl_3$ more strongly, whereas CH_2Cl_2 is the one preferred by cryptophane A.

Although the energy barriers for the complexation steps in aqueous and lipophilic media differ only slightly, the 'hydrophobic effect' seems to play a role all the same.

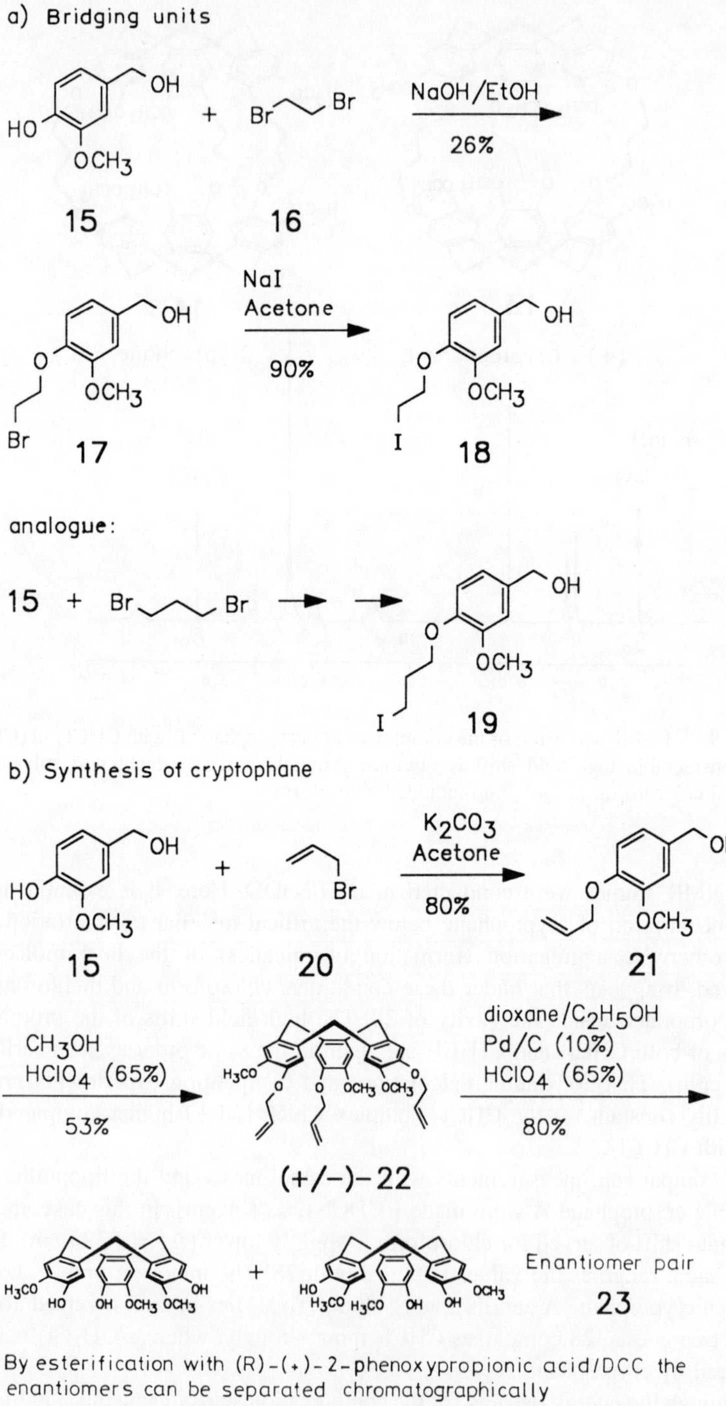

a) Bridging units

15 + 16 → NaOH/EtOH, 26%

17 → NaI, Acetone, 90% → 18

analogue:

15 + Br⌒⌒⌒Br → → 19

b) Synthesis of cryptophane

15 + 20 → K_2CO_3, Acetone, 80% → 21

→ CH_3OH, $HClO_4$ (65%), 53% → (+/−)-22 → dioxane/C_2H_5OH, Pd/C (10%), $HClO_4$ (65%), 80% →

Enantiomer pair 23

By esterification with (R)-(+)-2-phenoxypropionic acid/DCC the enantiomers can be separated chromatographically

Scheme 1. Synthesis of cryptophanes

separated diastereomeric esters $\xrightarrow[65-70\%]{LiB(C_2H_5)_3H,\ THF}$ 24 + 25

23 + 5-chloro-1-phenyltetrazole $\xrightarrow[90\%]{DMF/KO\text{-}t\text{-}Bu}$ $\xrightarrow[90\%]{Pd/C\ (10\%)/H_2,\ CH_2Cl_2/CH_3OH}$ $\xrightarrow[100\%]{BBr_3 \cdot CH_2Cl_2}$ 26 $\xrightarrow[\text{as described}]{\text{Enantiomeric separation}}$

18 (90%) $\xrightarrow{DMF/HCOOH}$ Cryptophane C/D

19 (36%) $\xrightarrow{HCOOH}$ Cryptophane E/F

24 + 18 $\xrightarrow{\text{1. Etherification, 2. Condensation, 3. Separation}}$ Cryptophane A $\xrightarrow[60\%]{\text{Li- diphenylphosphide, THF}}$ 27

Scheme 1. *(continued overleaf)*

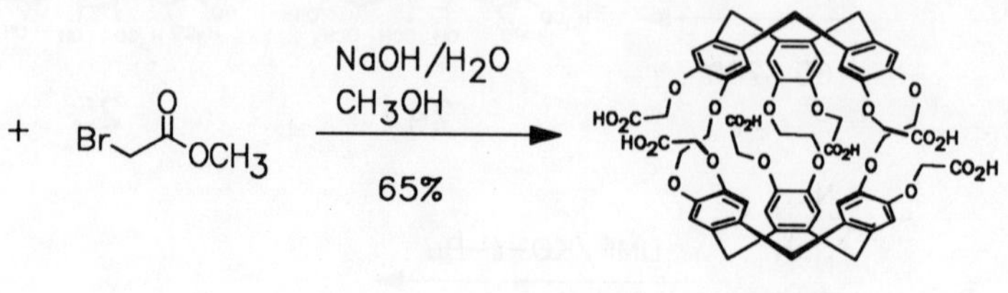

+ BrCH_2C(O)OCH_3 → ($NaOH/H_2O$, CH_3OH, 65%) → **28**

Scheme 1. (*continued*)

In this way it can be explained why in aqueous medium $CHCl_3$, which is larger than CH_2Cl_2 by 30%, is bound more strongly by **28**. This aspect also explains the difference in selectivity. Figure 8 shows the respective X-ray structures of the two DCM•cryptophane D complexes, in both cases looking in the direction of the C_3 axis. Details from a 200-MHz ^1H NMR spectrum of (+)-cryptophane C and racemic (±)-CHBrClF, recorded in $CDCl_3$, are shown in Figures 10 and 11. With increasing temperature, complexation increases, as shown by the broadened signals shifted to higher filed. At 329 K, a splitting into two doublets, corresponding to

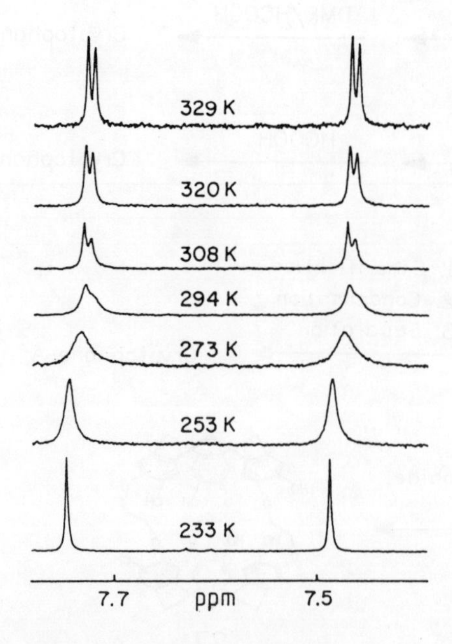

Figure 10. 200-MHz ^1H NMR spectrum of (±)-CHFClBr (**12**) at various temperatures and in the presence of (+)-cryptophane C (**10**)

the two diastereoisomeric complexes, is observed. The splitting into doublets (F–H coupling) is highlighted in Figure 11.

Cram *et al*. clamped CTV and calixarene rings on all sides with rigid bridges ('cavitands') [14]. The 'carcerands' obtained seem to incorporate small organic molecules and salts (e.g. DMF, THF, CsCl, Ar) even during the synthesis.

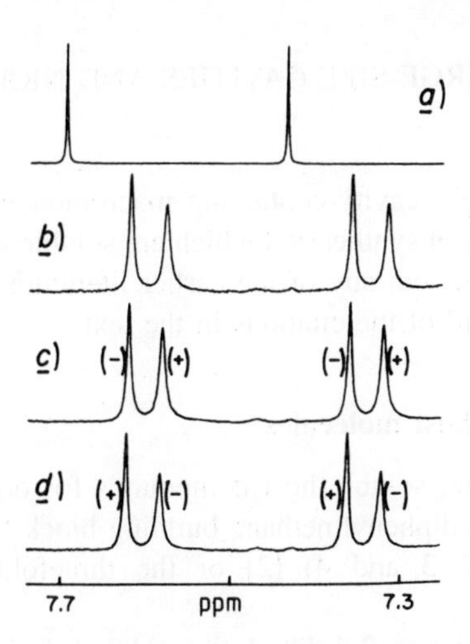

Figure 11. High-temperature (332 K) 200-MHz ^{1}H NMR spectrum in $CDCl_3$ of (*a*) (±)-CHFClBr; (*b*) (±)-CHFClBr (**12**) and (+)-cryptophane C; (*c*) (+)-CHFClBr (12) and (+)-cryptophane C; (*d*) (+)-CHFClBr (12) and (−)-cryptophane C

References

1. A.J.Ewins, *J. Chem. Soc*., **95**, 1486 (1909).
2. G.M.Robinson, *J. Chem. Soc*., **107**, 267 (1915).
3. A.S.Lindsey, *J. Chem. Soc*., 1685 (1965); A.Goldup, A.B.Morrison and G.W.Smith, *J. Chem. Soc*., 3864 (1965).
4. A.Lüttringhaus and K.C.Peters, *Angew. Chem*., **78**, 603 (1966); *Angew. Chem., Int. Ed. Engl*., **5**, 593 (1966).
5. V.K.Bhagwat, D.K.Moore and F.L.Pyman, *J. Chem. Soc*., 443 (1931).
6. H.Z.Zimmermann, R.Poupko, Z.Luz and J.Billard, *Z. Naturforsch., Teil A*, **40**, 149 (1985).
7. J.A.Hyatt, *J. Org. Chem*., **43**, 1808 (1978).
8. K.Frensch and F.Vögtle, *Liebigs Ann. Chem*., 2121 (1979).
9. J.Canceill, A.Collet, J.Gabard, F.Kotzyba-Hibert and J.-M.Lehn, *Helv. Chim. Acta*, **65**, 1894 (1982).
10. (a) J.Canceill, L.Lacombe and A.Collett, *J. Am. Chem. Soc*., **107**, 6993 (1985); (b) A.Collet, in *Inclusion Compounds* Vol.2 (J.L.Atwood, J.E.D.Davies and D.D.MacNicol, Eds.), Academic Press, London, 1984, p.97.
11. (a) J.Canceill, M.Cesario, A.Collet, J.Guilhem and C.Pascard, *J. Chem. Soc., Chem. Commun*., 361 (1986); (b) J.Canceill, M.Cesario, A.Collet, J.Guilhem, C.Riche and C.Pascard, *J. Chem. Soc., Chem. Commun*., 339 (1986).

12. J.Canceill, L.Lacombe and A.Collet, *J. Am. Chem. Soc.*, **108**, 4320 (1986).
13. J.Canceill, L.Lacombe and A.Collet, *J. Chem. Soc., Chem. Commun.*, 219 (1987); 583 (1988).
14. D.J.Cram *et al.*, *J. Am. Chem. Soc.*, **110**, 2554 (1988).

4.3.2 SYNTHETIC LARGE-SIZE CAVITIES AND NICHES FOR GUEST MOLECULES

In recent years, numerous cavity-containing macromonocyclic and macrobicyclic host compounds have been synthesized which are soluble in water, as enzymes are. These will now be discussed; any of the earlier literature not considered here can be consulted with the aid of the citations in the text.

4.3.2.1 Water-soluble host molecules

The first successful water-soluble host compounds for organic, neutral guests all contained either Koga's diphenylmethane building block (cf. **1**) [1], the analogous *o*-terphenyl system (cf. **3** and **4**) [2] or the threefold [*n.n.n*]paracyclophane framework (cf. **2**) [3].

When in an acidic solution 2,7-dihydroxynaphthalene is mixed with **2a–c** or **4**, a high-field shift of its aromatic protons is observed in the ^{1}H NMR; the degree of this shift depends on both the host added and the guest itself. Compounds **3** and **4**, for the first time, also complexed aliphatic guest molecules, such as cyclohexane-1,4-diol. Compound **3b**, like **2a**, accelerates the H–D exchange in 2,7-dihydroxynaphthalene.

Triphenylethane and triphenylbenzene frameworks have been used as spacers in the 30- and 40-membered carbobicyclic large-size cavities **5–7** [4]. As free dodecaacids, they are soluble in aqueous alkaline medium. In the case of **7a**, the inclusion of benzene, toluene and mesitylene into the host's cavity has been established by ^{1}H NMR spectroscopy. In these complexes, the guest's protons are markedly shifted to higher field, owing to the anisotropy effect of the host.

Vögtle *et al.* successfully prepared a bicyclic hexaamine (**8**) in which diphenylmethane units are used as bridges three times [5]. By protonating the amine nitrogens in acidic medium, the host compound becomes water soluble and then takes up diverse aromatic guest compounds. Their complexation was demonstrated by fluorescence and ^{1}H NMR spectroscopy. The geometry of the cavity allows a remarkably selective incorporation of flat, discotic guests, e.g. triphenylene, pyrene, fluoranthene, while partially hydrated aromatics, such as hexahydropyrene, acenaphthene and dodecahydrotriphenylene, are hardly complexed. Remarkable, too, is the selectivity of complexation in mixtures, such as phenanthrene–anthracene or 1,2-benzopyrene–chrysene. While the angularly built phenanthrene can be complexed, the linear anthracene does not bind to any extent. In the same way, chrysene, which deviates from the round geometry, is hardly bound at all, but 1,2-benzopyrene certainly is.

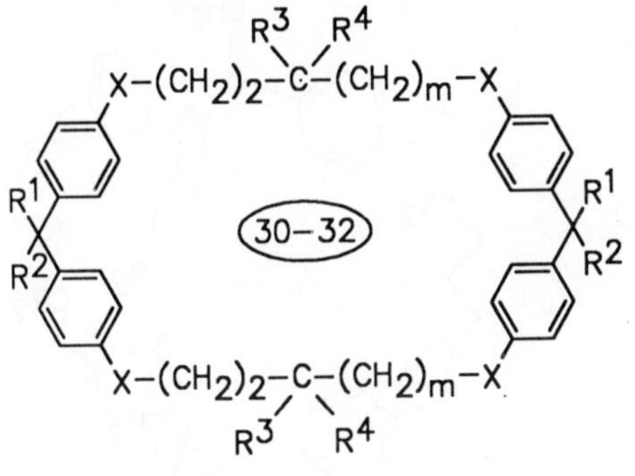

1

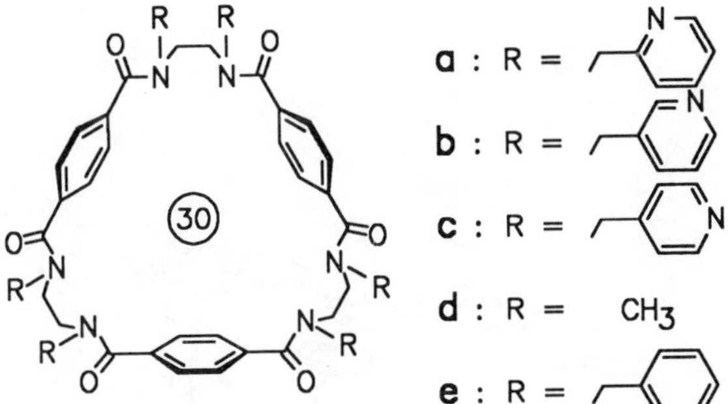

2

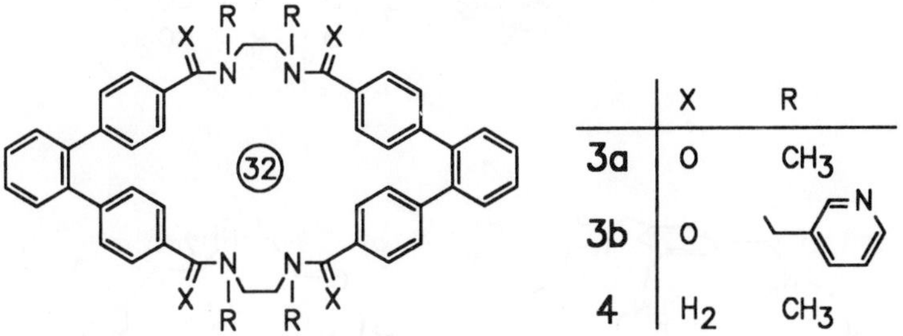

a : R =
b : R =
c : R =
d : R = CH_3
e : R =

	X	R
3a	O	CH_3
3b	O	
4	H_2	CH_3

Wilcox and Cowart prepared water-soluble macrobicyclic host cavities (**9**), derived from the Tröger base [6]. Their advantage lies in the relatively strong rigidity, with which the aromatic units are maintained in an angled position, thus forming a lipophilic pocket. With guests such as 2,4,6-trimethylphenol in aqueous medium, chemical shifts of the guest protons of up to $\Delta\delta = 2.33$ ppm were found.

Franke and Vögtle reported on the two isomeric host molecules **10** and **11**, which show different guest selectivities [7]. While the sterically unhindered cavity of the out/out isomer **10** is able to accept the spherical adamantane from an aqueous solution, the in/out isomer **11** is not. On the other hand, the flat naphthalene and the various dihydroxynaphthalenes are complexed by both hosts. From fluorescence

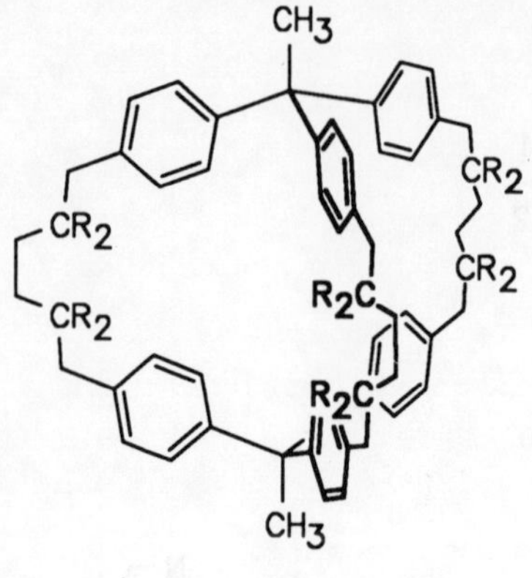

5

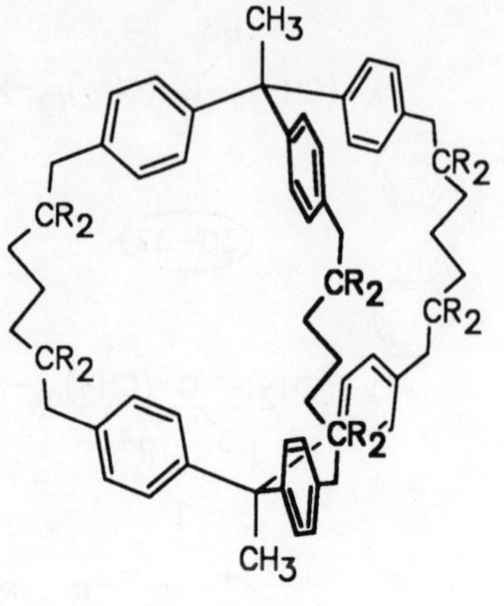

6

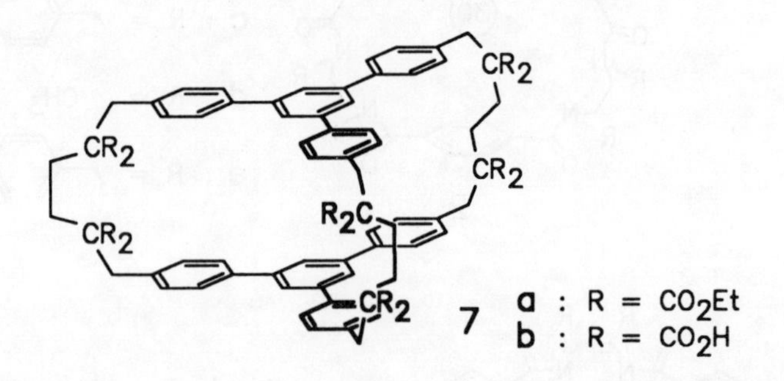
CR2
CR2
CR2
CR2
R2C
CR2
7
a : R = CO2Et
b : R = CO2H

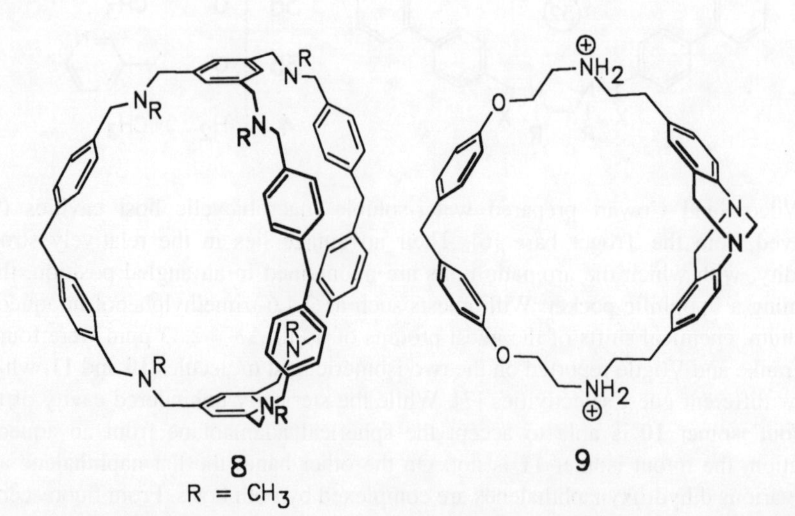
N R
N R
R N
N R
N R
R N
NH2
O
O
NH2
N
N
8
R = CH3
9

spectroscopy, a complex formation constant of pK = 4 was found for in/out isomer **11** and the fluoresceing 1,8-anilinonaphthalenesulphonic acid (1,8-ANS) as guest; the corresponding complex of the out/out isomer **10** with 1,8-ANS has pK = 4.6 and is therefore markedly more stable than the other complex. Molecular modelling calculations agree with the experimental findings [8].

out/out

10 : X, Y = C–CH_3
12 : X, Y = N
13 : X = N, Y = C–CH_3

out/in
11

a : R = H
b : R = $H_2^{\oplus}$

In the host molecules **12** and **13**, one and two triphenylethane units, respectively, are replaced with triphenylamine. Since the triphenylamine nitrogen is planar rather than pyramidal, owing to the delocalization of its lone pair into the aromatic rings, a flattened cavity is obtained, which exhibits a guest selectivity different from the triphenylmethane analogue. Accordingly, variously substituted dihydroxynaphthalenes, but not the spherical adamantane, can be complexed with it. The evidence was given by ^{1}H NMR and fluorescence spectroscopy. In addition, the triphenylamine unit, which electrochemically is easily oxidized to the radical cation, allows redox investigations on molecular cavities. Using cyclic voltammetry, large differences between the potentials of an open chain triphenylamine reference substance and the bicyclic host compounds **12** and **13** were found.

Compared with the macromoncyclic host molecules **15** and **16**, the association constant for **14** with 1,8-ANS is higher [9]. This is probably due to the fact that the framework of **14** is more rigid than those of **15** and **16**, the walls of the cavity therefore being better defined.

According to Shinkai *et al*., water-soluble calixarene carboxylic acids can discriminate between guest molecules, e.g. phenol blue, according to their size [9b]. Further information on the calixarenes themselves can be found in Gutsche's recent book [9c].

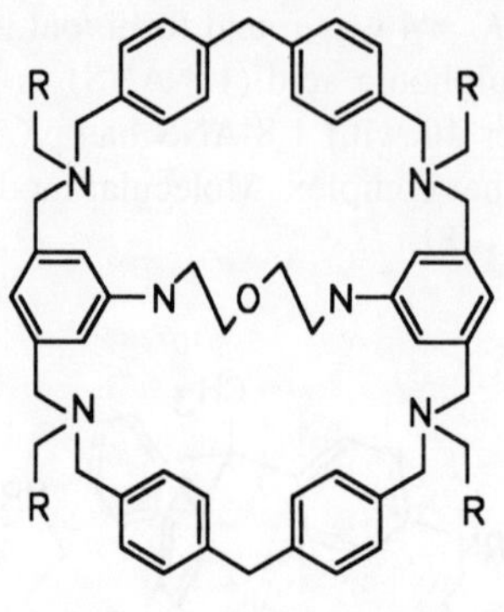

14

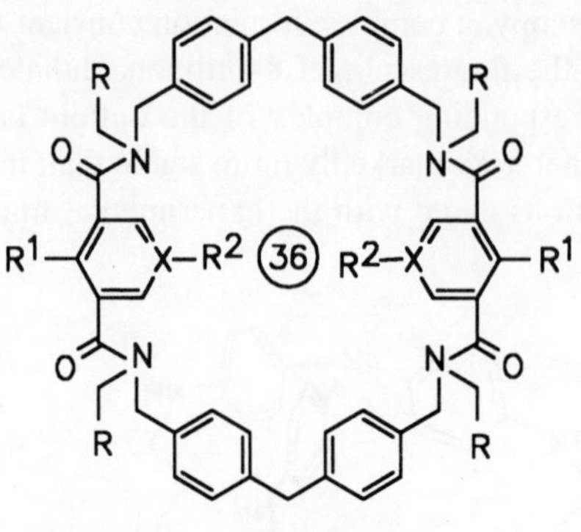

15,16
R = C_6H_5

	X	R^1	R^2
15a	N	H	–
15b	C	H	NO_2
15c	C	H	NH_2
15d	C	NO_2	H
16	C	NH_2	H

4.3.2.2 Host molecules with catalytic activity

Monocyclic host compounds with catalytic activity, different from the cyclodextrins, have been described by Schneider *et al*. [10], Cram [11], Mukarami and others. The monocyclic (**17**) and bicyclic (**18**) spiromolecules developed by Diederich *et al*. [12], which have a strong inclusion potential towards arenes and have proved themselves in numerous complexation experiments, have recently been supplemented by the catalytically active complexing agent **19** [13]. In order to restrict unproductive conformations, a catalytically active thiazolium group was introduced, using two amide bonds, into the previously known framework of **17**.

The kinetics of the benzoin condensation in the presence of **19** and, as references, **20** and **21**, were monitored using ^1H NMR spectroscopy by integrating the signals of both staring material and product. After 12 h, the signals of benzaldehyde had disappeared. Chromatographic work-up afforded, in addition to the unchanged catalysts **19**, **20** and **21**, benzoin in 93%, 74% and 27% yield, respectively. The comparatively high yield of 74% with **20** can be explained using CPK models, which show that here, too, a niche for the binding of benzaldehyde is present.

Measurements of the decrease in the NMR signals of the starting material proved the reaction to be of first order relative to benzaldehyde. The dependence of the starting reaction rate on the benzaldehyde concentration confirms that we are dealing with saturation kinetics. Compound **19** seems to be the most effective catalyst so far for the benzoin condensation [13].

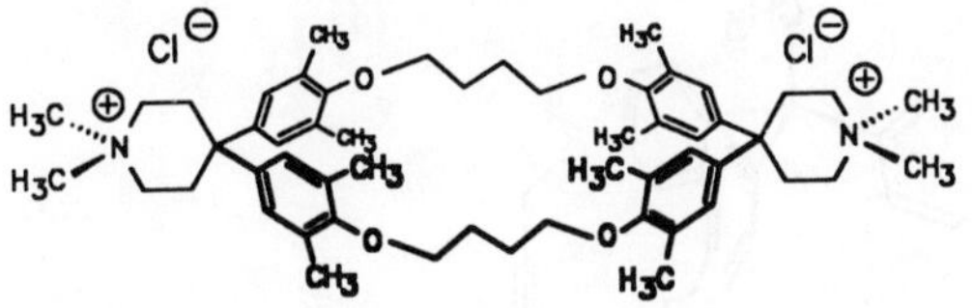

17

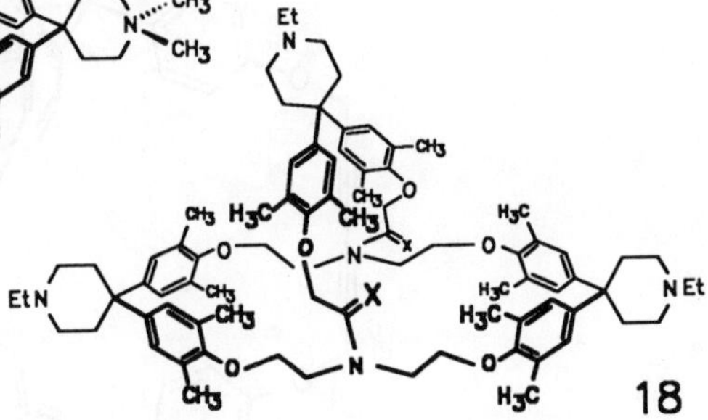

18

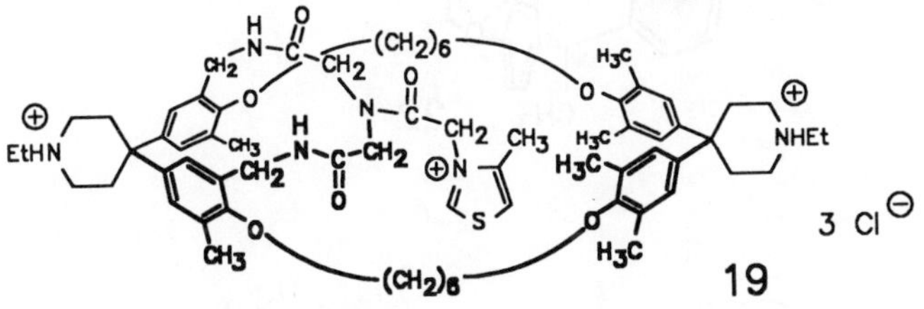

19

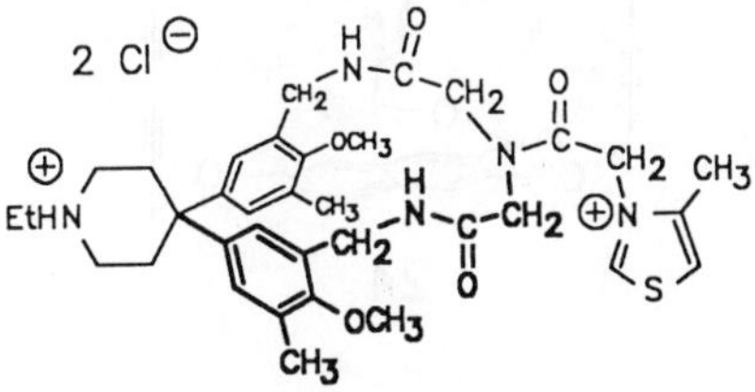

20

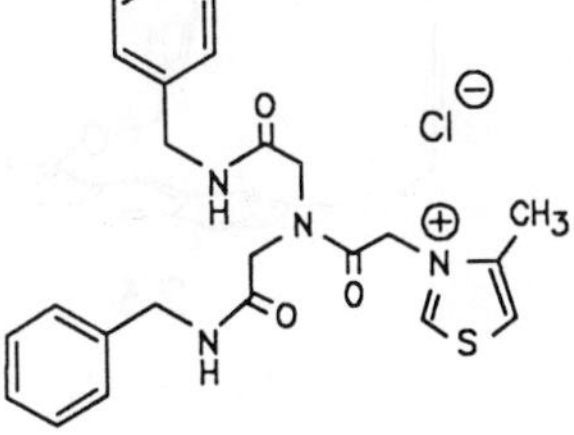

21

4.3.2.3 Chiral host molecules

In 1985, Breslow *et al*. successfully prepared a new bicyclic hexaether (**22**), which is built from two triphenylethane units linked by three diacetylene bridges [14]. The X-ray crystal structure analysis of the compound recrystallized from benzene showed it to be a 1 : 1 inclusion complex of benzene, where the host wraps itself helically around the guest.

The use of rigid diacetylene bridges was taken up by Whitlock *et al*. [15]. They succeeded in preparing two macrobicycles, **23** and **24**, in which two bridges are identical and the third is functionalized. They were isolated as isomers, both in the optically inactive *meso* form (**23**) and the active *d,l*-configuration (**24**).

Compound **24** is distinguished, apart from being chiral, by incorporating an aromatic donor centre into one bridge, the basicity of which is enhanced by an *N,N*-dimethylamino group in the *para* position. ^{1}H NMR spectroscopy (in $CDCl_3$) and X-ray crystallographic data confirmed the inclusion of 4-nitrophenol in **24**. Attempts to include also 2,4-dinitrophenol, failed. This seems to indicate that the

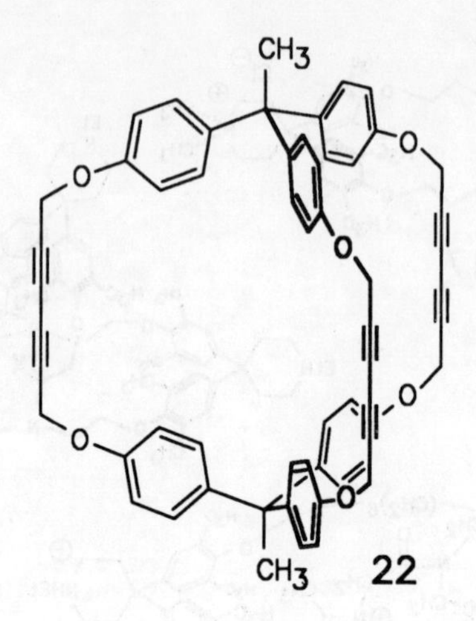

23

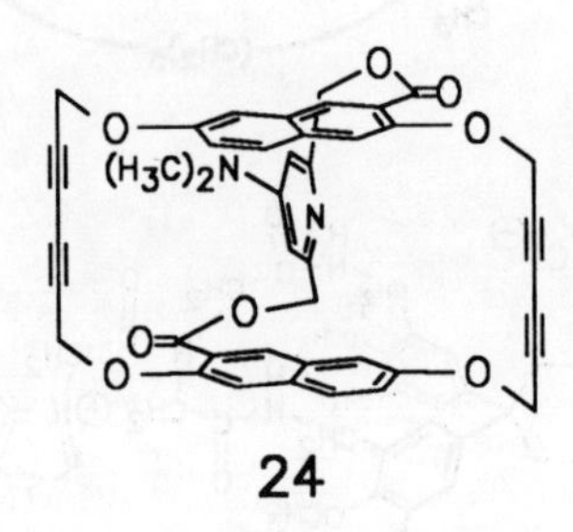

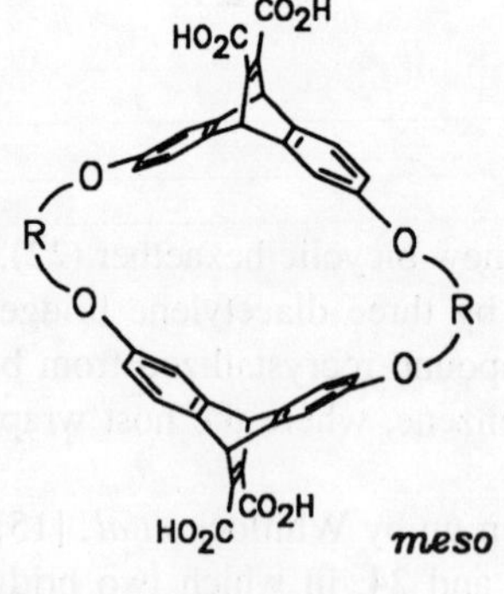

d,l

26

R = $(CH_2)_3$, $(CH_2)_4$, $(CH_2)_5$,

acid–base properties of host and guest have to be tuned exactly to each other. Hydrogen bonding between the pyridine nitrogen and the phenolic group pulls the guest into the cavity, where the aromatic ring will stay parallel to the two naphthalene systems [15].

The two water-soluble monocyclic molecules with *meso* and *d,l*-configurations, **25** and **26** are also isomeric and exhibit high affinity towards aromatic compounds such as anthracene and pyrene [16]. Intramolecular inclusions were established by UV spectroscopy (bathochromic shifts) and ^{1}H NMR spectroscopy (high-field shifts).

Diederich *et al*. described the macrocyclic, optically active host molecule (+)-**27** [17]. As a chiral building block lining the cavity it contains the non-natural alkaloid 4-phenyltetrahydroquinoline (−)-**28**. The formation of diastereoisomeric host–guest complexes of (+)-**27** with Naproxen® (**29**), a chiral cyclooxigenase inhibitor, was observed by ^{1}H NMR spectroscopy. The aromatic signals of the (*R*) and (*S*)-guests are shifted to different extents.

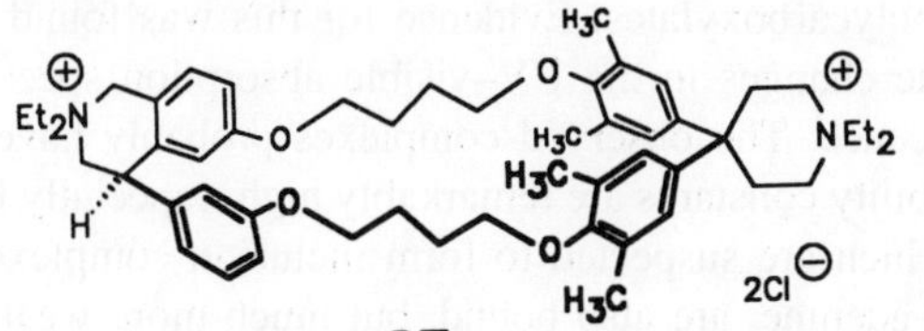

27

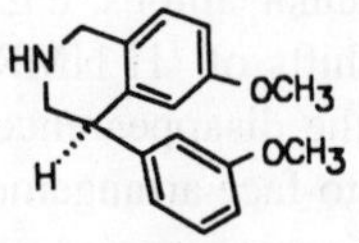

28

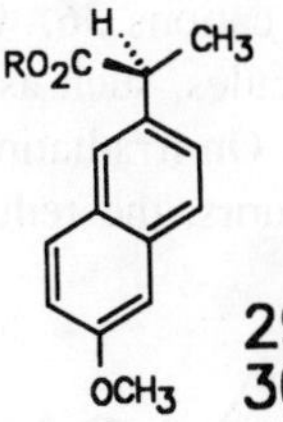

29: R = H
30: R = CH_3

4.3.2.4 Host molecules for the complexation of anionic guests

Compared with the complexation of neutral or positively charged guest molecules, the complexation of anions is a relatively little investigated area of bioorganic chemistry. In 1986, Lehn *et al*. [19] reported three bicyclic anion-complexing ligands which differ both in size and shape.

By protonating **31**–**33**, the corresponding polyammonium compounds are formed, where the quatenary nitrogen atoms act as binding sites for anions. There are two ways of binding anions: either by incorporation into the cavity or by coordination on the outside of the molecule. Thus, ligands **31** and **33** ought to be able to bind anions intramolecularly, whereas **32**, based on its size, would rather be binding on its surface. Although the size of the anion is important, there are also fundamental differences in the solid state and in solution. For instance, the nitrate complex of **31** in its solid state has six nitrate ions bound externally (X-ray crystal structure analysis). In solution, however, higher stability constants and further NMR data more likely indicate a 1 : 1 inclusion of a NO_3^- ion. The determination of complex formation constants gave far higher values for dianions (sulphate, oxalate, $S_2O_6^{2-}$) than for singly charged negative ions (NO_3^-, Cl^-).

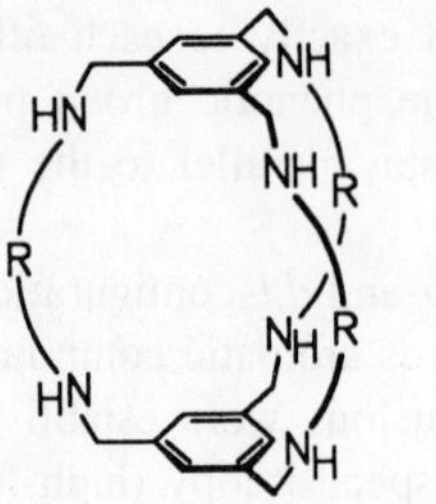

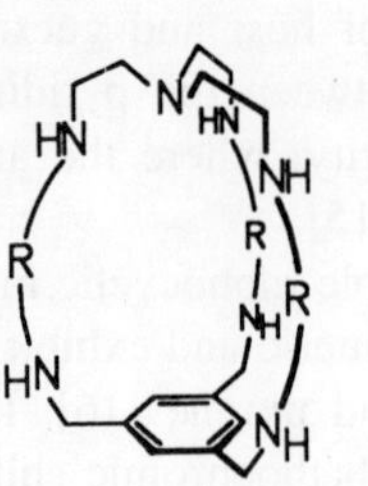

31: R = $-(CH_2)_3-$

32: R = $-CH_2CH_2OCH_2CH_2-$

33: R = $-CH_2CH_2OCH_2CH_2-$

Apart from these bicyclic compounds, Lehn *et al*. prepared different open-chain host molecules with 2,7-diazapyrenium units (DAP^{2+}) [20]: the cations **34** and **35**, as well as the niche-shaped bis(diazapyrenium) molecules **36**, bind different molecular anions, e.g. aromatic polycarboxylates. Evidence for this was found in the shifts of ^1H NMR signals, the changes in the UV–visible absorption spectra, and the disappearence of fluorescence. The observed complexes probably have a face-to-face arrangement. The stability constants are remarkably high, especially for bis(diazapyrenium) cations **36**, which are suspected to form inclusion complexes. Neutral guest molecules, such as adenine, are also bound, but much more weakly than anionic guests. On irradiating Me_2DAP^{2+} with visible light in the presence of several electron donors, the reduced salt Me_2DAP^+ is formed, which selectively cleaves DNA.

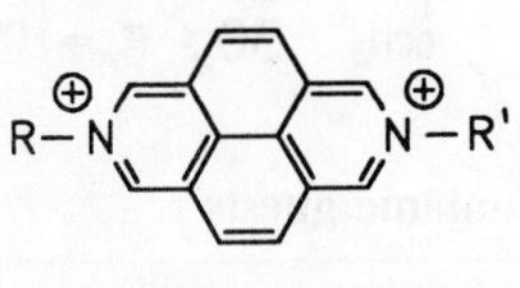

34: R , R' = CH_3

35: R = CH_3, R' = $C_6H_5CH_2$

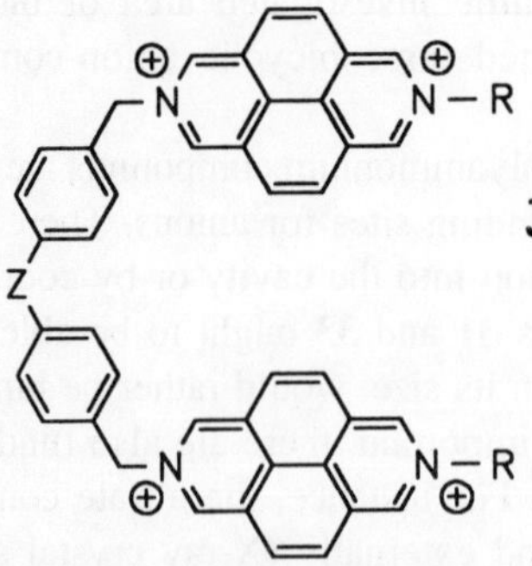

36 a: R = CH_3 , Z = CH_2

b: R = CH_3 , Z = O

c: R = CH_3 , Z = $(CH_3)_2C$

Pascal *et al*. prepared a smaller bicyclic host molecule (**37**) for anionic guests [21]. The model for this host compound was a sulphate-binding protein from *Salmonella typhimurium*, in which protons of the peptide bonds (O=C—NH) provide five of the eight binding sites. After extending 1,3,5-benzenetriacetyl chloride, this particular anion host was obtained without high dilution techniques in 11% yield.

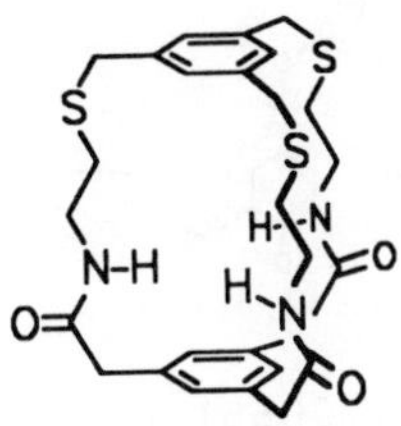

37

^{1}H and ^{19}F NMR spectra of **37** in the presence of tetrabutylammonium fluoride indicate an association between the host and a fluoride ion. Whether this is truly an intramolecular inclusion has not yet been determined.

Ditopic receptor molecules of type **38** were synthesized by Schmidtchen; they contain two distinct tetrahedral structural elements linked together by a *p*-xylene bridge [18].

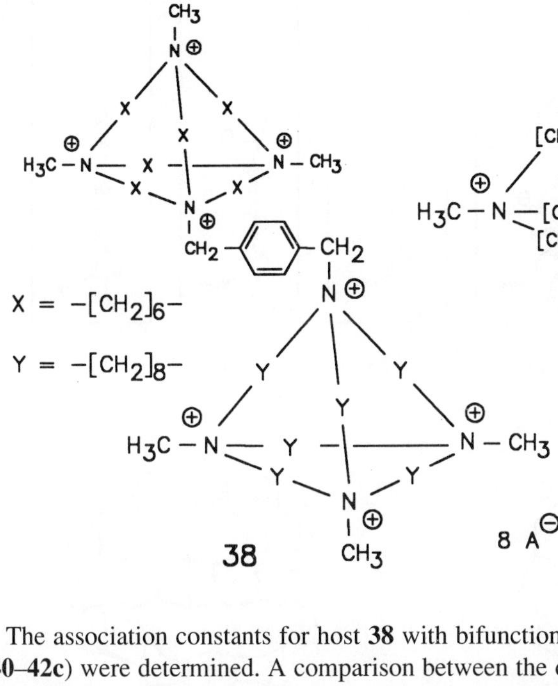

X = $-[CH_2]_6-$

Y = $-[CH_2]_8-$

38 $8\ A^{\ominus}$

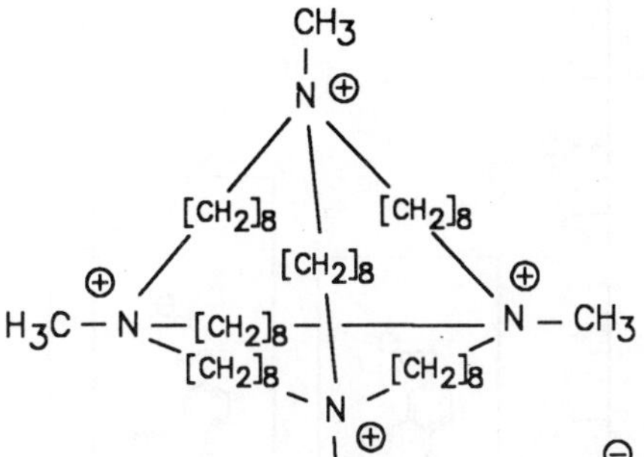

39 $4\ A^{\ominus}$

The association constants for host **38** with bifunctional guests of various lengths (**40**–**42c**) were determined. A comparison between the differences in selectivity for the different guests, which differ only in the spacing of the negative end-groups and not in their chemical character, shows a sudden leap in the cases of **42b** and **42c**. This behaviour suggests the participation of both structural elements of the

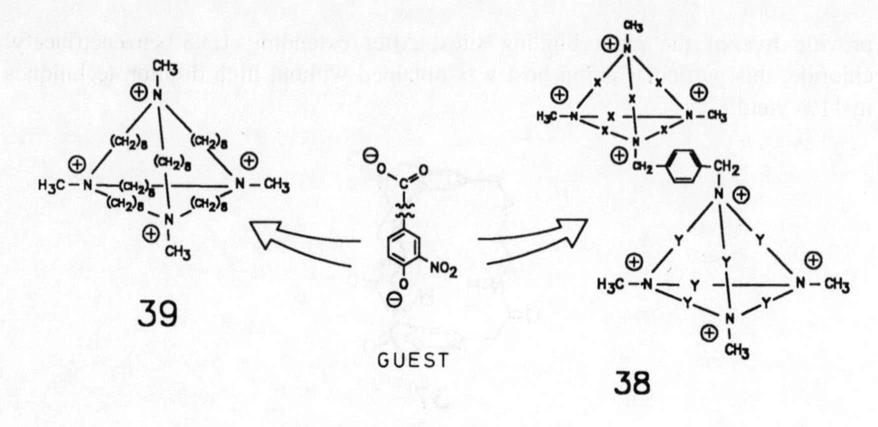
39
GUEST
38

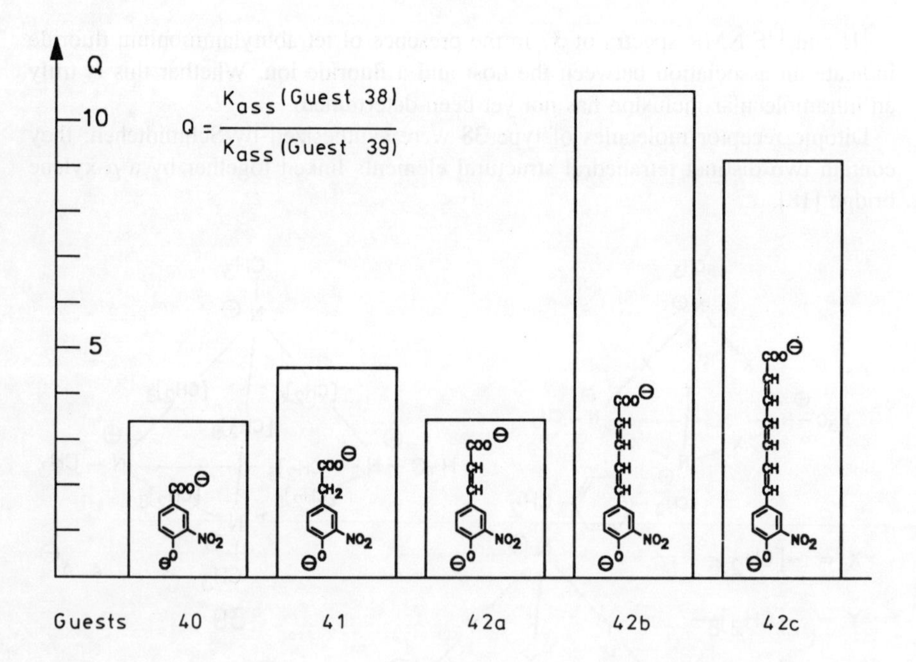
Q
10
5
Q = Kass (Guest 38) / Kass (Guest 39)
Guests
40
41
42a
42b
42c

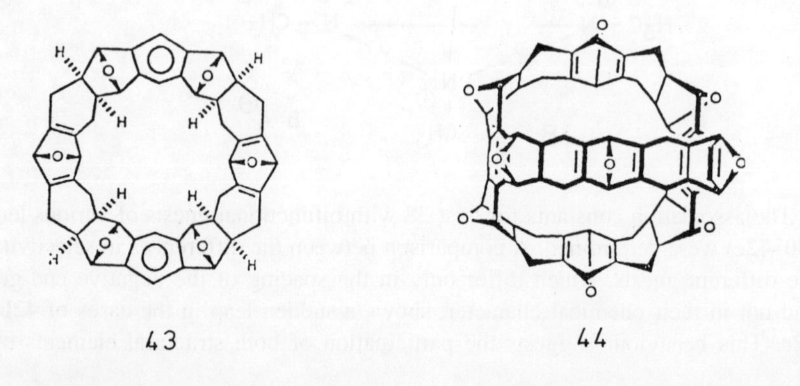
43
44

receptor in the complexation of these two guests, whereas the spacers in **40–42a** are too short for this. In comparison with the monotopic receptor **39**, the ditopic receptor **38** complexes three times more strongly.

The macrocycles kohnkene (**43**) and trinacrene (**44**), prepared by Stoddart *et al*. [21a] using multiple Diels–Alder reactions, form cavities; that of **43** contains, according to X-ray structure analysis, one molecule of water.

4.3.2.5 Open-chain molecular niches and pincers for the selective binding of complementary guests

The macrocyclic host molecules presented so far have their origin in investigations by Stetter and Roos, who in the early 1950s were engaged at the University of Bonn in the synthesis of nitrogen-containing ring systems. Their results later proved to be useful, although the original macrocycles only formed clathrates (cf. Chapter 5), but did not incorporate guests into the host's cavity. In subsequent work on these macrocycles by Vögtle *et al*. [23] and Saenger *et al*., the exact facts were settled in 1982 [24]. In the meantime, many workers throughout the world had thoroughly studied the synthesis of analogous macrocycles and had advanced the study of host–guest interactions dramatically, as described earlier.

In 1985, Rebek *et al*. chose to follow a different route [25]. They successfully prepared acyclic pincer-like molecules, such as **45** and **46**, which were meant to serve as receptor models (molecular niches) for guests of complementary size and shape. The binding is effected by hydrogen bonding from two carboxylic functions. In this, they follow the set-up in the active centre of aspartate proteinase and lysozyme, which also contain two carboxylic groups. Here, however, a less perfect niche is complemented by spacially oriented funtionalities:

The synthesis proceeds via a condensation reaction of *cis,cis*-1,3,5-trimethyl-cyclohexane-1,3,5-tricarboxylic acid and either acridine yellow or 2,7-diamino-3,6-dimethylnaphthalene. The distance between the two facing carboxylic groups is 800 pm in **45** and 550 pm in **46**. The aliphatic methyl groups are there to block any inversion of the cyclohexane chair conformation, while the aromatic CH_3 groups stop the rotation around the two C_{aryl}—N_{imide} bonds. According to their respective sizes, **45** binds pyridine and diazabicyclooctane, and **46** accepts diols, diamines and amino alcohols as guests. Owing to the proximity of host and guest, the aromatic protons pointing towards the guest show an average high-field shift of 0.5 ppm. The conversion of **45** into the corresponding diamide **47** allows the binding of piperazine diones (cf. **48**).

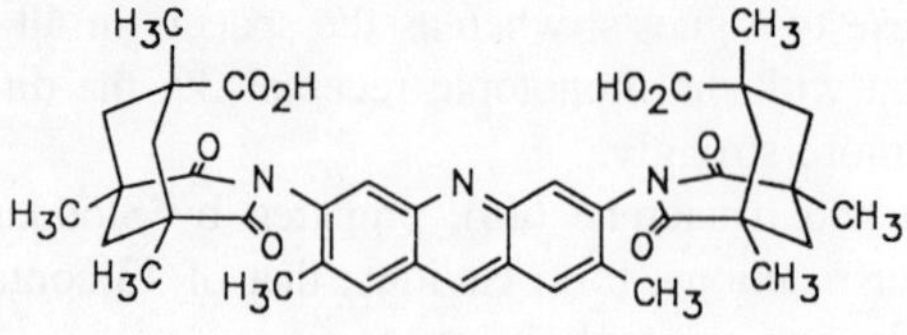

45

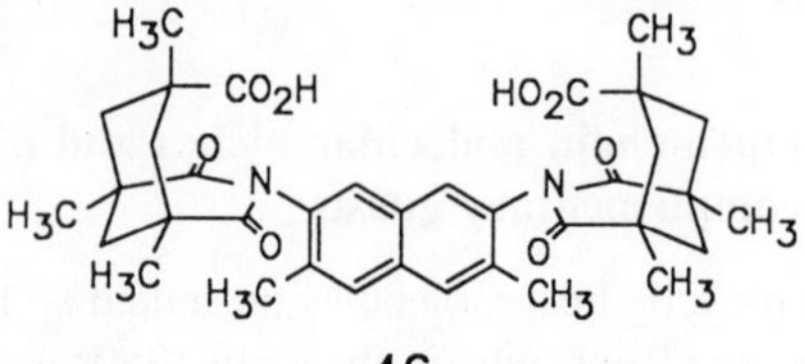

46

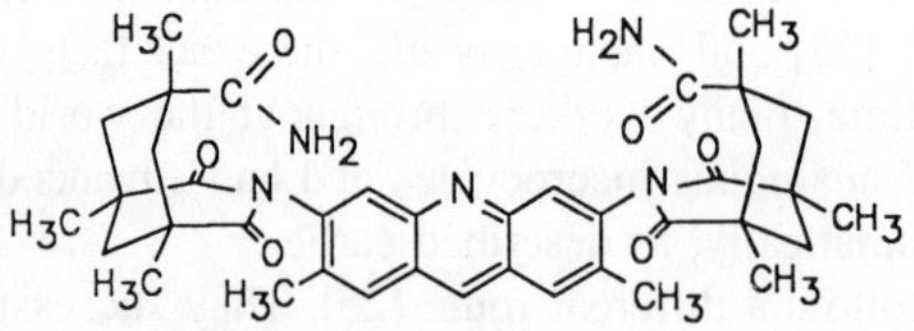

47

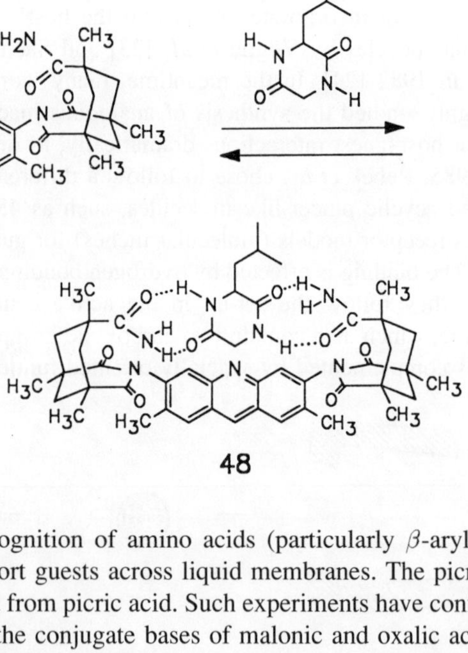

48

Apart from the selective recognition of amino acids (particularly β-arylamino acids), **45** is also able to transport guests across liquid membranes. The picrate of **45** is able to separate oxalic acid from picric acid. Such experiments have confirmed the high stabilization by **45** of the conjugate bases of malonic and oxalic acid.

The pincer molecule **49** differs from **45** and **46** by its lack of aromatic CH_3 groups. Because of this, a rotation around the C_{aryl}—N_{imide} bond is possible and this is stopped only by the addition of a dicarboxylic acid (cf. molecular complex **50**).

When using acridine-containing molecular pincers and suitable aromatic dicarboxylic acids, there is not only an internal proton transfer between guest and host (cf. **51** $\rightleftharpoons$ **52**), but also a π–π interaction, which also contributes to the stabilization of the complex (cf. **53**) [25c,26,27].

Ligand **54**, an EDTA analogue reported in 1981 by Leppkes and Vögtle [28], can also be considered to be a molecular pincer. Like ethylenediaminetetraacetic acid, but in spite of a larger distance between pincer tips [N(CH_2COO^-)$_2$], it favours Ca^{2+} ions.

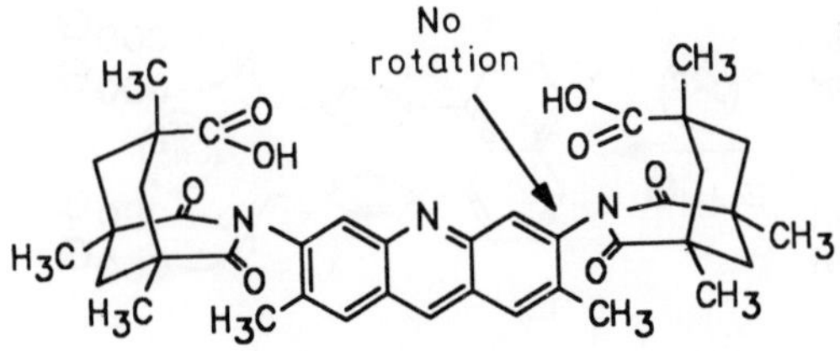

No rotation

45

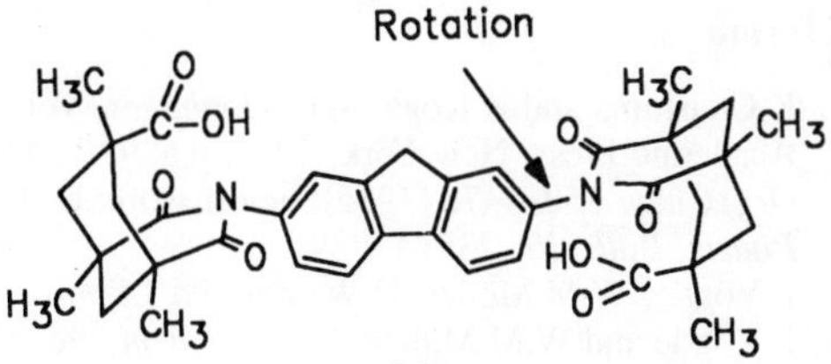

Rotation

49

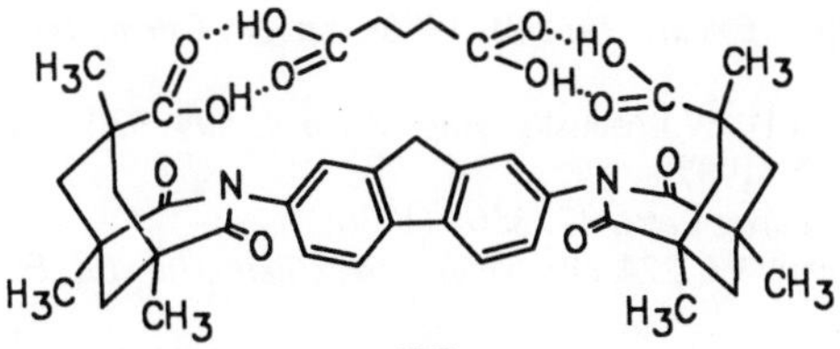

50

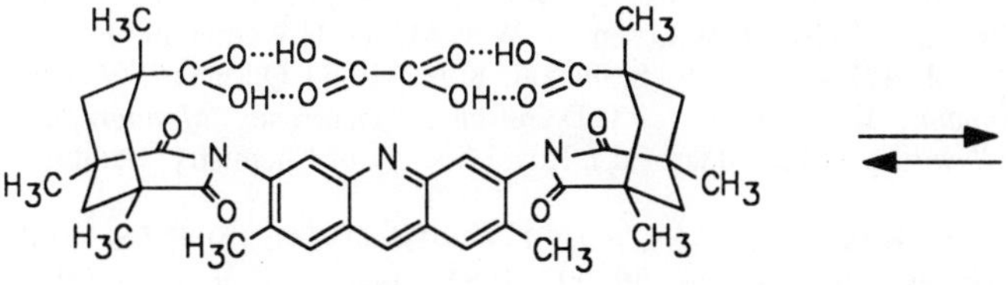

51

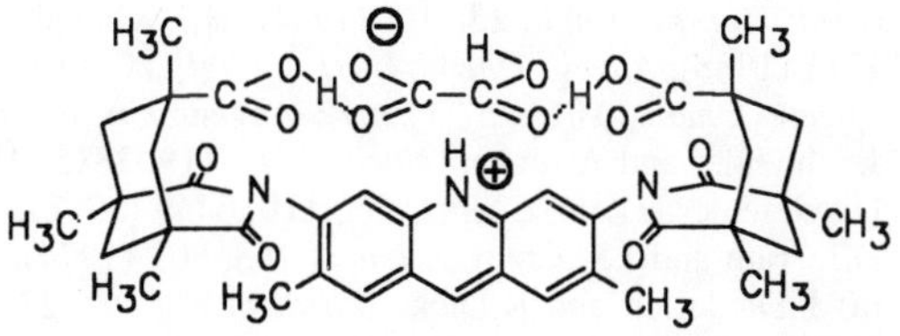

52

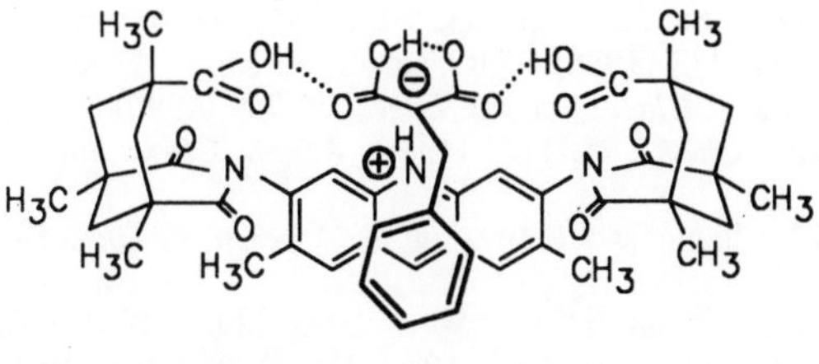

53

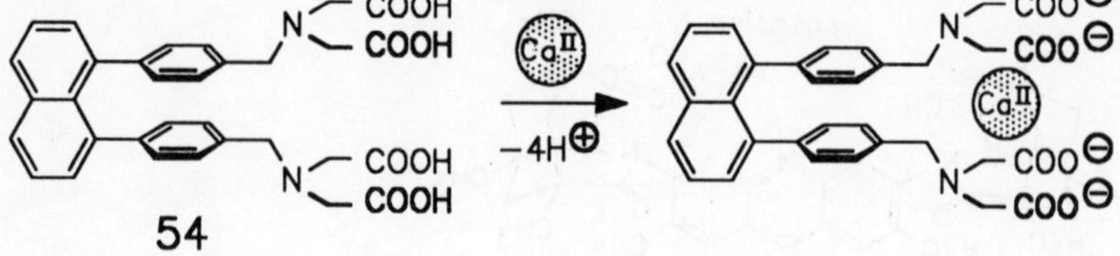

References

1. K.Odashima and K.Koga, in *Cyclophanes*, Vol.2 (P.M.Keehn and S.M.Rosenfeld Eds.), Academic Press, New York, 1983, p.629; K.Odashima, A.Ita, Y.Litaka and K.Koga, *J. Org. Chem.*, **50**, 4478 (1985); newer work: C.-F. Lai, K. Odashima and K. Koga, *Chem. Pharm. Bull.*, **37**, 2351 (1989).
2. F.Vögtle, W.M.Müller, U.Werner and J.Franke, *Naturwissenschaften*, **72**, 155 (1985).
3. F.Vögtle and W.M.Müller, *Angew. Chem.*, **96**, 711 (1984) *Angew. Chem., Int. Ed. Engl.*, **23**, 372 (1984).
4. T.Merz, H.Wirtz and F.Vögtle, *Angew. Chem.*, **98**, 549 (1986); *Angew. Chem., Int. Ed. Engl.*, **25**, 567 (1986).
5. F.Vögtle, W.M.Müller, U.Werner and H.-W.Losensky, *Angew. Chem.*, **99**, 930 (1987); *Angew. Chem., Int. Ed. Engl.*, **26**, 909 (1987).
6. C.S.Wilcox and M.D.Cowart, *Tetrahedron Lett.*, **27**, 5563 (1986).
7. J.Franke and F.Vögtle, *Angew. Chem.*, **97**, 224 (1985); *Angew. Chem., Int. Ed. Engl.*, **24**, 219 (1985).
8. cf. F.Vögtle, J.Franke, W.Bunzel, A.Aigner, D.Worsch and K.-H.Weissbarth, *Stereochemie in Stereobildern*, VCH, Weinheim, 1987, p.78.
9. (a) U.Werner, W.M.Müller, H.-W.Losensky, T.Merz and F.Vögtle, *J. Inclus. Phenom.*, **4**, 379 (1986); J.Franke, T.Merz, H.-W.Losensky, W.M.Müller, U.Werner and F.Vögtle, *J. Inclus. Phenom.*, **3**, 471 (1985); (b) S.Shinkai, K.Araki and O.Manabe, *J. Chem. Soc., Chem. Commun.*, 187 (1988); (c) C.D.Gutsche, *Calixarenes, Monographs in Supramolecular Chemistry* (J.F.Stoddart Ed.), Royal Society of Chemistry, Cambridge, 1989.
10. (a) H.J.Schneider and K.Philippi, *Chem. Ber.*, **117**, 3056 (1984); (b) H.J.Schneider, W.Müller and D.Güttes, *Angew. Chem.*, **96**, 909 (1984); *Angew. Chem., Int. Ed. Engl.*, **23**, 910 (1984); (c) H.J.Schneider and R.Busch, *Angew. Chem.*, **96**, 910 (1984); *Angew. Chem., Int. Ed. Engl.*, **23**, 911 (1984); H.J.Schneider and T.Blatter *Angew. Chem.*, **100**, 1211 (1988); *Angew. Chem., Int. Ed. Engl.*, **27**, 1163 (1988); H.-J.Schneider and I.Theis, *Angew. Chem.*, **101**, 757 (1989); *Angew. Chem., Int. Ed. Engl.*, **28**, 753 (1989); H.-J.Schneider and A.Junker, *Chem. Ber.*, **119**, 2815 (1986); H.-J.Schneider, D.Guttes and U.Schneider, *J. Am. Chem. Soc.*, **110**, 6449 (1988).
11. D.J.Cram and J.M.Cram, *Science*, **183**, 803 (1974).
12. (a) F.Diederich and K.Dick, *Tetrahedron Lett.*, **23**, 3167 (1982); (b) F.Diederich and K.Dick, *Angew. Chem.*, **95**, 730 (1983); *Angew. Chem., Int. Ed. Engl.*, **22**, 715 (1983); (c) F.Diederich and D.Dick, *J. Am. Chem. Soc.*, **106**, 8024 (1984); (d) F.Diederich and D.Griebel, *J. Am. Chem. Soc.*, **106**, 8037 (1984); (e) F.Diederich and K.Dick, *Angew. Chem.*, **96**, 789 (1984); *Angew. Chem., Int. Ed. Engl.*, **23**, 810 (1984); (f) for a review, see F.Diederich, *Angew. Chem.*, **100**, 372 (1988); D.R.Carcanague and F. Diederich, *Angew. Chem.*, **102**, 836 (1990); *Angew. Chem., Int. Ed. Engl.*, **29**, 769 (1990).
13. H.-D.Lutter and F.Diederich, *Angew. Chem.*, **98**, 1125 (1986); *Angew. Chem., Int. Ed. Engl.*, **25**, 1125 (1986).
14. D.O'Krongly, S.R.Denmeads, M.Y.Chiang and R.Breslow, *J. Am. Chem. Soc.*, **107**, 5544 (1985).
15. (a) S.P.Miller and H.W.Whitlock, *J. Am. Chem. Soc.*, **106**, 1492 (1984); (b) R.E.Sheridan and H.W.Whitlock, *J. Am. Chem. Soc.*, **108**, 7120 (1986); K.M.Neder

and H.W.Whitlock, *J. Am. Chem. Soc.*, **112**, 9412 (1990).
16. M.A.Petti, T.J.Shepodd and D.A.Dougherty, *Tetrahedron Lett.*, **27**, 807 (1986); see also T.J.Shepodd, M.A.Petti and D.A.Dougherty, *J. Am. Chem. Soc.*, **110**, 1983 (1988).
17. R.Dharanipragada and F.Diederich, *Tetrahedron Lett.*, **28**, 2443 (1987); 2443 (1987); R.Dharanipragda, S.B.Ferguson and F.Diederich, *J. Am. Chem. Soc.*, **110**, 1679 (1988); for a review, see F.Diederich, *Angew. Chem.*, **100**, 372 (1988); *Angew. Chem., Int. Ed. Engl.*, **27**, 362 (1988).
18. For a review, see F.P.Schmidtchen, *Nachr. Chem. Tech. Lab.*, **36**, 8 (1988).
19. D.Heyer and J.-M.Lehn, *Tetrahedron Lett.*, **27**, 5869 (1986).
20. A.J.Blacker, J.Jazwinski and J.-M.Lehn, *Helv. Chim. Acta*, **70**, 1 (1987); for a review, see J.-M.Lehn, *Angew. Chem.*, **100**, 91 (1988).
21. R.A.Pascal, J.Spergel and D.Van Engen, *Tetrahedron Lett.*, **27**, 4099 (1986).
21. (a) P.R.Ashton, N.S.Isaacs, F.H.Kohnke, G.Stagno d'Alcontres and J.F.Stoddart, *Angew. Chem., Int. Ed. Engl.*, **28**, 1261 (1989); J.F.Stoddart, *Nature (London)*, **334**, 10 (1988).
22. H.Stetter and E.-E.Roos, *Chem. Ber.*, **88**, 1390 (1955).
23. F.Vögtle, W.M.Müller and L.Rossa, unpublished work.
24. R.Hilgenfeld and W.Saenger, *Angew. Chem.*, **94**, 788 (1982); *Angew. Chem., Int. Ed. Engl.*, **21**, 781 (1982).
25. (a) J.Rebek, Jr, and D.Nemeth, *J. Am. Chem. Soc.*, **108**, 5637 (1986). (b) J.Rebek, Jr, B.Askew, N.Islam, M.Killoran, D.Nemeth and R.Wolak, *J. Am. Chem. Soc.*, **107**, 6736 (1985); (c) see also J.Rebek, Jr, B.Askew, P.Ballester and A.Costero, *J. Am. Chem. Soc.*, **110**, 923 (1988); J.Wolfe, D.Nemeth, A.Costero and J.Rebek, Jr, *J. Am. Chem. Soc.*, **110**, 983 (1988); J.B.Huff, B.Askew, R.J.Duff and J.Rebek, Jr, *J. Am. Chem. Soc.*, **110**, 5908 (1988); J.Rebek, Jr., *Top. Curr. Chem.*, **149**, 189 (1988); *J. Incl. Phenom.*, **7**, 7 (1989); cf. J.Rebek, Jr., *et al.*, *J. Am. Chem. Soc.*, **111**, 1082 (1989); overview: J.Rebek, Jr., *Angew. Chem.*, **102**, 261 (1990); *Angew. Chem., Int. Ed. Engl.*, **29**, 245 (1990).
26. See also T.R.Kelly and M.P.Maguiere, *J. Am. Chem. Soc.*, **109**, 6549 (1987).
27. S.C.Zimmerman and C.M.Van Zyl, *J. Am. Chem. Soc.*, **109**, 7894 (1987).
28. R.Leppkes and F.Vögtle, *Angew. Chem.*, **93**, 404 (1981); *Angew. Chem., Int. Ed. Engl.*, **20**, 396 (1981); *Chem. Ber.*, **115**, 926 (1982); F.Vögtle, H.Schäfer and C.Ohm, *Chem. Ber.*, **117**, 948, 955 (1984); cf. F.Vögtle, T.Papkalla, H.Koch and M.Nieger, *Chem. Ber.*, **123**, 1097 (1990).

5 Clathrate Inclusion Compounds

5.1 INTRODUCTION

Inclusion compounds have been known since the beginning of the last century. The fundamental feature of this type of compound is the fact that a cavity-containing host component incorporates one or several guest components, without any covalent bonding.

The oldest known 'compound' of this type is chlorine hydrate, $Cl_2 \cdot 6H_2O$, a species discovered by Humphrey Davy in 1810 and investigated by Michael Faraday in 1823. In 1949 von Stackelberg elucidated its inclusion structure [1].

After von Laue introduced X-ray crystal structure analysis, the conditions were ripe for a tempestuous phase in the evolution of 'inclusion chemistry'. In 1936, with this then new method, the physical chemist Kratky [2] succeeded in elucidating the structure of choleic acid. He also showed that deoxycholic acid forms typical inclusion compounds with fatty acids.

The term 'inclusion compound' was described by Cramer [3] in 1952 as follows:

> ... All these compounds namely have in common the ability to incorporate into the cavities of their own molecules, or within their lattices, other molecules of suitable size, spatially to enfold them, that is to hold them, though not by main or secondary valence forces, but mostly by physical imprisonment.

Thus, Cramer had already recognized that for the formation of inclusion compounds, no functional groups and no chemical reaction potential was needed. He wrote further:

> ... Rather, the essential requirements are:
> 1) The enclosing substance must possess cavities. Such cavities may be present either in the molecule, in large rings, for instance, or in the crystal lattice of the host compound. The corresponding cavity-containing lattice need not be stable as such, but is, in many cases, only formed on incorporating the guest substances.
> 2) The substance to be incorporated must find enough space in the cavities present ...

Most inclusion compounds were discovered purely by chance, by recrystallizing a compound, for example. Thus, Reddelien wrote in 1923 [4]:

> 1,2,5-Tris(biphenyl)benzene crystallizes from benzene with one molecule of solvent. The benzene-containing compound has the same crystal structure as the compound without solvent molecules. The benzene is held very strongly and is released only at elevated temperatures, without much change in the aspect of the crystals ...

At this stage, let us give a few examples of naturally occurring inorganic compounds, which possess cavities in their crystal lattices, and are therefore suitable as inclusion host compounds. Some of them are of industrial importance. Corresponding inclusions are known of zeolites, of argillaceous minerals, such as montmorillonite and kaolinite, of uranium mica, of graphite, and others.

One area of supramolecular chemistry is concerned with the creation of 'tailored' host compounds for potential guest molecules (cf. Section 1.1). A few fundamental results of clathrate chemistry (crystal lattice inclusion compounds) are presented below.

5.2 CLATHRATES

5.2.1 DEFINITION

The term clathrate was introduced by Powell [5] in 1948. Today, any crystalline lattice inclusion compound is called a clathrate; several molecules of the host compound form extramolecular cavities, where guest compounds can be incorporated. The occurrence of clathrates, therefore, is normally limited to the solid state, although clathrate formation in the liquid state has also been reported [5a].

Guest molecules are either incorporated into existing extramolecular cavities or induce, on crystallization, structures of the host lattice with guest-specific spaces. This organization of the clathrates is fundamentally different from molecular inclusion compounds, where the guest molecule is incorporated into the cavity of a single host molecule (Figure 1).

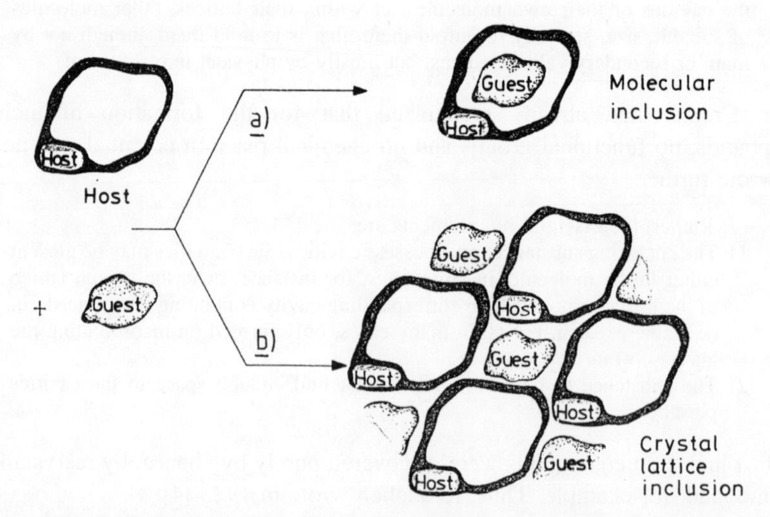

Figure 1. The difference between molecular and lattice inclusions: (*a*) Formation of a molecular complex, where a convex guest fits into the cavity of one host molecule; (*b*) inclusion of guest molecules into cavities between different host molecules in the crystal lattice (clathrate formation)

5.2.2 PRACTICAL USES OF CLATHRATES

The ease with which clathrates are prepared and the simplicity of their decomposition with solvents, make them interesting for industrial purposes. The reversibility of clathrate formation gives them some importance for technical processes. The following list gives a summary of possible applications of inclusion compounds:

— separation of mixtures (isomers, homologues);
— separation of racemates;
— 'solidification' of gases and liquids;
— stabilization of sensitive or toxic substances;
— polymerization inside inclusion channels (topochemistry);
— battery systems and organic conductors.

More detailed examples are given later on when specific clathrate hosts are discussed.

5.2.3 CLASSICAL CLATHRATE HOSTS

5.2.3.1 Water (gas hydrates)

Many gases are able to form crystalline hydrates with water at low temperatures and high pressures. Apart from chlorine hydrate, mentioned earlier, hydrates of ClO_2, SO_2, CO_2, N_2O, $CHCl_3$, and saturated and unsaturated hydrocarbons are known. Using X-ray crystal structure methods, von Stackelberg [6] in 1949 showed them to be cage inclusion compounds, where water acts as a clathrate host ('clathrand').

The H_2O molecules form polyhedral cages held together by hydrogen bonding. Compounds with polar bonds to hydrogen (hydrogen halides, alcohols, acids, amines, ammonia) and compounds with hydrogen bonding (aldehydes, ketones, nitriles) do not form gas hydrates. They have to be discarded as guest components, as they disturb the structure of the lattice and the cohesion of the water molecules. As with the cyclodextrins, incorporated guests are primarily lipophilic. The gas hydrates are not only of academic interest, as shown below.

In natural gas pipelines, unwanted formation of gas hydrates at low temperatures can occur and crystalline hydrates of the lower hydrocarbons may precipitate. The formation of the crystals can be stopped by drying the natural gas, or else the gas hydrates are destroyed and their formation is suppressed by the addition of ammonia. An interesting application consists in desalinating water with the help of propane hydrates [1]: when sea water is treated with propane at 8.5 °C, $C_3H_8 \cdot 17H_2O$ separates, 44 g of propane binding 306 g of water. After filtration of the crystals, the clathrates are decomposed by warming under pressure and thus salt-free water is obtained. Some doubt has now been thrown on these results.

5.2.3.2 Hydroquinone

As early as the mid-nineteenth century, Wöhler [7], Clemm [8] and Mylius [9] discovered crystalline inclusion compounds of hydroquinone (**1**). Thus, in 1849 Wöhler wrote:

> ... The colourless hydroquinone, $C_{12}H_6O_4$, has a peculiar ability to combine in two different proportions with hydrogen sulphide and thus to form two easily crystallized solids which apparently contain the added hydrogen sulphide as such ...

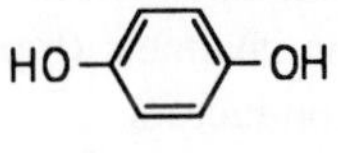

HO–⟨benzene ring⟩–OH

1

However, it was not before the mid-1940s that Palin and Powell [10] solved the crystal structure of the hydroquinone clathrates, using X-ray crystallography. The hydroquinone molecules, held by hydrogen bonding, form a three-dimensional network with cage-like cavities in the lattice. Incorporated guest compounds include SO_2, CH_3OH, HCl, H_2S, HCN and the noble gases Ar, Kr and Xe [10d].

5.2.3.3 Dianin

In 1914, the Russian chemist Dianin synthesized 4-(4-hydroxyphenyl)-2,2,4-trimethylchroman (**2**), which was later named after him [11]. He noticed that on recrystallization some organic solvents would stick to it. Powell and Wetters [12] demonstrated by X-ray crystal structure analysis that these host–guest compounds are of the clathrate type. Inclusions include, among others, Ar, SO_2, NH_3, benzene, ethanol and chloroform [13].

In its lattice, the clathrand forms a hexameric system of hydrogen bonding, which encloses the cavity in two directions. Figure 2 shows schematically a typical dianin-cage inclusion compound. The black dots symbolize the hydroxyl groups linked by hydrogen bonding and the cylinders represent the remainder of the dianin molecule. Figure 3 shows an example of an X-ray structure of a dianin clathrate.

In subsequent years, the influence of heteroatoms and substituents on the formation of clathrates was investigated by systematically introducing structural changes in dianin. As was demonstrated by Baker *et al*. [13b], and also by MacNicol *et al*.[13d], the inclusion behaviour is lost in the compounds **3a–c**. A similar effect was observed when substituents were varied in the sulphur analogues of dianin, **4a** and **4b** [13].

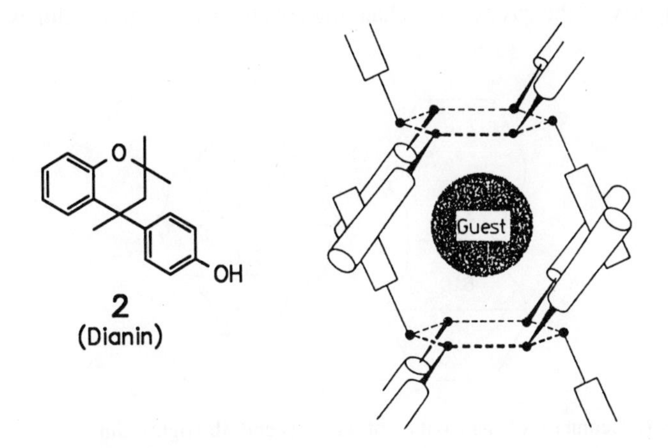

Figure 2. Cavity in the lattice of dianin (schematic)

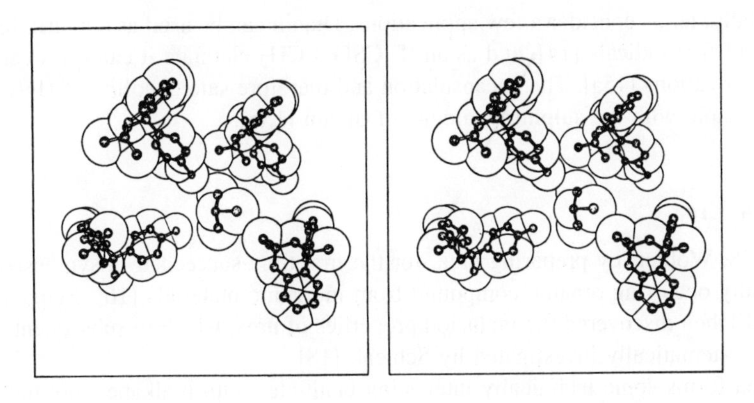

Figure 3. X-ray structure of the dianin•$CHCl_3$ clathrate

3a : $R^1 = R^2 = H$; $R^3 = CH_3$

3b : $R^1 = R^3 = H$; $R^2 = CH_3$

3c : $R^2 = R^3 = H$; $R^1 = CH_3$

4a : R = H

4b : R = CH_3

The introduction of only one additional methyl group leads to a dramatic change in the geometry of the cavity, thus changing the 'hour-glass' to a 'Chinese lantern' (Figure 4).

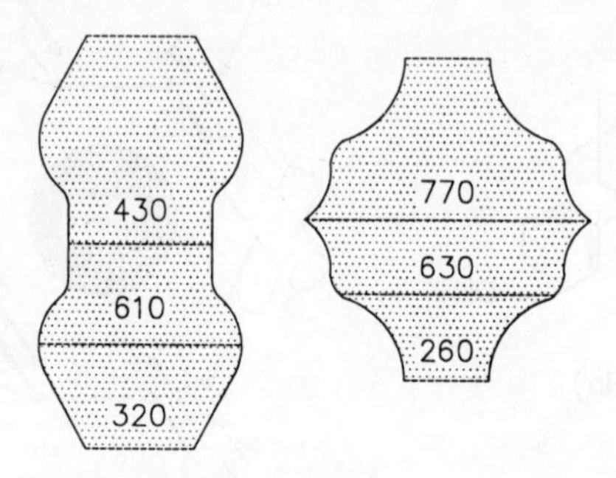

Figure 4. Geometry of the cavities of **4a** (left) and **4b** (right); dimensions in pm

Finally, let us consider a few applications. Dianin can be used as a matrix for the study of free radicals [14], and as an $(F_3CSO_2)_2CH_2$ clathrate it catalyses cationic polymerizations [15a]. The encapsulation and therefore safe handling of $Hg(CH_3)_2$ are possible with the sulphur analogue of dianin [15b].

5.2.3.4 Urea

In 1828, Wöhler, by preparing urea, for the first time succeeded in synthesizing a naturally occurring organic compound from inorganic materials [16]. Bengen [17] in 1940 then discovered the inclusion properties of urea, which in subsequent years were systematically investigated by Schlenk [18].

Urea forms some technically interesting clathrates with n-alkanes, normal fatty acids, halogenated hydrocarbons and many other guest compounds.

Normally, urea crystallizes tetragonally. The clathrates, though, which crystallize in the presence of hydrocarbons, have a hexagonal crystal structure. From X-ray crystal structure analysis it appears that in the lattice of 'hexagonal urea' channel-like cavities with a diameter of 520 pm are present, in which n-alkanes are preferentially incorporated (Figure 5). Urea is therefore suitable for the separation of linear and branched hydrocarbons (as clathrate guests) on an industrial scale.

A fundamental feature of the channels in the urea lattice is the fact that the urea molecules making them up are aligned either into left-handed or right-handed helices, which are like image and mirror image (see Figure 6). The achiral urea spontaneously crystallizes into two enantiomorphic lattices and can therefore be used to separate racemates into their enantiomers by clathrate formation.

In 1960, Brown and White [19] reported the polymerization of buta-1,3-diene in the lattice of urea and thiourea. Since then, a rich literature has been produced on 'inclusion polymerization' with the help of molecules forming channel-like cavities [20].

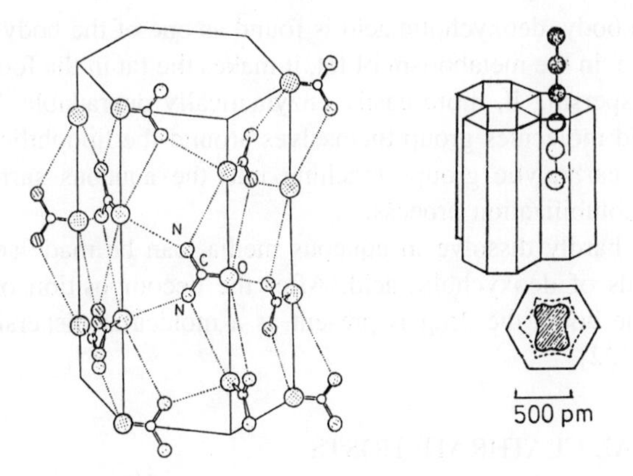

Figure 5. Channel structure in the urea lattice; on the right with a guest (n-alkane) and scale included

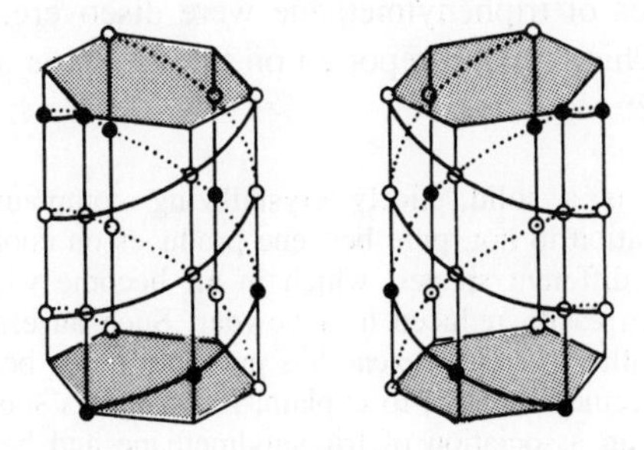

Figure 6. The two enantiomeric urea lattices (schematic)

5.2.3.5 Cholanic acids

The cholanic acids were discovered in 1916 by Wieland [21], but not until 1936 did Kratky [2] succeed in elucidating their structures.

Deoxycholic acid (**5**) alone crystallizes only with difficulty, but forms easily crystallized adducts with fatty acids, alkanes, xylene, naphthalene, benzoic acid, phenol, camphor, cholesterol and others. Kratky and Giacomello were able to show that the host lattice of deoxycholic acid contains channel-like cavities, in which the guest compounds can be incorporated [2].

HO — CO_2H — HO

5

In the human body, deoxycholic acid is found as one of the body's own steroids in the gall. There, in the metabolism of fat, it makes the fat in the food more soluble and thus, by dispersing it, more easily enzymatically degradable. To do this, the deoxycholic acid molecules group themselves around the lipophilic substance, the hydroxylic and carboxylic groups reaching into the aqueous surroundings, thus facilitating the solubilization process.

Drugs which hardly dissolve in aqueous media, can be made soluble as host–guest compounds of deoxycholic acid. After the decomposition of the inclusion compound in the body, the drug is present as a molecular dispersion and can be resorbed easily [22].

5.2.4 TRIGONAL CLATHRATE HOSTS

5.2.4.1 Triphenylmethane and its derivatives

The inclusion properties of triphenylmethane were discovered in the last century [23]. Kekulé and Franchimont [24] reported on the synthesis and the properties of this compound in 1872:

> Triphenylmethane is a solid, nicely crystallizing compound. It melts at 92.5 °C ... A solution in hot, pure benzene produces on cooling large, clear crystals of totally different shapes, which in air become white and opaque and which are then easily reduced to a powder. Such an efflorescence of a hydrocarbon crystallized from benzene has probably never been observed, so far, and initially seemed difficult to explain. Experiments soon demonstrated the crystals to be an association of triphenylmethane and benzene, and that they contain exactly one mole of benzene to every mole of triphenylmethane. The compound melts at 76 °C, slowly losing benzene, and then goes on to melt at 92.5 °C, like the triphenylmethane crystallized from alcohol ....

Anschütz in 1886 confirmed the 1 : 1 stoechiometry by heating the adduct to 100° and recording a loss in weight of 25.36% (theory: 24.22%) [25].

The inclusion of aniline was discovered by Lehmann in 1881 [26] and that of pyrrole was described by Hartley and Thomas in 1906 [27]. Both of these adducts also have a host–guest ratio of 1 : 1. In 1916, Norris described the adducts of triphenylcarbinol and triphenylchloromethane with both CCl_4 and acetone [28].

From the mid-1950s onwards, Driver [29] investigated derivatives of 4,4′-dihydroxytriphenylmethane with additional nitro, amino and bromo substituents and found them to be effective host compounds; they can incorporate a whole range of guest molecules with stoichiometries ranging from 2 : 1 to 1 : 2. These hydroxytriphenylmethane derivatives [30] build layered lattices, where the hydroxyl groups provide hydrogen bonding, and incorporate guests such as benzene, toluene and *p*-xylene. On the other hand, channel-like lattice cavities are also found, which can accommodate n-alkanes, n-alkenes, 2,2,4-trimethylpentane, diisobutene and squalene.

Triphenylmethane and its substituted derivatives, both in solution and in the solid state, all have the conformation of a three-bladed propeller (**6**). Triphenylmethane

6
X = H, Cl, OH

Table 1. Guest selectivities of triphenylmethane

Solvent mixture	Percentages of included guests	Overall host : guest stoichiometry
Benzene–aniline	35+65	1 : 2
Thiophene–aniline	45+55	1 : 1
Pyrrole–aniline	28+72	1 : 3
Benzene–thiophene	72+28	3 : 1
Benzene–pyrrole	38+62	1 : 2
Benzene–toluene*	80+20	5 : 1
Benzene–xylene	100+0	1 : 0

* On recrystallization from pure toluene, no inclusion is observed.

itself displays varying guest selectivities when recrystallized from solvent mixtures [31] (Table 1).

5.2.4.2 Trimesic acid (TMA)

Trimesic acid (benzene-1,3,5-tricarboxylic acid, **7**) is able, by itself or together with other molecules (e.g. water), to form extensive networks of molecules linked by hydrogen bonding, which as host lattices are suitable for the formation of channel inclusion compounds [32]. For example, six trimesic acid molecules build a hexagonal structure ('chicken-wire,' cf. Figure 7).

This type of structure is found in crystals of α-TMA [33], as well as in the isomorphic inclusion compounds $TMA \cdot 0.7H_2O \cdot 0.09HI_5$ ('$TMA \cdot I_5$', for short), $TMA \cdot 0.7H_2O \cdot 0.167HIBr_2$ and $TMA \cdot 0.7H_2O \cdot 0.103HBr_5$ ('$TMA \cdot Br_5$') [32]. These TMA–polyhalogen clathrates are prepared by recrystallizing TMA from water–HI_3, water–equimolar I^- and Br_2 and water–HBr_3, respectively. In the crystal, each of the hexagonal networks is penetrated by three identical frameworks in a chain-like manner (cf. Figure 8).

The polyhalogen ions (I_5^-, Br_5^-, IBr_2^-) are located in channel-like cavities, while the counter ions (H^+) are bound to water molecules, outside the channels and between the TMA networks. There is an analogy with the corresponding cyclodextrin inclusion compounds, where the metal cations are coordinated to water between the cyclodextrin molecules.

In a second group of clathrates, TMA and water form a network of molecules held together by hydrogen bonding (cf. Figure 9). Thus, on recrystallization of TMA from water, α-TMA and two hydrates, TMA•$3H_2O$ and TMA•$5/6H_2O$ [34] are obtained, the first hydrate easily losing water. By adding picric acid (PA), the more stable inclusion compound TMA•H_2O•2/9PA crystallizes.

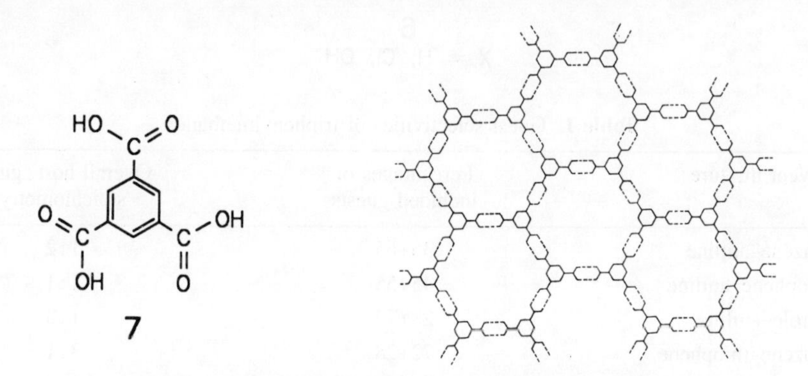

Figure 7. Hexagonal network of trimesic acid (TMA, **7**)

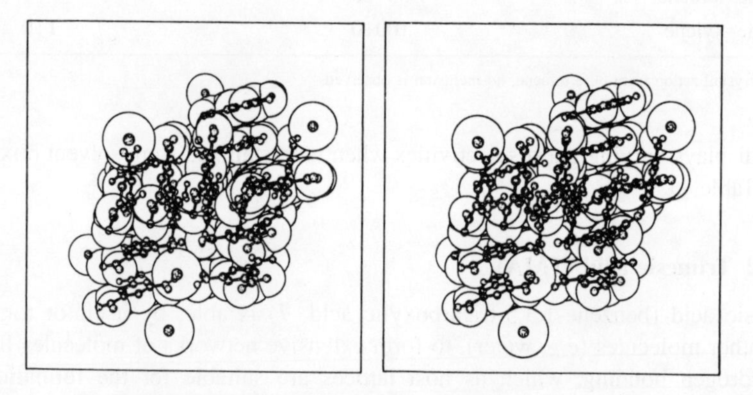

Figure 8. The 'TMA•$1/6Br_2$' clathrate (stereoview)

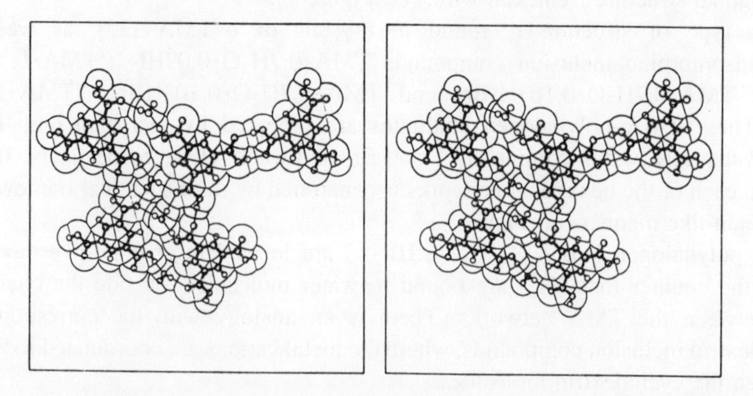

Figure 9. Structure of the TMA–H_2O network (stereoview)

Of a third kind of clathrate, where TMA and other molecules form a network, only one example is known: TMA•DMSO [35]. Here, the TMA molecules are woven into strips by individual hydrogen bonds of their carboxylic groups. Perpendicular to these, neighbouring TMA molecules form hydrogen bonds to the oxygen atoms of the dimethyl sulphoxide molecules (cf. Figure 10).

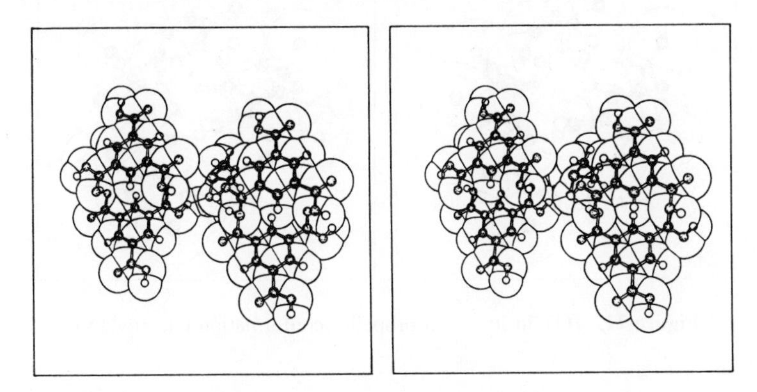

Figure 10. Structure of the TMA•DMSO clathrate (stereoview)

The TMA–DMSO adduct is special in that the DMSO molecules participate in the building of the cavities in the lattice, and also in the occupation of the channel-like gaps with parts of the molecule (sulphur and the two methyl groups).

5.2.4.3 Tri-*o*-thymotide (TOT)

Following Spallino's and Provenzal's [36] investigations in 1909 on the dehydration of 2-hydroxy-6-methyl-3-isopropylbenzoic acid (*o*-thymotic acid), Baker, Gilbert and Ollis succeeded in 1952 in separating, by fractional crystallization, the di-*o*-thymotide and tri-*o*-timotide (**8**) produced during this reaction [37]. During these recrystallizations, they found inclusions of n-hexane, ethanol, methanol, *m*- and *p*-xylene, benzene, carbon tetrachloride, chloroform and dioxane.

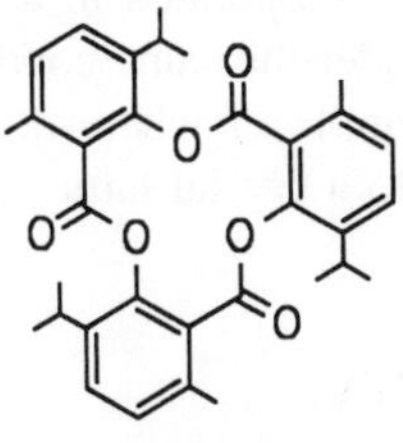

8

According to measurements of the dipole moment and NMR spectroscopic investigations, TOT adopts, in the solid state, a chiral conformation, i.e. all three carbonyl oxygen atoms are found on the same side of the twelve-membered ring [38] (Figure 11).

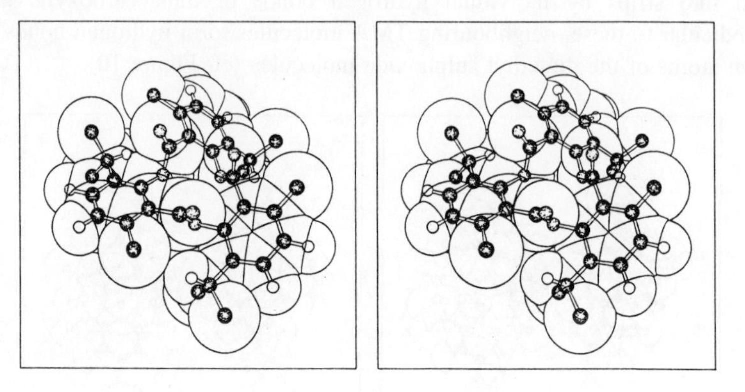

Figure 11. TOT in its chiral propeller conformation (stereoview)

An unusual aspect about the host lattice of TOT is its ability to form channel- and cage-like cavities, depending on the shape and size of the guest molecules. Channel inclusion compounds have monoclinic, triclinic or hexagonal structures, cage inclusion compounds are trigonal and guest-free TOT crystallizes orthorhombically.

An interesting property of TOT, which normally crystallizes as a racemate, is its spontaneous enantiomeric separation on forming inclusion compounds with n-hexane, benzene and chloroform [39].

If a racemic mixture of the guest component is used, then the guest enantiomers will be included to different extents. Thus, it was possible to separate racemic *sec*-butyl bromide into its antipodes using enantiomerically pure TOT. Arad-Yellin *et al*. [40] were able to show that the cage clathrates exhibit a higher enantiomeric selectivity than the channel inclusion compounds. This is understandable, since in a cage-like cavity the guest molecule is sheathed more fully, hence the higher 'chiral recognition factor.' Another interesting behaviour found by Arad-Yellin *et al*. [41] concerns the photochemical properties of the TOT clathrates of (*E*)- and (*Z*)-stilbene. In solution, both stereoisomers undergo (*E*)/(*Z*)-isomerization. However, whereas (*E*)-stilbene incorporated in a TOT lattice remains unchanged when irradiated, (*Z*)-stilbene, under the same conditions, isomerizes cleanly to the (*E*)-isomer. This is a good example of where a chemical compound changes its reactivity when incorporated into a crystal lattice.

5.2.4.4 Cyclotriveratrylene (CTV)

In 1915, Robinson reacted veratrol and formaldehyde under basic conditions and obtained a high-melting crystalline compound, $(C_9H_{10}O_2)_n$ [42]. The fact that this was the trimer (n=3) was only substantiated in 1963 by work by Lindsey [43], Erdtman *et al*. [44] and Goldup *et al*. [45].

The cyclotriveratrylene molecule (**9**) has C_{3v} symmetry. Lüttringhaus and Peters [46] were the first in 1966 to separate chiral cyclotriveratrylene derivatives into their enantiomers. Using NMR spectroscopy it was shown that CTV is found in a stable crown conformation (analogous to 10) (cf. Section 4.3.1).

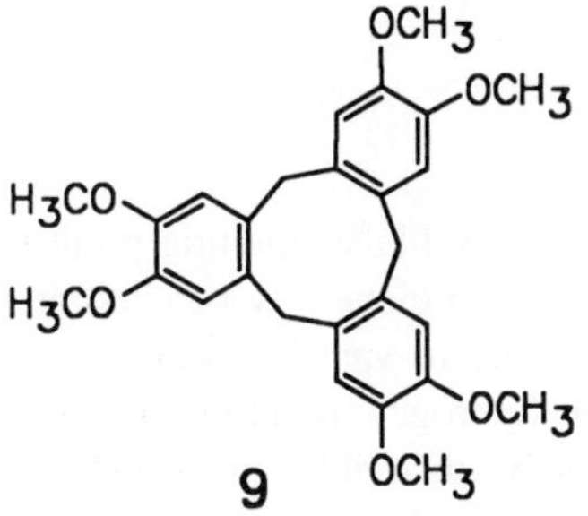

H_3CO H_3CO OCH_3 OCH_3 OCH_3 OCH_3

9

HO HO HO OH OH OH

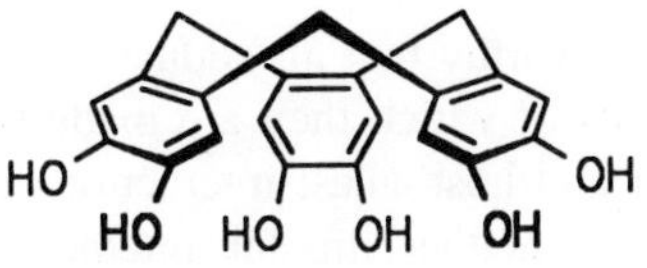

10

The multitudinous inclusion capabilities of cyclotriveratrylene were discovered by Bhagwat *et al.* [47]. Subsequently, clathrates of CTV with benzene, chlorobenzene, toluene, chloroform, acetone, carbon disulphide, acetic acid, thiophene, decalin, butan-2-one and ethanol were found. Derivatives of CTV such as cyclotricatechylene (**11**) and trithiacyclotriveratrylene (**12**) also show clathrand properties.

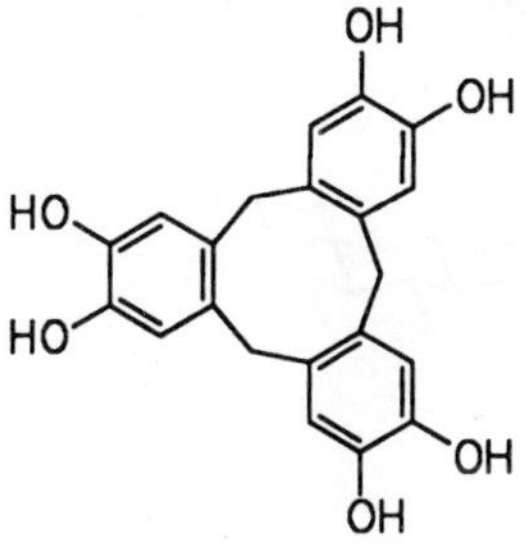

HO HO OH OH OH OH

11

H_3CO H_3CO S S S OCH_3 OCH_3 OCH_3 OCH_3

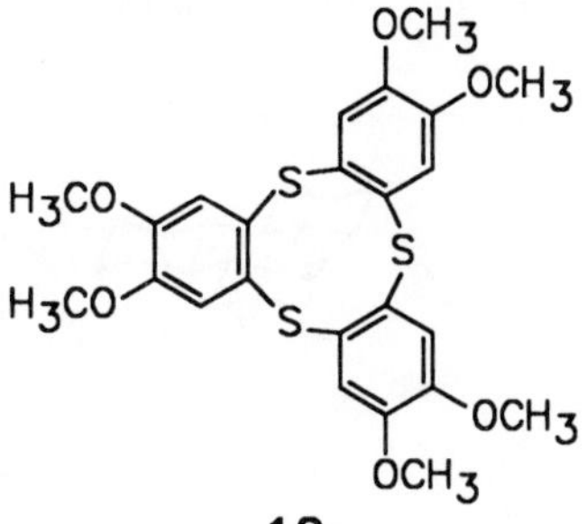

12

5.2.4.5 Perhydrotriphenylene (PHTP)

In 1923, Schrauth and Görig investigated the preparation of perhydro-9,10-benzophenantrene [48]; 40 years later, Farina hydrated dodecahydrotriphenylene, and from the mixture of stereoisomers obtained, isolated *trans–anti–trans–anti–trans–anti*-perhydrotriphenylene (all-*trans*-**13**) in 60% yield and investigated its properties [49].

PHTP forms many stable channel inclusion compounds with spherical, planar and linear guest molecules. Potential guests include aliphatic hydrocarbons and haloalkanes (n-heptane, 1-chloropentane), mono- and dicarboxylic acids (palmitic, stearic and adipic acids), alcohols, CCl_4, $CHCl_3$, cyclohexane, benzene and tetralin.

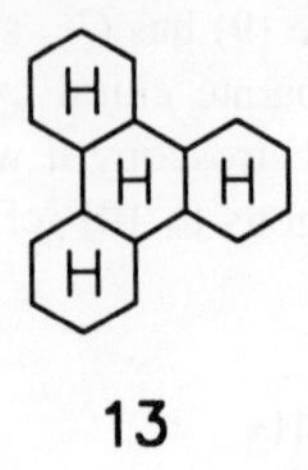

It is noteworthy that all adducts have a markedly higher melting point than the components of which they are made (thermodynamic stability of clathrates). The host–host and host–guest interactions are due only to van der Waals interactions, and contain neither dipolar interactions nor hydrogen bonding. In contrast to many other clathrands whose lattices are mostly invariable, the PHTP clathrates have many crystal conformations, depending on the type of guest molecule involved. Farina *et al*. discovered that PHTP forms stable inclusion compounds with linear macromolecules, such as polyethylene, (*E*)/(*Z*)-polybuta-1,4-diene and polyoxyethylene glycol.

All-*trans*-perhydrotriphenylene ('equatorial isomer') exists in two enantiomeric, optically active forms, both of which have been prepared (Figure 12). Crystals of either antipode offer to guest molecules an asymmetric environment. In such asymmetric channels (*E*)-penta-1,3-diene can be polymerized isotactically (i.e. the same configuration on all asymmetric carbon atoms) to polymethylbutadiene, obviously under the influence of the chiral structure of the host substance.

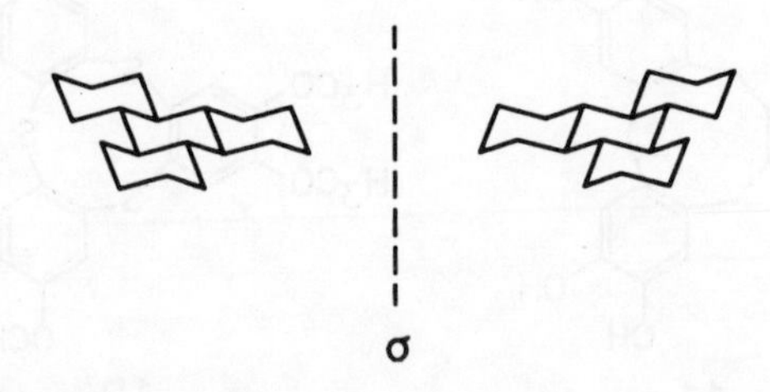

Figure 12. The enantiomers of all-*trans*-perhydrophenylene

5.2.4.6 Cyclophosphazenes

Searching for new polymers, Allcock in 1963 prepared a spirocyclophosphazene (**14**) from hexachlorocyclophosphazene and catechol [50]. On recrystallizing from benzene, he discovered the clathrand properties of tris(*o*-phenylenedioxy)-tricyclophosphazene (**14**). Further investigations showed the incorporation of aliphatic and aromatic hydrocarbons, alkenes, ethers, chloroalkanes, ketones, nitriles and alcohols. By X-ray crystal structure analysis Allcock found the guest-free compound to have a monoclinic or triclinic crystal lattice, while host–guest compounds crystallize hexagonally.

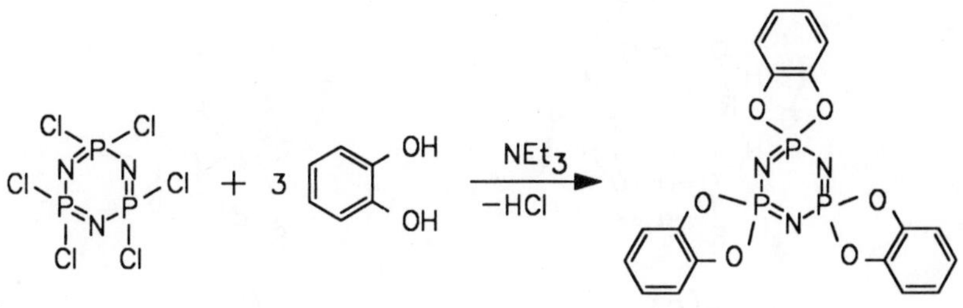

14

An unexpected discovery was the fact that the clathrand is also able to absorb liquids directly, such as n-heptane, ethyl acetate or chloroform. Even when brought into contact with only the vapour of an organic compound, the organophosphazene will incorporate it into its lattice. The guest molecules are embedded in channel-like cavities. Tris(*o*-phenylenedioxy)tricyclophosphazene is therefore suitable for 'inclusion polymerizations': Finter and Wegner in 1979 reported the polymerization of butadiene and vinyl chloride [50f].

Other active clathrands of the same cyclophosphazene type have been prepared and investigated, e.g. tris(2,3-naphthalenedioxy)cyclotriphosphazene (**15**) and tris(*o*-phenylenediamino)cyclotriphosphazene (**16**).

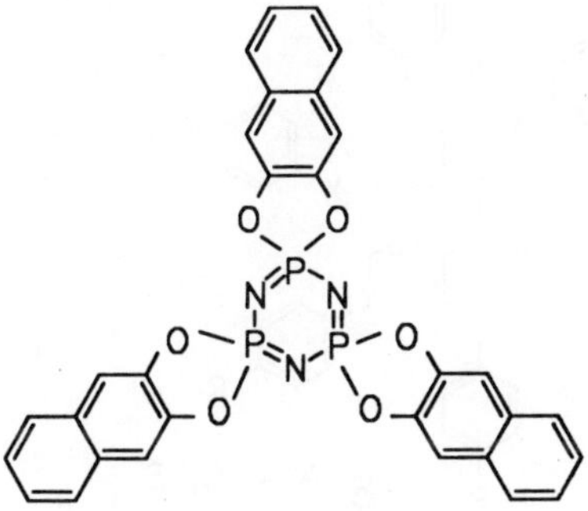

15

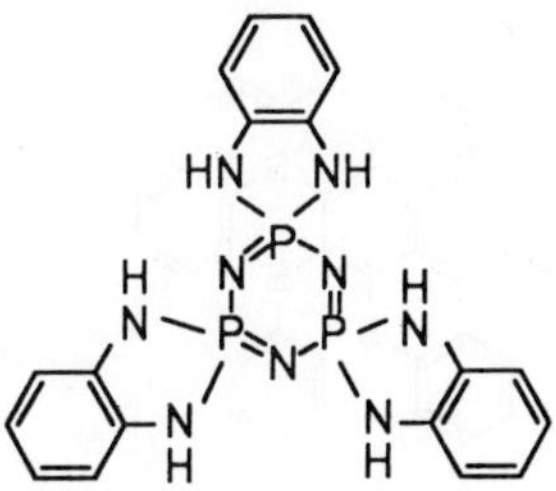

16

5.2.5 NEW CONCEPTS FOR THE CONSTRUCTION OF CLATHRATE HOSTS

5.2.5.1 'Hexahosts'

MacNicol recognized the structural importance of the system of hexameric hydrogen bonding, which in the lattice of dianin lines the cage cavities in two directions, and which is also present in phenol and hydroquinone inclusions [51]. Figure 13 shows that the dimensions in this hexameric bonding system are nearly the same as in a hexasubstituted benzene.

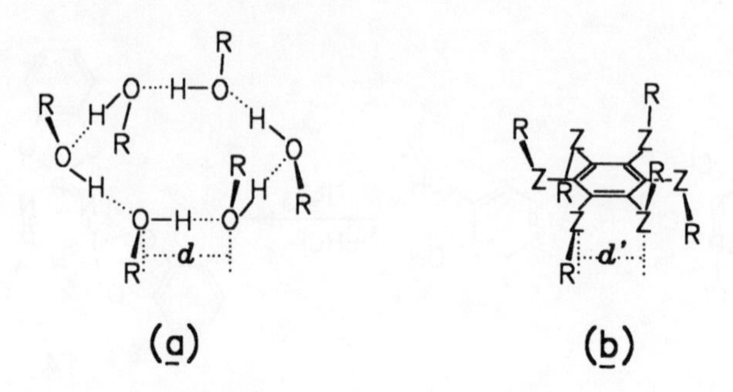

Figure 13. (*a*) Hexameric hydrogen-bonding system in the dianin lattice and (*b*) the measurements in a hexasubstituted benzene

Based on this analogy, MacNicol constructed the 'hexahosts' as a new type of host, e.g. **17** and **18**. Hexahosts form clathrates with, among others, acetone, dioxane, chloroform, carbon tetrachloride and derivatives of benzene. When recrystallizing from solvent mixtures, e.g. *o*- and *p*-xylene, different guest selectivities are found, depending on the structure of the host. Hexahosts have C_3 symmetry.

17   **18** X = O, S; R = H, Bu^t, Pr^i, OH

5.2.5.2 Organic onium salts

When attempting to recrystallize the azulene bisammonium compound **19**, Vögtle *et al.* in 1983 discovered that this substance would incorporate many types of solvents, usually in stoichiometric ratio [52]. With this, a new type of clathrate was found.

The many possible inclusions in these clathrands are based on general characteristics, observed in many clathrate hosts, such as bulk and limited conformational mobility and the stability of the hosts's ionic lattice. Moreover, the flexible onium side-arms are almost certainly responsible for the ability to incorporate guests of various shapes and sizes. X-ray crystal structure analysis

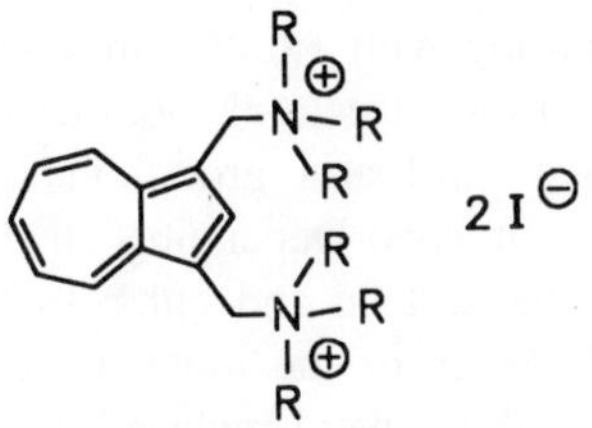

19

demonstrated the unusual adaptability ('induced fit') of host compound **19** to the steric requirements of the guest molecules.

Further organic onium clathrands, with modified guest selectivities, were prepared by systematically varying the following structural elements [53]:

— exchange of the alkyl rest on the ammonium nitrogen;
— modification of the onium side-arms;
— exchange of the anchoring group (azulene);
— variation of the counter ions.

From this work the following information was gained:

— similar alkyl residues on the ammonium nitrogen favour the clathrate formation, as do voluminous substituents in positions 1 and 3 on the anchoring group (azulene, benzene);
— exchange of iodide for the smaller bromide reduces the inclusion capability;
— extension or removal of the methylene spacer unit between the anchoring group and the ammonium nitrogen also has a negative effect.

5.2.5.3 Clathrands of the wheel-and-axle type

In 1968, Toda *et al*. reported on the clathrate properties of the diacetylene diol **20**, an elongated molecule with two voluminous groups on each of the terminal sp^3 carbons ('wheel-and-axle') [54].

20

This clathrand forms stoichiometric host–guest compounds with alkyl halides, arenes, alkenes, alkynes, aldehydes, ketones, esters, ethers, amines, nitriles, sulphoxides and sulphides. Essential structural characteristics, which favour clathrate inclusions, are the ability to form hydrogen bonding, the possibility for π-interactions to the arene rings and the basic linear structure of the diacetylene unit.

From this, Hart *et al*. developed a new concept for the structural design of host compounds [55]. Starting with 1,6-dihydroxy-1,1,6,6-tetraphenylhexa-2,4-diyne (**20**), he introduced various structural modifications. First, he replaced the hydroxy functionality with a third aryl group, i.e. phenyl, *p*-biphenyl and 4-methoxyphenyl as bulky substituents. Depending on the type of the aryl residue, inclusions of benzene, toluene, xylene and chloroform were observed. In order to determine whether the linearity of the molecule is of any importance to the inclusion behaviour, the sp carbons were replaced with sp^2 and sp^3 carbons (e.g. **21**). In spite of the great structural variations introduced, the new compounds still showed clathrand character.

$$Ph_3CCH_2CH=CHCH_2CPh_3$$

21

In a second step, heteroatoms or functional groups were included in the linear part of the molecule (cf. **22** – **24**). Whereas **22** and **23** incorporate toluene, a 1 : 1 clathrate is obtained with **24** and benzene. The phosphorus analogue 25 also forms clathrates with aromatic hydrocarbons [56].

$$Ph_3C-X-CH_2CH_2-X-CPh_3 \qquad Ph_3CCH=N-N=CHCPh_3$$

22 X = O, NH **23**

$$Ph_3CC(=O)-CH_2CH_2-C(=O)CPh_3$$

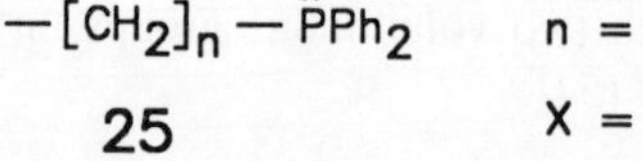

24

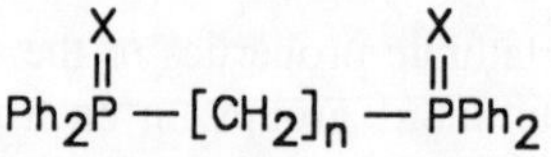

$$Ph_2P(=X)-[CH_2]_n-P(=X)Ph_2 \qquad n = 2, 3 \quad X = S, Se$$

25

Finally, the bulky trityl group was replaced with the triptycyl-(aryl) residue. Clathrand **26** forms clathrates with toluene and *m*- and *p*-xylene.

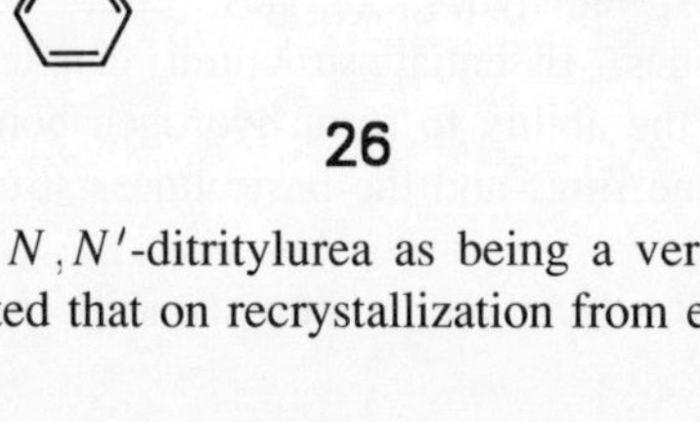

(triptycyl)–CH=N–N=CH–(triptycyl)

26

Hart *et al*. described *N*,*N*′-ditritylurea as being a versatile clathrand [57]. In 1925, Helferich [58] noted that on recrystallization from ethanol

... 2 molecules of crystal alcohol are retained.

The recipe for a successful clathrand design, therefore, seems to be to build an elongated molecule with bulky substituents on either end. By adding functional groups or heteroatoms, coordinative host–guest interactions, such as hydrogen bonding, are made possible. The bulky end-groups on the one hand prevent host–host hydrogen bonding and on the other favour the formation of intermolecular cavities in the lattice of longitudinally oriented molecules.

Toda *et al*. successfully prepared chiral 'wheel-and-axle' type clathrands and investigated their properties [59]. Thus, with optically active 1,6-dihydroxy-1,6-bis(*o*-halogenophenyl)-1,6-diphenylhexa-2,4-diyne (**27**), it was possible to separate the enantiomers of 3-methylcyclohexanone (**28**), 3-methylpentanone (**29**) and 5-methylbutyrolactone (**30**).

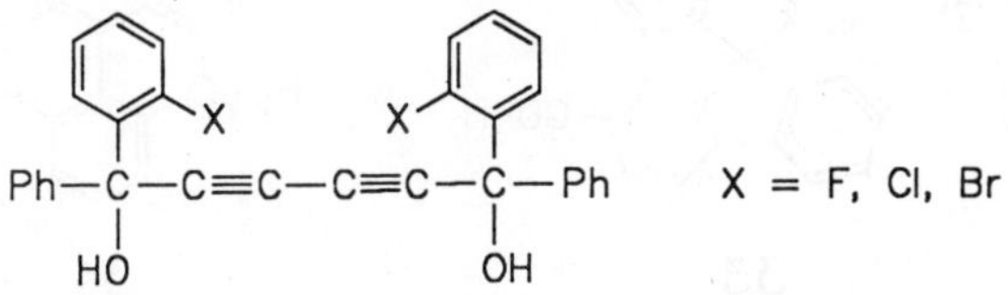

Ph—C(OH)(o-X-C₆H₄)—C≡C—C≡C—C(OH)(o-X-C₆H₄)—Ph   X = F, Cl, Br

27

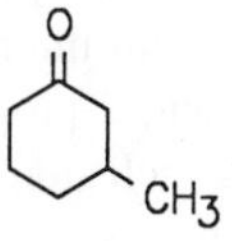

28 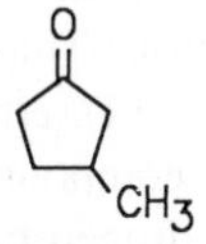   **29** 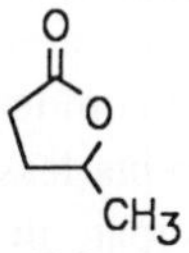   **30**

Toda also resolved optically active tertiary acetylene alcohols (propargyl alcohols, **31**) into their enantiomers, using brucine to form 1 : 1 clathrates (**32**; Scheme 1).

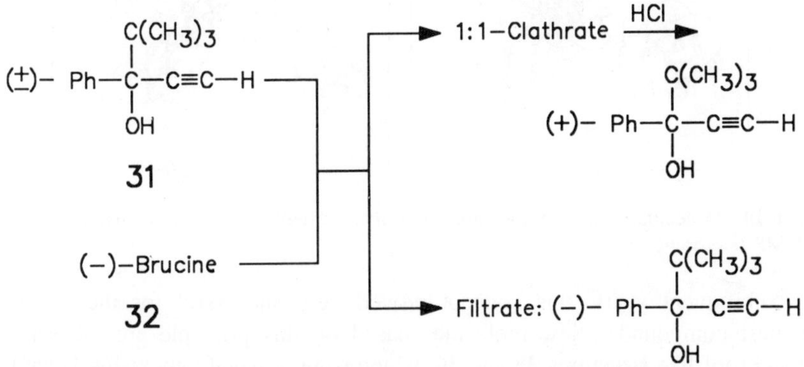

(±)- Ph—C(C(CH3)3)(OH)—C≡C—H **31** + (−)-Brucine **32** → 1:1-Clathrate —HCl→ (+)- Ph—C(C(CH3)3)(OH)—C≡C—H; Filtrate: (−)- Ph—C(C(CH3)3)(OH)—C≡C—H

Scheme 1. Resolution of racemates with the help of clathrates

5.2.5.4 Clathrates suitable for intermolecular interactions

Apart from a bulky skeleton, which is responsible for the formation of cavities, the host molecules of this type also contain functionalities suitable for selective hydrogen bonding [60]. In contrast to the classical hydroquinone and dianin clathrates, where hydrogen bonding between host molecules represents an essential structural characteristic of the lattice, in this new type of host, coordinative host–guest interactions are also found. A high selectivity in the clathrate formation can be expected from this combination of both topology and coordination.

Examples of molecules which follow this new concept ('coordination clathrates') are 9,9′-spirobifluorene-2,2′-dicarboxylic acid (**33**) and 1,1′-binaphthyl-2,2′-dicarboxylic acid (**34**).

HO$_2$C, CO$_2$H — **33** HO$_2$C, CO$_2$H — **34**

When recrystallized from polar solvents, such as alcohols, acids and amides, which are able to form hydrogen bonds to the host's carboxylic groups, the clathrates obtained are mostly stoichiometric. Thus, the spirofluorene derivative **33** incorporates dimethylformamide (Figure 14), ethanol and 2-propanol in a 1 : 1 ratio. However, **33** also accepts less polar compounds as inclusion partners, even if they are not able to participate in the hydrogen bonding.

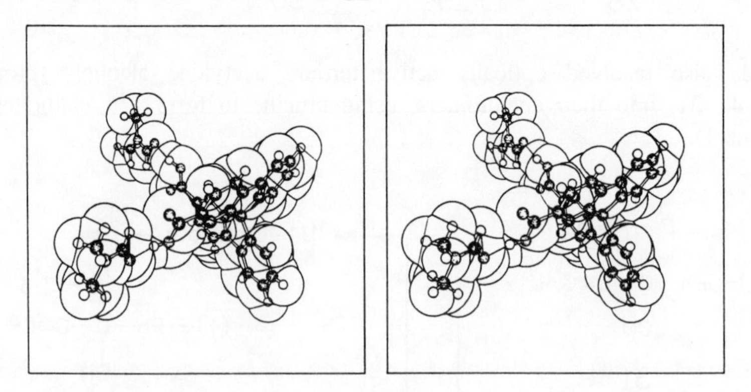

Figure 14. Molecular structure and molecular arrangement in the 1 : 1 inclusion compound **33**•DMF (stereoview)

In subsequent years, this concept proved very successful for the design of new host compounds. New molecules based on this principle are the scissors-like and roof-like structures **35** and **36**, which show a specific inclusion behaviour, depending on the basic skeleton and the type of functionalities attached [61].

35

35a: $R^1 = R^2 = H$
b: $R^1 = R^2 = Cl$
c: $R^1 = R^2 = CH_3$
d: $R^1 = Cl$, $R^2 = CN$
e: $R^1 = Cl$, $R^2 = COOH$
f: $R^1 = H$, $R^2 = COOH$

36

36a: $R^1 = R^4 = COOH$
$R^2 = R^3 = H$

36b: $R^1 = R^3 = COOH$
$R^2 = R^4 = H$

Inclusion compounds are not only of theoretical interest, but have also found various practical applications, e.g. in the separation of racemates with the help of urea adducts, something also successfully achieved with cyclodextrins. In pharmacology, inclusion compounds have long been of interest, first to protect drugs from autoxidation and other forms of decomposition, and second to effect rapid resorption by the body.

In Japan, industrial applications of cyclodextrin inclusion compounds currently command a great deal of interest. Apart from the fixation of volatile fragrances and drugs, there is particularly the hope that inclusion compounds of pesticides will be safer and easier to handle.

Finally, a reminder that smokers turn inclusion compounds of menthol cigarettes into 'blue haze.'

References

1. W.Schlenk, Jr, *Chem. Unserer Zeit*, **3**, 120 (1969).
2. O.Kratky and G.Giacomello, *Monatsh. Chem.*, **69**, 427 (1936).
3. F.Cramer, *Angew. Chem.*, **64**, 437 (1952).
4. G.Reddelien, Z. *Angew. Chem.*, **36**, 515 (1923).
5. H.M.Powell, *J. Chem. Soc.*, 61 (1948).

5a. J.L.Atwood, J.E.D.Davies and D.D.MacNicol (Eds), *Inclusion Compounds*, Academic Press, New York, 1984.

6. (a) M.v.Stackelberg, *Naturwissenschaften*, **86**, 327, 359 (1949); (b) M.v.Stackelberg and H.R.Müller, *Naturwissenschaften*, **38**, 456 (1951); **39**, 20 (1952).
7. F.Wöhler, *Liebigs Ann. Chem.*, **69**, 297 (1849).
8. A.Clemm, *Liebigs Ann. Chem.*, **110**, 357 (1859).
9. F.Mylius, *Ber. Dtsch. Chem. Ges.*, **19**, 999 (1886).
10. (a) D.E.Palin and H.M.Powell, *J. Chem. Soc.*, 208 (1947); (b) D.E.Palin and H.M.Powell, *Nature (London)*, **156** 334 (1945); (c) D.E.Palin and H.M.Powell, *J. Chem. Soc.*, 571, 815 (1948); (d) H.M.Powell, *J. Chem. Soc.*, 298, 300, 468 (1950).

11. A.P.Dianin, *J. Russ. Phys. Chem. Soc.*, **46**, 1310 (1914) [*Chem. Zentralbl.*, **I**, 1063 (1915)].
12. H.M.Powell and B.D.Wetters, *Chem. Ind. (London)*, 256 (1955).
13. (a) W.Baker, A.J.Floyd, J.F.W.McOmie, G.Pope, A.S.Weaving and J.H.Wild, *J. Chem. Soc.*, 2010 (1956); (b) W.Baker, J.F.W.McOmie and A.S.Weaving, *J. Chem. Soc.*, 2018 (1956); (c) J.L.Flippen, J.Karle and I.L.Karle, *J. Am. Chem. Soc.*, **92**, 3749 (1970); (d) A.D.U.Hardy, J.J.McKendrick and D.D.MacNicol, *J. Chem. Soc., Chem. Commun.*, 972 (1974).
14. L.K.Kispert and J.Pearson, *J. Phys. Chem.*, **76**, 133 (1972).
15. (a) J.E.Kropp, M.G.Allen and G.W.B.Warren, *Ger. Offen.*, 2 012 103 [*Chem. Abstr.*, **74**, 43074 (1971)]; (b) R.J.Cross, J.J.McKendrick and D.D.MacNicol, *Nature (London)*, **245**, 146 (1973).
16. F.Wöhler, *Ann. Phys.*, **12**, 253 (1828).
17. H.Bengen, *Angew. Chem.*, **63**, 207 (1951).
18. (a) W.Schlenk, Jr, *Liebigs Ann. Chem.*, 1145 (1973); (b) W.Schlenk, Jr, *Liebigs Ann. Chem.*, **565**, 204 (1949); (c) W.Schlenk, Jr, *Angew. Chem.*, **62**, 299 (1950); (d) W.Schlenk, Jr, *Fortschr. Chem. Forsch.*, **2**, 92 (1951).
19. J.F.Brown and D.M.White, *J. Am. Chem. Soc.*, **82**, 5671 (1960).
20. (a) H.Clasen, *Z. Elektrochem.*, **60**, 983 (1956); (b) A.Colombo and G.Allegra, *Macromolecules*, **4**, 579 (1971); (c) M.Miyata and K.Takemoto, *Angew. Makromol. Chem.*, **55**, 191 (1976); (d) Y.Chatani and S.Kuwata, *Macromolecules*, **8**, 12 (1975); (e) M.Farina, *Makromol. Chem.*, **4**, 21 (1981).
21. H.Wieland and H.Sorge, *Hoppe-Seylers Z. Physiol. Chem.*, **97**, 1 (1916).
22. K.H.Frömming, *Pharm. Unserer Zeit*, **2**, 109 (1973).
23. J.E.D.Davies, P.Finocchiaro and F.H.Herbstein in *Inclusion Compounds*, Vol.2 J.L.Atwood, J.E.D.Davies and D.D.MacNicol, (Eds.), Academic Press, New York, 1984 p.418.
24. A.Kekulé and A.Franchimont, *Ber. Dtsch. Chem. Ges.*, **5**, 906 (1872).
25. R.Anschütz, *Liebigs Ann. Chem.*, **235**, 208 (1887).
26. O.Lehmann, *Z. Kryst. Min.*, **5**, 472 (1881).
27. H.Hartley and N.G.Thomas, *J. Chem. Soc.*, 1013 (1906).
28. J.F.Norris, *J. Am. Chem. Soc.*, **38**, 702 (1916).
29. (a) J.E.Driver and S.F.Mok, *J. Chem. Soc.*, 3914 (1955); (b) J.E.Driver and T.F.Lai, J. Chem. Soc., 3219 (1958).
30. G.B.Barlow and A.C.Clamp, *J. Chem. Soc.*, 393 (1961).
31. P.Finocchiaro, A.Recca, F.A.Bottino and E.Libertini, *Gazz. Chem. Ital.*, **109**, 213 (1979).
32. J.E.D.Davies, P.Finocchiaro and F.H.Herbstein, in *Inclusion Compounds*, Vol.2 (J.L.Atwood, J.E.D.Davies and D.D.MacNicol (Eds.), Academic Press, New York, 1984, p.407; F.H.Herbstein, M.Kapon and G.M.Reisner, *Acta Crystallogr., Sect.B* **41**, 348 (1985).
33. D.J.Duchamp and R.E.Marshy, *Acta Crystallogr., Sect.B*, **25**, 5 (1969.
34. F.H.Herbstein and R.E.Marsh, *Acta Crystallogr., Sect.B*, **33**, 2358 (1977).
35. F.H.Herbstein, M.Kapon and S.Wassermann, *Acta Crystallogr., Sect.B*, **34**, 1613 (1978).
36. R.Spallino and R.Provenzal, *Gazz. Chim. Ital.*, **39**, II, 325 (1909).
37. W.Baker, B.Gilbert and W.D.Ollis, *J. Chem. Soc.*, 1443 (1952).
38. D.J.Williams and D.Lawton, *Tetrahedron Lett.*, 111 (1975).
39. (a) A.C.D.Newman and H.M.Powell, *J. Chem. Soc.*, 3747 (1952); (b) J.E.D.Davies, W.Kemula, H.M.Powell and N.O.Smith *J. Inclus. Phenom.*, **1**, 3 (1983).
40. R.Arad-Yellin, B.S.Green and M.Knossow, *J. Am. Chem. Soc.*, **102**, 1157 (1980).
41. R.Arad-Yellin, S.Brunie, B.S.Green, M.Knossow and G.Tsoucaris, *J. Am. Chem. Soc.*, **101**, 7529 (1979).
42. G.M.Robinson, *J. Chem. Soc.*, 267 (1915).
43. (a) A.S.Lindsey, *Chem. Ind. (London)*, 823 (1963); (b) A.S.Lindsey, *J. Chem. Soc.*, 1685 (1965).

44. H.Erdtman, F.Haglid and R.Ryhage, *Acta Chem. Scand.*, **18**, 1249 (1964).
45. A.Goldup, A.B.Morrison and G.W.Smith, *J. Chem. Soc.*, 3864 (1965).
46. A.Lüttringhaus and K.C.Peters, *Angew. Chem.*, **78**, 603 (1966); *Angew. Chem., Int. Ed. Engl.*, **5**, 593, (1966).
47. V.K.Bhagwat, D.K.Moore and F.L.Pyman, *J. Chem. Soc.*, 443 (1931).
48. W.Schrauth and K.Görig, *Ber. Dtsch. Chem. Ges.*, **56**, 2024 (1923).
49. (a) M.Farina, in *Inclusion Compounds*, Vol.2, J.L.Atwood, J.E.D.Davies and D.D.MacNicol, (Eds.), Academic Press, New York, 1984 p.69; (b) M.Farina, G.Allegra and G.Natta, *J. Am. Chem. Soc.*, **86**, 516 (1964); (c) M.Farina and G.D.Silvestro, *J. Chem. Soc., Chem. Commun.*, 842 (1976); (d) M.Farina, *Tetrahedron Lett.*, 2097 (1963); (e) G.Allegra, M.Farina, A.Immirzi, A.Colombo, U.Rossi, R.Broggi and G.Natta, *J. Chem. Soc., B*, 1020 (1967).
50. (a) H.R.Allcock, *Acc. Chem. Res.*, **11**, 81 (1978); (b) H.R.Allcock, M.T.Stein and E.C.Bissell, *J. Am. Chem. Soc.*, **96**, 4795 (1974); (c) H.R.Allcock and L.A.Siegel, *J. Am. Chem. Soc.*, **86**, 5140 (1964); (d) H.R.Allcock, R.W.Allen, E.C.Bissell, L.A.Smeltz and M.Teeter, *J. Am. Chem. Soc.*, **98**, 5120 (1976); (e) H.R.Allcock and M.T.Stein, *J. Am. Chem. Soc.*, **96**, 49 (1974); (f) J.Finter and G.Wegner, *Makromol. Chem.*, **180**, 1093 (1979); (g) H.R.Allcock, *J. Am. Chem. Soc.*, **85**, 4050 (1963).
51. (a) D.D.MacNicol, J.J.McKendrick and D.R.Wilson, *Chem. Soc. Rev.*, **7**, 65 (1978); (b) D.D.MacNicol and D.R.Wilson, *J. Chem. Soc., Chem. Commun.*, 494 (1976); (c) D.D.MacNicol, D.R.Wilson, *Chem. Ind. (London)*, 84 (1977); (d) D.D.MacNicol, A.D.U.Hardy, D.R.Wilson, *Nature (London)*, **266**, 611 (1977); (e) D.D.MacNicol, A.D.U.Hardy and D.R.Wilson, *J. Chem. Soc., Perkin Trans. 2*, 1011 (1979).
52. F.Vögtle, H.G.Löhr, H.Puff and W.Schuh, *Angew. Chem.*, **95**, 425 (1983); *Angew. Chem., Int. Ed. Engl.*, **22**, 409 (1983).
53. (a) H.G.Löhr, H.P.Josel, A.Engel, F.Vögtle, W.Schuh and H.Puff, *Chem. Ber.*, **117**, 1487 (1984); (B) H.G.Löhr, F.Vögtle, W.Schuh and H.Puff, *J. Inclus. Phenom.*, **1**, 175 (1983); (c) H.G.Löhr, F.Vögtle, W.Schuh and H.Puff, *J. Chem. Soc., Chem. Commun.*, 924 (1983); (d) F.Vögtle, H.G.Löhr, J.Franke and D.Worsch, *Angew. Chem.*, **97**, 721 (1985); *Angew. Chem., Int. Ed. Engl.*, **24**, 727 (1985).
54. (a) F.Toda and K.Akagi, *Tetrahedron Lett.*, 3695 (1968); (b) F.Toda, D.L.Ward and H.Hart, *Tetrahedron Lett.* **22**, 3865 (1981).
55. H.Hart, L.T.W.Lin and D.L.Ward, *J. Am. Chem. Soc.*, **106**, 4043 (1984).
56. J.D.H.Brown, R.J.Cross, P.R.Mallison and D.D.MacNicol, *J. Chem. Soc., Perkin Trans. 2*, 993 (1980).
57. H.Hart, L.T.W.Lin and D.L.Ward, *J. Chem. Soc., Chem. Commun.*, 293 (1985).
58. B.Helferich, L.Moog and A.Junger, *Ber. Dtsch. Chem. Ges.*, **58**, 872 (1925).
59. (a) F.Toda, K.Tanaka, T.Omata, K.Nakamura and T.Oshima, *J. Am. Chem. Soc.*, **105**, 5151 (1983); (b) F.Toda, K.Tanaka and H.Ueda, *Tetrahedron Lett.*, **22**, 4669 (1981); (c) F.Toda, K.Tanaka, H.Ueda and T.Oshima, *J. Chem. Soc., Chem. Commun.*, 743 (1983).
60. (a) M.Czugler, J.J.Stezowski and E.Weber, *J. Chem. Soc., Chem. Commun.*, 154 (1983); (b) E.Weber and M.Czugler, Coll. Abstr. 2nd International Symposium on Clathrate Compounds and Molecular Inclusion Phenomena, Parma, Italy, 1982, p.82; (c)E.Weber, I.Csöregh, B.Stensland and M.Czugler, *J. Am. Chem. Soc.*, **106**, 3297 (1984).
61. (a) M.Czugler, E.Weber and J.Ahrent, *J. Chem. Soc., Chem. Commun.*, 1632 (1984); (b) E.Weber, I.Csöregh, B.Stensland and M.Czugler, *J. Am. Chem. Soc.*, **106**, 3297 (1984); (c) E.Weber, in *Molecular Inclusion and Molecular Recognition—Clathrates I* (E.Weber, Ed.), *Top. Curr. Chem.*, **140**, 1, Springer-Verlag, Berlin-Heidelberg-New York, 1987; (d) E.Weber and M.Czugler, in *Molecular Inclusion and Molecular Recognition—Clathrates II* (E. Weber, Ed.), *Top. Curr. Chem.*, **149**, 45, Springer-Verlag, Berlin-Heidelberg-New York, 1988; (e) E.Weber, *J. Mol. Graphics*, **7**, 12 (1989); E.Weber, I.Csöregh, J.Ahrendt, S.Finge and M.Czugler, *J. Org. Chem.*, **53**, 5831 (1988); (f) E.Weber, M.Hecker, I.Csöregh and M.Czugler, *J. Am. Chem. Soc.*, **111**, 7866 (1989); (g) E.Weber, M.Hecker, I.Csöregh and M.Czugler, *Mol. Cryst. Liq. Cryst.*, **187**, 165 (1990).

6 Directed Crystal Formation with Tailored Additives

6.1 INTRODUCTION

The effect of impurities or additives on the crystallization process may also be considered as a host–guest interaction. All areas, from stereochemistry to materials science, benefit from investigations of the influence that tailored inhibitors have on both the growth and dissolution of organic crystals. In organic crystals, that have been grown in the presence of growth-inhibiting additives, there is a connection between the crystal structure and the crystal habit obtained under those conditions. The knowledge of such relationships can be used to produce crystals with wanted habits, to separate conglomerates of enantiomeric compounds or crystals and, as will be shown later, to determine directly or indirectly the absolute configuration of chiral molecules. In addition, as will also be discussed, a new model for the spontaneous creation of optical activity in Nature can be derived from these findings. In analogy with crystal growth, the reverse process, i.e. the dissolution of organic crystals in the presence of additives, also produces etching figures on selected crystal planes, from which certain conclusions can also be drawn [1].

6.2 HISTORICAL ASPECTS

In antiquity, the habit of crystals, owing to their diversity and harmony, already roused the interest of man, and of naturalists in particular. Kepler even suggested crystals might be built of very small, indivisible units. In 1848, Pasteur for the first time demonstrated the relationship between a crystal's habit and the symmetry of the compound of which it is made. He separated the two enantiomers of sodium ammonium tartrate on the basis of the crystals' asymmetry. Later, it appeared that the habit of a crystal depends not only on the structure of the crystallizing molecule, but also on experimental parameters during the crystallization, such as solvent, temperature, oversaturation and impurities. For instance, $PbCl_2$, which normally crystallizes centrosymmetrically, is obtained in a chiral crystal habit of 222 symmetry when crystallized in the presence of dextrin. The use of additives for the generation of a given crystal habit is widespread in industry, although the underlying mechanism is hardly known. In the industrial praparation of large NaCl crystals for IR and Raman spectroscopy, for instance, small amounts of Pb^{2+} or

SO_4^{2-} ions are present as crystallization aids. In another area, oligosaccharides have a certain effect on the shape and speed of crystallization of cane-sugar from molasses.

6.3 ENANTIOSPECIFIC SYNTHESIS IN CRYSTALS

For a better understanding of the following discussion, it is important to know a few things about the photodimerization of molecules in the crystal. Enantiomerically pure chiral compounds can be obtained from achiral starting materials by reactions in which the crystal lattice is a contributory factor. It is necessary, however, that single crystals precipitate from the solution of the starting materials. Figure 1 illustrates this, taking as an example the preparation of chiral cyclobutane polymers from achiral dienes which form ideally packed crystals.

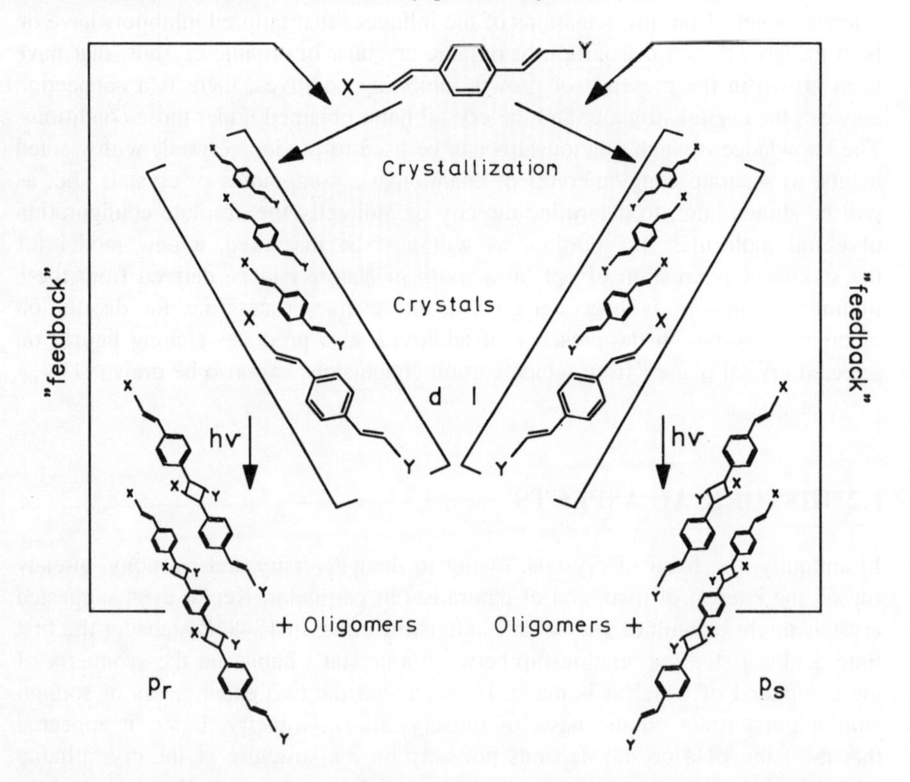

Figure 1. Formation of chiral cyclobutane derivatives (enantiomers p_r and p_s) from achiral dienes, suitably packed in chiral crystals (*d* and *l* symbolize corresponding enantiomeric lattices) [1]

The 'feedback' results in an inhibition of the crystallization. In these chiral crystals (depicted in large square brackets in Figure 1), the molecules are arranged in such a way that neighbouring carbon—carbon double bonds can photodimerize.

On irradiation of a single crystal with UV light, dimers, trimers and oligomers of only one chirality (p_r or p_s) are formed. In an achiral solvent, equal amounts of *d*- and *l*-crystals are produced, therefore allowing no asymmetric induction in the product mixture. However, an asymmetric induction is observed when the crystalline phases are grown in the presence of dimers, trimers or oligomers of one given chirality (either p_r or p_s). The additive inhibits the crystallization of the enantiomorphic phase from which it has been obtained, thus producing a large excess of the other phase (see Section 6.4). The experiment shows that, indeed, *d*-crystals (*l*-crystals) are obtained from a solution which has been 'spiked' with a p_s (p_r) additive ('chirality reversal', Figure 2).

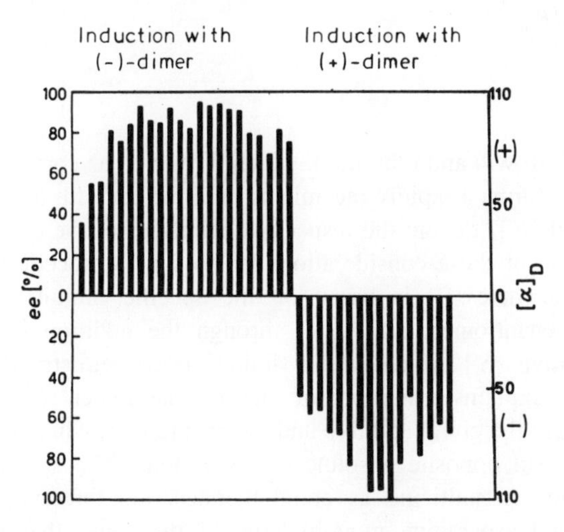

Figure 2. Enantiomeric excesses (ee) of dimers, formed after the crystallization of monomers in the presence of chiral dimers as additives, using different crystallization conditions (cf. Figure 1)

Here, the stereochemical similarity between the additive and the respective enantiomorphic crystal of the monomer proved to be of importance. Parameters such as temperature and concentration have only a quantitative influence on the observed asymmetric induction.

6.4 SEPARATION OF ENANTIOMERS BY CRYSTALLIZATION IN THE PRESENCE OF ADDITIVES

Based on the described stereochemical similarity between the additive and the enantiomorphic monomer crystal, it was proposed that a chiral dimer (trimer, oligomer) replaces two (three, several) monomers at the crystallization centre of a crystal, thus hindering the crystal's growth so dramatically that the crystallization equilibrium is strongly shifted towards the unaffected phase. Figure 3 illustrates this

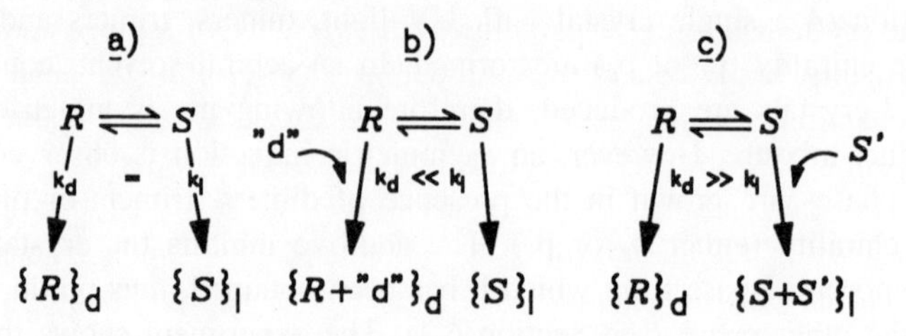

Figure 3. (*a*) Spontaneous crystallization of a racemate (*R,S*) to an $\{R\}_d/\{S\}_l$ conglomerate. (*b*) Crystallization in the presence of additive '*d*' (products of the $\{R\}_d$ phase formed under lattice control); preferential crystallization of $\{S\}_l$. (*c*) Crystallization in the presence of any chiral additive S' (with stereochemical 'affinity' to S); preferential crystallization of $\{R\}_d$ [1]

notion of crystal growth and inhibition, which proved to be correct. In solution, the achiral monomer forms a rapidly racemizing equilibrium with the conformations R and S; $\{R\}_d$ and $\{S\}_l$ denote the respective chiralities of the crystalline phases.

The application of these considerations led to a new procedure for the kinetic resolution of racemates. It is based on the specific inhibition of the crystal growth of one enantiomer, e.g. $\{S\}_l$, through the addition of small amounts of a chiral additive $\{S\}'$ which has a similar stereochemistry (Figure 3c). This mechanism also explains a number of kinetic enantiomer separations observed fortuitously, where the chiral additive and the enantiomer, which crystallized faster in its presence, had opposite absolute configurations. The scheme was checked with amino acids crystallizing as conglomerates, i.e. threonine, glutamic acid hydrochloride and asparagine monohydrate, in the sense that the crystallizing enantiomer was predicted beforehand. In this case, several other amino acids were used as chiral additives for the enantiomeric separation.

The quality of such separations is variable and depends on the type and concentration of the growth-inhibiting additive. In the best of cases [e.g. (*R,S*)-glutamic acid•HCl + (*S*)-lysine] a complete resolution is achieved (100% ee). Compared with the unaffected crystals, the growth of the affected ones is retarded by several days. In all systems mentioned, the additive was found in the inhibitedly grown crystals in amounts ranging between 0.05% and 1.5% (by weight); in the respective enantiomorphic crystals, none or only traces of it are found.

The changed crystal habit of the affected crystals gives positive proof of growth inhibition by adsorption. A crystal's habit is a function of the relative rates at which its faces grow. Since only the speed of growth of those faces which have adsorbed a growth-inhibiting additive will be changed, it is to be expected that crystals grown with and without the additive will have different habits. Accordingly, in the course of enantiomeric separations crystals with a markedly changed habit appear following the crystallization of the uninhibited enantiomer. By selecting crystals of each type and analysing their composition, it was shown that they were indeed composed of the uninhibited and the affected component, respectively. For instance, the crystallization of asparagine in the presence of various additives leads to different habits. In the case of (*R*,*S*)-threonine and (*S*)-glutamic acid, the growth of the (*S*)-enantiomer is inhibited so extensively that it is only obtained as a fine powder covering the well grown crystals of the (*R*)-enantiomer.

Directed crystallizations of this kind allow the modification of Pasteur's experiment and its adaptation to systems which show spontaneous racemate separation, but do not form hemihedral faces, where the enantiomorphic crystals, therefore, cannot be differentiated by their crystal habit.

Another possibility for the visual differentiation of enantiomorphic crystals consists in crystallizing a conglomerate of otherwise colourless crystals in the presence of a chiral, coloured additive. If, for instance, a solution of (*R*,*S*)-glutamic acid•hydrochloride is made to crystallize in the presence of the yellow *N*-(2,4-dinitrophenyl)-(*S*)-lysine, the colourless crystals of (*R*)-glutamic acid•hydrochloride will precipitate first, followed by the crystals of the (*S*)-enantiomer, coloured yellow by the additive.

The described stereochemical relationship between crystals and growth-inhibiting additives offer a new possibility for the comparative determination of the absolute configuration. Here, the additive influences the enantiomer with the same absolute configuration, exclusively.

6.5 THE DIRECTED EFFECT ON THE CRYSTAL HABIT

The drastic morphological changes produced by the presence of additives during the growth of organic crystals reveal a great measure of specific interaction between the additive and the structurally different faces of the crystal. The morphological changes are therefore in direct relation to the adsorption/inhibition processes proceeding on a molecular level. Generally, the size of a face will increase, compared with other faces in the crystal, when its growth is inhibited perpendicularly to itself and its rate of growth in this direction is small (Figure 4). The changes in the size ratio of faces grown on crystals with and without the presence of a growth-inhibiting additive therefore allow the identification of the affected faces and the corresponding crystallographic direction. In all cases investigated so far, there is a stereochemical relationship between the structure of the affected crystal and the molecular structure of the inhibiting molecule.

It is assumed that the additive is only adsorbed on such faces where the modified molecular part of the additive is pointing outwards. The adsorbed additive thus prevents the addition of further substrate molecules, reduces the rate of growth perpendicularly to this face and in the process causes a relative enlargement of the face, in comparison with the others. A similar mechanism is observed in the change of habit of cane-sugar crystals when oligosaccharides have been added.

Consequently, a systematic variation of the crystal habit with specific additives ought to be possible, additives which adsorb on chosen faces, inhibiting their growth in a predictable manner. The procedure was demonstrated with benzamide, which crystallizes from ethanol in plate-like {001} crystals growing along the *b*-axis

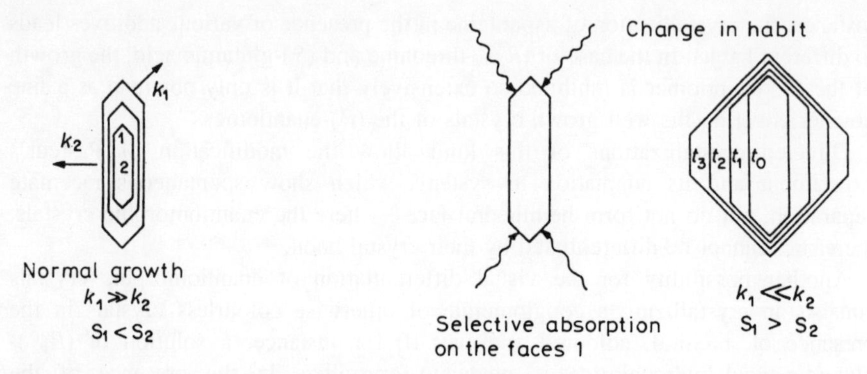

Figure 4. Formation of crystal faces S in relation to the rate of growth k perpendicular to it at given time intervals t. Left, habit under normal conditions; right, change in habit through the inhibition of the rate of growth of the faces 1

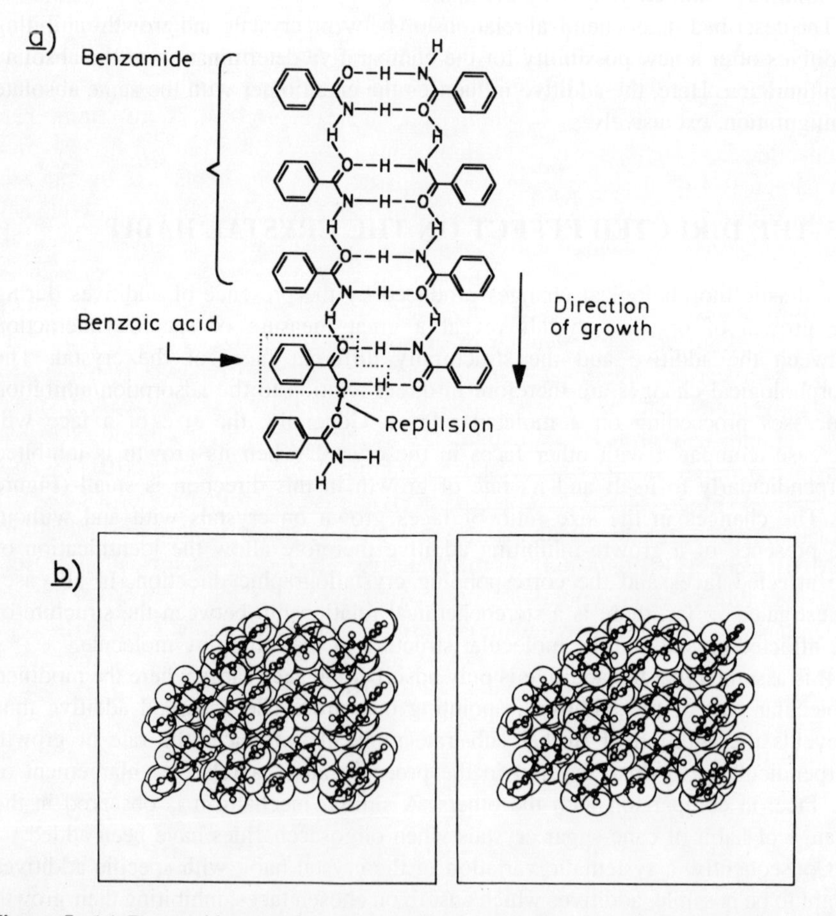

Figure 5. (*a*) Benzamide crystal structure with the strip-like system of hydrogen bonding extending in the *b*-direction; at the end a molecule of benzoic acid has been included. (*b*) Projection of the benzamide packing along the *b*-axis (the *b*-axis is perpendicular to the page) [1]

(Figure 5a) [2]. The packing (cf. Figure 5b) is characterized by a typical strip-like structure of cyclic dimers held by hydrogen bonding, which are joined by NH· · ·O bonds along the *b*-axis.

A growth inhibition of the {011} faces along the *b*-axis is achieved with benzoic acid as an additive. Now, if a benzamide molecule is replaced with a molecule of benzoic acid (cf. Figure 5a), at the end of the strip, in the direction of *b*, the last NH· · ·O bond is replaced with a repelling O· · ·O interaction; the resulting growth inhibition in this direction leads to rod-like crystals grown along the *a*-axis.

A growth inhibition along the *a*-axis is brought about by the addition of 2-methylbenzamide to the crystallizing solution. The crystals then grow as rods along the *b*-axis. The additive 2-methylbenzamide is easily inserted into the hydrogen bonding network without interfering with the crystal growth in the direction of *b*. The *ortho*-methyl group, however, occupies the (100) face and in doing so inhibits the growth in the direction of *a* in which the dimer pairs are stacked (Figure 5b). With larger amounts of 2-methylbenzamide, increasingly thinner crystal plates are obtained, since the methyl groups are interfering with the already weak van der Waals interactions between the layers of phenyl residues in the direction of *c*.

Similar changes in the crystal habit could be induced in other primary amides, and also in any other organic compound, such as carboxylic acids, amino acids, dipeptides and steroids. The crystallographic changes observed when crystallizing asparagine•H_2O in the presence of aspartic acid are explained by the same mechanism. The experimental results and their interpretation allow both explanation and prediction of changes in the crystal habit. We shall not discuss the details of how to determine simultaneously the absolute configurations of both a chiral molecule and the additive used [1].

Let us now discuss a number of crystal structures of glycine, grown without and with various additives. Pure glycine crystallizes from water in the shape of bipyramids with the *b*-axis perpendicular to the base (Figure 6a). Except for the amino acid proline, all natural (*S*)-amino acids induce drastic changes in habit on the *b* side of the crystals: Pyramids with a prominent $(0\bar{1}0)$-face are produced (Figure 6b). All (*R*)-amino acids induce the formation of a large (010) face (Figure 6c) and, finally, a racemic additive leads to the crystallization of {010} plates [2] (Figure 6d). In the plate-like glycine crystals usually 0.02–0.2% of the additive is incorporated. HPLC Analysis of the incorporated amino acids demonstrated a completely enantioselective inclusion along the *b*-axis.

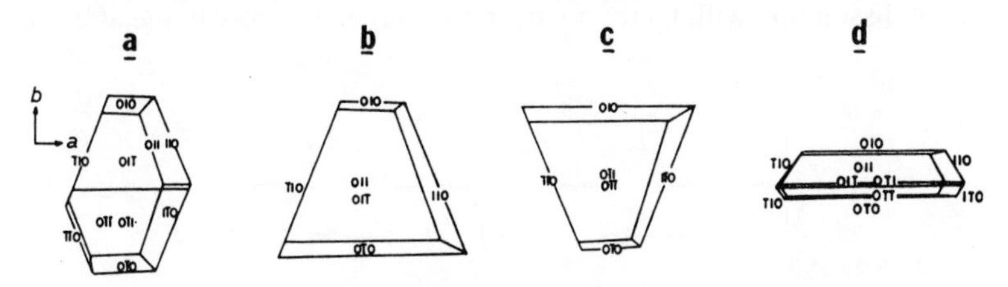

Figure 6. Glycine crystals obtained with the addition of various additives: (*a*) without additve; (*b*) with (*S*)-α-amino acids; (*c*) with (*R*)-α-amino acids; (*d*) with (*R,S*)-amino acids

Finally, let us outline a possible explanation for the spontaneous emergence of optical activity within the frame of evolution. Assuming that the first plate-like crystal of glycine emerging from the primitive ocean has its (010) face oriented towards the interface, it will grow incorporating (*S*)-amino acids exclusively, thus accumulating the (*R*)-enantiomers in solution. The excess of hydrophobic amino acids thus produced will subsequently orientate the crystals formed at the interface with their (010) faces towards that interface, thus further increasing the excess of the (*R*)-enantiomers in solution. In this way, a model has been found to explain the emergence and subsequent replicative propagation of optical activity in an originally symmetrical world.

6.6 TAILORED ETCHING ADDITIVES FOR ORGANIC CRYSTALS

Growth and dissolution of crystals are opposite processes, which can be induced by changing the degree of saturation of a solution. This means that an inhibitor to the growth of a particular crystal face must, in principle, also influence the rate of dissolution of that same face. In order to check on this interaction, the etching of crystals in the presence of tailored growth inhibitors was systematically investigated. For instance, glycine plates with well formed {010} faces were etched in non-saturated solutions containing additional amino acids in various amounts. In the presence of enantiomerically pure (*R*)-alanine in solution, only on the (010) faces do well formed etchings appear. These etched pits have a two-fold symmetry with the surface edges parallel to the *a*- and *c*-axes of the crystal. The enantiotopic $(0\bar{1}0)$ face is etched regularly and therefore behaves as if in an unsaturated solution of glycine. As expected, (*S*)-alanine induces pits on the $(0\bar{1}0)$ faces and in the presence of racemic alanine both {010} faces are etched.

The dissolution of a given surface, as is well known, begins at exit points of a fault. There, the interface to the solvent consists of several types of surface structures (Figure 7). Dissolution starts at this centre, radiating in the various directions at the same relative rates as during crystal growth; under such conditions, the crystal's structure, overall, is retained. However, if the solution contains an additive which selectively binds to a given face of the crystal, the rates of dissolution in the various directions will be different from that in the pure solvent. The rate of dissolution will, therefore, decrease perpendicularly to the affected face,

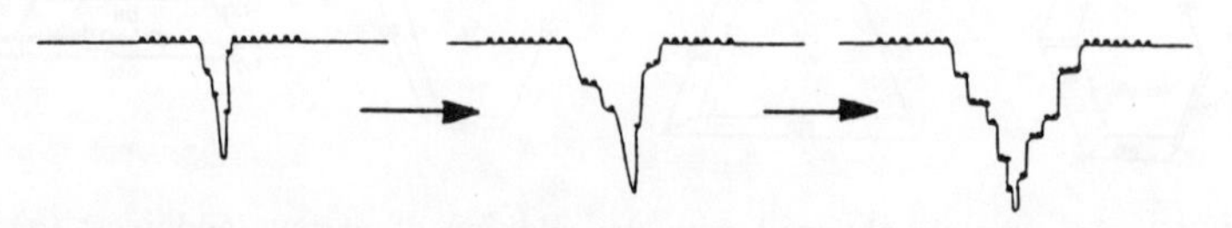

Figure 7. Formation of etching pitches where faults in the crystal reach the surface, an additive binding to different extents on the various faces

as opposed to an unaffected face, and at faults on the first face etched pits will appear.

6.7 THE 'RESORCINOL PROBLEM'

In 1949, Wells found that in aqueous solution α-resorcinol crystals (spatial group $Pna2_1$) will grow exclusively along the c-axis. The crystals (Figure 8) are bound at one end by phenyl-rich (011) and $(0\bar{1}1)$ faces and at the other end by hydroxyl-rich $(01\bar{1})$ and $(0\bar{1}\bar{1})$ faces (Figure 8, bottom). For a long time, the absolute direction of growth along the polar axis, relative to the packing of the molecules, could not be determined [3].

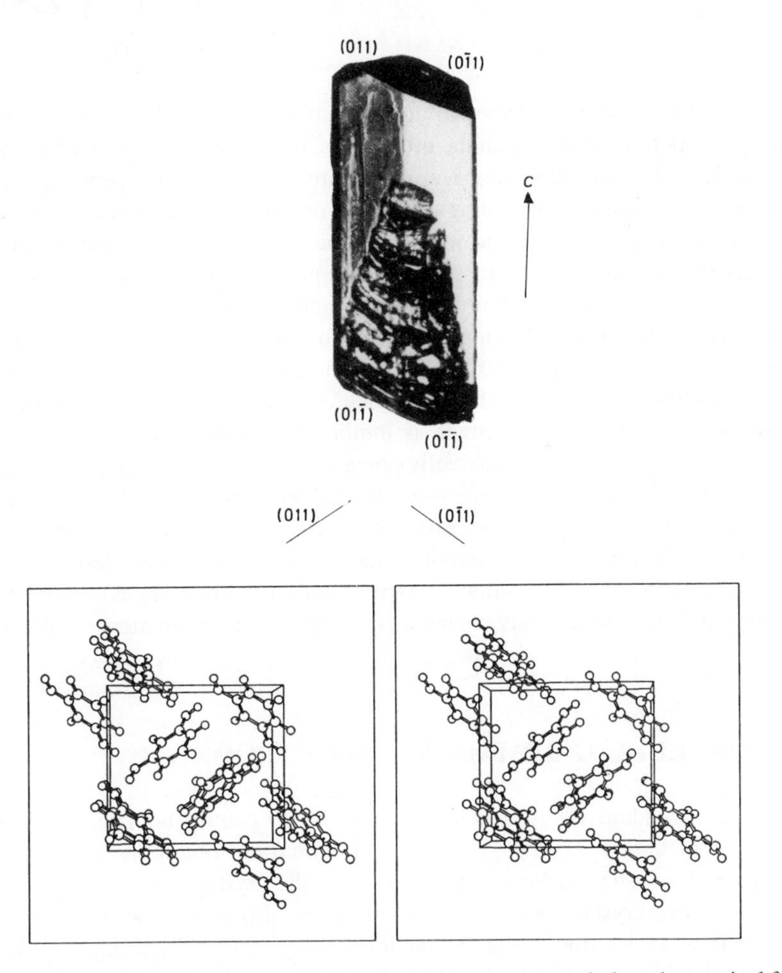

Figure 8. Top: typical α-resorcinol crystals grown from aqueous solution; the terminal faces $\{011\}$ [i.e. (011) and $(0\bar{1}1)$] and $\{0\bar{1}\bar{1}\}$ [i.e. $(01\bar{1})$ and $(0\bar{1}\bar{1})$] are shown, in addition to the faces parallel to the phenyl-rich faces $\{011\}$ and the hydroxyl-rich faces $\{0\bar{1}\bar{1}\}$ (according to [1]). Bottom: Packing of the resorcinol molecules (stereoview along the a-axis)

Therefore, it was not known which end of a given crystal is phenyl-rich and which is hydroxyl-rich. According to Wells, the growth along the *c*-axis was due to a marked adsorption of water at the hydroxyl-rich $\{0\bar{1}\bar{1}\}$ face, whereby its growth was retarded in comparison with the $\{010\}$ faces.

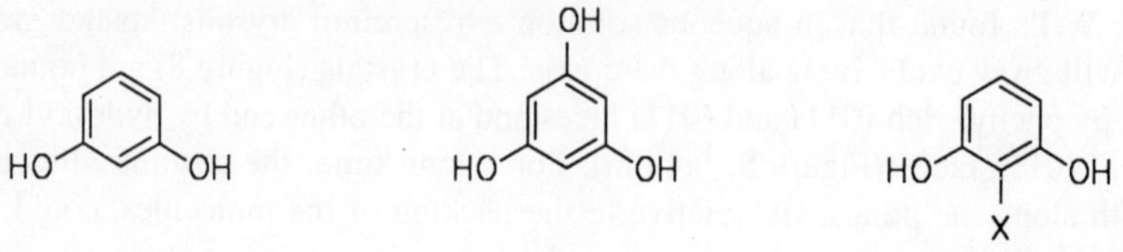

1

2

3: X=OH; 4: X=CH_3
5: X=COOH

In order to determine whether in aqueous solution the hydroxyl-rich faces are really growing more slowly than the other faces, the absolute direction of growth of resorcinol crystals was determined, using additives **1–5** and, independently, X-ray crystal structure analysis according to the Bijvoet method. According to these and in contrast to Wells' idea, the growth in aqueous solution is occurring primarily at the hydroxyl-rich $\{011\}$ faces. It seems that the preferential growth of the hydroxyl-rich $\{0\bar{1}\bar{1}\}$ faces is brought about by the inhibited growth of the phenyl-rich $\{011\}$ faces, because the latter have a higher affinity to water than the hydroxyl-rich faces. It also turned out, that in water it is better to call the faces 'acidic' and 'basic', rather than 'phenyl-rich' and 'hydroxyl-rich', respectively. Wells' interpretation, according to which a face's growth is inhibited by strong adsorption of solvent molecules, has proved to be essentially correct; however, in the case of resorcinol, under the incorrect assumption that water is preferentially adsorbed at the hydroxyl-rich face, he drew the wrong conclusion concerning the direction of growth.

The method for the investigation of surface–solvent interactions described above is the basis for a general quantitative explanation of the role played by the crystal structure and the surface–solvent interactions in the determination of the crystal habit.

6.8 CONCLUDING REMARKS AND OUTLOOK

Using specific additives, it is therefore possible to control both the growth and dissolution of organic crystals. Here, it is important to bring a given additive to a determined face, and thus to a chosen centre of the crystal.

Furthermore, crystallization with additives is also a method with which to investigate weak intermolecular interactions in crystals by observing the changes in crystal habit induced, e.g. the O· · ·O repulsion mentioned in relation to the benzamide–benzoic acid system.

The effect of tailored additives is probably not restricted to the growth process, but also extends to the aggregation of molecules to form higher molecular structures. This stage precedes the germ formation and structurally already corresponds to the finished crystal.

So far, additives have been limited to molecules similar in structure to the host molecules. In the future, this similarity will be reduced. This will offer the possibility of investigating the effect of the solvent on the crystal habit. Such studies could also lead to a better understanding of crystallization processes on structured biological interfaces, such as the formation of bones, teeth and shells. Corresponding investigations might also help answer the question as to why no ice is formed in fishes which live at temperatures below the freezing point of water. All in all, improvements of the old and up to now mostly empirical technology of modifying crystal growth with 'impurities' and tailored additives will open up a host of new aspects in theoretical and applied chemistry [4].

References

1. For a review, see L.Adachi, Z.Berkovitch-Yellin, I.Weissbuch, J.van Mil, L.J.W.Shimon, M.Lahav and L.Leiserowitz, *Angew. Chem.*, **97** 476 (1987); *Angew. Chem., Ind. Ed. Engl.*, **26**, 1171 (1987); cf. also L.J.W.Shimin, M.Vaida, L.Addachi, M.Lahav and L.Leiserowitz, *J. Am. Chem. Soc.*, **112**, 6215 (1990); I.Weissbach, G.Berkovic, L.Leiserowitz and M.Lahav, *J. Am. Chem. Soc.*, **112**, 5874 (1990).
2. The symbol $\{hkl\}$ stands for a complete set of equivalent planes of symmetry. For example, in the case of serine, $\{011\}$ indicates the (011), $(0\bar{1}1)$, $(01\bar{1})$ and (011) planes. The symbol (hkl) gives only a specific (hkl) plane [1].
3. A.F.Wells, *Phil. Mag.*, **37**, 184 (1949).
4. Crystal structure of silver halides used for photography: J.J.Marchesi, *Photographie*, **10**, 108 (1990).

7 Photoresponsive Host–Guest Systems: Organic Switches Based on Azobenzene

7.1 INTRODUCTION

At first, research in host–guest chemistry mostly dealt with the synthesis of host compounds which displayed either a high affinity or a pronounced selectivity for given types of guests. Given today's knowledge, however, the desire arises to have in hand host compounds whose complexation behaviour and selectivity can be controlled reversibly 'from outside', and which, therefore, can act as 'chemical switches'.

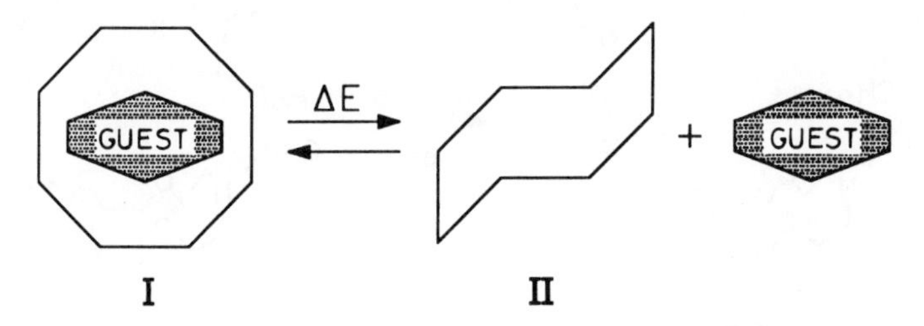

Figure 1. Principle of a switchable host molecule (II)

A chemical switch should be built in such a way that a host molecule in position I would, after changing into position II (Figure 1), lose its complexing ability, or else markedly change its guest selectivity. Such chemical switching processes can be effected either with a change in constitution (opening or closing of the cavity), or without one. In the latter case, a change in conformation or configuration (rotation of a C—C bond, isomerization of double bonds) is responsible for the change in guest selectivity. The following examples will illustrate this.

a. Nagai *et al*. prepared a number of β-cyclodextrin derivatives (**1a–c**) which have two complementary nucleic bases [adenine (**2**) and thymine (**3**)] attached on side-chains [1]. The complexing ability of **1a–c** for the sodium salt of adamantanecarboxylic acid decreases dramatically when the pH is increased; this is due to the formation of an enolate which breaks the hydrogen bonding between the nucleic bases, thus leaving adenine free to compete with the guest substrate for complexation.

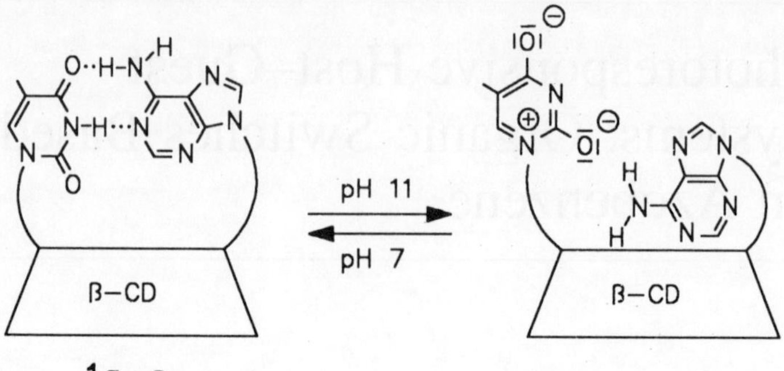

1a–c

(varying substituents at the cyclodextrin core)

b. The redox system consisting of the two components thiol and disulphide has been known for some time; it is important in biochemistry. It was therefore obvious to use this system to open and close a cavity, which was done with the molecules **4a** and **4b** [2].

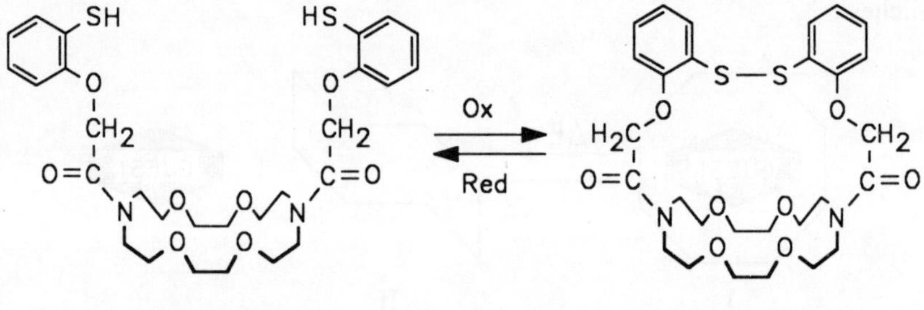

4a 4b

c. The complexes of crown compounds investigated by Kaifer *et al*. are examples of compounds which can be switched without changing their constitution [3]. The observation was that on electrochemical reduction, the stability of the alkali metal cation complexes of **5** and **6** increased by several orders of magnitude, indicating additional binding forces in the reduced groups. Since the electrochemical reduction of these complexes is reversible, a system such as this is also suitable as a chemical switch.

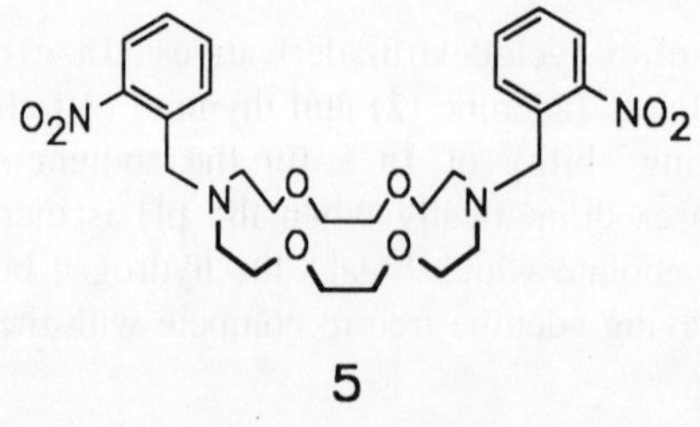

5 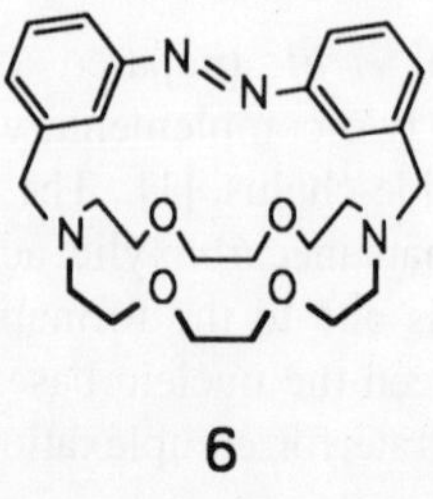 6

With the host molecules discussed so far, the corresponding switching process is always associated with drastic changes in the reaction medium (use of oxidants, changes in pH, etc.). A more elegant solution consists in isomerizing double bonds with light, a process which can be reversed thermally (Figure 2).

$$\underset{(E)}{R\text{-}X{=}Y\text{-}R} \;\underset{h\nu''}{\overset{h\nu'}{\rightleftarrows}}\; \underset{(Z)}{R\text{-}X{=}Y\text{-}R}$$

Figure 2. Isomerization of double bonds

Stilbene (X = Y = CH, R = Ph) is the classical example [4]. The photoisomerization is induced by the promotion of an electron to a higher molecular orbital; in most cases this is a π–π^* transition. With increasing conjugation, a bathochromic shift of the absorption band is observed [5]. Excitation wavelength in conjugated polyenes are given in Table 1.

Table 1. Excitation wavelengths in conjugated polyenes

$H_3C\text{-}[CH{=}CH]_n\text{-}CH_3$

n	Excitation wavelength
2	227 nm
3	263 nm
4	352 nm
5	413 nm

In analogy with the C=C double bond, other double bonds, such as P=C [6], P=P [7] and N=N, can also be isomerized. In the case of azobenzene, a stilbene analogue, even visible light is sufficient to effect its isomerization.

7.2 AZOBENZENE AS A PHOTOSWITCH

In 1834, Mitscherlich described azobenzene for the first time [8]. From an alkaline solution of nitrobenzene in alcohol, he isolated a compound which grew as red crystals, and for which he found the empirical formula $C_{12}H_{10}N_2$, i.e. azobenzene (**7**). In subsequent years, several procedures for its synthesis were developed and patented [9].

However, it was not until 1937, i.e. over 100 years later, that Hartley recognized the influence of light on the configuration of the N=N double bond, when he exposed an acetone solution of azobenzene to light [10]. Subsequently, a systematic investigation of their interconversion and the properties of the two configurational isomers began. It should be pointed out that the photoisomerization always leads to a photostationary state (PSS), which is dependent on the wavelength used. Thus, when irradiating at $\lambda = 313$ nm, approximately 80% (Z)-isomer [(Z)-**7**] is obtained, whereas at $\lambda = 365$ nm the yield is only 40% [11a]. Further, a dependence of the PSS on the polarity of the solvent has been observed [11b]. The thermal reisomerization of (Z)-**7** and its derivatives has been thoroughly investigated. As solids, the (Z)-isomers are relatively stable, but in solution, with increasing temperature and decreasing polarity of the solvent, they isomerize more or less rapidly to the (E)-isomer [12]. Thus, the half-life of (Z)-**7** varies from 94 h in carbon tetrachloride to 600 h in aqueous solution. From these experiments, an activation energy of 90–100 kJ/mol has been calculated.

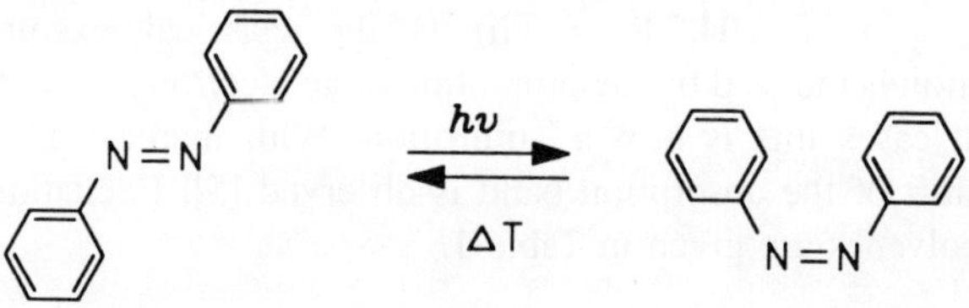

N=N $\xrightarrow{h\nu}$ / $\xleftarrow{\Delta T}$ N=N

(E) – 7 (Z) – 7

For the mechanism of the thermal reisomerization, two models have been proposed (Figure 3). For the rotation mechanism, a π–π^* transition is postulated, which should allow a rotation around the N=N double bond. However, today's views seem to favour the inversion mechanism, which is associated with an n–π^* transition. Indeed, the reisomerizations of a number of azobenzene derivatives show no dependence on pressure or solvent, something expected of a bipolar transition state in the rotation mechanism [13]. Compounds **8–10**, in which the N=N double

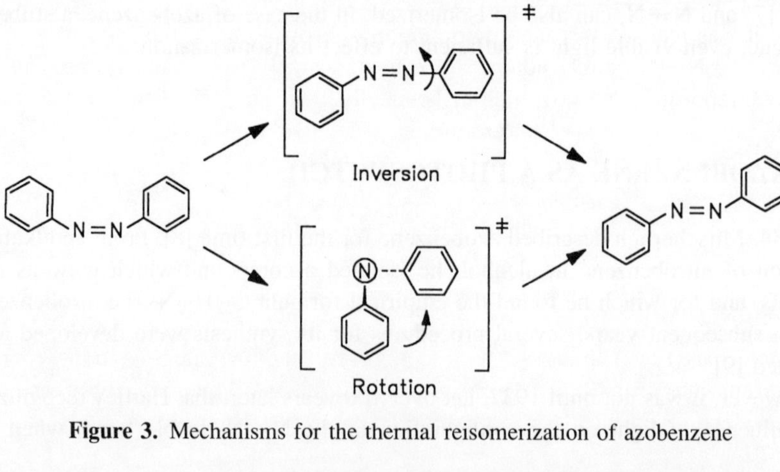

N=N ‡

Inversion

N=N N=N

N ‡

Rotation

Figure 3. Mechanisms for the thermal reisomerization of azobenzene

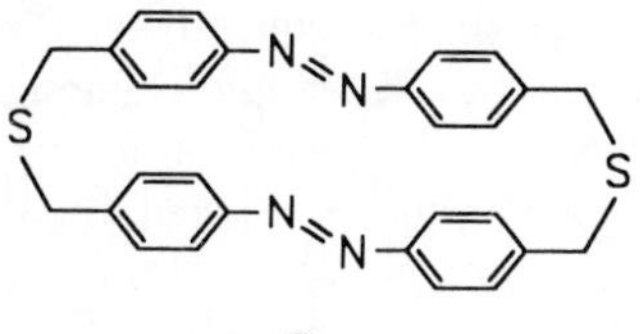

8

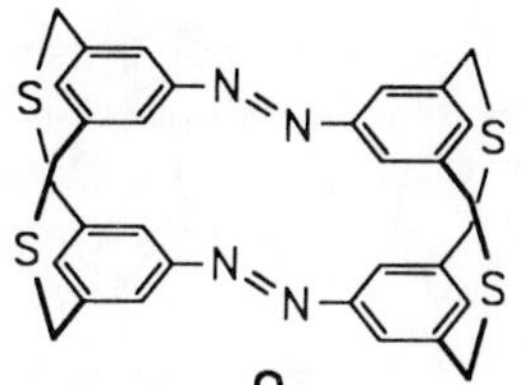

9

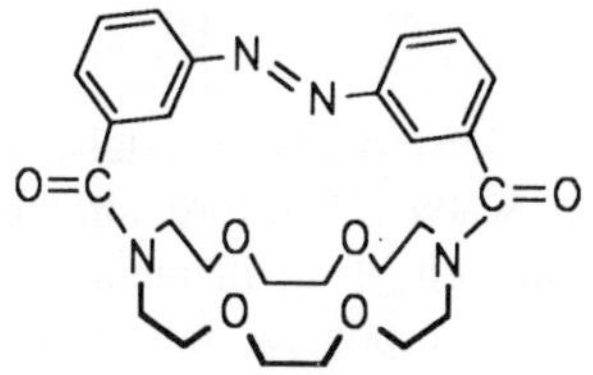

10

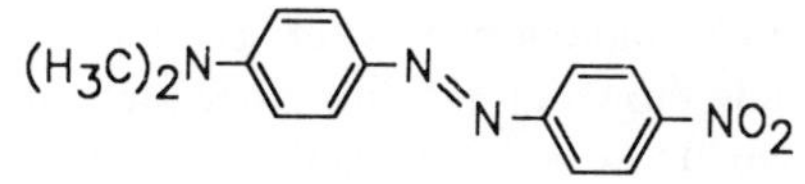

11

bond cannot rotate, something confirmed by thermodynamic calculations [14], were used for further investigations.

With 'push–pull' systems, such as 4-dimethylamino-4′-nitroazobenzene (**11**), a distinct solvent dependence was observed, which the authors ascribed to a rotation mechanism. However, based on thermodynamic calculations, an inversion mechanism is also being discussed for the same example [15]. So far, it has not yet been possible to determine unequivocally which mechanism is actually operating.

7.3 PHOTOCONTROLLED HOST SYSTEMS BASED ON AZOBENZENE

Most azobenzene-containing photresponsive host systems are based on crown ethers combined with some azobenzene moiety [16]. Compounds of this type are also known as 'ion-selective crown ether dyes', in which on complexation considerable changes in their UV–visible absorptions are occasionally observed [17]. In the early 1980s, it was realised that the conformation of crown ethers could be changed by isomerizing N=N double bonds. Since then, scarcely one month passes without someone reporting on a novel photochemically switchable molecule.

We shall consider some examples in more detail, all of which belong to any one of four classes:

azophanes	azocryptands
azocrowns	azocyclodextrins

7.3.1 AZOPHANES

The molecules **12a**, **12b** and **13** represent some of the first attempts to incorporate azobenzene into crowns; they were prepared from 2,2′-dihydroxyazobenzene and oligoethylene glycol dichlorides or ditosylates [18].

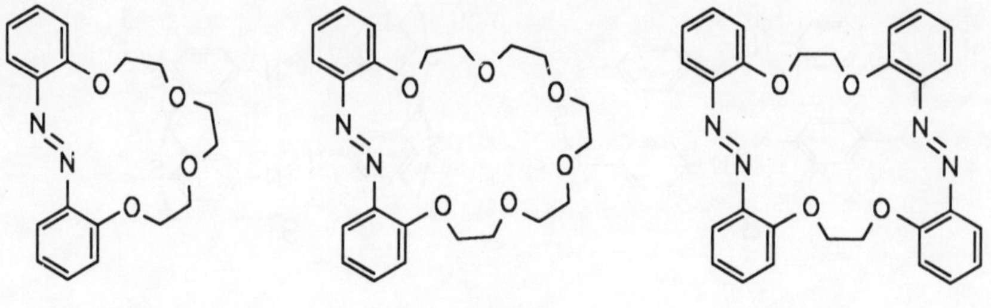

12a 12b 13

These chromoionophores (cf. Section 2.2) form reddish orange crystals, which, like simple crown ethers, are moderately soluble in water. Therefore, similarly to the crowns, **12a** and **12b** can be used for the extraction of alkali metal cations, such as Na^+ and K^+. Nevertheless, their use as photoresponsive host molecules seems pointless, since on irradiation ($\lambda > 300$ nm), photodecomposition with only little isomerization [20% (*Z*)-isomer] is observed. The open-chain analogue 2,2′-dimethoxyazobenzene, isomerizes to at least about 60%. In addition, the thermal reisomerization is slow, the half-life being of the order of days. Here, steric hindrance in the molecule seems to be too high to afford a completely reversible system. A further aspect becomes apparent: the ability to extract the cations mentioned earlier decreases to the same extent as the percentage of the (*E*)-isomer, from which it can be deduced that the (*Z*)-isomer has no appreciable complexing ability.

In order to reduce the steric strain occurring during isomerization, Shinkai *et al*. increased the size of the macrocycle and also moved the linkage to the azobenzene from the 2- and 2′-positions to the 4- and 4′-positions [19]. As a result, he obtained molecules **14a–c**.

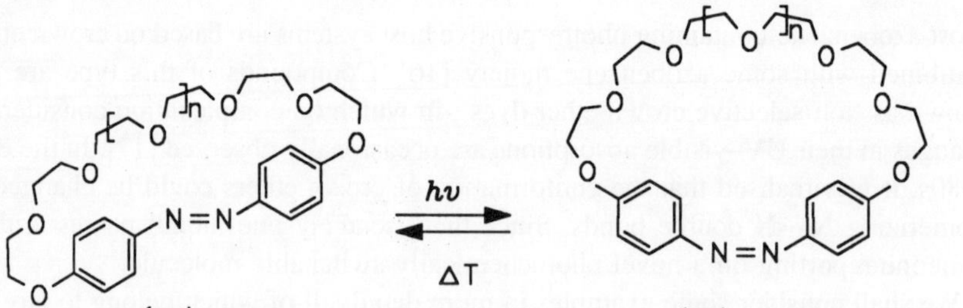

(*E*) − 14 $\underset{\Delta T}{\overset{h\nu}{\rightleftharpoons}}$ (*Z*) − 14

a : n = 1

b : n = 2

c : n = 3

These compounds are of interest in more than one way. First, the synthesis seemed to be demanding; the authors described it as one of the most difficult they had ever carried out. After several unsuccessful attempts at cyclization by bridging 4,4′-dihydroxybenzene with suitable 1,ω-dihalides, the oxidative azo coupling of the corresponding diamines was used, as shown in the reaction scheme. In contrast to the open-chain diamines, the ethylene protons of the cyclic compounds are shifted upfield with decreasing ring size; this is explained by an anisotropy effect of the azobenzene moiety. The expectations concerning the isomerization possibilities were completely fulfilled. Thus, a photostationary state is reached after only 30 s of irradiation with a filtered mercury high-pressure lamp (330 nm $< \lambda <$ 380 nm), producing a 70–80% excess of the (Z)-isomer. The percentage of the (Z)-isomer is usually estimated using the decrease or increase in the (E)-band in the UV spectrum (363 nm for **14a** and **b**, 361 nm for **14c**).

2 NaO–C₆H₄–NO_2 + CH_2CH_2Cl–$[OCH_2CH_2]_{n+4}$–Cl → CH_2CH_2O–C₆H₄–NO_2 / $[OCH_2CH_2]_{n+4}$–O–C₆H₄–NO_2

H_2 / Pd/C → CH_2CH_2O–C₆H₄–NH_2 / $[OCH_2CH_2]_{n+4}$–O–C₆H₄–NH_2

O_2/*t*–BuOK → 14a–c

A complete reisomerization can also be achieved, both thermally and photochemically ($\lambda >$ 460 nm). The (E)-isomers show no extracting ability for alkali metal cations, which can be explained with the stretched arrangement of the polyethylene glycol units. On isomerization from (E) to (Z), the distance between the *para* positions decreases from 0.9 to 0.55 nm [20]. This leads to the formation of rings similar to crown ethers. The phenolic oxygens do not participate, and therefore cannot contribute to any complexation. Formally, **14a** corresponds to [15]crown-5, **14b** to [18]crown-6 and **14c** to [21]crown-7. Indeed, similar complexation behaviours are observed: (Z)-**14a** binds preferentially Na^+, as well as K^+ and Rb^+, but not Cs^+; (Z)-**14b** prefers K^+, but also binds Na^+, Rb^+ and Cs^+. Finally, (Z)-**14c** has the strongest complexing ability for all the cations mentioned, with a preference for Rb^+, but also with the lowest selectivity. As a consequence of the fairly stable complexes formed, the thermal reisomerization is inhibited by the presence of alkali metal cations, those bound most strongly having the greatest effect. The photochemical reisomerization, however, cannot be inhibited in this way.

Compounds **14a–c** could be called 'all-or-nothing' switches, since the (E)-isomer has no complexing ability, whereas the (Z)-isomer can be used to extract alkali metal cations.

7.3.2 AZOCROWNS

The synthesis of azophanes in which the azo group is part of the ring can, on occasion, be fairly tedious. It is easier to prepare molecules which contain azobenzene as a side-chain (cf. Section 2.2). The azobis(crown ether) **16** is an outstanding example of this and is obtained by the reduction of 4′-nitrobenzo[15]crown-6 (**15**); azobis(crowns) with other ring sizes have been prepared in the same way.

15 —Zn→ 16

When **16** is isomerized in the presence of alkali metal cations, a marked dependence of the isomerization kinetics on the type of cation used is observed. The position of the photostationary state can be shifted in favour of the (*Z*)-isomer and additionally the rates of the thermal and, in contrast to **14a–c**, that of the photoreisomerization can be reduced, as shown in Table 2.

Table 2. Effect of cations on the photochemistry of azobis(crown ether) **16** [22,23]

Cation	(*Z*)-isomer in photostationary equilibrium	Relative rate of thermal reisomerization	Relative rate of photochemical reisomerization
None	52	1.00	3.10
Na^+	55	0.89	1.70
K^+	57	0.36	0.76
Rb^+	98	0.11	0.15
Cs^+	92	0.18	–*

* Not available

It is obvious that Rb^+ has the largest influence on the photochemistry, since its ionic radius is the most favourable for complexation. During the (*E*)→(*Z*) isomerization, a configuration is obtained in which the two crown rings are stacked one above the other. The cations will then be incorporated between the two rings, so that a ‘sandwich structure’ is obtained (cf. Figure 4). Depending on the ionic radius, a more or less stable bonding builds up, the energy of which will add to the amount already needed for reisomerization. Na^+ being a comparatively small cation, it will be incorporated at the centre of one of the rings and will, therefore, have only a small effect on the isomerization process. This is demonstrated by the fact that with an increasing degree of isomerization, the ability of **16** to extract a cation from water increases in the order $K^+ < Rb^+ < Cs^+$, but diminishes for Na^+. The extent of the increase is strongly dependent on the hydrophilicity of the counter ion. Since both the (*E*)- and (*Z*)-isomers are nearly insoluble in water, the higher the hydrophilicity, the higher the concentration at the interface will be, thus enhancing the degree of complexation. Based on these properties, it seems natural to use **16** as a light-controlled transport system for ions (Figure 4).

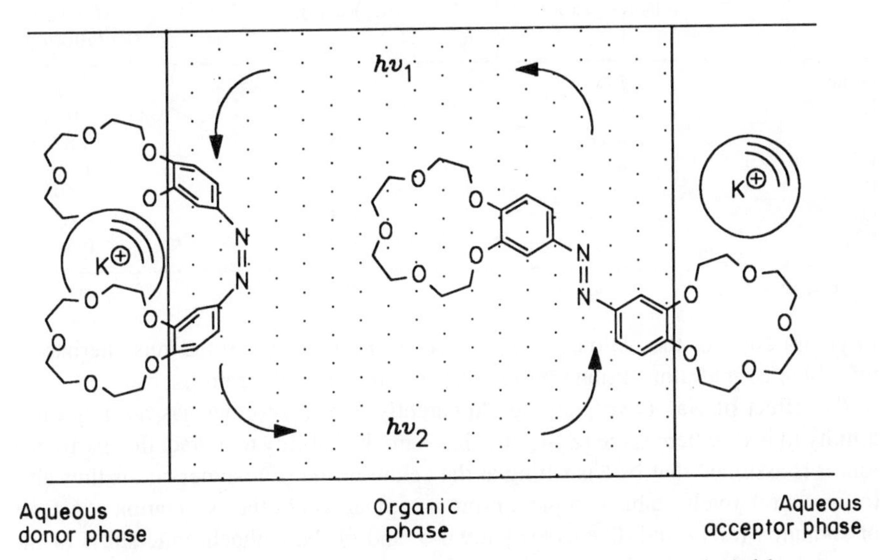

Figure 4. Light-controlled ion transport with azobis(crown ether) **16**

As it turns out, the highest transport rates are obtained with the **16**–K^+ system. The (Z)-**16**•Rb^+ complex is so stable that the release of ions into the aqueous phase is associated with a sizable energy barrier. On the other hand, the (Z)-**16**•Na^+ complex is too labile to effect any enrichment in the organic phase; on the contrary, Na^+ has its largest transport rate with the non-isomerized (E)-isomer. Transport processes therefore require a complex of 'moderate' stability, something other investigations on crown ethers and cryptands have confirmed [24].

In **16**, the principle of a light-controlled transport system consists in the ability to isomerize the two crown ether rings into a favourable complexing position. Another way was chosen for ligand **17**, where the coordination sites can be blocked photochemically in an intraannular fashion [25], as shown below:

(E) – 17 + M⊕ ⇌ (hv' / hv'') (Z) – 17

Table 3. Effect of cations on the photochemistry of azocrown ether **17**

Cation	Relative rate of thermal reisomerization	Relative extractability, (*E*)-isomer	Relative extractability, (*E*)–(*Z*) equilibrium
None	1.00	—	—
Li^+	1.20	0	0
Na^+	0.17	0	0.67
K^+	0.28	0.85	1.00
Rb^+	—*	0.49	0.69
Cs^+	0.78	0.52	0.63
NH_4^+	—*	0.35	0.52

* Not available.

In contrast to **16**, the cation has an effect on the rate of thermal reisomerization (cf. Table 3), but almost none on the photochemical equilibrium.

The effect of Na^+ is very strong. Apparently, the (*E*)-isomer has no noticeable affinity to Na^+, whereas the (*Z*)-isomer has a marked ability to extract that particular ion. It is assumed that Na^+ is sitting at the centre of the (*Z*)-isomer and is thus able to form a relatively stable complex. From a comparison of the association constants of sodium with **17** and dibenzo[18]crown-6, and of the bathochromic shifts of the $n \rightarrow \pi^*$ band, a coordination of one of the nitrogen atoms of the diazo moiety with the metal has been deduced. On the other hand, Cs^+, K^+ and Rb^+ should, in agreement with studies on other crown ethers [26], rest on top of the ring (cf. Figure 5) and therefore have weaker bonding to the ring oxygens. This would also

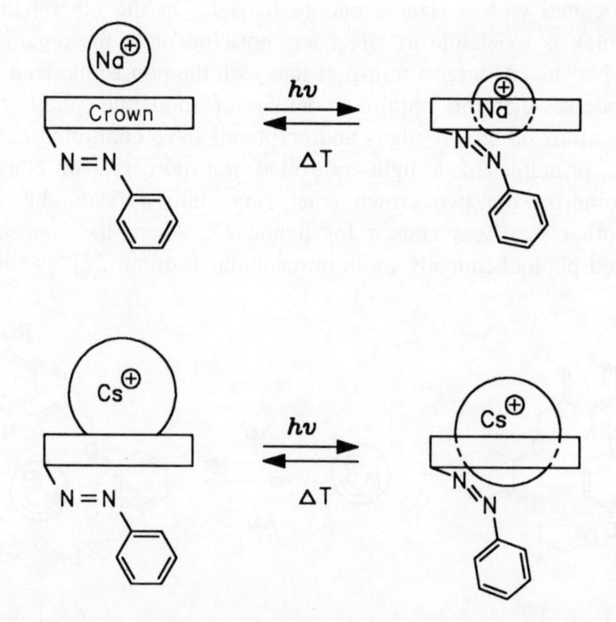

Figure 5. Complexation of Na^+ and Cs^+ by azocrown ether **17**

explain the extracting ability of (*E*)-**17** for the larger alkali metal cations.

The ionic radius of Li^+, however, is too small for complexation by **17**. Remarkably, the rate of thermal reisomerization is increased in the presence of Li^+; unfortunately, the authors did not give any explanation for this effect. The fact that ions such as NH_4^+, which are able to form hydrogen bonds to the oxygen atoms, have a greater affinity for the (*Z*)- than for the (*E*)-isomer, leads to the assumption that a change in configuration of the azo double bond also induces a change in conformation. Based on its selectivity for Na^+, **17** has, like **16**, been successfully used for the transport of Na^+ through organic membranes.

Both ion carriers, **16** and **17**, allow only passive transport, i.e. the transport of ions from a higher to a lower concentration, since it is based purely on a complexation equilibrium. Active ion transport, however, requires the ion carrier to have different complexation properties in the 'donor' and 'acceptor' phases. An example of this has been realized with the preparation of the azocrown ether **18**; not only can its complexation behaviour be changed photochemically, but also the amino function on the side-chain is pH dependent [27].

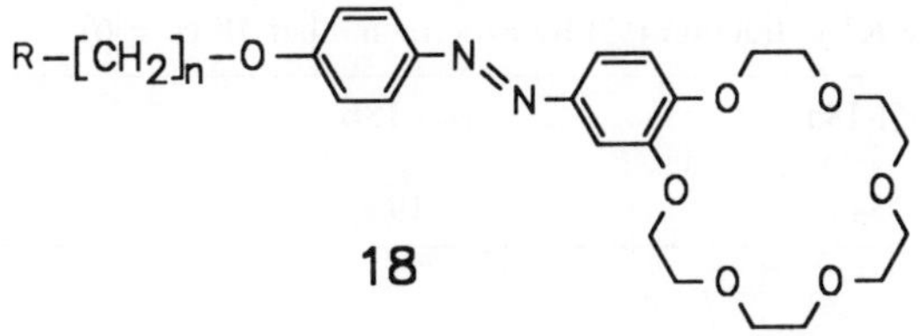

$R-[CH_2]_n-O-$

18

a: $R = NH_2$
b: $R = NH_3^{\oplus}$

As can be seen in Table 4, the protonated (*Z*)-isomers (**18b**) generally have a lesser tendency for thermal reisomerization than the unprotonated (*Z*)-amino compounds (**18a**). However, when an excess of K^+ is present, there is no appreciable difference in the isomerization kinetics of (*Z*)-**18a**, and (*Z*)-**18b** is no longer observed. This is explained by the fact that in an acidic medium intramolecular blocking of the crown ether ring by the ammonium group formed occurs. The strongest effect is observed with (*Z*)-**18a** ($n = 6$). It is also remarkable that, compared to (*Z*)-**18a** ($n = 4$) and (*Z*)-**18a** ($n = 10$), the unprotonated (*Z*)-isomer of **18a** ($n = 6$) shows a distinctly smaller tendency for reisomerization, the

Table 4. Effect of K^+ on the thermal reisomerization of azocrown ether **18**

Isomer	(*Z*)-isomer in photostationary equilibrium (%)	Reisomerization without K^+	Reisomerization with an excess of K^+
(*Z*)-**18a** ($n = 4$)	63	1.00	1.03
(*Z*)-**18a** ($n = 6$)	72	0.27	0.33
(*Z*)-**18a** ($n = 10$)	61	0.96	0.94
(*Z*)-18b ($n = 4$)	77	0.46	0.98
(*Z*)-18b ($n = 6$)	80	0.16	0.32
(*Z*)-18b ($n = 10$)	63	0.60	0.91

effect of K^+ being not very pronounced either. From this, the authors deduced that a chain length of $n = 6$ offered the best conditions for intramolecular complexation by both the amino and ammonium moieties.

Hence ammonium and potassium ions, in addition to the amino group and K^+, compete for complexation by the ring oxygens of the crown ether. As mentioned earlier, if a large excess of cations is present, no great difference in the rate of thermal reisomerization of (*Z*)-**18a** ($n = 6$) and (*Z*)-**18b** ($n = 6$) is observed.

Extraction experiments with (*Z*)-**18a** ($n = 6$) and (*Z*)-**18b** ($n = 6$) confirmed their suitability as a transport system for K^+ (Table 5). (*Z*)-Isomer **18b** ($n = 6$), when protonated, has a distinctly lower extracting ability than amino compound **18a** ($n = 6$). It was therefore obvious to build an active transport system for K^+, which would make use of a basic K^+-donor phase [formation of the equilibrium (*E*)-**18a** $\rightleftharpoons$ (*Z*)-**18a**] and an acidic K^+-acceptor phase [formation of the equilibrium (*Z*)-**18b** $\rightleftharpoons$ (*E*)-**18b**]. A back-transport, and therefore the build-up of an equilibrium, would be impossible.

Table 5. Relative K^+ extraction (%) by azocrown ether **18** ($n = 6$)

(*E*)-**18a**	(*Z*)-**18a**	(*E*)-**18b**	(*Z*)-**18b**
100	34	19	3

Surprisingly, in first experiments with a high concentration of ion carrier, (*E*)-**18a** ($n = 6$) itself, together with its protonated form, was already able to effect an active transport. After isomerization to the corresponding (*Z*)-isomers, comparable transport rates were observed. This is attributed to an intermolecular blocking by the ammonium moiety, something which has also been observed in other pH-dependent transport systems [28]. Therefore, with high carrier concentrations the system is pH-controlled (cf. Figure 6a). Only with low concentrations are there marked differences in transport rates for (*E*)/(*Z*)-**18a** ($n = 6$) and (*E*)/(*Z*)-**18b** ($n = 6$), so that an active transport system for K^+ becomes photochemically controllable (cf. Figure 6b).

Recent investigations (osmometric determinations of molecular weight) have demonstrated that *all* (*E*)-isomers of **18b** ($n = 4$, 6 and 10), and also the (*Z*)-isomer ($n = 4$), are found as 'pseudocyclic dimers,' and not as polymeric aggregates. (*Z*)-**18b** ($n = 6$) and (*Z*)-**18b** ($n = 10$), however, form cyclic monomers: the authors speak of 'tail-biting monomers' (cf. Figure 7) [29]. This has been substantiated by conductivity measurements, as with an increasing content of (*Z*)-isomer the conductivity of **18b** ($n = 6$) and **18b** ($n = 10$) increases, but not in the case of **18b** ($n = 4$). By analogy, cations which are complexed by **18a** or **18b**, are able to interfere with both intramolecular and intermolecular aggregation.

In this example, one should be aware that the active transport is pH-dependent, whereas the (*E*)–(*Z*)-isomerization is only able either to accelerate or to inhibit the transport process.

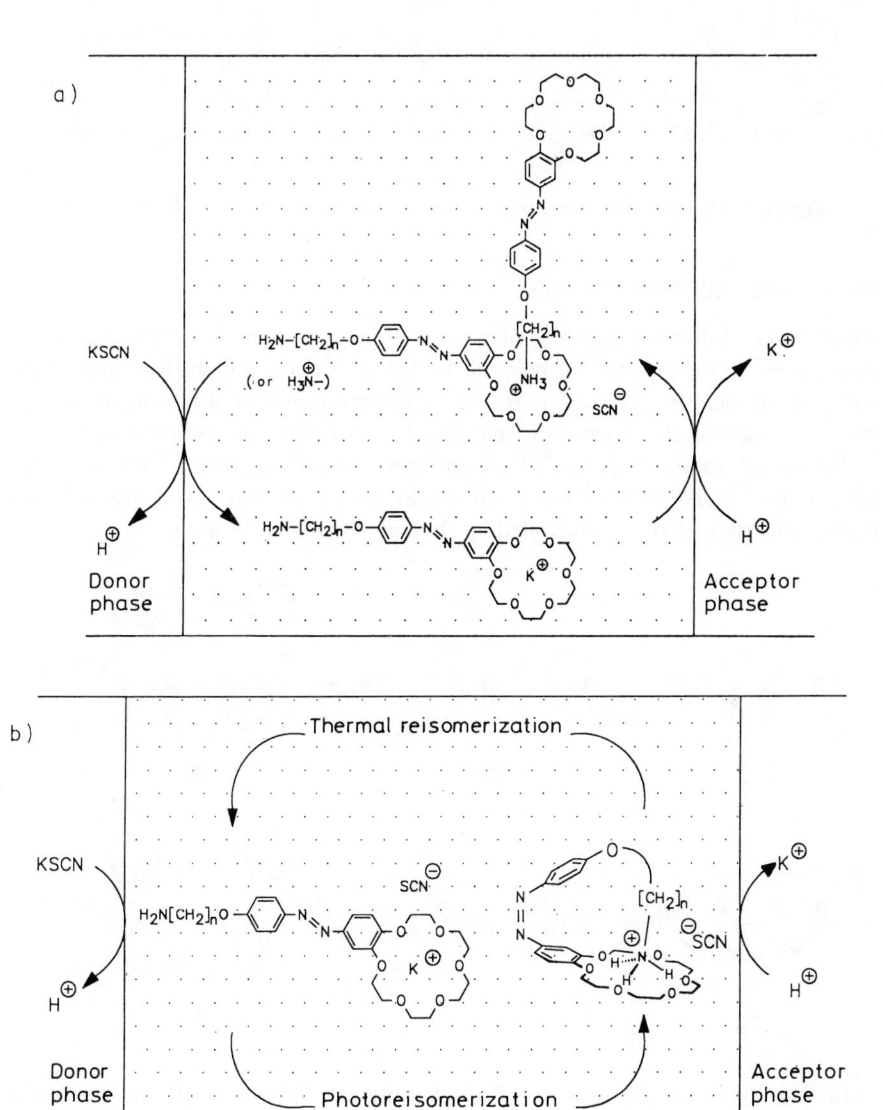

Figure 6 (a) Active transport of K^+ with a high concentration of **18**. (b) Active transport of K^+ with a low concentration of **18**

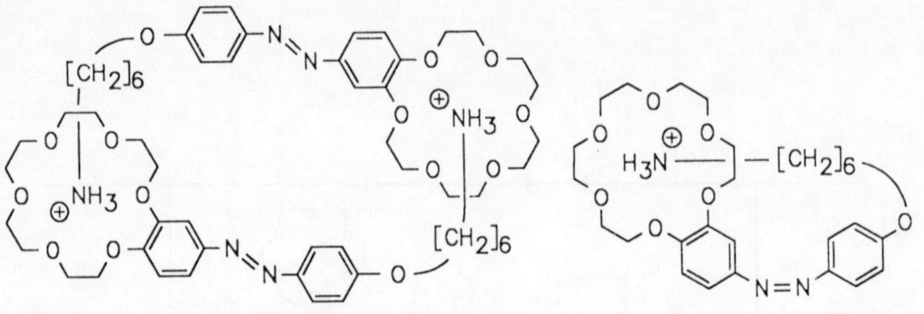

Figure 7. Pseudocyclic dimer and 'tail-biting' monomer of azocrown ether **18b**

7.3.3 SOME AZOCRYPTANDS

Another possibility for building photoresponsive host–guest systems consists in fitting a photoisomerizable azobenzene bridge of suitable length to a macrocycle with known inclusion properties. (*E*) (*Z*) Isomerization should drastically change the steric strain in the macrocycle and thus have an effect on its conformation.

The complexation properties of crown ethers have been known for many years [30]. It was therefore an obvious choice to bridge an azacrown ether with an azobenzene derivative, as illustrated by the two examples shown.

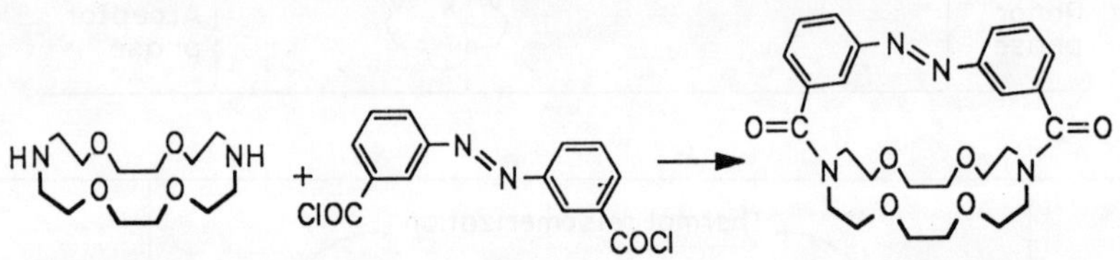

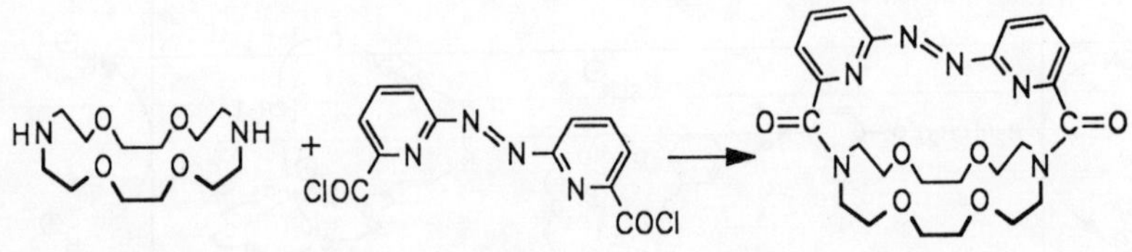

In both syntheses, 1,10-diaza-4,7,13,26-tetraoxa[18]crown-6 (**19**) is used as a starting material, a crown well suited for the complexation of alkali metal cations [31]. The use of **19** not only offers synthetic advantages (simple cyclization of an acid chloride with an amine under high-dilution conditions), but should also, compared with oxacrowns, offer some advantages related to the complexation of ions.

Thus, the introduction of nitrogen atoms into crown ethers usually affords improved solubility in water and renders the complexation pH-dependent. Because of the rapid inversion of the nitrogens, the flexibility of the ligand is also increased. The higher polarizability is also more likely to lead to complexes with small, highly charged heavy metal cations, at the same time decreasing the affinity towards alkali metal cations.

In 1979, Shinkai *et al*. presented the first photoswitchable crown ether (**10**), prepared from diazacrown ether **19** and 3,3′-bis(chlorocarbonyl)azobenzene (**20**) in 23% yield [32]. As determined by UV spectroscopy [decrease of the (*E*)-band at 324 nm], isomerization leads to a photostationary state with ca 60% (*Z*)-isomer. On thermal reisomerization, the original UV spectrum is recovered quantitatively. From temperature-dependent measurements, an activation energy of approximately 80 kJ/mol was obtained, which is slightly lower than the value for azobenzene (**7**; 90–100 kJ/mol) [12]. Mechanistic investigations established that the thermal reisomerization of (*Z*)-**10** (Figure 8) is pressure-dependent, which points to an inversion mechanism. As a consequence of steric strain, a rotation mechanism is probably inhibited; its activation energy must be much higher. Electronic considerations and measurements of the quantum yield of the photochemical reisomerization substantiate the fact that the $\pi \rightarrow \pi^*$ transition, which is the basis of a rotation mechanism, is less likely than the $n \rightarrow \pi^*$ transition associated with the inversion [14c].

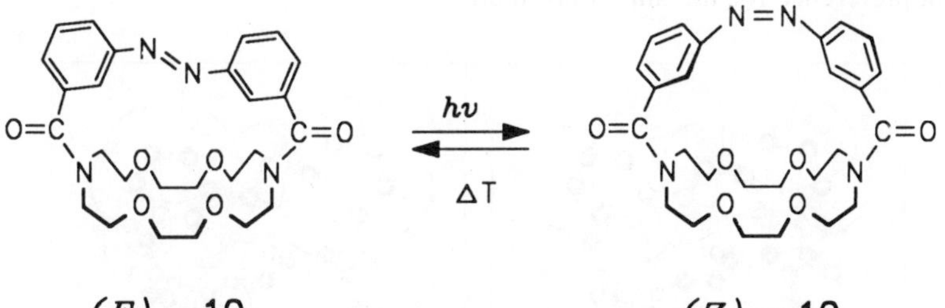

Figure 8. (*E*)–(*Z*)-Isomerization of azocryptand **10**

Similarly to the simple azocrowns, the thermal reisomerization of **10** can be inhibited by the addition of cations (K^+), except that Na^+ and Li^+ have no appreciable effect. On the other hand, organic ammonium ions generally have a strong effect. CPK models of (*E*)-**10** and (*Z*)-**10** show that in the (*E*)-isomer the azo bridge is planar and perpendicular to the plane of the crown ring, whereas in the (*Z*)-isomer the two benzene rings are skewed, affording a larger cavity.

These results are in accord with those from extraction experiments. Thus, K^+ and Rb^+ are more easily extracted with (*Z*)-**10**, whereas Li^+ and Na^+ have a higher affinity to the (*E*)-isomer. Surprisingly, the (*E*)-isomer also extracts ammonium salts, although these inhibit the thermal reisomerization more strongly than K^+. This points to interactions between the ions in the transition state. The stability of the (*E*)- and (*Z*)-cryptates alone cannot explain these findings.

When the azobenzene bridge in **10** is replaced with 2,2′-azopyridine, cryptand **22** is obtained, which by analogy with **10** is available from the reaction of acid chloride **21** and diazacrown ether **19** [34]. The two pyridine nitrogen atoms make it much more water-soluble than **10**, and as basic centres they are easily protonated. Here, we have to bear in mind that the (*Z*)-isomers of 2,2′-azopyridine and its derivatives have a lower basicity than the (*E*)-isomers. These findings are related to the stabilization of the protonated (*E*)-isomer by intramolecular hydrogen bonding to one of the nitrogen atoms in the azo moiety, which in the (*Z*)-isomer is impossible. From this, but possibly also from the electron-withdrawing effect of the pyridinium nitrogens, a pH dependence of the photostationary state results: with decreasing pH, the stability of (*E*)-2,2′-azopyridine increases. However, no such effect was observed for cryptand **22** between pH 1 and 7. In strongly acidic solution, however, an irreversible change in the UV spectrum occurs, which must be ascribed to cleavage of the amide bond.

Pyridine and bipyridine can be used as complexing ligands for heavy metals [35]. CPK models indicate that in **22**, the position of the 2,2′-azopyridine bridge must also be planar and perpendicular to the crown ring. This was confirmed by X-ray crystal structure analysis (Figure 9) [36]. The distance between the two pyridine nitrogens is 70.9 pm, and two oxygens are pointing into the cavity, the other two outwards. The bond lengths in the crown ether ring correspond to those in [18]crown-6 [37]. In contrast to the unsubstituted crown, the C—O bonds in **22** show a preference for the anti conformation.

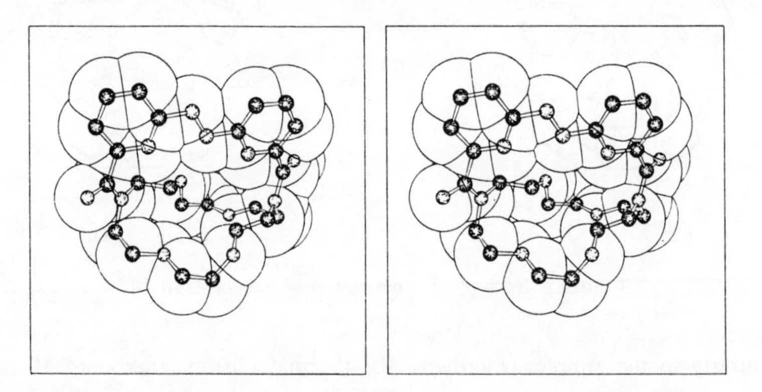

Figure 9. Stereoview of azocryptand **22** (X-ray crystal structure)

Since the azopyridine bridge must severely limit the conformational mobility of the crown ether ring, it must be assumed that there are only very small differences in conformation between **22** itself and one of its complexes. The (*E*)-isomer must, therefore, have stronger interactions with the cations than the (*Z*)-isomer, since in that case the coordination ability must be distinctly lower, owing to the lack of coplanarity of the pyridine rings.

Indeed, the rates of thermal reisomerization in the presence of Cu^{2+} and Pb^{2+} are increased by a factor of 26 and 11, respectively, without shifting the position of the photostationary state. When extracting metal salts into an organic phase with

Table 6. Relative extractability of metal cations: azocryptands **10** and **22** as their (*E*)-isomers and in the photostationary equilibrium (PSS)

Cation	(*E*)-10	(*E*)–(*Z*)-**10** (PSS)	(*E*)-**22**	(*E*)–(*Z*)-**22** (PSS)
Na^+	14.8	13.5	39.8	35.6
K^+	15.8	27.2	22.0	37.8
Rb^+	—	1.1	17.2	10.8
Cs^+	—	—	8.9	5.1
$Ca2^+$	—*	—*	6.7	5.3
$Ba2^+$	—*	—*	8.1	11.6
$Cu2^+$	1.9	2.2	9.5	1.0
$Ni2^+$	1.0	0.4	0.6	—
$Co2^+$	3.0	2.6	3.7	—
$Hg2^+$	1.9	2.3	3.1	—
$Pb2^+$	7.9	4.3	10.8	8.8

* Not available

22, some similarities to **10** are observed (cf. Table 6). However, whereas there is almost no difference between the two isomers of **10** in their complexation behaviour towards heavy metal cations, (*E*)-**22** has a markedly higher affinity for Cu^{2+}, Ni^{2+}, Co^{2+} and Hg^{2+} than its (*Z*)-isomer. Pb^{2+}, on the other hand, seems to be extracted equally well by both isomers, the cause of which must be the formation of stable crown ether complexes. Complexes of similar stability have also been observed for diazacrown ether **19** [38]. These results indicate that in the (*E*)-isomer, coordination of the cation by the pyridine nitrogen atoms occurs, whereas in the (*Z*)-isomer this is made more difficult, or even impossible, by the lack of coplanarity. However, since the (*Z*)-isomers of **10** and **22** also have an occasionally strong propensity to complex heavy metal cations, the influence of the crown on the complexation should not be neglected.

7.3.4 AZOCYCLODEXTRIN

The two azocryptands **10** and **22** demonstrate impressively that the combination of a crown ether of known complexation properties with an azobenzene can lead to a photoswitchable host–guest system. The disadvantage of azocrowns and azocryptands consists generally in their only being able to bind metal cations, and in some rare cases a few organic ammonium ions [20,32b,39].

It therefore seems logical to modify other host structures. Here, the cyclodextrins appear promising, since they are known for their many inclusion compounds [40] (see Chapter 4). What is more, most bridged cyclodextrins also possess considerable inclusion properties, although significant differences between them and the free cyclodextrins have been observed [41].

A good example of such a modified cyclodextrin is **23**, which was prepared from β-cyclodextrin and 4,4′-bis(chlorocarbonyl)azobenzene [42]. This host displays remarkable complexation properties towards organic guest molecules, which were studied with the help of circular dichroism (CD). The (*E*)–(*Z*) mixture in a

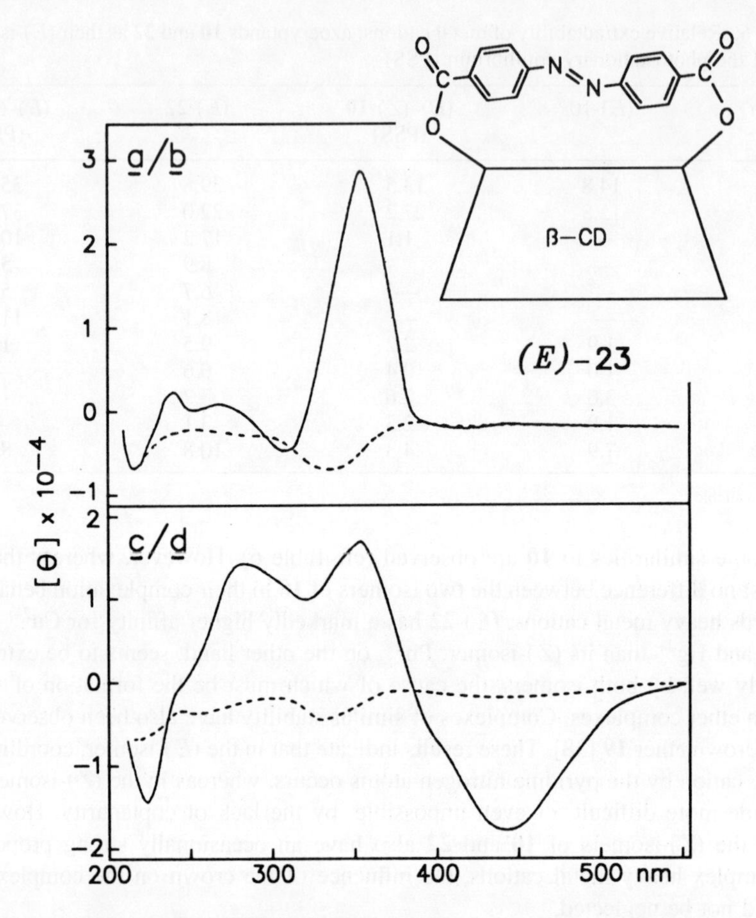

Figure 10. Circular dichroism of azocyclodextrin **23**. (*a*) (———), (*E*)-**23**; (*b*) (- - -), (*E*)-**23** with a 2000-fold excess of cyclohexanol; (*c*) (———), (*E*)–(*Z*)-**23** in PSS; (*d*) (- - -), (*E*)/(*Z*)-**23** in PSS with a 2000-fold excess of cyclohexanol

photostationary state and the (*E*)-isomer have distinctly different CD traces, as can be seen in Figure 10. In the presence of guest molecules, the CD bands lose some of their intensity. From the concentration dependence of the ellipticity, conclusions can be drawn on both the stoichiometry and the association constant of the inclusion compound.

In the first instance, 4,4′-bipyridine, cyclohexanol, toluene, methoxybenzene and the configurationally isomeric terpenes nerol (**24**) and geraniol (**25**) were used as guests. For all guest molecules, except 4,4′-bipyridine, a 1 : 2 stoichiometry is observed with (*Z*)-**23**, but with the (*E*)-isomer only a 1 : 1 ratio is found. Besides the complexes of (*Z*)-**23** are more stable than those of the corresponding β-cyclodextrin (with the exception of toluene), whereas the (*E*)-isomer shows lesser binding properties. This is particularly evident for 4,4′-bipyridine, with which, in contrast to β-cyclodextrin, (*E*)-**23** is completely unable to form a complex. It is also interesting that the 1 : 1 (*Z*)-**23**•geraniol complex is more stable than the corresponding nerol compound, but that in the 1 : 2 complexes the situation is reversed.

The different complexation properties of the (*E*)- and (*Z*)-isomers are attributed to the dependence of the cavity's size on the configuration of the azo double bond. It is assumed that the cavity of (*Z*)-**23** is deeper than that of (*E*)-**23**, which is thought to have a shallow, but broader, cavity. This would also explain the fact that, in contrast to (*E*)-**23**, (*Z*)-**23** is forming 1 : 2 host–guest complexes, where the first guest molecule to enter the cavity is bound more strongly than the second, since it penetrates more deeply into the hydrophobic cavity of cyclodextrin (Figure 11). The different rates of hydrolysis of *p*-nitrophenyl acetate in the presence of (*E*)- or (*Z*)-**23** also points in that direction. Thus, a higher rate of hydrolysis is observed with the (*E*)-isomer than with the (*Z*)-isomer, but with the latter forming the more stable complex. Because of the shallower cavity, the substrate finds a better orientation in (*E*)-**23**, which promotes a faster hydrolysis than in the (*Z*)-isomer. This, then, is an example of a photocontrolled catalysis!

Apart from 1 : 1 and 1 : 2 complexes, a few cases are known where a non-stoichiometric concentration dependence of the CD spectra is observed [42b]. This

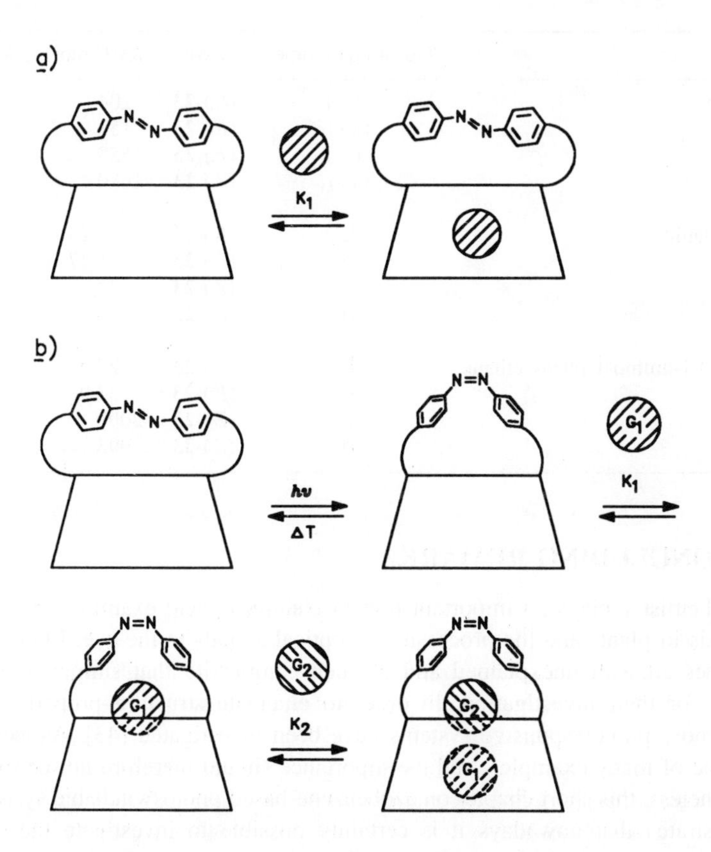

Figure 11. The formation of host–guest complexes with (*a*) (*E*)-**23** and (*b*) (*Z*)-**23** (G_1 = first guest; G_2 = second guest)

is attributed to an aggregation of guest molecules on the outside of the host, which beyond a certain concentration leads to a change in conformation in **23**.

The inclusion behaviour towards chiral guests is of particular interest. Here, the question is whether a chiral discrimination of enantiomers can be observed. For the elucidation of this question, the optical isomers of, among others, carvone, phenylalanine and *N*-acetyl-1-amino-1-phenylethane were used as guests [42c]. They form 1 : 2 host–guest complexes with both the (*E*)- and (*Z*)-isomers, the (*Z*)-complex being distinctly more stable, which agrees with other investigations. Moreover, differences in the association constants of the enantiomeric guest molecules were observed, where the (*E*)- and (*Z*)-isomers show exactly opposite enantioselectivities (cf. Table 7). It is assumed that interactions of the substrate with the wall or the rim of the cyclodextrin are at the origin of this selectivity. Even though the enantiodifferentiation of chiral guests with azocyclodextrin (**23**) complexes is relatively poor, this can nonetheless be called photocontrolled chiral recognition by complexation.

Table 7. Enantiodifferentiating complexation of chiral guest molecules with azocyclodextrin **23**

Guest	Guest enantiomer	Host	K_2 (l/mol)	K(L)/K(D)
Carvone	L(−)	(*E*)-**23**	204	1.29
	D(+)	(*E*)-**23**	158	
	L(−)	(*Z*)-**23**	1550	0.92
	D(+)	(*Z*)-**23**	1680	
Phenylalanine	L	(*E*)-**23**	1.2	0.82
	D	(*E*)-**23**	1.47	
	L	(*Z*)-**23**	45	1.22
	D	(*Z*)-**23**	37	
N-Acetyl-1-amino-1-phenylethane	L	(*E*)-**23**	27.5	0.73
	D	(*E*)-**23**	37.9	
	L	(*Z*)-**23**	2090	2.23
	D	(*Z*)-**23**	993	

7.4 CONCLUDING REMARKS

Photochemistry plays an important role in Nature, typical examples being photosynthesis in plants and the processing of optical signals in the eye. Many of these processes are still unexplained and of such complexity that simpler models are needed for their investigation. In order to elucidate structure–property relationships, more photoresponsive systems have been investigated [43]. Azobenzene is only one of many examples, and its importance should therefore not be overrated. Nevertheless, this short chapter on azobenzene-based photoswitchable systems [44] demonstrates that nowadays it is certainly possible to investigate the effect of light on supramolecular assemblies — in the sense of a 'supramolecular photochemistry' [45].

The first photoswitchable host–guest system containing azobenzene was published only in 1979. The examples of applications presented here, a cross-section of research work in the last 10 years, show that in this field, a number of fundamental discoveries are yet to be made.

References

1. K.Nagai, S.Ukai, K.Hayakawa and K.Kanematsu, *Tetrahedron Lett.*, **26**, 1735 (1985).
2. (a) S.Shinkai, K.Inuzuka and O.Manabe, *Chem. Lett.*, 747 (1983); (b) S.Shinkai, K.Inuzuka K.Hara, T.Stone and O.Manabe, *Bull. Chem. Soc., Jpn.*, **57**, 2150 (1984).
3. (a) D.A.Gustowski, V.J.Gatto, A.Kaifer, L. Echegoyen, R.E.Godt and G.W.Gokel, *J. Chem. Soc., Chem. Commun.*, 923 (1984); (b) A.Kaifer, D.A.Gustowski, L.Echegoyen, V.J.Gatto, R.A.Schults, T.P.Cleary, C.R.Morgan, D.M.Goli, A.M.Rios and G.W.Gokel, *J. Am. Chem. Soc.*, **107**, 1958 (1985).
4. (a) J.Saltiel, W.L.Chang, E.D.Megarity, A.D.Rousseau, P.T.Shannon, B.Thomas and A.K.Uriarte, *Pure Appl. Chem.*, **41**, 559 (1975); (b) J.Saltiel, J.T.D.Agostino, E.D.Megarity, L.Metts, K.R.Neuberger, M.Wrighten and O.-C.Zafiriou, *Org. Photochem.*, **3**, 1 (1973); (c) J.Saltiel, *Surv. Prog. Chem.*, **2**, 239 (1964); (d) A.Schönberg, *Preparative Organic Photochemistry*, Springer, New York, 1968, p.56.
5. F.Bohlmann and H.-J.Mannhardt, *Chem. Ber.*, **89**, 1307 (1956).
6. (a) M.Yoshifuji, K.Toyota and N.Inamoto, *Tetrahedron Lett.*, **26**, 1727 (1985); (b) M.Yoshifuji, K.Toyota, N.Inamoto, K.Hirotsu and T.Higuchi, *Tetrahedron Lett.*, **26**, 6443 (1985).
7. M.Yoshifuji, T.Hashida, N.Inamoto, K.Hirotsu, T.Horiuchi, T.Higuchi, K.Ito and S.Nagase, *Angew. Chem.*, **97**, 230 (1985).
8. E.Mitscherlich, *Ann. Pharm.*, **12**, 311 (1834).
9. *Beilsteins Handbuch der Organischen Chemie*, Vol. 16, System-Nr. 2092, Springer, Berlin, Heidelberg, New York.
10. G.S.Hartley, *Nature (London)*, **140**, 281 (1937).
11. (a) I.Hausser, *Naturwissenschaften*, 315 (1949); (b) M.Frankel and R.Wolovsky, *J. Chem. Phys.*, **23** 1367 (1955); (c) G.Zimmermann, L.-Y.Chow and U.-J.Paik, *J. Am. Chem. Soc.*, **80**, 3528 (1958).
12. (a) G.S.Hartley, *J. Chem. Soc.*, 633 (1938); (b) G.S.Hartley and R.J.W.LeFevre, *J. Chem. Soc.*, 531 (1939); (c) R.J.W.LeFevre and J.Northcott, *J. Chem. Soc.*, 867 (1953); (d) P.Bortolus and S.Monti, *J. Chem. Phys.*, **83**, 648 (1979).
13. (a) T.Sueyoshi, N.Nishimura, S.Yamamoto, S.Hasegawa, *Chem. Lett.*, 1131 (1974); (b) P.Haberfield, P.M.Block and S.M.Lux, *J. Am. Chem. Soc.*, **97**, 5804 (1975); (c) T.Asano, *J. Am. Chem. Soc.*, **102**, 1205 (1980); (d) T.Asano, T.Yano and T.Okada, *J. Am. Chem. Soc.*, **104**, 4900 (1982); (e) J.P.Otruba, III and R.G.Weiss, *J. Org. Chem.*, **48**, 3448 (1983); (f) T.Asano and T.Okada, *Chem. Lett.*, 695 (1987); (g) N.Nishimura, T.Sueyoshi, H.Yamanaka, E.Imai, S.Yamamoto and S.Hasegawa, *Bull. Chem. Soc., Jpn.*, **49**, 1381 (1976).
14. (a) S.Shinkai, Y.Kusano, K.Shigematsu and O.Manabe, *Chem. Lett.*, 1301 (1980); (b) D.Gräf, H.Nitsch, D.Ufermann, G.Sawatzki, H.Patzelt and H.Rau, *Angew. Chem.*, **94**, 385 (1982); (c) H.Rau and E.Lüddecke, *J. Am. Chem. Soc.*, **104**, 1616 (1982); (d) H.Rau, *J. Photochem.*, **26**, 221 (1984).
15. (a) P.D.Wildes, J.G.Pacifici, G.Irick, Jr, and D.G.Whitten, *J. Am. Chem. Soc.*, **93**, 2004 (1971); *Angew. Chem., Int. Ed. Engl.*, **21**, 706 (1982); (b) T.Asano and T.Okado, *J. Org. Chem.*, **49**, 4387 (1984); (c) T.Asano and T.Okado, *J. Org. Chem.*, **51**, 4454 (1986); (d) N.Nishimura, T.Tanaka and Y.Sueishi, *J. Chem. Soc., Chem. Commun.*, 903 (1985).
16. (a) S.Shinkai and O.Manabe, *Top. Curr. Chem.*, **121**, 67 (1984); *Host Guest Complex Chemistry, Macrocycles* (F.Vögtle and E.Weber, Eds.), Springer, Berlin, 1985; (b) S.Shinkai, *Pure Appl. Chem.*, **59**, 425 (1987).

17. (a) J.P.Dix and F.Vögtle, *Angew. Chem.*, **90**, 893 (1978); *Angew. Chem., Int. Ed. Engl.*, **17**, 857 (1978); (b) T.Yamashita, H.Nakamura, M.Takagi and K.Ueno, *Bull. Chem. Soc., Jpn.*, **53**, 1550 (1980); (c) J.P.Dix and F.Vögtle, *Chem. Ber.*, **113**, 457 (1980); (d) T.Kaneda, K.Sugihara, H.Kamiya and S.Misumi, *Tetrahedron Lett.*, 4407 (1981); (e) K.Nakashima, S.Nakatsuji, S.Akiyama, T.Kaneda and S.Misumi, *Chem. Lett.*, 1781 (1982).
18. M.Shiga, M.Takagi and K.Ueno, *Chem. Lett.*, 1021 (1980).
19. (a) S.Shinkai, T.Minami, Y.Kusano and O.Manabe, *Tetrahedron Lett.*, 2581 (1982); (b) S.Shinkai, T.Minami, Y.Kusano and O.Manabe, *J. Am. Chem. Soc.*, **105**, 1851 (1983).
20. S.Shinkai, Y.Honda, T.Minami, K.Ueda, O.Manabe and T.Tashiro, *Bull. Chem. Soc. Jpn.*, **56**, 1700 (1983).
21. (a) S.Shinkai, T.Ogawa, Y.Kusano and O.Manabe, *Chem. Lett.*, 283 (1980); (b) S.Shinkai, T.Nakaji, T.Ogawa, K.Shigematsu and O.Manabe, *J. Am. Chem. Soc.*, **103**, 111 (1981).
22. S.Shinkai, T.Ogawa, Y.Jusano, O.Manabe, K.Kikukawa, T.Goto and T.Matsuda, *J. Am. Chem. Soc.*, **104**, 1960 (1982).
23. S.Shinkai, K.Shigamatsu, M.Sato and O.Manabe, *J. Chem. Soc., Perkin Trans. 1*, 2735 (1982).
24. (a) Y.Kobuke, K.Hanji, K.Horiguchi, M.Asada, Y.Nakayama and J.Furakawa, *J. Am. Chem. Soc.*, **98**, 7414 (1976); (b) M.Kirch and J.-M.Lehn, Angew. Chem., 87, 542 (1975); (c) J.D.Lamb, J.J.Christensen, J.L.Nielsen, B.W.Asay and R.M.Izatt, *J. Am. Chem. Soc.*, **102**, 6820 (1980).
25. (a) S.Shinkai, K.Miyazaki and O.Manabe, *Angew. Chem.*, **97**, 872 (1985); (b) S.Shinkai, K.Miyazaki and O.Manabe, *J. Chem. Soc., Perkin Trans. 1*, 449 (1987).
26. (a) F.Vögtle and E.Weber, *Host Guest Complex Chemistry*, Vols I–III, *Top. Curr. Chem.*, **98**, (1981), **101** (1982), **121** (1984); (b) F.Vögtle and E.Weber, *Biomimetic and Bioorganic Chemistry*, Vols.I–III, *Top. Curr. Chem.*, **128** (1985), **132** (1986), **136** (1986).
27. (a) S.Shinkai, M.Ishihara, K.Ueda and O.Manabe, *J. Chem. Soc., Chem. Commun.*, 727 (1984); (b) S.Shinkai, M.Ishihara, K.Ueda and O.Manabe, *J. Inclus. Phenom.*, **2**, 111 (1984).
28. H.Tsukube, *J. Membr. Sci.*, **14**, 155 (1983).
29. S.Shinkai, T.Yoshida, K.Miyazaki and O.Manabe, *Bull. Chem. Soc. Jpn.*, **60**, 1819 (1987).
30. (a) C.J.Pedersen, *J. Am. Chem. Soc.*, **89**, 2495, 7017 (1967); (b) F.Vögtle, H.Sieger and W.M.Müller, *Top. Curr. Chem.*, **98**, 107 (1981); (c) F.Vögtle, W.M.Müller and W.H.Watson, *Top. Curr. Chem.*, **125**, 131 (1984).
31. B.Dietrich, J.-M.Lehn and J.P.Sauvage, *Tetrahedron Lett.*, 2885, 2889 (1969).
32. (a) S.Shinkai, T.Ogawa, T.Nakaji, Y.Kusano, O.Manabe, *Tetrahedron Lett.*, 4569 (1979); (b) S.Shinkai, T.Nakaji, Y.Nishida, T.Ogawa and O.Manabe, *J. Am. Chem. Soc.*, **102**, 5860 (1980).
33. T.Asano, T.Okado, S.Shinkai, K.Shigematsu, Y.Kusano and O.Manabe, *J. Am. Chem. Soc.*, **103**, 5161 (1981.
34. (a) S.Shinkai, T.Minami, T.Kouno, Y.Kusano and O.Manabe, *Chem. Lett.*, 499 (1982); (b) S.Shinkai, T.Kouno, Y.Kusano and O.Manabe, *J. Chem. Soc., Perkin Trans. 1*, 2741 (1982).
35. (a) F.Kober, *Grundlagen der Komplexchemie*, Salle, Sauerländer, Frankfurt-Aarau, 1979; (b) W.R.McWhinnie and J.D.Miller, *Adv. Inorg. Radiochem.*, **12**, 135 (1969).
36. H.L.Ammon, S.K.Bhattacharjee, S.Shinkai and Y.Honda, *J. Am. Chem. Soc.*, **106**, 262 (1984).
37. E.Maverick, P.Seiler, W.B.Schweizer and J.D.Dunitz, *Acta Crystallogr., Sect. B*, **36**, 615 (1980).
38. J.D.Lamb, R.M.Izatt, P.A.Robertson and J.J.Christensen, *J. Am. Chem. Soc.*, **102**, 2452 (1980).

39. E.Weber, in *Synthesis of Macrocycles: the Design of Selective Complexing Agents* (R.M.Izatt and J.J.Christensen, Eds.), Wiley, New York, 1987, p.337.
40. (a) M.L.Bender and M.Komiyama, *Cyclodextrin Chemistry*, Springer, Berlin, 1978; (b) J.Szejtli, *Cyclodextrins and Their Inclusion Complexes*, Akadémiai Kiadó, Budapest, 1982.
41. (a) J.Emert and R.Breslow, *J. Am. Chem. Soc.*, **97**, 670 (1975); (b) T.Tabushi, K.Shimokawa, N.Shimzu, H.Shirakata and K.Fujita, *J. Am. Chem. Soc.*, **98**, 7855 (1976); (c) R.Breslow, *Science*, **218**, 532 (1982); (d) S.Shinkai, M.Yamada, T.Stone and O.Manabe, *Tetrahedron Lett.*, 3501 (1983); (e) A.P.Croft and R.A.Bartsch, *Tetrahedron*, **39**, 1417 (1983); (f) A.Ueno, and T.Osa, *J. Inclus. Phenom.*, **2**, 555 (1984); (g) J.Franke, T.Merz, H.-W.Losensky, W.M.Müller, U.Werner and F.Vögtle, *J. Inclus. Phenom.*, **3**, 471 (1985); (h) K.Kano, H.Matsuomoto, Y.Yoshimura and S.Hashimoto, *J. Am. Chem. Soc.*, **110**, 204 (1988).
42. (a) A.Ueno, H.Yoshimura, R.Saka and T.Osa, *J. Am. Chem. Soc.*, **101**, 2779 (1979); (b) A.Ueno, R.Saka and T.Osa, *Chem. Lett.*, 841, 1007 (1979); 29 (1980); (c) A.Ueno, R.Saka, K.Takahashi and T.Osa, *Heterocycles*, **15**, 671 (1981); (d) A.Ueno, K.Takahashi and T.Osa, *J. Chem. Soc., Chem. Commun.*, 94 (1981).
43. H.-W.Losensky, H.Spelthann, A.Ehlen, F.Vögtle and J.Bargon, *Angew. Chem.*, **100**, 1225 (1988); *Angew. Chem., Int. Ed. Engl.*, **27**, 1189 (1988); K.H.Neumann and F.Vögtle, *J. Chem. Soc., Chem. Commun.*, 520 (1988); J.Schmiegel, U.Funke, A.Mix and H.-F.Grützmacher, *Chem. Ber.*, **123**, 1397 (1990); J.Schmiegel and H.-F.Grützmacher, *Chem. Ber.*, **123**, 1749 (1990); N.Tamaoki, K.Koseki and T.Yamaoka, *Angew. Chem.*, **102**, 66 (1990); *Angew. Chem., Int. Ed. Engl.*, **29**, 105 (1990); *Tetrahedron Lett.*, **31**, 3309 (1990).
44. See, for example, the molecular cage of thioindigo derived from a podand: M.Irie and M.Kato, *J. Am. Chem. Soc.*, **107**, 1024 (1985).
45. V.Balzani, *Supramolecular Photochemistry*, Reidel, Dordrecht, 1987.

8 Liquid Crystals

8.1 INTRODUCTION*

We are used to describing the substances around us as either solid, liquid or gaseous; the temperatures, at which substances change from one state into another, are characteristic, tabulated values. Whereas the boiling point depends strongly on the pressure, the pressure dependence of the solid-to-liquid transition is only of the order of 1 °C per 1000 atm, and therefore negligible [1]. The determination of the melting point is, in spite of highly developed spectroscopic and chromatographic techniques, still one of the most important methods for the determination of the purity and identity of an organic compound. It is therefore not surprising that in 1888, the Austrian botanist Reinitzer [2] took great care when describing the peculiar melting behaviour of cholesteryl acetate **1a** and cholesteryl benzoate **1b** (Figure 1):

> Concerning the melting point, there was a significant discrepancy from Schulze's data. He found it to be 150–151°. But in spite of continued and careful purification, I was only able to obtain 145.5° (146.6°, corrected). I noticed, however, that in the process, the substance was not melting to a clear, transparent liquid, which I first thought to be a sign of impurity, although microscopic and crystallographic investigations did not reveal any sign of dissimilarity. Indeed, on close scrutiny, it appeared that at higher temperatures, the turbidity suddenly disappears. This happens at 178.5° (180.6°, corrected). At the same time, I found that on cooling, the substance which had been heated to this high temperature displayed similar colour effects as have already been described for the acetate. It was mainly the manifestation of the existence of two melting points, as it were, and the occurrence of the colour effects which gave me the idea that here and in the acetate case, a physical isomerism must be present, which is why I asked Prof. Lehmann in Aix-la-Chapelle to take a close look at these things . . . .

8.1.1 80 YEARS OF RESEARCH

For years, the physicist Lehmann attempted to prove that a substance not only occurred in different crystal forms, but could also exist in different liquid states. At first, however, he agreed with Reinitzer, according to whom the turbid molten mass consisted of

* Cooperation by W.Calaminus (during doctoral thesis work) and by H.P.Schwenzfeier, A.Schröder, P.Windscheif, Ch.Seel and other former co-workers is appreciated.

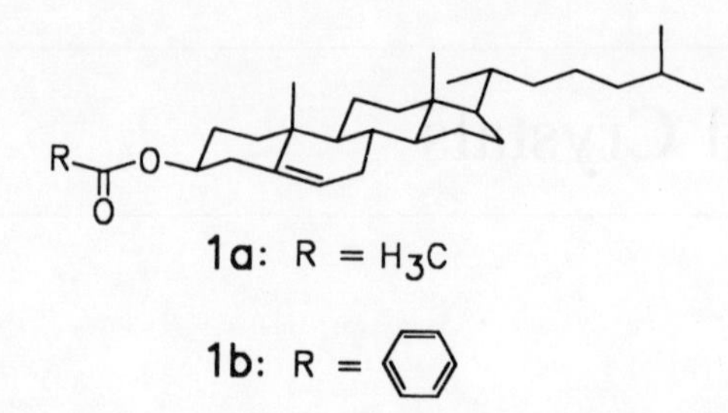

Figure 1. The Austrian botanist Reinitzer first observed the strange melting behaviour of the two cholesteryl esters **1a** and **1b**. Lehmann recognized that they displayed typical properties of both crystals and liquids, and because of this coined the term 'liquid crystals'

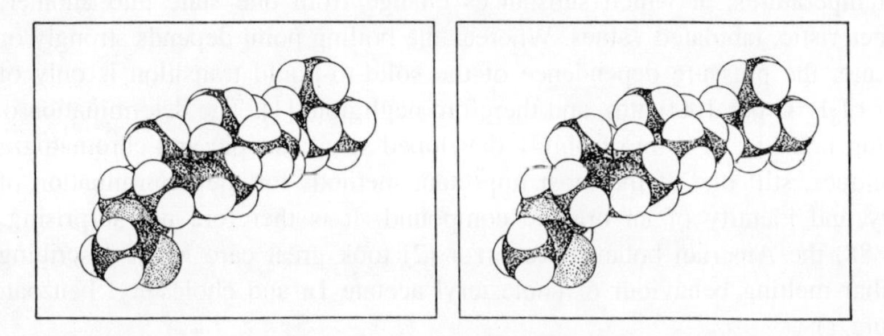

Figure 1a. Stereoview of a CPK model of **1a**, i.e. (10*R*)-3*c*-acetoxy-10*r*,13*c*-dimethyl-17*c*-[(*R*)-1,5-dimethylhexyl]-(8*c*H,9*t*H,14*t*H)-Δ^5-tetradecahydro-1*H*-cyclopenta[*a*]phenanthrene [3]

> a paste of doubly refracting crystals and an isotropic liquid, physically isomeric modifications namely, as there is no apparent reason for a chemical isomerism [4].

Based on further investigations he was able, one and a half years later, to rule out the possibility of pollution of the molten substance by minuscule crystals, which could have caused the observed double refraction of light. Initially, however, Lehmann was unable to leave the familiar concept behind, whereby crystalline order is only possible in the solid state, and ascribe to the observed liquid the properties typical of crystals.

Thus, the circumlocutions used for the observed samples range from 'flowing crystals' to 'seemingly liquid' and 'crystalline liquids' [4,5,7,8]. The systematic investigations that followed, led to an increase in the number of liquids studied from approximately 35 in 1903 [9,10] to over 100 in 1907 [11]. With the increasing number of liquid-crystalline substances available, the physical analyses also became more precise and more versatile (cf. Figures 2 and 2a). Thus, Lehmann and Vorländer found different, consecutive liquid-crystalline states in the same substance [9–12]. This was comparable to the different crystal modifications into which certain solids, such as tin, change with increase in temperature. The term 'polymorphism' ('allotropy' for elements), introduced in 1822 by Mitscherlich, was adopted. Vorländer was able to demonstrate that for certain compounds, a liquid-

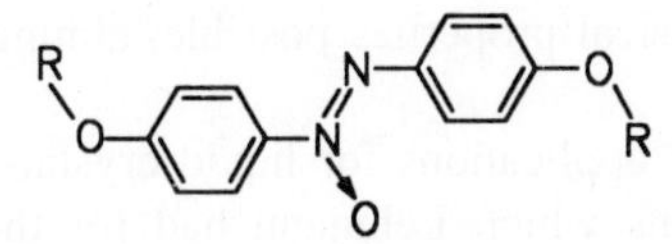

2a: R = CH_3
2b: R = C_2H_5
2c: R = $C_{11}H_{23}$

Figure 2. The short-chain 4,4′-alkoxyazoxybenzenes **2a** and **2b**, prepared by Gattermann, melt at 117 and 137 °C, respectively, and form clear, anisotropic nematic liquid melts, which become isotropic at 135 and 168 °C [7,8]. With longer alkoxy chains, these compounds also start to display smectic phases. Thus, 4,4′-diundecyloxyazoxybenzene (**2c**) melts at 80.8 °C to form, over a range of 41.1 °C, a smectic s_C phase

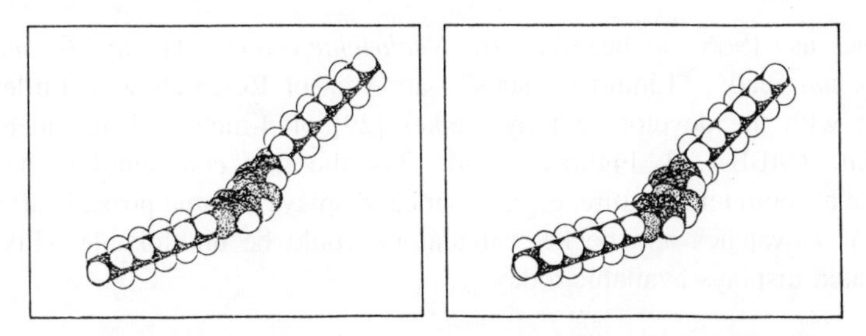

Figure 2a. CPK model of 4,4′-diundecyloxyazoxybenzene **2c** [6] (stereoview)

crystalline phase could only be observed after careful undercooling of the isotropic liquid melt, affording a so-called monotropic metastable phase. He was also able to show that mixtures of compounds which otherwise are not liquid crystalline, are able to form liquid-crystalline phases [14].

Shortly after the turn of the century, the effects of magnetic and electrical fields on liquid crystals were studied by, among others, Mauguin and Björnstahl [15]. Encouraged by von Laue, van der Lingen (1913) and Hückel (1921) conducted the first experiments in which X-rays were applied to liquid-crystalline phases [16,17]. The first useful results, however, were only obtained in 1923 by de Broglie and Friedel [18–20]. At the time, mostly owing to the lack of a general model for the structural organization in a liquid crystal, a bewildering multitude of terms and expressions were in use. In 1922, Friedel therefore suggested for the classification of liquid-crystalline phases the now commonly used terms *smectic*, *nematic* and *cholesteric*. From their optical appearances and X-ray diffraction patterns, he deduced the structures of some crystalline liquids [21]. In order to avoid any confusion between the structure of molecules and the structure of a liquid-crystalline phase, Friedel in 1931 introduced the term texture [22], which today is mostly used for the characterization of the appearance of a liquid-crystalline phase by polarization microscopy. Friedel also introduced the term 'mesomorphism', which was meant to replace the contradictory expression 'liquid crystal'. However, the latter describes the nature of these compounds so well that it prevailed.

With all this, the basics had been acquired, and in subsequent years the number of liquid crystals studied increased from approximately 1000 in the early 1940s to about 2000 by the mid-1950s [9,23]. This made comprehensive comparisons between molecular structure and physical properties possible, aiming at deducing a basic structural principle.

Almost immediately, technological applications for liquid crystals were looked for, but at first the great expectations which Lehmann had for them were not fulfilled, and liquid crystals remained a mere curiosity.

In *The Chemical Crystallography of Liquids*, Vorländer in 1924 observed [26,27]:

> I have certainly been asked, whether the liquid-crystalline substances might be used industrially. I see no possibilities for it. And yet, to the knowledgeable, they offer so much [26,27].

As late as 1966, a headline in *Nachrichten aus Chemie, Technik und Laboratorium* read: ' "Liquid Crystals"—an Area of Research with Little Use?' [28]. But with the development by Kelker [29] of 4-methoxybenzylidene-4′-n-butylaniline (MBBA, **3**; Figures 3 and 3a), the first compound to be liquid crystalline at room temperature, electro-optical displays became possible after 1962 [30,31]. Wristwatches and pocket calculators would be unthinkable without the sophisticated displays available today.

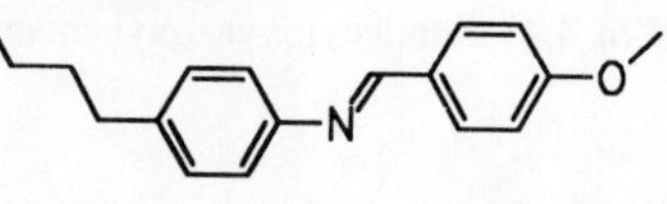

3 "MBBA"

Figure 3. With 4-methoxybenzylidene-4′-n-butylamine (**3**), Kelker in 1969 prepared the first compound to be liquid crystalline at room temperature. This Schiff's base with the acronym MBBA has a melting point of 21 °C, and below 45 °C it forms a nematic phase

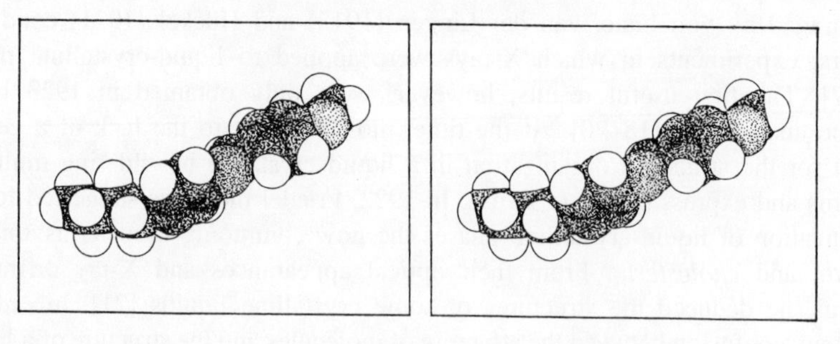

Figure 3a. Stereoview of the CPK model of 4-methoxybenzylidene-4′-n-butylamine (MBBA, **3**) [25]

8.2 MOLECULAR ORDER—AND THE CONSEQUENCES

We find the state of highest order which molecules can adopt in crystals. Here the molecules, being fixed in all three dimensions and unable to rotate, occupy rigid, regular positions in the lattice*. The smallest periodic unit in a crystal lattice is the unit cell, which was postulated by the French mineralogist Haüy (1743–1822) and established by von Laue (1879–1960) [33–36]. Provided the crystal growth is unhindered, this arrangement causes the formation of the ideal crystal form, with planar crystal faces and constant, characteristic angles of intersection [37]. In such single crystals, the unit cells are linearly arrayed in rows and columns, and this is called ideal long-range order. However, as a rule, more or less extensive disorders or structural faults, induced by impurities or other external factors, occur in the lattice.

As a consequence of the fixed order of the molecules, various physical properties are direction dependent, and this is termed *anisotropy*. Anisotropic physical properties of crystals include their cleavage, their polarizability and the velocity of the propagation of light. Thus an expert is able, after careful examination and with simple tools, to cleave the hard, anisotropic diamond crystal along certain crystal faces (Figure 4).

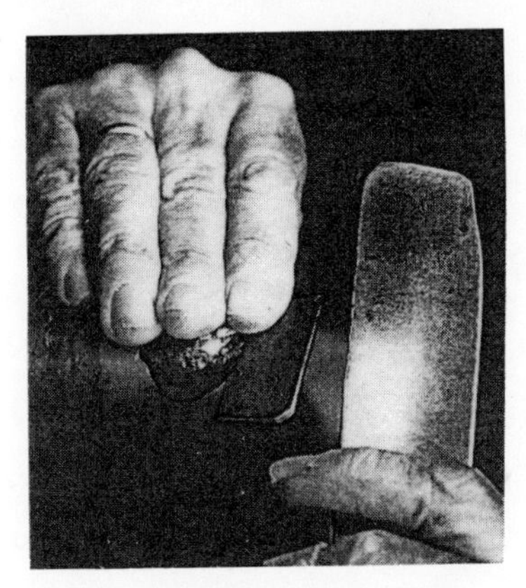

Figure 4. With simple tools, an expert cleaves an anisotropic diamond crystal [38]

The anisotropy of optical properties manifests itself in what is called *birefringence* (double refraction in crystals) [39,40]. Bartholinus in 1669 observed that an object, which was viewed through a crystal of calcite, appeared as a double-image. This observation was explained by Huygens in 1678 using the wave theory: light rays coming from each point on the object are, on passing through the crystal, split into two rays, which are each subject to different diffractions (Figure 5; see also Figure 6). After the discovery of linearly polarized light by Arago (1807), it was recognized that both rays are linearly polarized, and that their planes of vibration are perpendicular to each other. The direction of polarization is determined by the crystallographic structure of the crystal (Figure 7).

* Solids, in which only weak intermolecular interactions are present, leading to easy migration processes within the crystal and, thus, to plastic deformations, are called plastic crystals [32,33]. Since both their heats of fusion and vaporization are so low (cf. camphor), they produce a large molar depression of the freezing point, which in cryoscopy can be used for the determination of the molar mass.

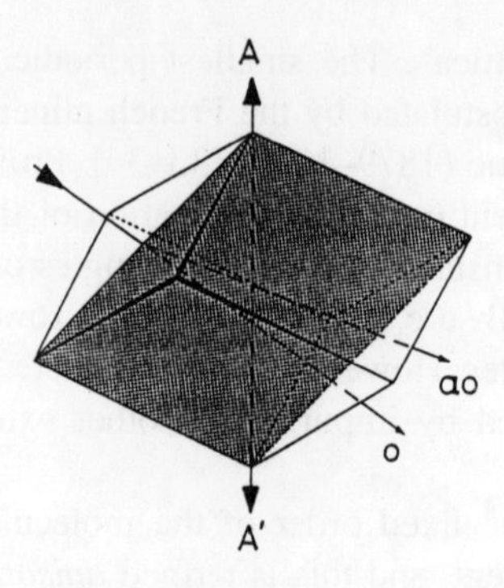

Figure 5. The birefringence in calcite. The plane defined by the incident ray and the axis AA′ is the ray's 'main section'

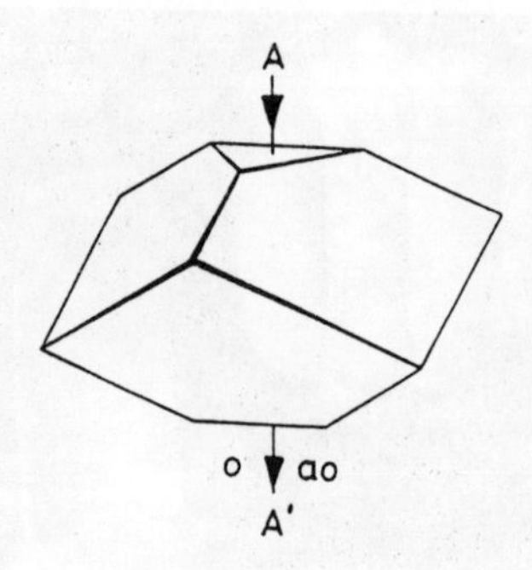

Figure 6. In the optically uniaxial calcite crystal, the optic and crystallographic axes (C_3 symmetry) are coincident

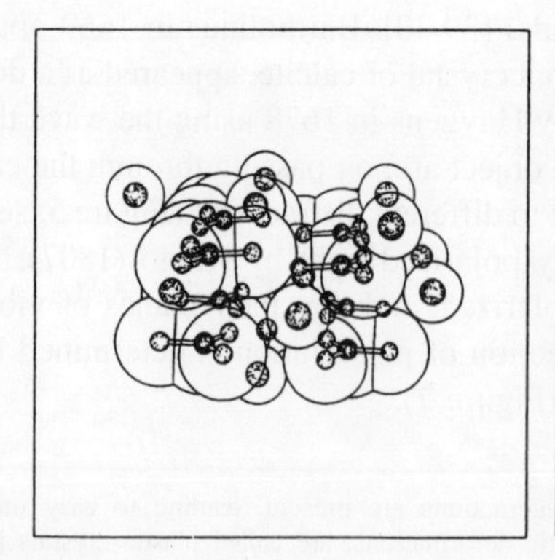
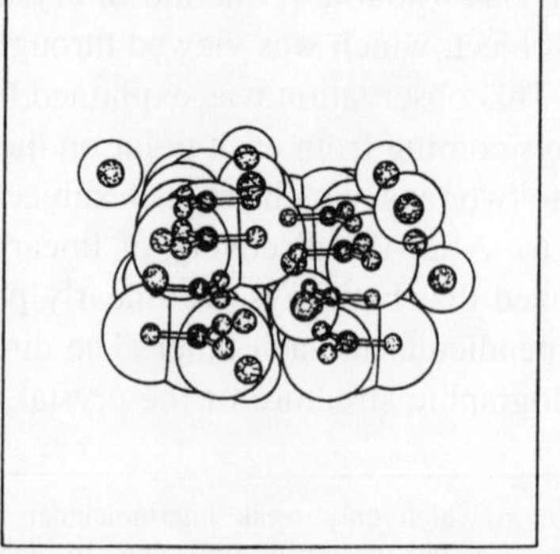

Figure 7. Stereoview of the cleavage rhombohedron of calcite ($CaCO_3$) with the optic axis AA′ (C_3 symmetry) [41]

The direction-dependent refractive indices are due to the different velocities of propagation of light in the crystal. If, from a point within an uniaxial crystal, distances are measured which are proportional to the velocity of propagation of each of the linearly polarized components, wavefronts are obtained corresponding to a sphere and an ellipsoid, respectively. The sphere and the ellipsoid touch at two points, which together define a direction, called the *optic axis*. This is a distinguished direction in the crystal, in which both rays have the same velocity. A ray's component for which the velocity of light is the same in all directions in the crystal is called an *ordinary ray*; it oscillates perpendicularly to the plane formed by its direction of incidence and the optic axis. The ray oscillating perpendicularly to it is called the *extraordinary ray* [39].

If the difference between the refractive indices of the ordinary and extraordinary rays is larger than zero, it is said that the crystal has a positive optical behaviour (Figure 8) [42,43]. In Figure 9, Fresnel's construction of birefringence is demonstrated for two simple cases. Crystals, when examined with a simple microscope, only show their natural colour but if, as was done by Sorby in 1858,

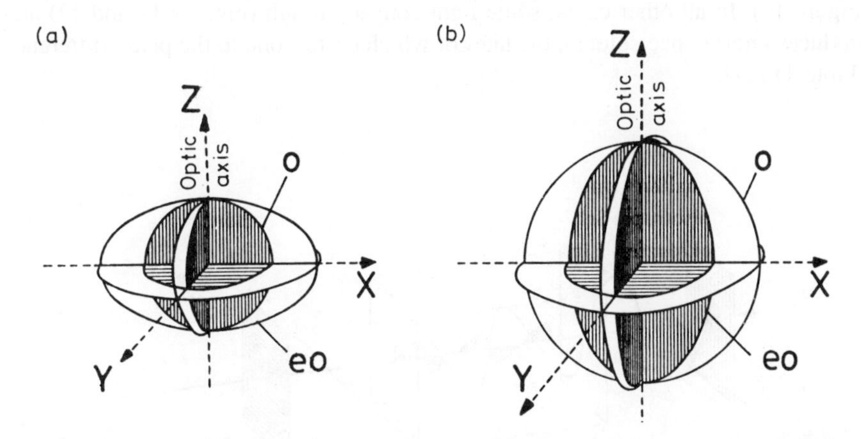

Figure 8. Optical planes in crystals. (*a*) Calcite (negative uniaxial) $c_{eo} = 1.116c_o$; $n_{eo} = 1.4864 < n_o = 1.6583$; $\Delta n = -0.1719$; c = velocity of light; o = ordinary ray. (*b*) Quartz (positive uniaxial). $c_{eo} = 0.993c_o$; $n_{eo} = 1.5553 > n_o = 1.5442$; $\Delta n = +0.0111$; eo = extraordinary ray; n = refractive index

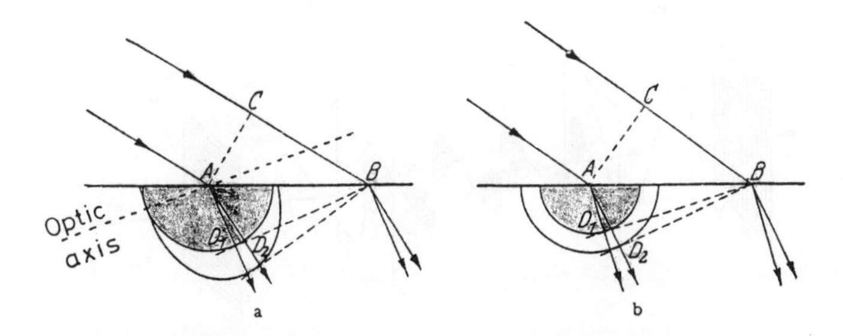

Figure 9. Fresnel's construction of birefringence

two polarizing filters whose planes of polarization are perpendicular to each other are placed before and after the sample, anisotropic materials will shine with bright interference colours. Isotropic solids and liquids, and also the free field of vision, remain dark.

This behaviour in the polarization microscope is also due to the double refraction of anisotropic materials. If linearly polarized light is incident at right-angles to a sample face, which is parallel to the optic axis of the anisotropic compound, the ray will be separated into two components, oscillating perpendicularly to each other. Under these conditions, the two components are not refracted, but keep moving in the same direction, although with different phase velocities.

After they have passed through the sample, the second polarizer (analyser) will only let pass the component which is oscillating in its plane of polarization. Owing to the different phase velocities incurred during the passage through the anisotropic sample, a phase difference results as shown in Figures 10–12, which illustrate the behaviour of light on passing through an anisotropic sample placed between crossed polarizing filters. If the phase difference is zero, as is formally also the case after the passage through an isotropic medium, then no light passes the analyser (Figure 10). In all other cases, some light comes through (Figures 11 and 12) and produces interference colours, the hues of which correspond to the phase difference (Table 1) [37].

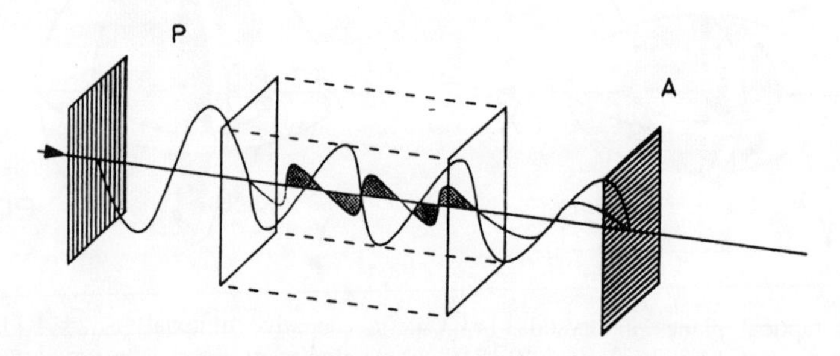

Figure 10. Quenching by interference of the ordinary and the extraordinary rays

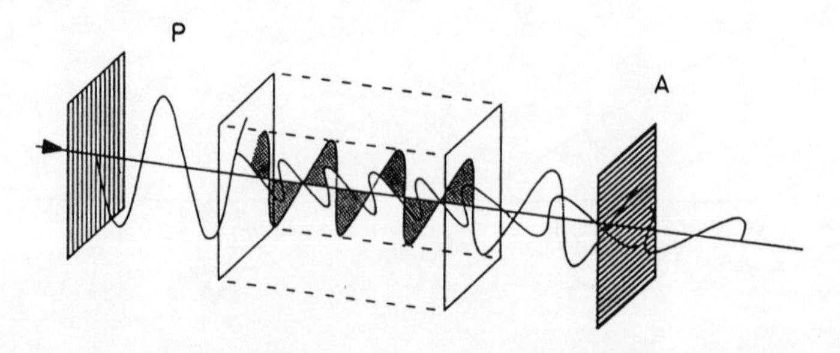

Figure 11. The partial rays of shorter waves are not quenched and pass through the analyser

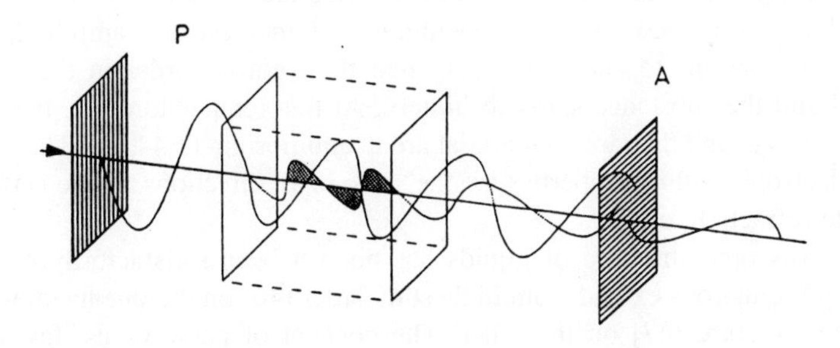

Figure 12. If the anisotropic sample is thin enough, rays of light will reach the analyser, even in a situation as in Figure 10

Table 1. The interference colours usually obtained with daylight as a function of the phase difference

Phase difference (nm)	Order	Interference colours between crossed polarizers	Interference colours between parallel polarizers
0	First order	Black	Light white
40		Dark grey	White
97		Lavender–grey	Yellowish white
218		Grey	Brown–yellow
259		White	Light red
306		Light yellow	Indigo blue
430		Brown–yellow	Grey–blue
536		Red	Pale green
565		Purple	Light green
575	Second order	Violet	Greenish yellow
589		Indigo blue	Golden
664		Azure	Orange
728		Greenish blue	Brownish orange
747		Green	Light carmine
866		Greenish yellow	Violet
910		Yellow	Indigo blue
998		Vivid orange–red	Greenish blue
1101		Deep red–violet	Green
1128	Third order	Light bluish violet	Yellowish green
1151		Indigo blue	Yellow
1258		Greenish blue	Flesh-coloured
1334		Sea-green	Brown–red
1426		Greenish yellow	Grey–blue
1534		Carmine	Green
1621		Dull purple	Dull sea-green
1652		Violet–grey	Yellowish green

8.2.1 MOLECULAR MOTION IN THE CRYSTAL

Warming a crystal induces the molecules to oscillate at their sites in the lattice. Eventually, a continued rise in temperature will increase the amplitude of the particles' oscillations to such an extent, that the rigorous order in the crystal is disrupted and the substance starts to liquefy. At this temperature, i.e. the melting point, the crystal and the molten material are in equilibrium [1,44,45]. The resulting liquid is isotropic and its properties are the same in all directions. There is therefore no double refraction, either.

The microscopic structure of liquids has not yet been satisfactorily explained. Model representations extend from highest disorder [46] on the one hand, to nearly crystalline structure [47] on the other. The concept of paracrystals, favoured by Hosemann, is somewhere in between [35,36,48]; here, the microscopic structure is thought to consist of crystalline regions of varying size, where the lattice is somewhat 'blurred,' while mobile particles make the liquid fluid. The only small difference in density between the two states seems to indicate that a liquid is more similar to a solid, than to a gas. In isotropic liquids, only short-range order has been detected and its three-dimensional range is too short to be optically recognized (Figure 13). During the solid-to-liquid transition, the molecular long-range order, which ensured the periodical array in the crystal over macroscopic distances, disappears. The difference between crystal and liquid is, in that sense, only a question of degree.

In about 5% of all organic compounds, the thermal energy is, at first, not sufficient to transform the three-dimensional, periodic array in the crystal into an isotropic liquid [50]. Thus, intermolecular interactions, strong enough to guarantee

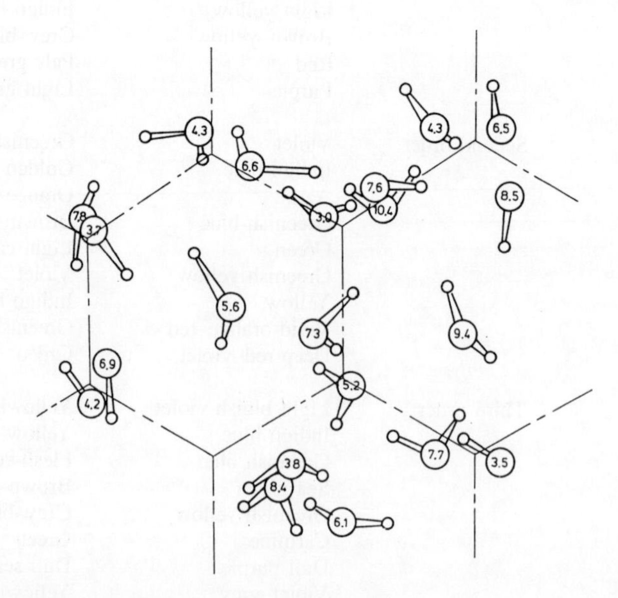

Figure 13. A Monte-Carlo simulation of the orientation of molecules in liquid water reveals short-range order within small volumes

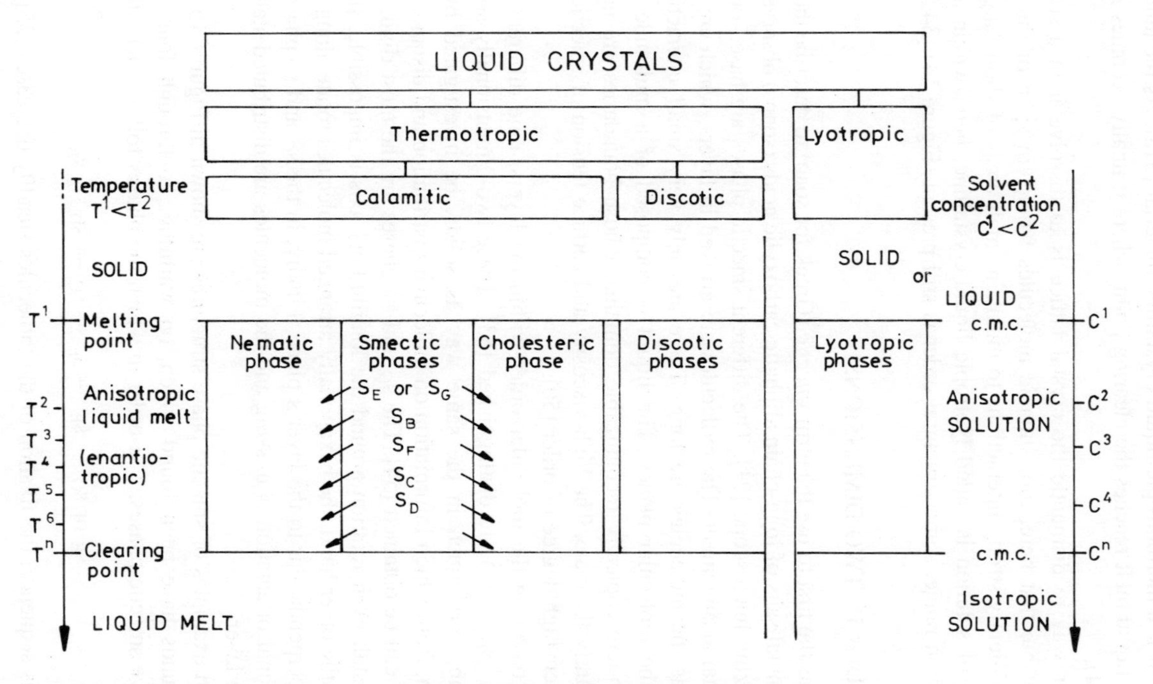

Figure 14. Classification of liquid crystals and their possible phase transitions

a characterizable molecular order, still subsist in the liquid melt. This long-range order must be over 400 nm wide, since it can be detected with visible light (birefringence) [51].

This type of liquid melt displays the typical characteristics of both a liquid and a crystal—it is a thermotropic liquid-crystalline material [51a]. As the temperature rises, this liquid melt reaches the clearing point, where it finally becomes isotropic (Figure 14).

Another way to dismantle the crystal lattice is to dissolve it in a solvent. It is conceivable that here, too, suitable molecules would, in spite of the solvent, retain sufficiently strong interactions to maintain small areas of short-range order. This type of solution is called lyotropic liquid-crystalline; here, within a certain concentration range, some structural order is still present (Figure 14) [34,52].

8.2.2 ORDER IN TWO DIMENSIONS

Friedel was the first to use the term *smectic* (Greek for soap) to describe the texture of alkali metal salts of long-chain aliphatic carboxylic acids (soaps) observed under the polarizing microscope [21]. The different smectic phases are based on a two-dimensional arrangement. The molecules are arrayed in layers, which are roughly as thick as the molecules are long. There are only very small interactions left between the molecular planes. The liquid is composed of a multitude of such ordered, microscopically recognizable domains, whose boundaries are not rigid, but constantly fluctuate. The high viscosity and surface tension of smectic phases reflects their high degree of order [50].

Today, based on the molecular order within the layers, nine different smectic phases (s_A, s_B, ..., s_I) are distinguished [53]. These two-dimensionally structured liquids can, very much in the same way as solids, be investigated by X-ray diffraction, from which information on molecular distribution and distances within the layers can be obtained [54]. The s_G and s_E phases are the most rigid; here, as in the crystal, even rotation around the longitudinal axis is impossible. In the s_B phase, on the other hand, the hexagonally arranged molecules rotate along the axis which is perpendicular to the layer's plane. Finally, in the s_C and s_A phases, even the hexagonal orientation is missing and the molecules are distributed statistically within the layer.

Selected examples of smectic phase structures are shown in Figures 15–18.

Compounds have been found which, on warming, go through four or more consecutive smectic phases. The order in which the phases follow each other is

$$s_E \text{ or } s_G \rightarrow s_B \rightarrow s_F \rightarrow s_C \rightarrow s_D \rightarrow s_A$$

Within this sequence, the fixation of the molecules steadily decreases [55].

8.2.3 ORGANIZED BUT MOBILE MOLECULES

Friedel called a second type of thermotropic, liquid-crystalline phases *nematic* (Greek for thread); between crossed polarizing filters these show thread-like

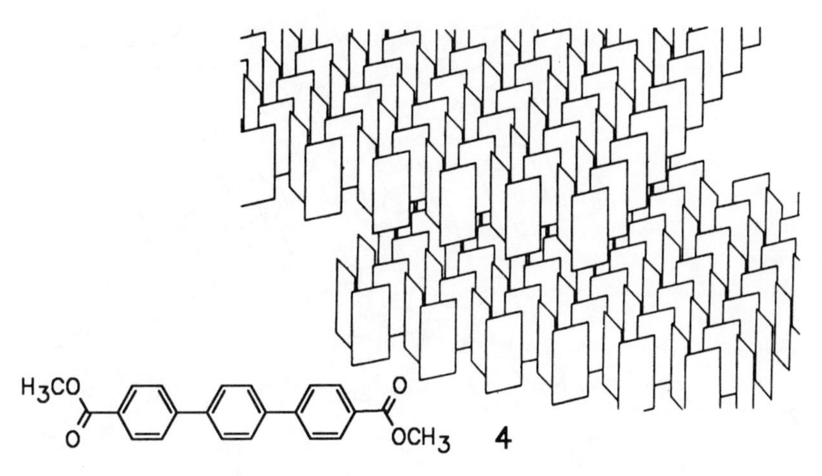

Figure 15. Molecular array in an s_E phase, as observed with diethyl *p*-terphenyldicarboxylate (**4**). The almost crystalline structure allows no rotation of the molecules

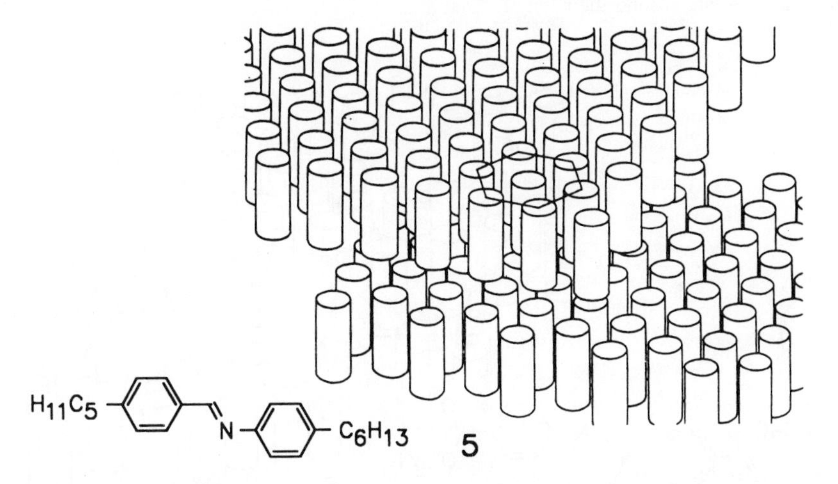

Figure 16. Molecular array in an s_B phase. The molecules are perpendicular to the layer's plane, each at the centre of a hexagonal system, and are free to rotate around their longitudinal axis. 4-Pentylbenzylidene-4′-hexylaniline (**5**) is such an example

irregularities in the liquid melt and are less viscous than smectic phases [21]. Particularly striking are the highly mobile drops, sparkling with bright colours, and the so-called schlieren texture (centred texture) [42,51,53].

The nematic phase consists of molecules which are parallel to each other and free to move in all three directions, but can only rotate along their longitudinal axis (Figure 19). An ideally parallel orientation of the molecules, however, is not possible, owing to their thermal motion. Using the statistical mean value of the angle, θ, describing the departure of the molecules' axes from the ideal orientation,

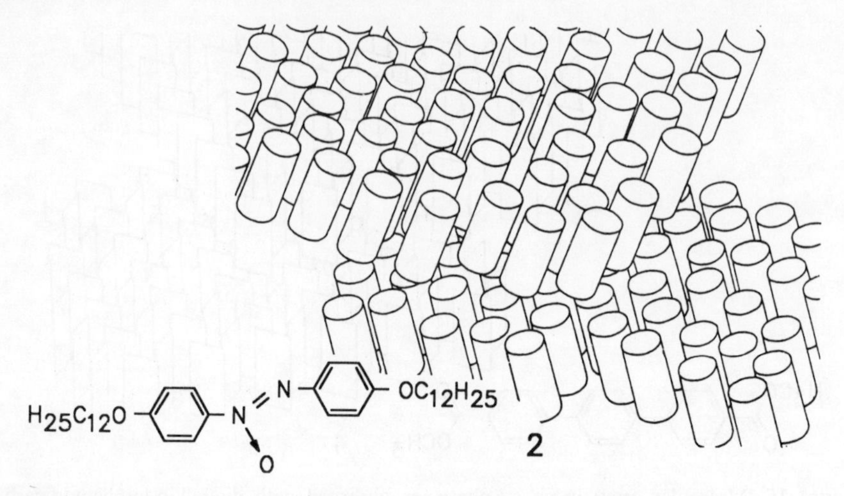

Figure 17. The molecular array in an s_C phase, which occurs with 4,4′-didodecoxyazoxybenzene (**2**). The molecules are distributed statistically within a layer, all tilted to one side and free to rotate around their longitudinal axis

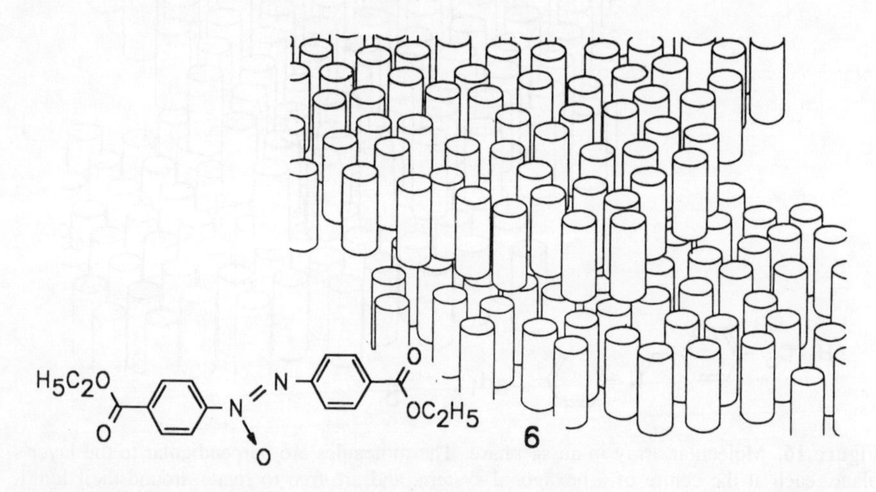

Figure 18. Molecular alignment in an s_A phase. The molecules are perpendicular to the layer's boundary, rotate around their longitudinal axis and move freely within their layer. An example is 4,4′-diethoxycarbonylazoxybenzene (**6**)

Tsvetkov in 1938–39 introduced a definition for the degree of order, S, in a nematic phase [56]:

$$S = 1 - \frac{3}{2}\sin^2\theta$$

Accordingly, ideally organized nematic phases have a degree of order $S = 1$ and an isotropic liquid melt has $S = 0$.

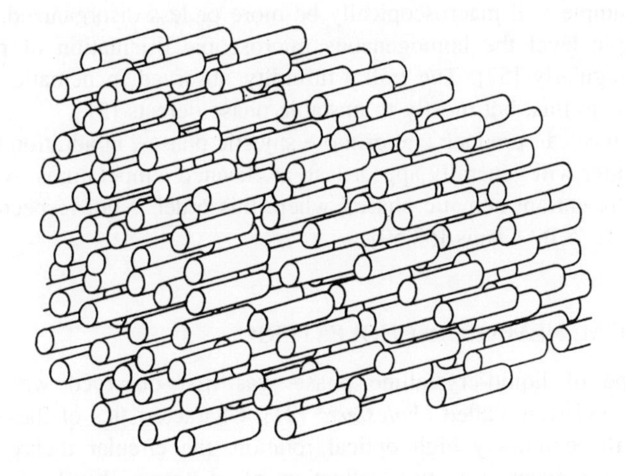

Figure 19. Phase structure of a nematic liquid crystal

The average deviation of the molecules' orientation from the favoured direction can be determined by X-ray diffraction or NMR investigations. These measurements can only be carried out on nematic phases, which have been organized over a macroscopic area, either by electric or magnetic fields, i.e. with an additional energy expenditure. The S-values of nematic phases calculated from such experiments are between 0.3 and 0.8 (Figure 20). When an additional outside orientation is not used, even the smallest disturbances, e.g. variations of density due to unavoidable inhomogeneities in temperature, produce strong internal bending and torsional deformations, and also currents and swirls. Without particular measures,

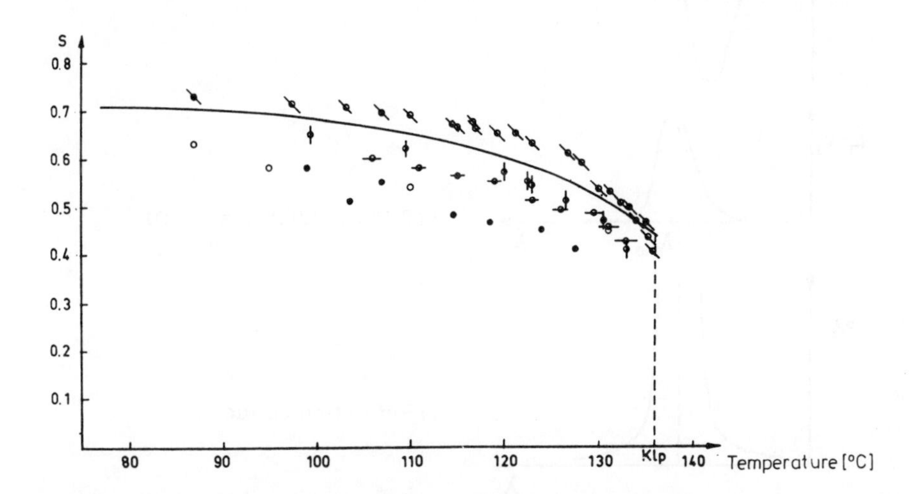

Figure 20. Degree of order S of 4,4′-dimethoxyazoxybenzene (**2a**) from (•) IR spectra, (○) UV spectra, (ϕ) diamagnetic susceptibility, (-○-) refractive index at 546 nm, (⦰) proton resonance. The solid curve was calculated from theory [42]

a nematic sample will macroscopically be more or less disorganized, because on a macroscopic level the homogeneous microscopic orientation of molecules is changing irregularly [57]. The milky turbidity observed in nematic liquid melts is due in part to this, but mostly to unstable phase defects [51].

If a substance can produce one or more smectic phases, in addition to a nematic phase, the latter will generally appear last, at elevated temperatures. A rare case of a so-called 're-entrant nematic phase,' where this order is not respected, was first observed in 1975 by Cladis [58,59].

8.2.4 HELICAL NEMATIC = CHOLESTERIC

A third type of liquid-crystalline phase was first observed with cholesteryl compounds, and were called *cholesteric* [21]. Characteristics of these cholesteric phases are the extremely high optical rotation, the circular dichroism and the temperature-dependent, selective reflection of a narrow band of wavelengths (Figure 21), producing the colour effects observed many years earlier by Reinitzer [60]. Friedel recognized the close relationship between nematic and cholesteric phases, considering the latter to be a special case of the nematic state [21]. This assumption was supported by some of Vorländer's experiments, where he was able to induce a cholesteric behaviour in nematic phases by

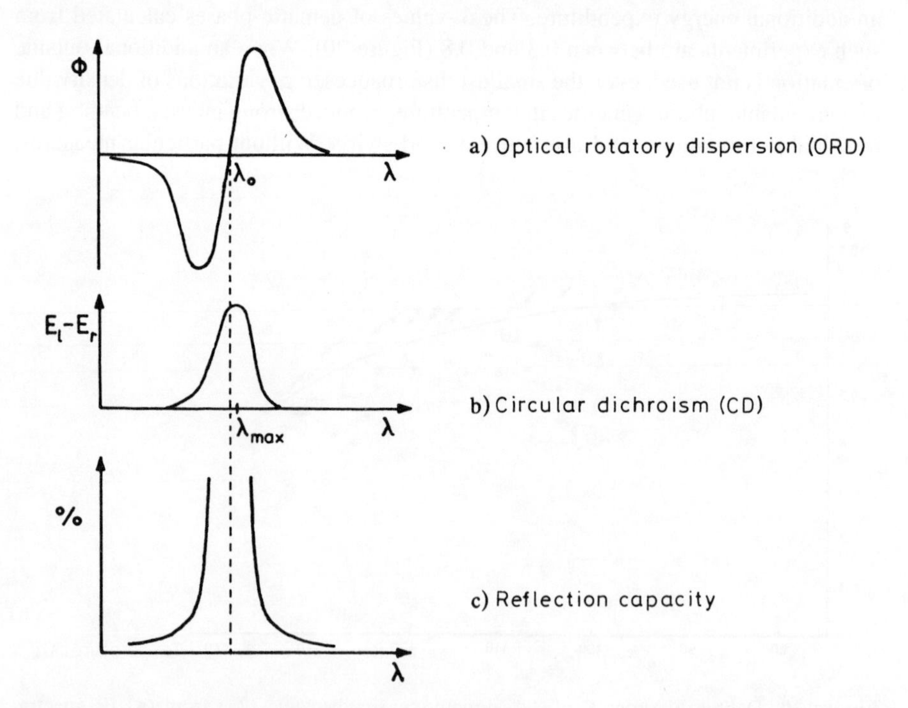

Figure 21. The optical properties of cholesteric phases are strongly dependent on the wavelength of the irradiated light

adding chiral, as such not liquid-crystalline compounds to a nematic phase [61]. Because X-ray diffraction experiments did not reveal any differences between cholesteric and nematic phases, their structure and the complex optical phenomena remained unexplained for some time [27]. In 1933, Oseen assumed that molecular chirality, as a prerequisite for cholesteric behaviour, was imposing a chiral, helical superstructure on a nematic phase [62]. The molecules are organized in a kind of parallel order but, unlike in nematic phases, their preferred orientation varies continually in a direction perpendicular to the molecular axis, i.e. from layer to layer. Thus, a chiral, helical phase structure is formed, in which the pitch ranges from 200 to 2000 nm. Like the absolute configuration of chiral molecules, the dextrorotatory and laevorotatory phase-helices cannot be established by X-ray crystal structure analysis [63].

Enantiomers form phases with the same pitch, but with opposite senses. Nematic phases can be considered to be cholesteric phases with an infinite pitch.

De Vries in 1951 demonstrated that, with this model, the temperature-dependent, selective and circular-polarizing reflection of light on cholesteric phases could be explained [60,65,66]. The highest specific molar rotation ever recorded, ca 18 000 °/mm, was obtained with some cholesteric liquid crystals; for comparison, the value for active quartz is about 20–24 °/mm.

Such a high rotation can only be attributed to the chiral helical structure of the cholesteric phase (Figure 22), and not to the chiral molecules themselves [60,61,64]. The light reflected from a cholesteric phase has a wavelength which, in a narrow range, is proportional to the pitch of the helix; depending on the rotatory sense of the cholesteric phase, it could, for example, be laevorotatory, while its dextrorotatory component is let through. In contrast to metallic reflections, there is no phase change and no inversion of the rotatory sense (Figure 23) [59,61,62].

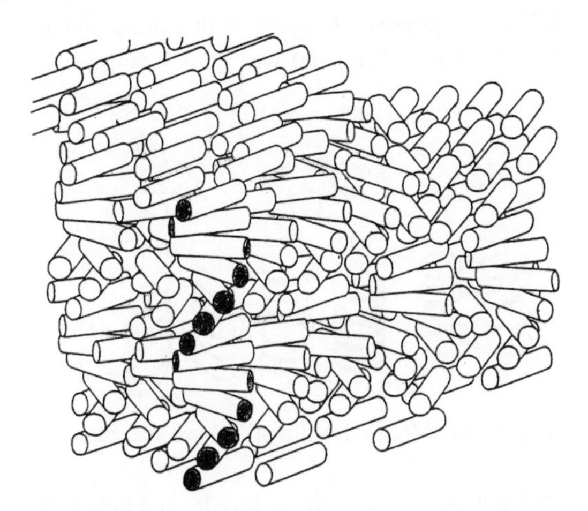

Figure 22. In cholesteric phases, the uniform change of the molecules' favoured orientation describes a helix

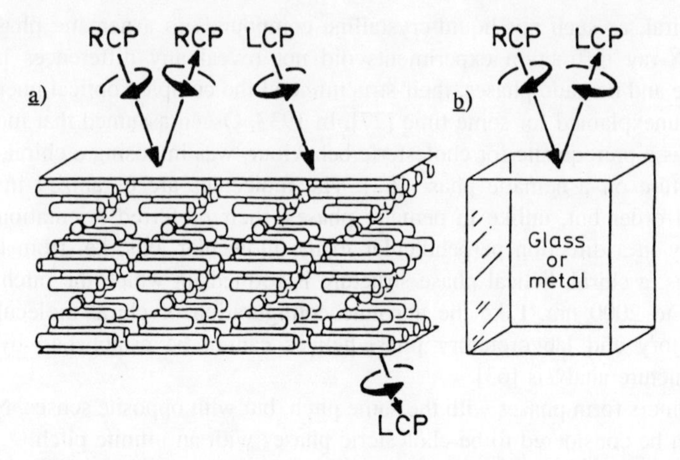

Figure 23. Circular polarized light behaves differently on the surface of a cholesteric liquid crystal than on the surface of glass or metal (RCP = right circular polarized, LCP = left circular polarized)

8.2.5 STACKED MOLECULES

The three 'classic' thermotropic liquid-crystalline phases are made of elongated and linearly built molecules, as shown by Vorländer. He also recognized the possibility of a structure, where

> molecular plates would lay in bundles, face-to-face, one on top of another, much like a voltaic pile, resulting in an optically anisotropic formation [67].

His investigations with star-, cross- and plate-like molecules, however, went against this particular hypothesis. It was only in 1977 that liquid-crystalline phases built from similar, disc-like molecules were detected [68]. They are called *discotic* (Greek for disc), whereas for phases which are built from elongated molecules, the collective name *calamitic* (Greek for tube) was coined (Figure 14).

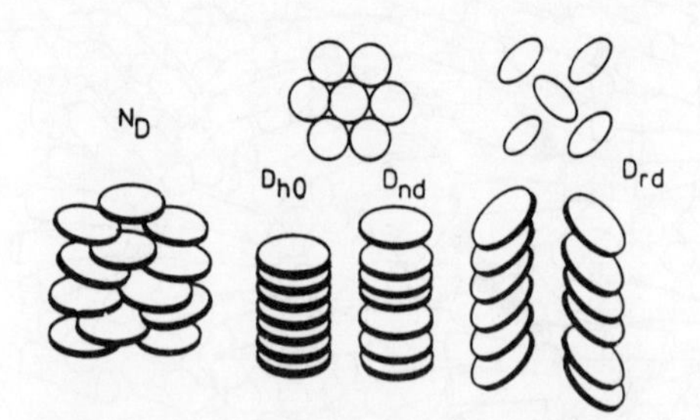

Figure 24. Models for discotic phase structures

Since 1961, a 'carbonaceous' mesophase has been known, which consists of large, disc-shaped polycondensed aromatics with a molecular mass of around 2000 [69].

An X-ray crystal structure analysis of the mesophase of hexakis(n-octanoyloxy)benzene (**7c**, Figures 25 and 25a), one of the first discotic liquid crystals, revealed a hexagonal stack of molecules [68].

Among the 100 or so compounds which have been characterized as discotic liquid-crystalline, there are also examples of molecules, whose shapes are not flat and discotic, but conical or prismatic, allowing them to stack one inside another (Figures 26 and 26a) [73,74]. Comprehensive investigations of the various discotic liquid-crystalline phases are still lacking.

8.2.6 STRUCTURED SOLUTIONS

A *lyotropic system* was probably observed for the first time in 1854 by Virchow, who reported it in an essay 'On the Widespread Occurrence of a Substance Similar to Marrow in Animal Tissues', but without describing it as such [75]. Again, it was Lehmann who recognized the anisotropic behaviour of aqueous solutions of ammonium oleate, calling it liquid-crystalline [76].

Lyotropic systems consist of at least two components: a solvent, mostly water, and an amphiphilic compound with a polar, hydrophilic 'head' and a long hydrophobic 'tail'. Examples are soaps, sphingomyelin, which is found in cell membranes, and the sulphonic acid **9** (Figures 27 and 27a). By mixing both components within certain concentration and temperature ranges, micellar or laminar molecular aggregates are formed; Oswald in 1931 already pointed

	R	Melting point (°C)	Clearing point (°C)
7a	C_5H11	68.3	86.0
7b	C_6H_{13}	80.2	83.6
7c	C_7H_{15}	79.4	81.5

Figure 25. Chandrasekhar in 1977 introduced the first discotic liquid crystals of low molecular weight, the hexakis(n-alkanoyloxy)benzenes (**7a–c**). When the length of the side-chains exceeds five carbon atoms, on melting the compounds change into a mesophase consisting of stacks of the plate-like molecules [68]

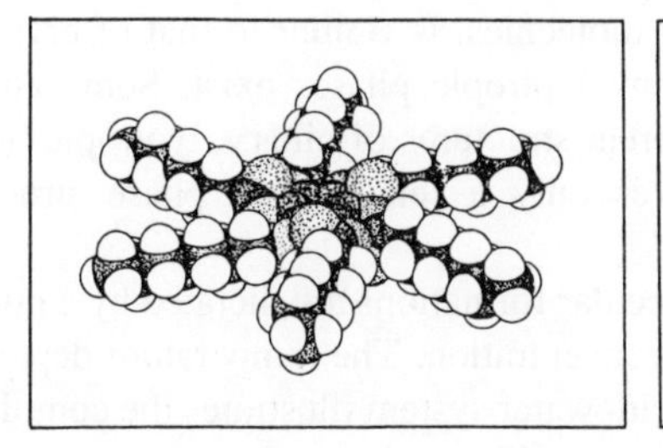

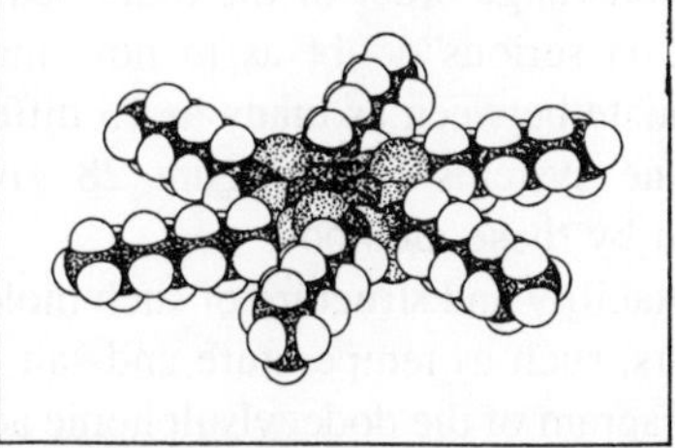

Figure 25a. CPK Model of hexakis(n-octanoyloxy)benzene (**7c**) [71] (stereoview)

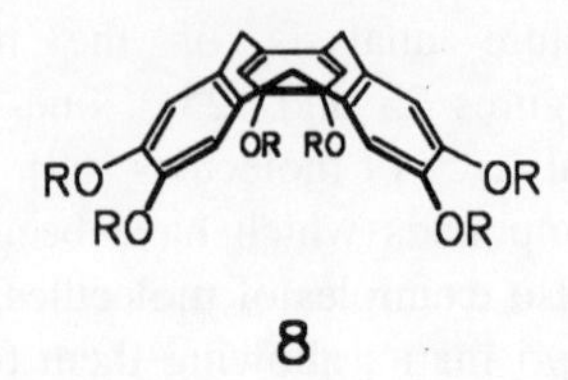

Figure 26. Discotic mesophases have been observed not only with flat disc-like molecules, but also with conical molecules. In 1985, several so-called pyramidal liquid-crystalline phases of 2,3,7,8,12,13-hexaalkoxy-5,10,15-dihydrotribenzo[*a*,*d*,*g*]cyclononenes (**8**) were observed [72,74]

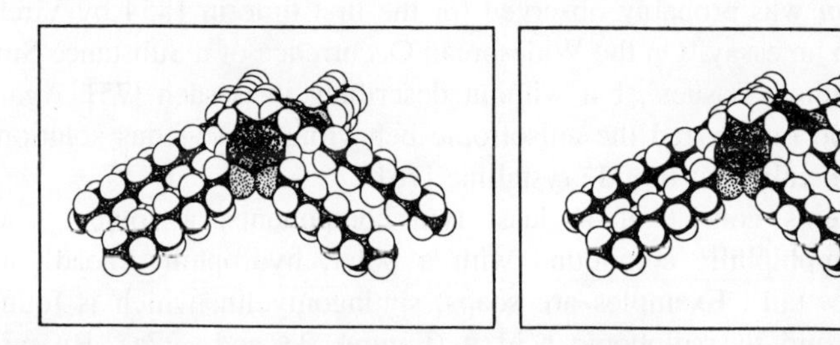

Figure 26a. Conformation of 2,3,7,8,12,13-hexahexyloxy-5,10,15-dihydrotribenzo[*a*,*d*,*g*]-cyclononene (**8**, R = hexyl) [75] (stereoview)

out their close relationship to colloidal solutions [77]. On either increasing the solvent concentration or raising the temperature, lyotropic liquid-crystalline systems reversibly change into anisotropic solutions. An important parameter is called the critical micellar concentration (c.m.c., cmc), which is defined as being the concentration at which system properties, such as surface tension, density and turbidity, change, due to micelle formation or changes in aggregate structure (Figure 14).

Hess's [78] and Hartley's [79] model for the structure of these aggregates was later confirmed by X-ray diffraction experiments, which relate to first investigations by Stauff in 1939 [80,81]. In addition, the size of the molecular aggregates, which determine the properties of lyotropic liquid crystals, allowed investigations with the electron microscope once suitable techniques had been developed [82,83]. It turned out that the long-range order of these molecular aggregates, in spite of the only small short-range order of the individual molecules, is similar to that of crystals.

There is serious doubt as to how many lyotropic phases exist. Some authors differentiate between as many as 15 different structures of binary, lyotropic liquid-crystalline systems [54]. Figure 28 gives only a selection of phase structures analysed by these methods.

The stability and structure of such molecular formations are dictated by a number of factors, such as temperature and salt concentration. The temperature-dependent phase diagram of the dodecylsulphonic acid–water system illustrates the complexity of the phase transitions which may be present (Figure 29) [86].

The use of lyotropic liquid crystals is far behind the technological applications of

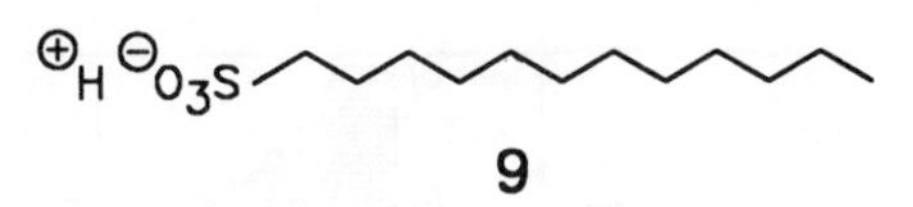

Figure 27. Numerous amphiphilic compounds form, together with solvents or solvent mixtures, characteristic aggregates. For instance, dodecylsulphonic acid (**9**) with water can form lyotropic liquid crystals, depending on concentration and temperature (cf. Figure 28)

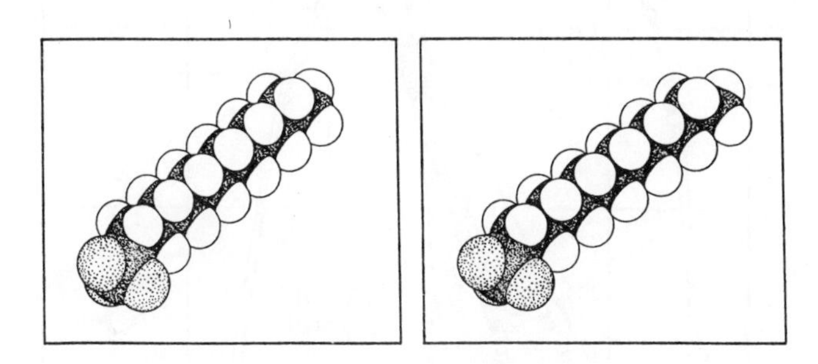

Figure 27a. The anion of dodecylsulphonic acid (**9**) [84] (stereoview)

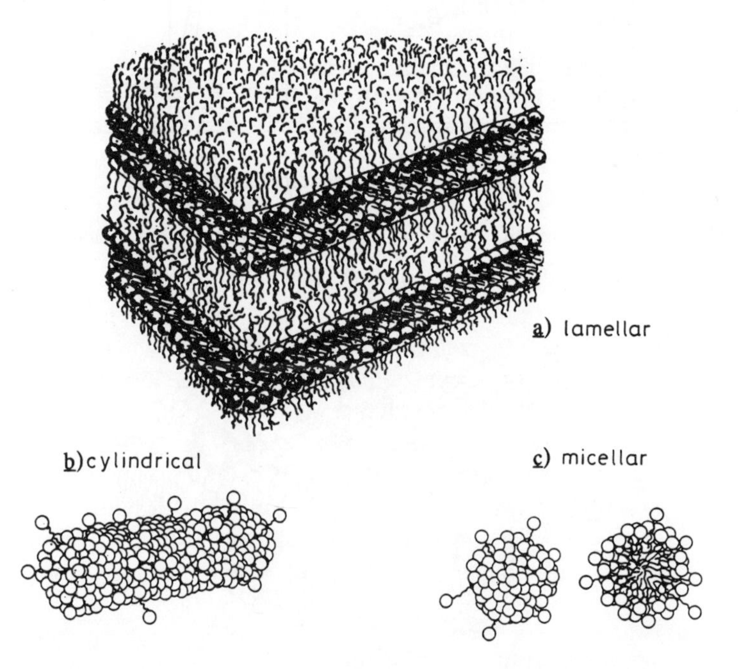

Figure 28. Lyotropic liquid-crystalline structures exist in three general forms: (*a*) lamellar double layers (neat or G-phase); (*b*) hexagonal, cylindrical structure (middle or M_1 phase); (*c*) cubic micellar structures (viscous or V_1 phase). Moreover, numerous variants are conceivable, some of which have already been detected

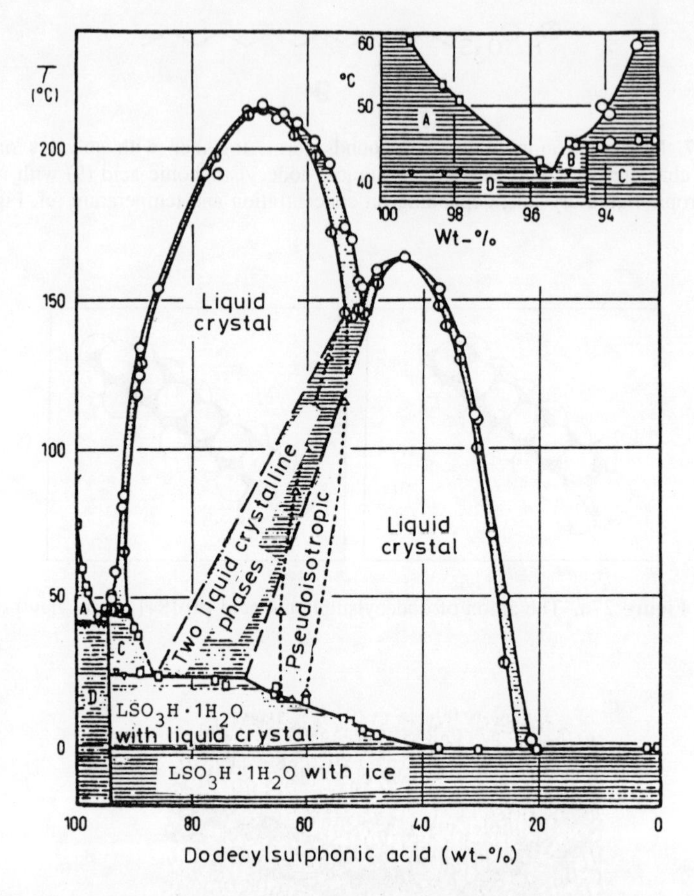

Figure 29. Phase diagram of the binary lyotropic dodecylsulphonic acid–water system [85,86]

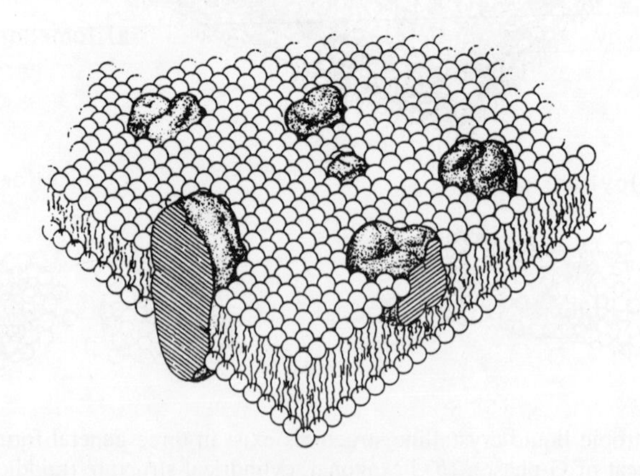

Figure 30. A cell membrane may be considered as a 'liquid mosaic,' a lyotropic liquid-crystalline lipid matrix studded with globular proteins [88]

thermotropic mesophases, even though they belong to some of the most important structures in biological systems, and were recognized as such very early [24].

The liquid-crystalline state is uniquely suited for the formation of complex molecular arrangements, in which order and mobility are combined. These structures possess variable permeability, varying flow behaviour and direction-dependent properties, in addition to the ability to dissolve other compounds without losing their mesogenic character; all these factors are very important for the functions of cells. Such organized molecules should form an ideal medium for catalytic processes of the metabolism; thus, lyotropic liquid crystals are being discussed as models for biological membranes [86,87] (see Figure 30).

8.3 CHEMICAL STRUCTURE ELEMENTS IN LIQUID CRYSTALS

Since the discovery of liquid-crystalline phases, it has been the aim of many preparative-organic chemists to discover the relationships between the molecular structure of a compound and its liquid-crystalline properties. Above all, the work of Vorländer [12,67,89], Weygand [90], Kast [91] and Gray [92] was concerned with defining general structural features for liquid-crystalline compounds, and related those to the corresponding physical properties.

By comparing the molecular structures of a large number of calamitic–thermotropic liquid crystals, a basic principle emerges, which can roughly be summarized in four points:

1. A thin, elongated molecular shape is necessary, especially with inflexible molecular frameworks. The length should be at least 1.3–1.4 nm.
2. Branched or angular molecular frameworks reduce or prevent the formation of liquid-crystalline regions.
3. Since the orientation of the molecules in liquid-crystalline phases is brought about by dispersive forces, a high anisotropy of polarizability is necessary; this is promoted by permanent dipoles or easily polarizable groups.
4. In order to avoid mere metastable, monotropic liquid-crystalline phases, not too high a melting point is preferable.

From the same comparative investigations, it emerges that liquid crystals can be built from typical structural elements [54] (Figure 31). They usually consist of rigid, linear or linearly linked building blocks S, polar, or easily polarizable functional groups P linking the rigid parts and end-groups E, which are mostly unbranched aliphatic chains extending the molecule and lowering the melting point [93].

By far the most often used rigid unit S, is the *para*-substituted benzene ring. Figure 32 gives an arbitrary selection of other rigid structural elements used in liquid crystals. Noticeably, these are mostly aromatics or unsaturated polycyclics and heterocyclics, many containing easily polarizable π-electrons. The saturated rings or ring systems appearing in the first liquid crystals remained for a long time unconsidered as basic structural principles, because the elucidation of cholesterol's

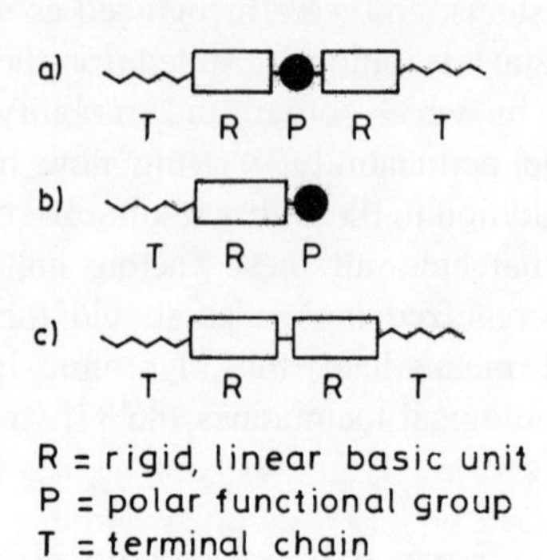

Figure 31. Molecules of liquid-crystalline compounds are built from typical building blocks (R, P, T)

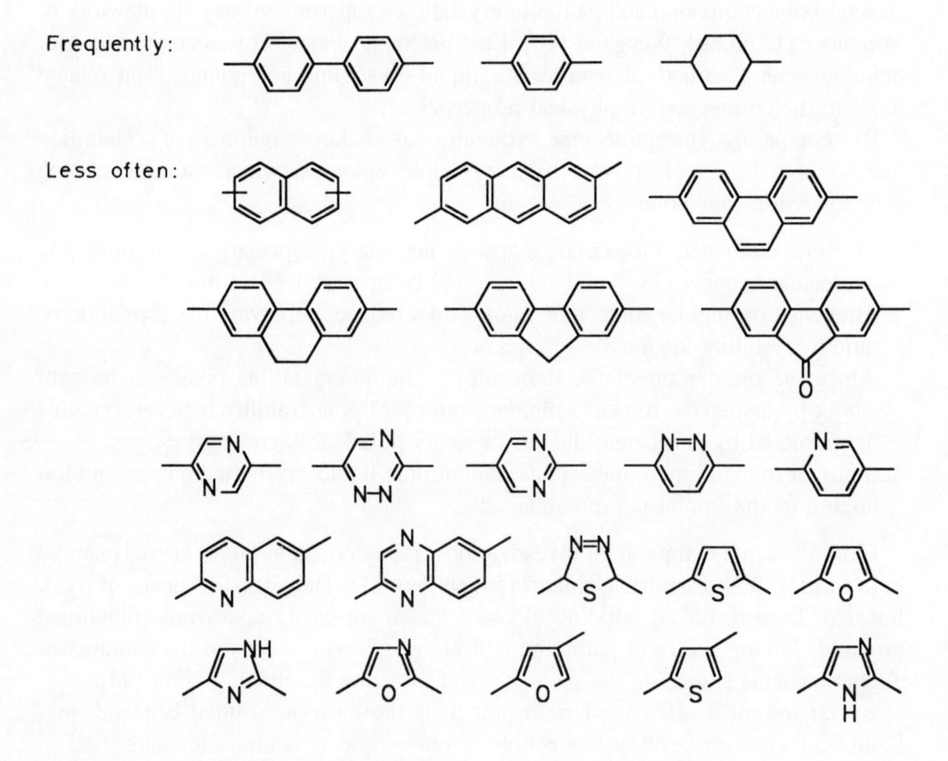

Figure 32. A selection of rigid building blocks used for the preparation of rigid frameworks in liquid crystals

constitution by Windaus and Wieland [94] only succeeded in 1932, while the X-ray structure analysis of cholesteryl iodide by Crowfoot came only in 1945 [95].

Not so many years ago, nobody doubted that the total hydrogenation of aromatic liquid-crystalline compounds would lead to the loss, or at least to a reduction, of the mesomorphic domain [96,97]. However, the occurrence of liquid-crystalline phases with *trans*-1,4-substituted cyclohexane derivatives demonstrates that saturated, conformatively mobile rings can have the necessary rigid linear geometry for mesomorphic behaviour. The latter is of greater importance for the thermal stability of liquid-crystalline phases than easily polarized π-electrons of aromatic or unsaturated systems [92]. Polarizable groups between two aromatic units also seemed essential for mesogenic behaviour. Azomethine, azo, azoxy and ester groups are the four most widely used polar building blocks.

As shown by various examples in Figure 33, there are only few completely linear functional groups, such as the triple bond. Far more frequently, the building blocks are angular, allowing a parallel arrangement of the two halves of the molecule [54]. Thus, azomethines (Schiff's bases) and liquid-crystalline stilbenes are found in their elongated (E)-configuration (Figure 34) [54,98].

As demonstrated by spectroscopic investigations of azomethines, except for the rod-like shape of the molecule, there is no need for the rigid, mostly flat building

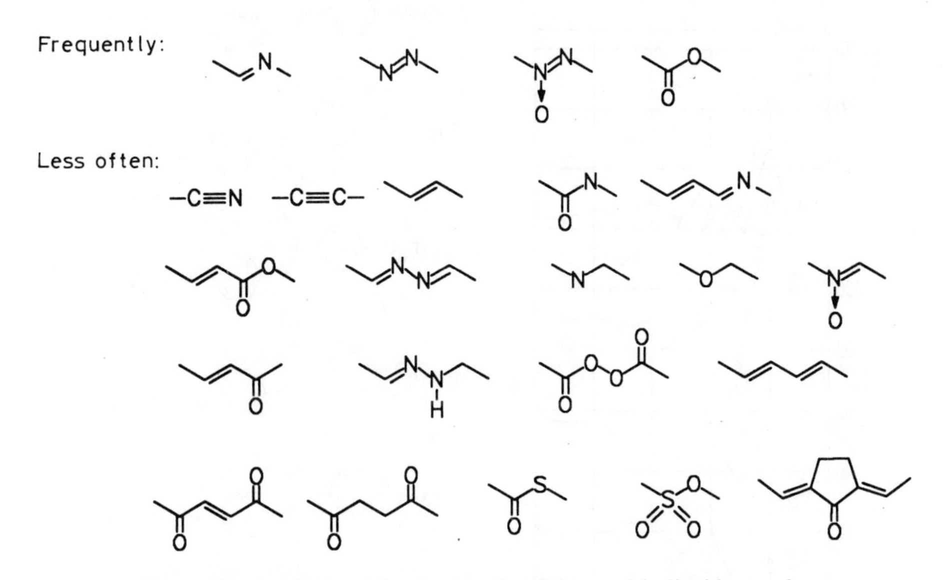

Figure 33. A selection of polar functionalities used in liquid crystals

Figure 34. The elongated (E)-configuration of stilbenes and Schiff's bases is well suited for liquid-crystalline properties

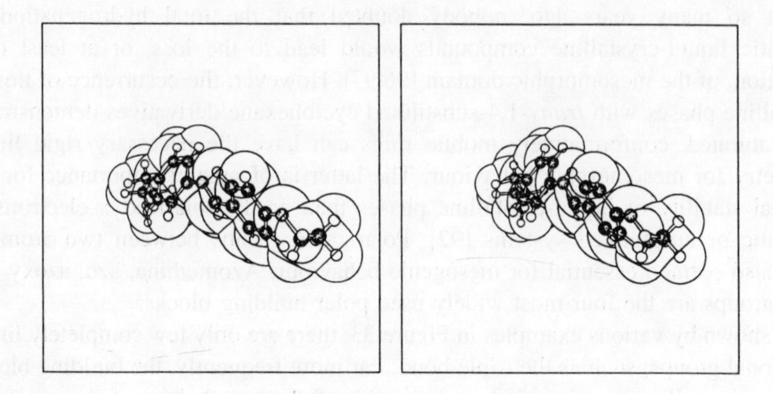

Figure 35. Spectroscopic investigations have shown that Schiff's bases in liquid crystals are not planar. The stereoview of 4-ethoxybenzylidene-4′-n-butylaniline (EBBA, **10**) shows the twist of the aromatic moieties [3] (X-ray crystal structure)

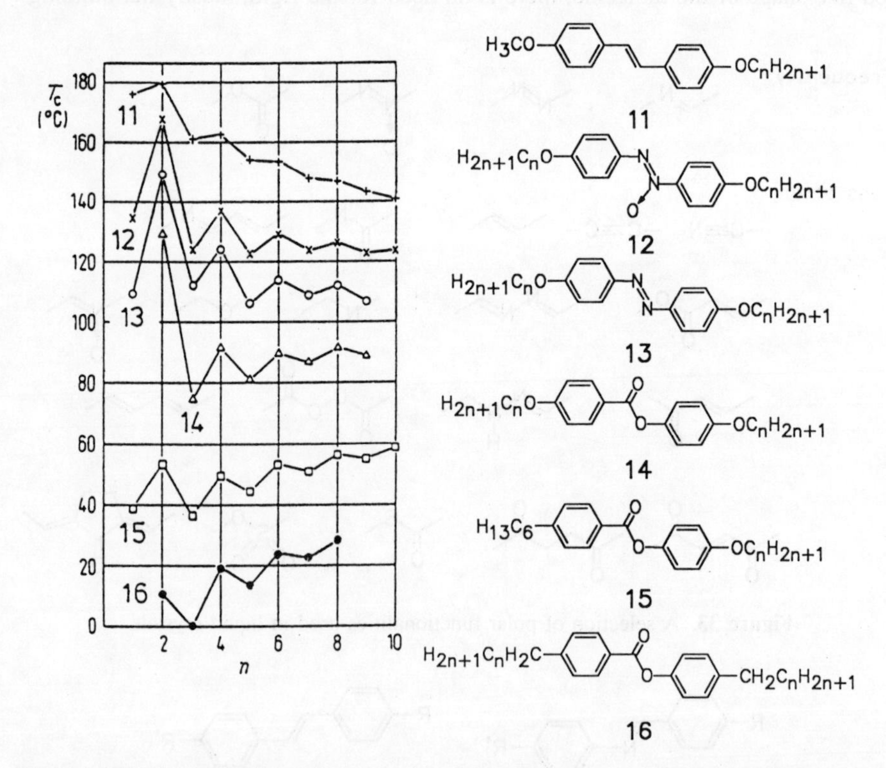

Figure 36. The clearing-point temperatures of analogous compounds are mostly determined by the type of central polar group present. Homologues have alternating clearing points

blocks to be in a planar alignment [99]. Indeed, such a twisted conformation has been detected in the solid state by X-ray crystallography, e.g. in the case of 4-ethoxybenzylidene-4′-n-butylaniline (**10**, EBBA), a liquid-crystalline homologue of MBBA (Figure 35) [100].

It has been shown that (*E*)-stilbenes with non-planar conformations have liquid-crystalline phases at lower, technologically more useful temperatures [98]. The influence of individual structural units on the formation of mesomorphic phases can only be judged on the basis of a comparison of the melting and clearing points of analogously built compounds, which differ in only one structural element. Thus, in Figure 36 the clearing points of homologous nematic substances in which various different polar central building blocks have been used are plotted [101].

The replacement of benzene with a heteroaromatic ring not only has an effect on the clearing point, but can also change the intermolecular interactions to such an extent, that additional liquid-crystalline phases may appear (Figure 37) [102]. As Gray *et al.* demonstrated with n-pentylcyanobiphenyl (**17**a) and pentyloxycyanobiphenyl (**18**), the transfer of the polar functional group to a terminal position (cf. Figure 31b) is possible without loss of liquid-crystalline properties [103]. Indeed, today's technologically used liquid crystals are often built according to this model and contain terminal cyano, ester or carboxylic acid groups (Figure 38) [104].

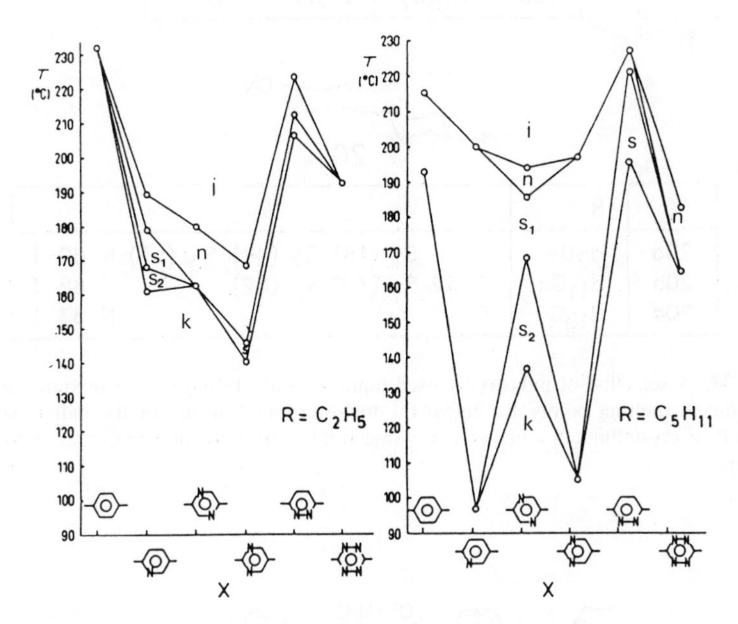

Figure 37. Variation of the central ring (X) in compounds of the following type

R–⟨benzene⟩–[X]–⟨benzene⟩–R

leads, apart from considerable changes in phase transition temperatures, to different phase sequences

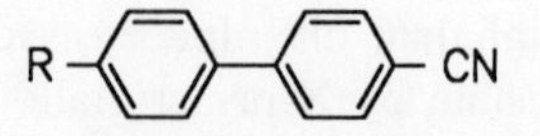

	R	
17a	$H_{11}C_5$	C 22.5 N 35.0 I
17c	$H_{15}C_7$	C 28.5 N 42.0 I
18a	H_5C_2O	C 48.0 N 69.0 I
18b	H_7C_3O	C 102.0 N (90.5) I
18c	$H_{11}C_5O$	C 71.5 N (64.0) I

R CN

19

	R	
19b	H_3C	C 38 N (−25) I
19d	H_7C_3	C 42 N 46 I
19e	$H_{11}C_5$	C 30 N 55 I
19c	$H_{15}C_7$	C 30 N 57 I

R CN

20

	R	
20a	H_7C_3	C 58 S_1 (18) S_2 (44) S_3 (57) N 80 I
20b	$H_{11}C_5$	C 62 S_1 (43) S_2 (52) N 85 I
20c	$H_{15}C_7$	C N 83 I

Figure 38. A selection of industrially used liquid crystals. In most cases mixtures are used whose mixed melting points are below the melting points of any of the individual components (C = crystalline; N = nematic; S = smectic; I = isotropic; in parentheses, monotropic transition)

O···H–O

O–H···O

21

Figure 39. Through formation of dimers, even short carboxylic acids, such as 4-propyloxybenzoic acid (**21**), reach the molecular length necessary for liquid crystals

Polar moieties, because of their chemical reactivity, are often only of limited use. The drawbacks of some such groups are given in Table 2 [105].

As demonstrated by hydrocarbons **22**, **23**, and **24** (Figure 40), a permanent dipole is not absolutely necessary for the stabilization of mesomorphic phases, s_B phases in this case [106]. On the contrary, this kind of example supports the theory according to which dispersive interactions are responsible for the order in liquid crystals [107].

Table 2. The reactivity of polar groups leads to unwanted side-effects in commercial applications

Type of compound	Group	Disadvantage
Schiff's base	—CH=N—	Hydrolytic cleavage
Stilbene	—CH=CH—	UV stability
Azo compound	—N=N—	Oxidation, isomerization
Ester	—COOR	Nucleophilic attack
Alkynes	—C≡C—	UV stability
Azoxy compounds	—N=N(O)—	Yellowing

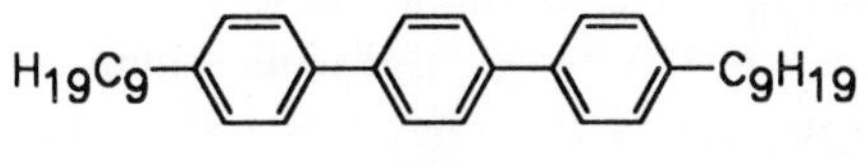

$H_{19}C_9$—⟨benzene⟩—⟨benzene⟩—⟨benzene⟩—C_9H_{19}

22

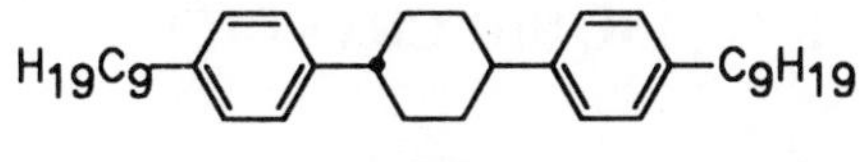

$H_{19}C_9$—⟨benzene⟩—⟨cyclohexane⟩—⟨benzene⟩—C_9H_{19}

23

$H_{19}C_9$—⟨cyclohexane⟩—⟨cyclohexane⟩—⟨cyclohexane⟩—C_9H_{19}

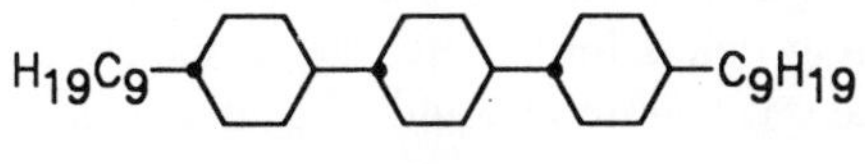

24

Figure 40. Permanent dipoles are not necessary for the formation of mesomorphic phases

Terminal n-alkyl or n-alkoxy chains are primarily used to lower the melting point. At the same time, however, they have an effect on the polarization anisotropy of the molecule. The alternating of the clearing points in homologous series, as seen in Figure 36, is explained in this way. In Figure 41, the inscribed longitudinal molecular axis corresponds to the direction of highest polarizability.

Adding an atom to the basic skeleton increases the polarizability along the main axis more than that at a right-angle to it. The addition of a further atom at a tetrahedral angle of 109° increases the amount of cross-polarization more than that along the molecular axis. Prolongation by a third atom then has the reverse effect again. According to the Maier–Saupe theory [107], the difference between the two

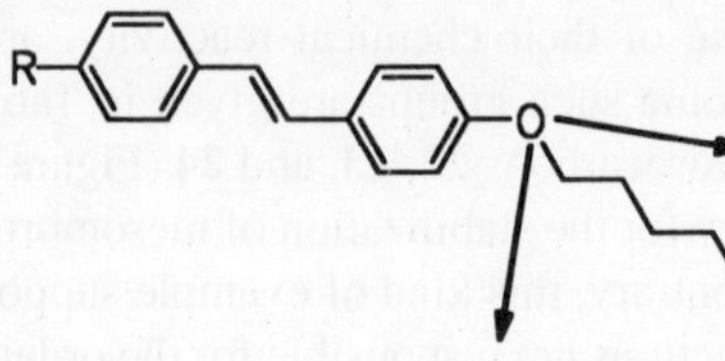

Figure 41. The progressive extension of terminal chains leads to a change in polarization anisotropy

polarizabilities is a measure of the value of the clearing point. With odd-numbered terminal chains, the polarizability difference is increased, in addition to the clearing point, whereas even-numbered chains increase the cross-polarization, which lowers the polarization anisotropy, which in turn leads to a lower clearing point. This model corresponds exactly to the observed behaviour in homologous series. As can be expected, the effect is more pronounced with two variable side-chains than with only one [107].

Concerning the molecular geometry of compounds which together with solvents form lyotropic systems, similar criteria as for calamitic–thermotropic liquid crystals apply. In addition to the interactions between the amphiphilic molecules themselves, considerable interactions with the solvent are observed.

Because to date only a few examples of discotic liquid crystals are known, the search for generalized structural criteria has only just begun [108].

8.4 APPLICATIONS OF LIQUID CRYSTALS

8.4.1 FROM A LABORATORY CURIOSITY TO A COMMODITY

In first applications, cholesteric liquid crystals were used for the relative optical determination of melting points [66]. Here, the temperature-dependent change in colour of light reflected on cholesteric liquid-crystalline phases is used. An increase in temperature will decrease the pitch of the phase's helical structure [109]. In addition, the wavelength of light reflected on such phases is dependent on the same pitch. Therefore, an increase in temperature will decrease the wavelength of the reflected light, i.e. on purely cholesteric phases a change of colour from red to green and blue is observed (Figure 42) [110]. On addition of nematic liquid crystals the colours may change differently, hence calibration of the colour–temperature correlation is required [27]. The advantages of this method lie in the minimum amounts of time and effort required, and in the fact that the temperature distribution on surfaces is made visible. In order to avoid reflections and seeing the surface's proper colour, it is first blackened, after which a layer of a suitable liquid crystal is applied [27]. For protection against mechanical stress and chemical reactions, these two layers can be combined to form a film, which also makes handling easier [111].

Using a number of mixtures, it is today possible to cover a temperature range from about −20 °C to +250 °C [112]. In order to obtain reproducible results,

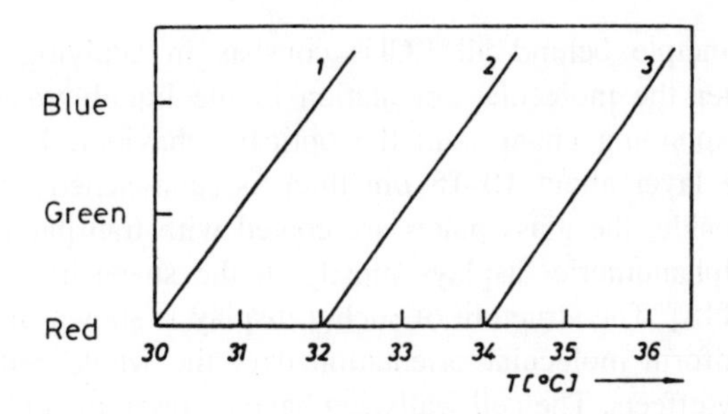

Component	Weight ratios for		
	1	2	3
1b (cholesteryl benzoate)	50	50	50
1c (cholesteryl oleate)	300	300	300
1d (cholesteryl chloride)	40	40	40
1e (cholesteryl pelargonate)	80	125	125

Figure 42. Mixtures of cholesteric liquid crystals can be used for the optical determination of temperatures

it is necessary to observe and illuminate the sample surface at right-angles. The achievable temperature resolution is dependent on the width of the range observed. Thus, within a range of 3–4 K, a distinguishable temperature difference of 0.2 K is possible; in a range of 0.1 K, the precision improves to an excellent 0.007 K. The limit of spatial temperature resolution is of the order of approximately 2 μm and the resolution in time reaches a limit of 30 ms. In this respect, temperature measurement with liquid crystals is surpassed by no other method [113].

Applications in medical areas, such as the detection of disorders in the blood circulation and of breast cancer, are as varied as those in non-destructive material testing, e.g. in measurements of stress and tension in components [114]. Because thermometers based on liquid crystals are cheap and toxicologically harmless, and therefore safe to use, they are ideal for the measurement of body temperature or as an 'intelligent egg timer' [115]. In advertising, interesting possibilities arise, e.g. when body heat makes a brand-name appear in various colours.

8.4.2 LIQUID CRYSTAL DISPLAYS

It was in the guise of liquid crystal displays (LCD), built into wristwatches and pocket calculators, that liquid crystals made their appearance in everyday life. The ever increasing miniaturization of electronic devices, and the concurrent reduction in their consumption of electricity, made new approaches to the conversion of electrical impulses into optical signals necessary. This is why, from 1960 onwards, Williams [116] and Heilmeier *et al*, [117] at the Radio Corporation of America (RCA), and Allen [105] in England, started investigations into the possible use of liquid crystals in electrooptical displays.

The basic principle behind all LCDs consists in applying an electric field to disturb or alter the molecular orientation in the liquid-crystalline layer, thus producing corresponding changes in the optical behaviour. For this purpose, a liquid-crystalline layer about 10–15 μm thick is sandwiched between two glass plates. On the inside, the glass plates are coated with transparent SiO_2 or In_2O_3 electrodes, in alphanumeric displays mostly in the shape of a so-called seven-segment array [118]. The structure of such a display is shown in Figure 43.

The basic, uniform molecular orientation over the whole cell is produced by surface boundary effects. The cell walls can have an orienting effect on molecules, in much the same way as a molecule within the liquid-crystalline layer is oriented by interactions with its neighbours. Through appropriate pretreatment of these walls, the arrangement of the molecules in a nematic phase can be set to be either perpendicular (homeotropic) or tangential (homogeneous) to them. Intermolecular interactions propagate this outer order 50–100 μm into the interior of the layer so that, even without any energy expenditure, a durable, uniformly oriented layer, some 150–200 μm deep, can be achieved. With two homogeneously orienting plates, a sandwiched nematic phase can be twisted, if the orientations of the two parallel surfaces are set at an angle to each other. A maximum twist of 90° over the depth of a liquid-crystalline layer can then be achieved (Figure 44) [48,51].

If the deformation of the oriented liquid-crystalline phase is elastic, so that it can

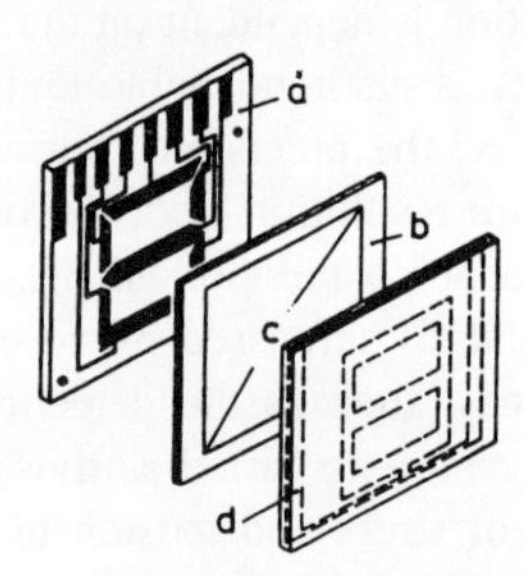

Figure 43. Structure of a digital, seven-segment liquid crystal display: a = glass plate with electrode segments; b = spacer frame, with c = liquid crystal layer; d = glass plate with transparent counter electrode

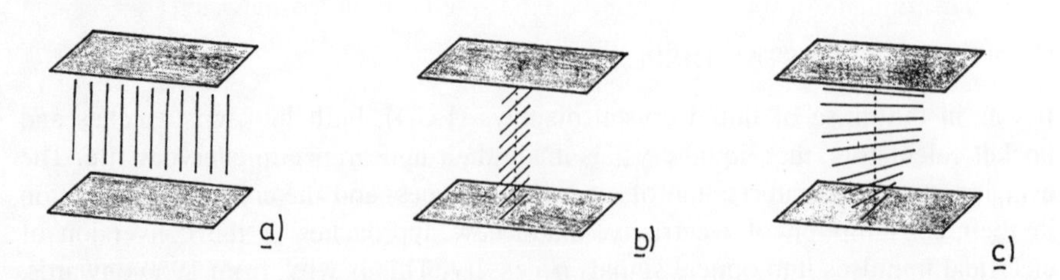

Figure 44. Liquid-crystalline phases are pre-oriented by boundary effects, without any energy consumption: (*a*) homeotropic alignment; (*b*) homogeneous arrangement; (*c*) homogeneously twisted orientation

be described by Hooke's law, one speaks of field effects or normal deformations; with dynamic effects or abnormal deformations, currents also occur (see below) [43,119].

The first liquid crystal displays were based on dynamic light scattering. Such cells usually contain homogeneously oriented nematic phases with a negative anisotropy of the dielectric constant. If a voltage of 20–30 V is applied to the electrodes, a charge transport and hence hydrodynamic currents will occur, thus turning the molecules into the field's direction, even though the dielectric moment is counteracting this. The previously structurally uniform and optically clear liquid crystal now consists of countless birefringent domains, a few micrometres across, on whose surfaces light is strongly diffracted, the layer now appearing milky white (Figure 45) [27,50,118,119].

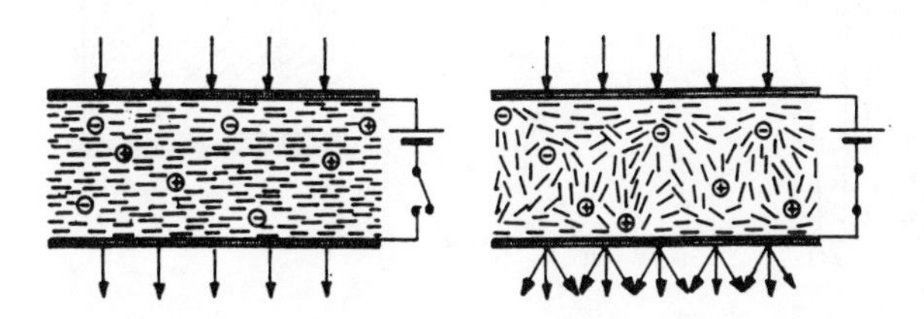

Figure 45. In early liquid crystal displays, the uniform, homogeneous orientation of the layer was often disturbed by hydrodynamic currents, causing light scattering

Displays using only the field effect (Freedericksz effect [120]) have almost completely replaced these diffraction cells, because in spite of their lower power consumption, they give better contrasts and have a longer life-span. Among these are displays which working according to the DAP mode (Deformation of Aligned Phases). The cell structure is the same as before, but without a voltage applied the liquid-crystalline phase is homeotropic, i.e. the molecules are perpendicular to the layer boundary. Moreover, the front and back planes are fitted with crossed linear polarizers. Without a voltage, incident light will pass unaffected, and the display will remain dark. A voltage of 3–6 V leads to a deformation of the original homeotropic orientation. The degree to which the molecular axis deviates from the field's direction, and with it the amount of birefringence, depend on the voltage applied and the distance from the orienting surfaces (Figure 46). The light transmittance of the set-up can be controlled within a wide range by varying the applied voltage. When white light is used, the transmittance also depends on the wavelength, and the Newtonian interference colours are observed (cf. Figures 10–12 and Table 1) [27,43,118,119].

Colour displays can also be made by incorporating pleochroitic dyestuffs into the nematic phase. Pleochroitic dyes have different light absorptions, and hence colours, along distinct axes [121]. Because of the uniform orientation of the nematic phase, the added dye molecules become organized in the same way. In a homogeneous liquid crystal orientation, the cell displays the colour produced across the main molecular axis. A nematic phase with a positive anisotropic dielectric constant, on applying an electric field to it, is reoriented homeotropically, thus turning the dye molecules around, presenting the differently coloured longitudinal axis to the viewer (Figure 47) [27,43,118,119]. Examples of pleochroitic dyes (**25–28**) are shown in Figure 48.

Today's most widely used display elements are based on the deformation of twisted nematic cells (TNC), a technology developed by Schadt and Helfrich [122]. The plane of vibration of polarized light, passing through this twisted liquid crystal layer, follows its 90° twist.

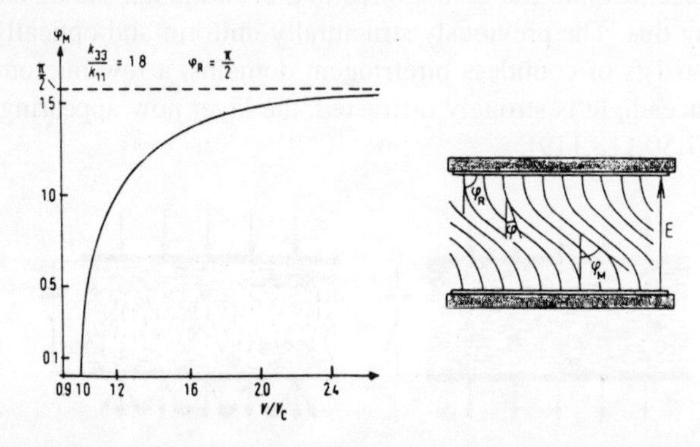

Figure 46. In cells where a homeotropic phase is deformed, the deformation angle ϕ_M at the centre depends on the applied voltage V (V_c is an applied basic voltage, which does not induce any deformation)

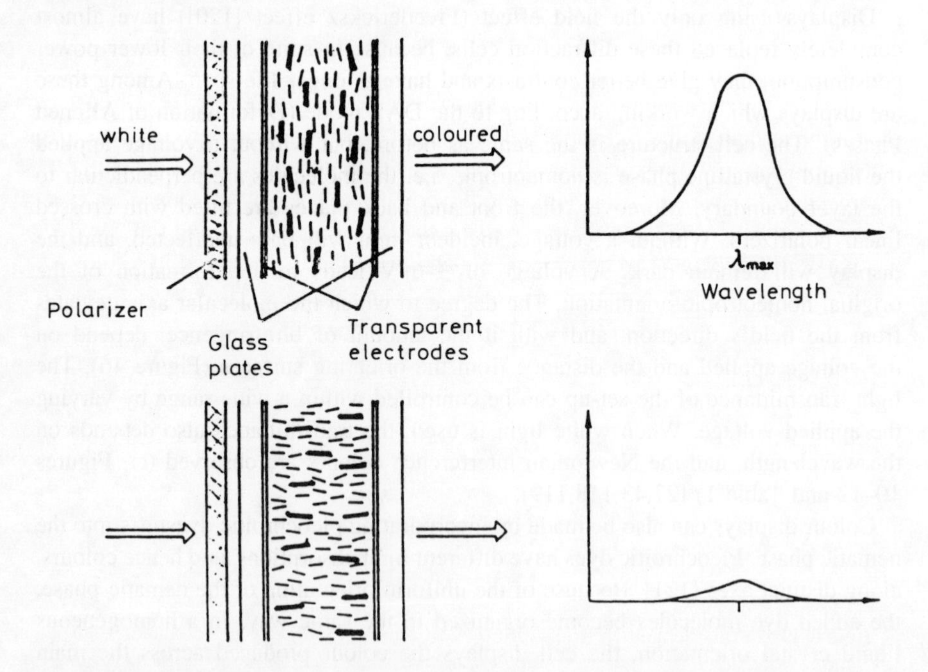

Figure 47. Pleochroitic dye molecules follow the changes in orientation of their host liquid-crystalline phase. Their anisotropic light absorption can be used in colour LCDs

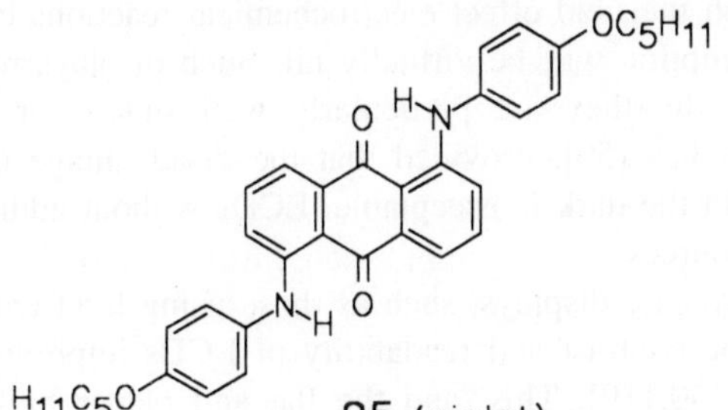

25 (violet) 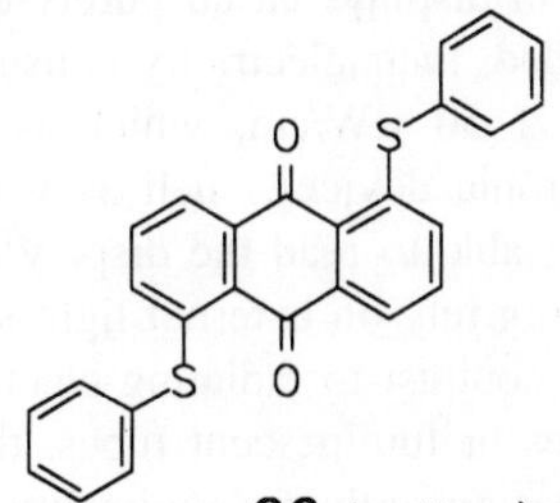 **26** (yellow)

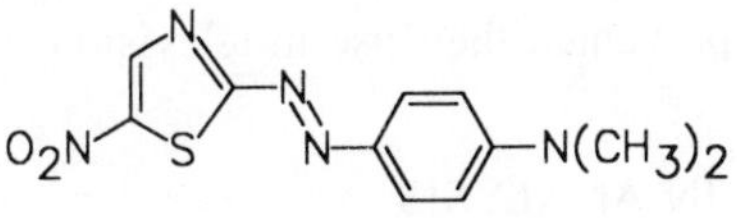

27 (blue)

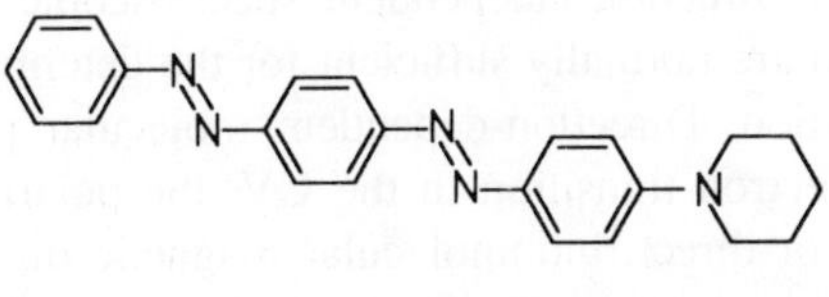

28 (red)

Figure 48. Examples of pleochroitic dyes

At rest, this type of system, packed between two crossed polarizers, is transparent to light. Again, if the liquid-crystalline phase has a positive dielectric anisotropy, the application of a voltage will produce a homeotropic rearrangement of the molecules. The plane of vibration of incident polarized light cannot be twisted any more, hence stopping it from passing the second polarizer, and thus producing a darkened display [27,43,50,118,119]. The principle is illustrated in Figure 49.

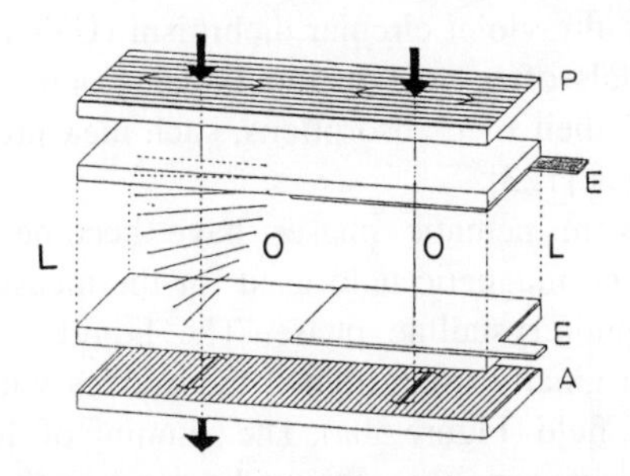

Figure 49. In this Schadt–Helfrich cell, the left part, without applied voltage, is transparent to light, whereas on the right an applied voltage has rearranged it homeotropically, rendering it opaque. P = polarizer; A = analyser; E = electrode; L = liquid crystal; O = preferred orientation

If in displays based purely on the field effect electrochemical reactions can be avoided, their electricity consumption will be virtually nil. Such displays require only 1–50 μW/cm, which is why they are particularly well suited for small electronic devices, such as watches [50], provided that the disadvantage of not being able to read the display in the dark is acceptable. LCDs without additional lighting rely on external light sources.

In contrast to radiating electrooptic displays, such as those using light-emitting diodes or luminescent tubes, the contrast and readability of LCDs improve with increasing ambient luminosity [50,119]. This, and the flat and planar build are the reasons behind the ever more widespread use of liquid crystals in computer displays. Here, their slow response, compared with conventional cathode-ray tubes, of about 10 ms, which prevented their use in television sets, is of no importance.

8.4.3 APPLICATIONS IN ANALYSIS

In recent years, liquid crystals have become of increasing interest in various areas of chemical analysis. Direction-independent spectroscopic methods (UV–Visible, IR, NMR, ORD, CD) are normally sufficient for the determination of an unknown compound's constitution. Direction-dependent molecular properties, such as the polarization of an electron transition in the UV, the polarization of a vibrational transition in the IR, or direct, intramolecular magnetic dipole–dipole interactions in the NMR, are averaged out by the random orientation of the molecules, which, incidentally, makes these methods practicable. However, if direction-dependent molecular properties are to be measured, all the molecules have to be uniformly aligned in the same way. This can be achieved by dissolving the compound in a nematic phase; depending on the molecular geometry, the concentration can be as high as 30%, without destroying the mesomorphic state. With wall effects, electric or magnetic fields, the liquid crystal molecules, and thus the solute, can be uniformly oriented.

IR and UV spectra recorded parallel and perpendicular to the longitudinal molecular axis display the same bands as 'non-polarized spectra,' but the intensities of analogous bands are different. These phenomena are called infrared circular dichroism (IRCD) and ultraviolet circular dichroism (UVCD), respectively.

Because liquid crystals often contain functional groups, such as C=N, C=O and C=C, which have their own absorptions, such measurements are undisturbed only within small ranges [123].

NMR investigations in nematic phases have become valuable applications [63,50,124,125,126]. The magnetic field used for the measurement is at the same time orienting the liquid-crystalline phase. The liquid crystals with a positive magnetic anisotropy originally used aligned themselves with their molecular axis parallel to the magnetic field (Figure 50a). The spinning of the sample, usually used to remove residual inhomogeneities, cannot be used in this case, since it would disturb the orientation of the molecules. A loss in resolution is the consequence (Figure 52). This problem can be avoided by using liquid crystals with a negative magnetic anisotropy (Figure 51) [127].

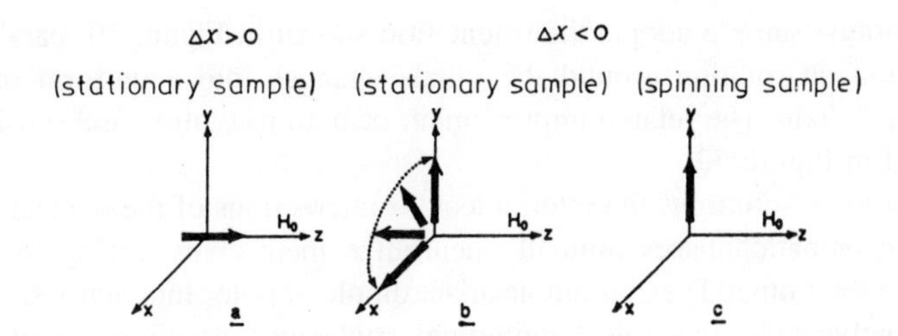

Figure 50. In contrast to molecules with $\Delta\chi > 0$ (*a*), those with $\Delta\chi < 0$ (*b*) align themselves orthogonally to the direction of the magnetic field. Spinning the sample aligns all the molecules parallel to the *y*-axis (*c*)

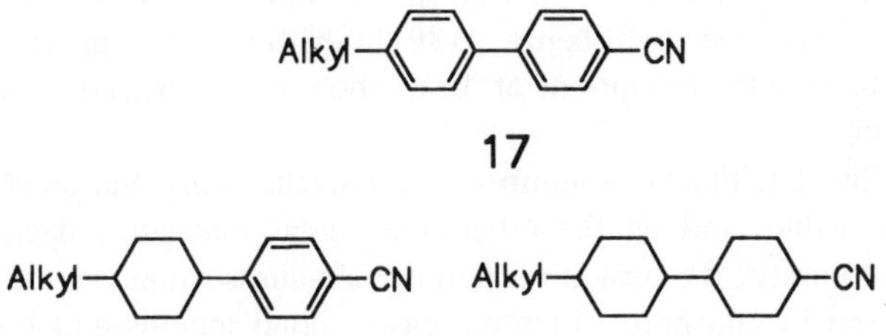

Alkyl–(C₆H₄)–(C₆H₄)–CN

17

Alkyl–(C₆H₁₀)–(C₆H₄)–CN   Alkyl–(C₆H₁₀)–(C₆H₁₀)–CN

19   **20**

Figure 51. The liquid-crystalline nitriles **17**, **19** and **20** have a negative magnetic anisotropy

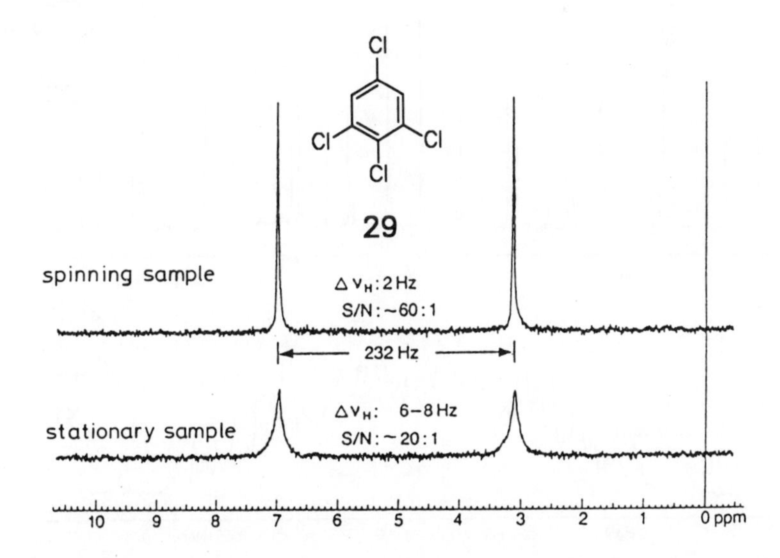

Figure 52. The ^{1}H NMR spectra of 1,2,3,5-tetrachlorobenzene, recorded in a nematic phase with negative magnetic anisotropy; the spinning sample is far better resolved

The immobile sample adopts the orientation shown in Figure 50, parallel to the *xy*-plane, and on rotation around the *y*-axis changes into a uniform orientation parallel to this axis. The related improvement of both resolution and sensitivity are exemplified in Figure 52.

As in isotropic solutions, the intermolecular interactions of the nuclear magnetic moments in nematic phases mutually neutralize themselves, owing to the rapid molecular vibrations. The intramolecular dipole–dipole interactions, however, remain largely intact, as the longitudinal molecular axes are lined up with the nematic phase's orientation. Owing to dipole coupling, the two protons of tetrachlorobenzene (**29**) experience a magnetic field H, which, depending on the orientation of their spins, will differ from the original field H_0 by $\pm\Delta H$. In proton NMR spectra, the magnitude of the line resolution ($\Delta\nu$) directly corresponds to the spin–spin coupling constant. Because, as a rule, shielding effects in nematic phases are different from those in isotropic solutions, other chemical shifts are observed. The proton spectra of benzene (**30**) [128] and cyclooctatetraene (**31**) [129] (Figure 53) illustrate the limitations of the method: it is restricted to molecules with only few protons.

On the one hand, with a large number of signals the interpretation of the spectrum becomes impossible, and on the other, the signal intensities decrease. For the last reason mentioned, the proton-containing nematic solvent does not appear, but remains covered by the noise. In some cases, bond lengths and bond angles of dissolved molecules have been determined with great precision using this method. For example, the molecular geometry of cyclobutane (**32**) has been investigated with this technique [116a,b]. Together with other, non-measurable quantities, the fine structure shown in Figure 55 results (Figures 54 and 55).

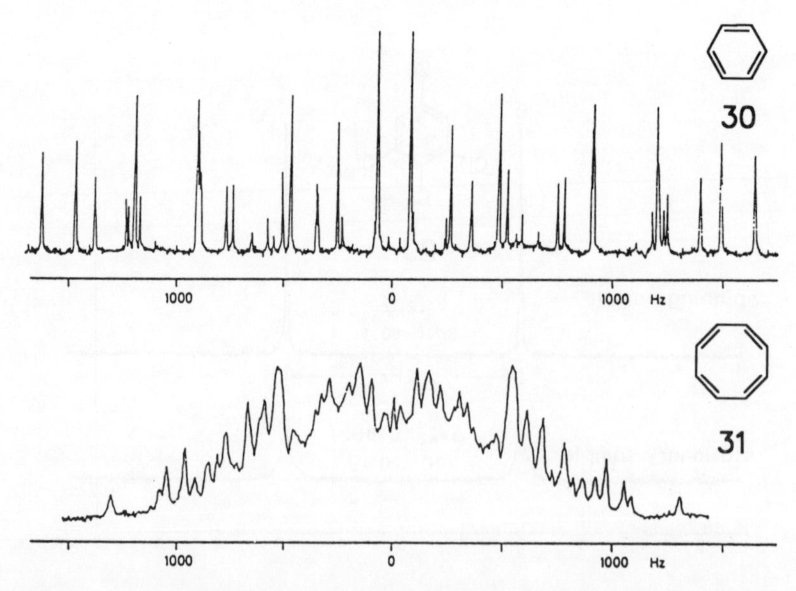

Figure 53. In the ^{1}H NMR spectra of benzene (**30**) and cyclooctatetraene (**31**) in nematic solvents, direct dipole–dipole nuclear interactions lead to complex signal patterns

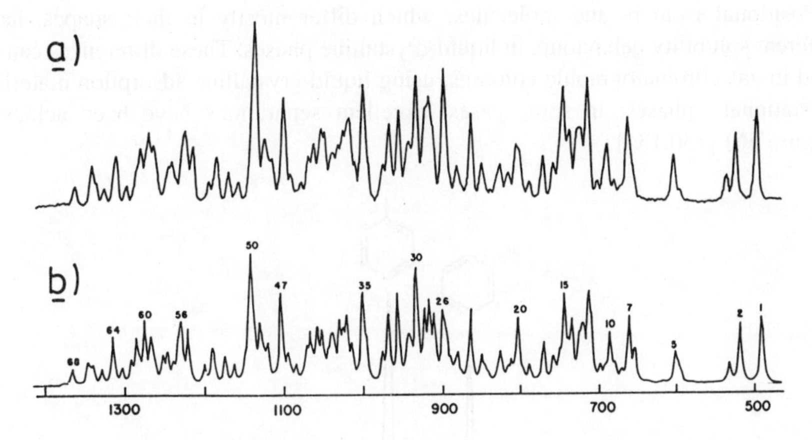

Figure 54. Low-field part of the ^1H NMR spectrum of cyclobutane in *p,p'*-di-n-hexyloxyazobenzene at 80 °C. (*a*) Experimental spectrum; (*b*) computer simulation

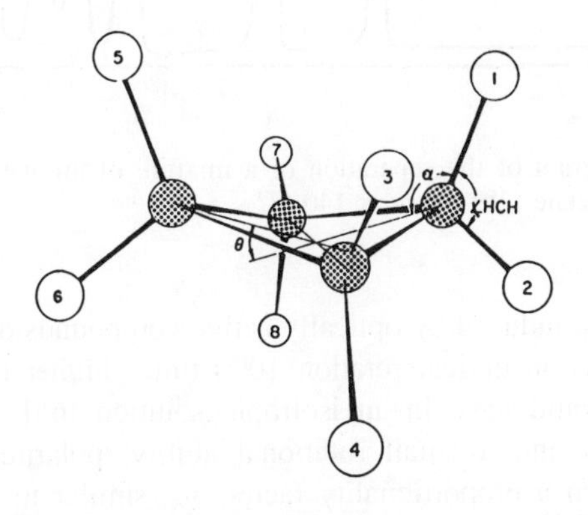

	Assumptions		
	1	2	3
	$C—H^1 = C—H^2$	$C—H^1 = C—H^2 + 4$ pm	HCHO = 109.47°
$C—H^1$	113.3	112.29	112.33
$C—H^2$	113.3	108.3	108.5
CHC	108.1	109.5	109.47
α	122.1	118.5	118.7
θ	27.0	22.7	22.9

Figure 55. Structural data for cyclobutane, obtained from ^1H NMR measurements in a nematic phase (bond lengths in pm; in all cases, a C—C bond length of 154.8 pm was assumed)

Positional isomers and molecules, which differ mostly in their shapes, have different solubility behaviours in liquid-crystalline phases. These differences can be used in gas chromatographic columns, using liquid-crystalline adsorption materials as stationary phases; in some cases, excellent separations have been achieved (Figure 56) [130,131].

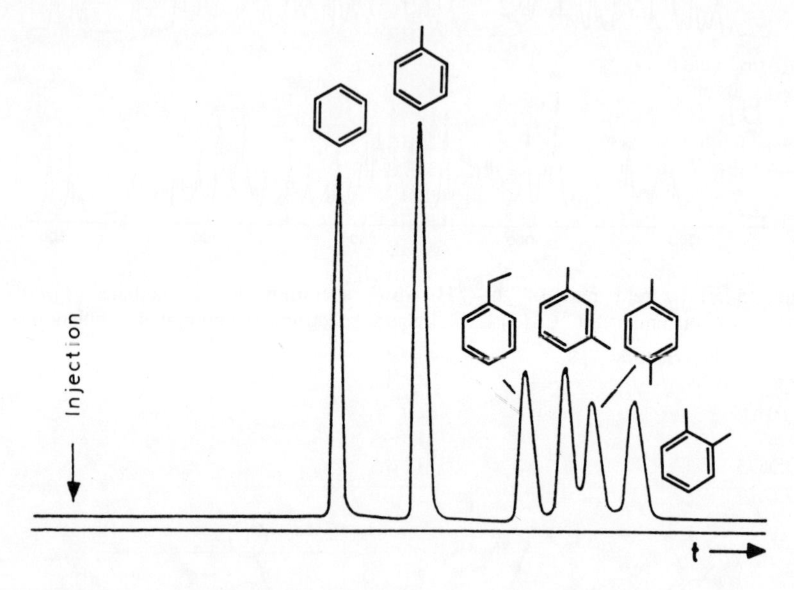

Figure 56. Chromatogram of the separation of a mixture of aromatic hydrocarbons on a 4,4′-diethoxyazoxybenzene (**2b**) phase at 140 °C

Cholesteric phases, induced by optically active compounds dissolved in nematic phases, often possess an optical rotation 1000 times higher than that which the active compound would have in an isotropic solution [65]. If, owing to small amounts of material and/or small rotational ability, polarimetric investigations become difficult, then a proportionality factor β_M, similar to the optical rotation $[\alpha]_D$, can be obtained using induced-cholesteric phases. The concentration c should be kept below 10 mol-% in order to ensure an inverse proportionality to the observed helix pitch, p:

$$p^{-1} = \beta_M \cdot c$$

In addition, induced-cholesteric phases offer a simple way of determining racemization barriers and racemization kinetics. The main advantage over other methods consists in the very small amounts of material required (<10 mg), which do not even need to be weighed precisely (in polarimetric methods, weighing errors are the main source of errors).

Table 3 leads to several general considerations:

Table 3. Rotatory power β_M of chiral compounds

Chiral compound	Nematic phase	β_M ($\mu m^{-1} mol^{-1}$)
33 (−)-Verbenone	MBBA	0.1
34 (+)-2-Octanol	EBBA	0.5
	ZLI 304†	0.5
	MBBA	0.8
35 D-Amyl benzoate	MBBA	1.0
36 (−)-Carvone	EBPA	1.6
	EBBA	1.9
37 (−)-Menthol	MBBA	2.2
36 (−)-Carvone	MBBA	2.3
	EBPA	3.2
38 CBAC	EBBA	3.3
	MBBA	3.5
1d Cholesteryl chloride	MBBA	9.3
1f Cholesteryl propionate	EBBA	13.5
	MBBA	15.0
39 α-Phenylethylamine	MBBA	15.3
40 PAC	MBBA	17.1
41 MBAC	ZLI 304†	17.8
	MBBA	19.4
42 DDCO	MBBA	32

		R	R'
3	MBBA	CH_3	$n\text{-}C_4H_9$
10	EBBA	C_2H_5	$n\text{-}C_4H_9$
43	EBPA	C_2H_5	$n\text{-}C_5H_{11}$

† Mixture of azoxybenzenes (Merck)

a. The most pronounced twist is induced by the inherently chiral compound DDCO (**42**), the least by the centrochiral verbenone (**33**).
b. Chiral compounds which themselves form mesophases, or have a corresponding shape, produce a greater twist than do spherical molecules.
c. The surprisingly strong rotatory power β_M displayed by α-phenylethylamine (**39**) is probably due to an amine exchange with the nematic Schiff's base MBBA (**3**).
d. There is no simple relationship between the optical rotation $[\alpha]_D$, the rotatory power β_M, and the type of chirality of the guest molecule.

With some non-steroidal cholesteric liquid crystals [133], or in cases where the mechanism of the induced cholesteric twisting of nematic phases is known, the absolute configuration of the chiral component can be deduced from the observed optical behaviour.

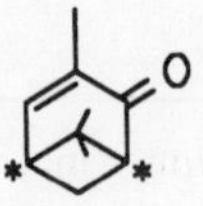

33 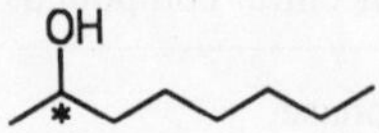 34 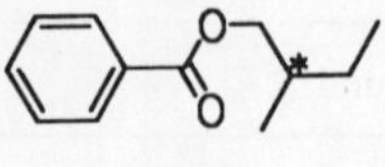 35

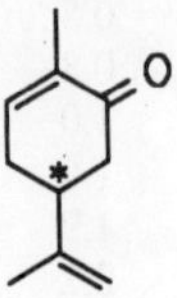

36 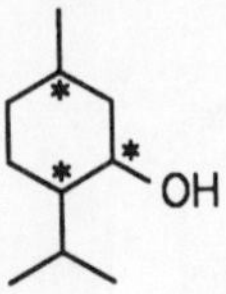 37 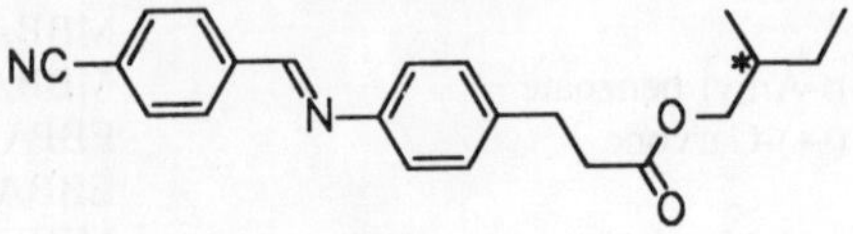 38

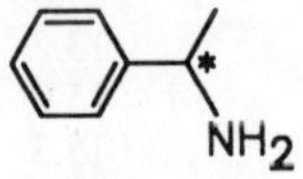

39 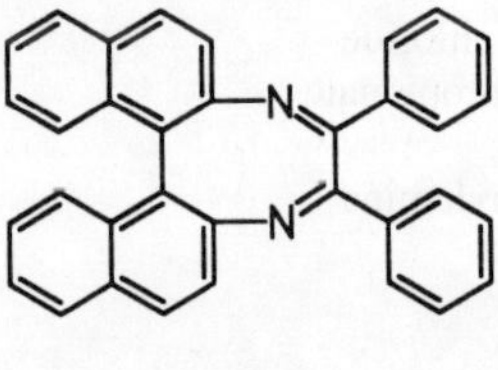 42

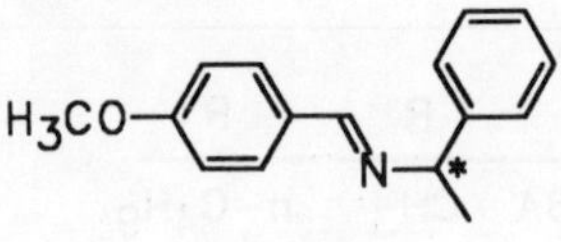

40 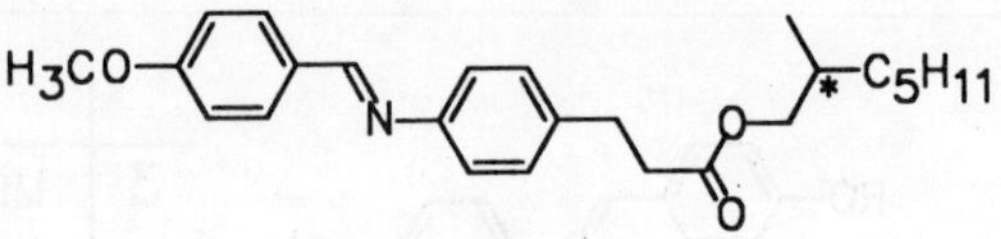 41

8.5 RECENT DEVELOPMENTS

8.5.1 POLYMERIC LIQUID CRYSTALS

The molecular mass of calamitic liquid crystals lies between 200 and 1000 mass units, while discotic liquid-crystalline compounds have masses of about 500 to 2000. As early as 1923, Vorländer answered the question of whether 'infinitely' long molecules would lose their mesogenic properties in the negative [67]. However, it was only 50 years later that the Kevlar fibre (poly-*p*-phenyleneterephthalamide, **44**; DuPont), the first liquid-crystalline fibre, was discovered. The theoretical basis, however, was laid down by Flory in the 1950s [134].

Macromolecules in the solid state have only minuscule crystalline domains, which are separated by more or less amorphous intervals (Figure 57) [36]. In solution or in the melt, they form disordered coils. In order to produce liquid-crystalline molecular arrangements, the polymer must consist of mesogenic units, capable of ordering intermolecular and intramolecular interactions, and spacers contributing the necessary flexibility. When alternating mesogenic groups and spacers form the basic polymer thread, this is called a main-chain polymer [139].

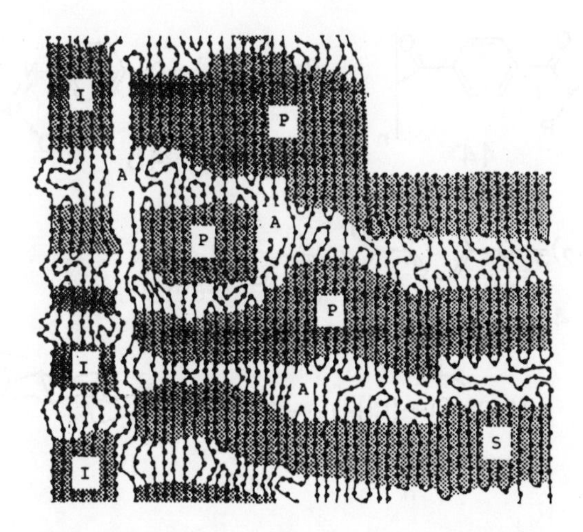

Figure 57. Model of a linear polymer with high molecular weight. A = amorphous domain; S = similar to a single crystal; I = isolated ultrafibril; P = paracrystalline layer

In the side-chain polymers developed by Finkelmann and Ringsdorf, the liquid-crystalline groups are attached to the main chain by flexible spacers (Figure 58) [135,137,138]. With these polymers, the distinction between thermotropic and lyotropic [136] liquid crystals is also made. Basically, they behave as do the low-mass liquid crystals and form the corresponding nematic, smectic, cholesteric and various lyotropic phases. Obviously, the spacers must satisfy one basic requirement, i.e. they must allow an undisturbed organization of the aligned or cross-linked mesogenic units (Figure 59).

With thermotropic liquid-crystalline polymers, the liquid-crystalline order can be frozen by cooling the liquid melt below the vitrification temperature. In this manner, polymeric glasses with liquid-crystalline texture can be obtained. Thus, undercooled cholesteric polymeric liquid crystals in the form of sheets can be used as selective reflectors and circular polarizers (cf. Figure 23) [135].

Such a pseudo-cholesteric structure is nothing new in Nature. As discovered by Michelson in 1911 and confirmed by Gaubert in 1924, the scintillating, often metallic, colour effects observed on beetles are due to the selective reflection of laevorotatory polarized light [140]. As seen under the electron microscope, these reflections are produced by cholesteric-helical structures in the outer 5–20 μm of the skin [50,131].

Mesomorphous polyesters with chain elements of poly-*p*-hydroxybenzoic acid (**49**) (Figure 60), of which Vorländer investigated the basic type [67], are of technological interest. By either pressing or drawing the liquid melt, fibres are produced, which retain a certain nematic molecular orientation along their main axis. This temperature-resistant fibre, because of its fibril-type structure, has a high tensile strength. Here, too, a comparison with biological fibrous structures in muscles and tendons is possible.

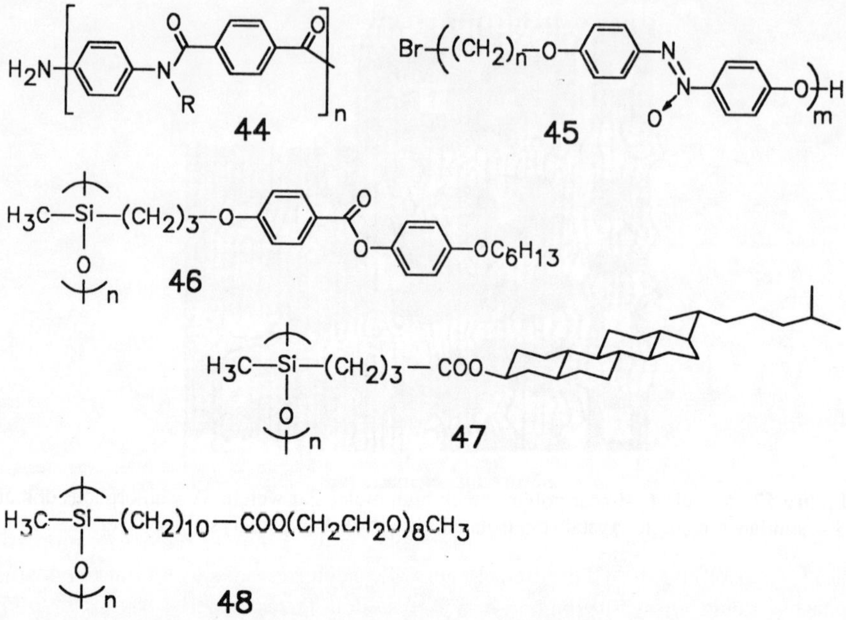

Figure 58. Examples of liquid-crystalline polymers

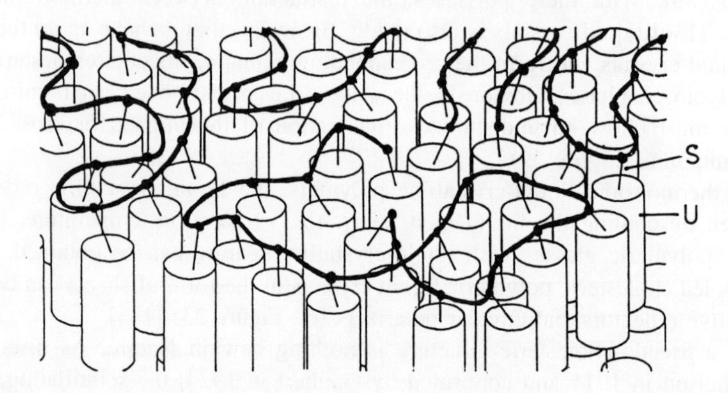

Figure 59. Either the spacer units (S) or the basic polymer must allow a strainless interaction of the liquid-crystalline subunits (U) if a mesophase is to be formed

8.5.2 FERROELECTRIC LIQUID CRYSTALS

The technical applications of liquid crystals were first limited to nematic and cholesteric phases. It was only in 1972 that Kahn first presented a display which used a liquid crystal forming a smectic A phase [141]. Special smectic phases today promise a new generation of quickly switching displays.

From symmetry considerations, Harvard physicist Meyer in 1974 deduced

49

DP*	Number of spacer units	Phase behaviour†
1	13.8	C 342 a 360 I
2	20.5	C 185 a_1 222 a_2 288 I
3	28.4	C 180 a_1 203 a_2 257 I
4	34.6	C 121 a_1 211 a_2 245 I

* DP = Degree of polymerization of the polyglycol units.

† C = crystalline; I = isotropic; a, a_1, a_2 = Liquid-crystalline phases (without characterization).

Figure 60. Nematic polyester

that smectic C phases of chiral molecules must have ferroelectric properties [142]. According to Clark and Lagerwall, such ferroelectric liquid-crystalline phases should allow substantially shorter switching times and posses a 'memory effect' [143]. Ferroelectric phenomena are the electric analogue of ferromagnetic properties and manifest themselves in the spontaneous electric polarization of a given macroscopic sample. Whether such a polarization actually occurs depends on the symmetry of the molecules and their arrangement. Only if the electric polarization remains unchanged (invariant) by symmetry operations of the medium does spontaneous polarization of the sample occur [144,145].

For paramagnetic nematic phases we refer to the literature [147].

Ester **50**, which forms an s_C phase, will illustrate this. As seen in Figure 61, the

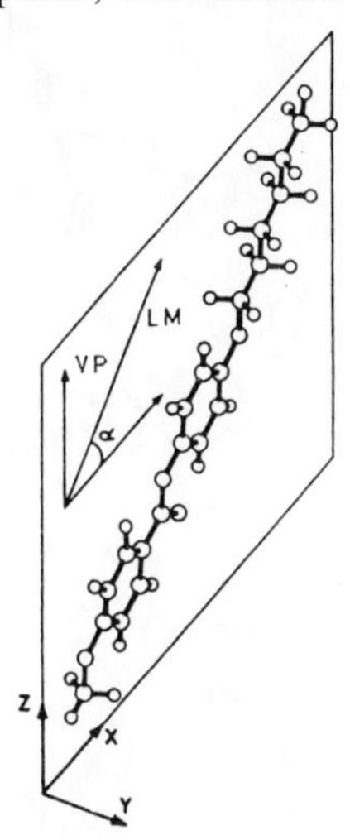

Figure 61. Orientation of ester **50**, which forms an s_C phase. VP = vector perpendicular to the layer boundary; LM = direction of the longitudinal molecular axis

longitudinal molecular axis is tilted towards the layer boundary. Reflection on a plane σ_{xz} containing the molecular axis, and also rotation around the y-axis (C_{2y}), produce identical molecules and identical layers. The symmetry operations of the medium are, therefore, σ_{xz} and C_{2y}.

A dipole P, associated with a molecule, can be resolved into three components, P_x, P_y and P_z, along the coordinates. The application of the symmetry operations on these dipole components give:

$$C_{2y} \quad \begin{array}{l} P_x \rightarrow -P_x \\ P_y \rightarrow P_y \\ P_z \rightarrow -P_z \end{array} \qquad \sigma_{xz} \quad \begin{array}{l} P_x \rightarrow P_x \\ P_y \rightarrow -P_y \\ P_z \rightarrow P_z \end{array}$$

A change in polarity is observed for the reflection of P_x and P_z and the rotation of P_y. Accordingly, ordinary s_C phases are not ferroelectric.

An s_C phase of a chiral molecule, such as ester **51** (Figure 62), is denoted by s_C^*.

Here, the reflection on the xz-plane is no longer a symmetry operation. The reflection preserves the layer structure, but changes the configuration of the chiral molecule (Figure 62). The remaining C_{2y}-axis leaves the sign of the dipole component P_y unchanged:

$$C_{2y} \quad \begin{array}{l} P_x \rightarrow -P_x \\ P_y \rightarrow P_y \\ P_z \rightarrow -P_z \end{array}$$

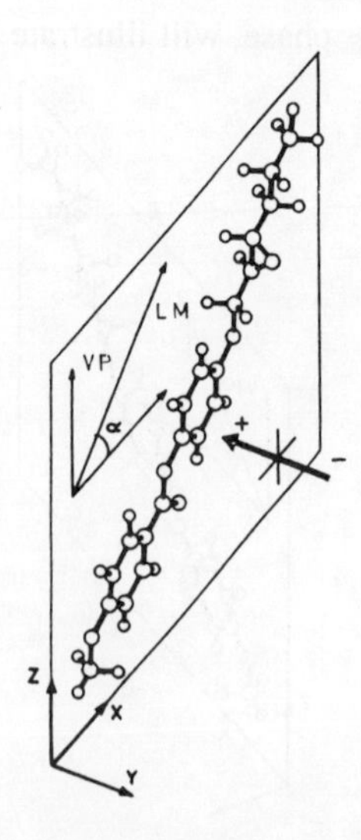

Figure 62. Orientation of chiral ester **51**, which forms an s_C^* phase. VP = vector perpendicular to the layer boundary; LM = direction of the longitudinal molecular axis

If this component points sideways out of the molecule, then the necessary condition for ferroelectric behaviour is given [144].

The cause of ferroelectricity in S_C^* phases is not found in the reciprocal interactions of the lateral dipoles adding to a macroscopic polarization, but rather in the molecular arrangement in smectic phases. An external field can switch a ferroelectric liquid crystal [146] between two stable states (Figure 63). The similarly directed molecular dipoles align themselves in the direction of the external field, adopting one of two stable states. This could be used for optical information storage, opening new fields of application. Moreover, the switching process is faster than in today's displays and could therefore be of interest for television screens [144,145].

For paramagnetic nematic phases we refer to the literature [147].

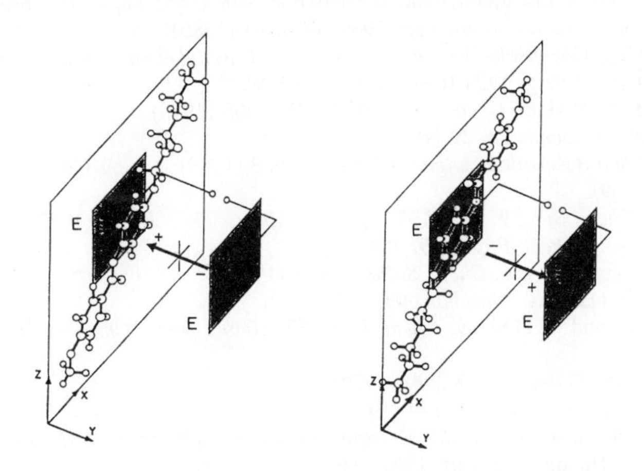

Figure 63. The two stable molecular alignments of ferroelectric liquid crystals can be used in electrooptic storage and display systems (E = electrode)

References

1. H.G.Eberle, *Kontakte (Darmstadt)*, **74**, 37 (1974).
2. F.Reinitzer, *Monatsh. Chem.*, **9**, 421 (1888).
3. Computer diagrams based on the following X-ray structure data: P.Sawzik, B.M.Craven, *Acta Crystallogr., Sect.B*, **35**, 895 (1979); G.Ferguson and J.Ball, *Structure Reports for 1979*, **45B**, 577 (1984).
4. O.Lehmann, *Ber. Dtsch. Chem. Ges.*, **41**, 3774 (1908).
5. O.Lehmann, *Z. Phys. Chem.*, **4**, 462 (1889).
6. Calculated using the X-ray data for 4,4′-dimethoxyazoxybenzene of W.R.Krigbaum, Y.Chatani and P.G.Barber, *Acta Crystallogr., Sect.B*, **26**, 97 (1970); W.B.Pearson and J.Trotter, *Structure Reports for 1970*, **35B**, 69 (1974).
7. L.Gattermann and A.Ritschke, *Ber. Dtsch. Chem. Ges.*, **23**, 1738 (1890).
8. O.Lehmann, *Z. Phys. Chem.*, **5**, 427 (1890).
9. C.Weygand, *Ber. Dtsch. Chem. Ges.*, **76**, Abt. A, 41 (1943).
10. D.Vorländer, *Ber. Dtsch. Chem. Ges.*, **39**, 803 (1906).
11. D.Vorländer, *Ber. Dtsch. Chem. Ges.*, **40**, 1415 (1907).
12. D.Vorländer, *Ber. Dtsch. Chem. Ges.*, **40**, 4527 (1907).
13. D.Vorländer, *Z. Phys. Chem.*, **57**, 357 (1907).
14. D.Vorländer and A.Gahren, *Ber. Dtsch. Chem. Ges.*, **40**, 1966 (1907).

15. Y.Björnstahl, *Ann. Phys.*, **56**, 161 (1919).
16. St.van der Lingen, *Ver. Dtsch. Phys. Ges.*, **15**, 913 (1913).
17. E.Hückel, *Phys. Z.*, **22**, 561 (1921).
18. M.deBroglie and E.Friedel, *C. R. Acad. Sci.*, **176**, 738 (1923).
19. E.Friedel, *C. R. Acad. Sci.*, **180**, 269 (1925).
20. K.Hermann and A.H.Krummacher, *Z. Kristallogr.*, **79**, 134 (1931).
21. G.Friedel, *Ann. Phys.*, **18**, 273 (1922).
22. G.Friedel and D.Friedel, *Z. Kritsallogr.*, **79**, 1 (1931).
23. 'Flüssigkristalle,' in *Römpps Chemie Lexikon*, Franckh'sche Verlagsbuchhandlung, Stuttgart, 8. Aufl., Band 2.
24. O.Lehmann, *Flüssige Kristalle und die Theorien des Lebens*, J.A.Barth, Leipzig, 1906 [*Chem. Abstr.*, **3**, 208 (1909)].
25. Structure calculated using X-ray data for 4-ethoxybenzylidene-4′-n-butylaniline (10), von J.Howard, A.J.Leadbetter and M.Sherwood, *Mol. Cryst. Liq. Cryst.*, **56**, 271 (1980); G.Ferguson, *Structure Reports for 1980*, **46B**, 94 (1985).
26. D.Vorländer, *Chemische Kristallographie der Flüssigkeiten*, Akademische Verlagsgesellschaft, Leipzig, 1924 [*Chem. Abstr.*, **19**, 9549 (1925)].
27. H.Kelker and R.Hatz, *Chem. -Ing. -Tech.*, **45**, 1005 (1973).
28. *Nachr. Chem. Techn.* **14**, 29 (1966).
29. H.Kelker and B.Scheurle, *Angew. Chem.*, **81**, 903 (1969); *Angew. Chem., Int. Ed. Engl.*, **8**, 884 (1969).
30. G.Elliot, *Chem. Ber.*, **9**, 213 (1973) [*Chem. Abstr.*, **79**, 36043m (1973].
31. R.Williams, *J. Chem. Phys.*, **39**, 384 (1963).
32. J.Timmermans, *J. Phys. Chem. Solids*, **18**, 1 (1961).
33. B.Böttcher, D.Gross, *Umschau*, **69**, 574 (1969).
34. G.H.Brown and W.G.Shaw, *Chem. Rev.*, **57**, 1049 (1957) [*Chem. Abstr.*, **52**, 5071d (1958)].
35. R.Hosemann, *Endeavour*, **32**, 99 (1970).
36. R.Hosemann, *Umschau*, **72**, 749 (1972).
37. R.Mosebach, in *Methoden der Organischen Chemie (Houben–Weyl–Müller)*, 4.Aufl., Band III/1, Thieme, Stuttgart, 1962, p.625.
38. *Geo 1986(3)*, 28/29, Verlag Gruner & Jahr, Hamburg, 1986.
39. L.Bergmann and C.Schäfer, *Experimentalphysik, Band III, Optik*, 5.Aufl., De Gruyter, Berlin, 1972.
40. (a) W.H.Westphal, *Physik*, 25./26.Aufl., Springer, Berlin, 1970; (b) C.Gerthsen, H.O.Kneser and H.Vogel, *Physik*, 11.Aufl., Springer, Berlin, 1974.
41. Completed using structural data of P.P.Ewald and C.Hermann, *Strukturbericht 1913–1928, der Z. Kristallographic*, Ergänzungsband, 1931.
42. H.Sackmann and D.Demus, *Fortschr. Chem. Forsch.*, **12**, 349 (1969).
43. D.Demus and G.Pelzl, *Z. Chem.*, **21**, 1 (1981).
44. P.W.Atkins, *Physical Chemistry*, 2nd ed., Oxford University Press, Oxford, 1984.
45. D.Turnbull, *Sci. Am.*, **212**, 38 (1965).
46. (a) J.D.Bernal, *Nature (London)*, **183**, 141 (1959); (b) J.D.Bernal, *Nature (London)*, **185**, 68 (1960).
47. R.Kaplow, S.L.Strong and B.L.Averbach, *Phys. Rev. A*, **138**, 1336 (1965).
48. *Umschau*, **79**, 362 (1979).
49. S.A.Rice, *Structure of Amorphous and Liquid Water, Top. Curr. Chem.*, **60**, 109 (1975).
50. R.Steinsträsser and L.Pohl, *Angew. Chem.*, **85**, 706 (1973); *Angew. Chem., Int. Ed. Engl.*, **12**, 617 (1973).
51. D.Demus and L.Richter, *Textures of Liquid Crystals*, Verlag Chemie, Weinheim, 1978; (a) G.W.Gray (Ed.), *Thermotropic Liquid Crystals*, Vol.22, Wiley, New York, 1987; (b) More recent reviews of liquid crystals: D.Erdmann, *Kontakte (Darmstadt)*, No.2, (1988) 3; U.Finkenzeller, *Kontakte (Darmstadt)*, No.2, 7 (1988); E.Pötsch, *Kontakte (Darmstadt)*, No.2, 15 (1988); R.Eidenschink, *Kontakte*, 15 (1979/1); H.Finkelmann,

Angew. Chem., **100**, 1019 (1988); *Angew. Chem., Int. Ed. Engl.*, **27**, 987 (1988).
52. S.Friberg, *Naturwissenschaften*, **64**, 612 (1977).
53. G.H.Brown and P.P.Crooker, *Chem. Eng. News*, **61**, 24 (1983).
54. H.Kelker and R.Hatz, *Handbook of Liquid Crystals*, Verlag Chemie, Weinheim, 1980 [*Chem. Abstr.*, **92**, B 172906r (1980)].
55. A.Loesche, *Wiss. Z. Karl-Marx-Univ., Leipzig, Math. Nat. Reihe*, **25**, 60 (1976) [Chem. Abstr., 86, 198053m (1977)].
56. V.N.Tsvetkov, *Acta Physicochim. U.S.S.R.*, **16**, 132 (1942) [*Chem. Abstr.*, **37**, 22387 (1943)].
57. A.Sauper and W.Maier, *Z. Naturforsch., Teil A*, **14**, 816 (1961).
58. P.D.Cladis, *Phys. Rev. Lett.*, **35**, 48 (1975).
59. G.Heppke, R.Hopf, B.Kohne and K.Praefcke, *Z. Naturforsch., Teil B*, **35**, 1384 (1980).
60. Hl.deVries, *Acta Crystallogr.*, **4**, 219 (1951).
61. D.Vorländer and F.Janecke, *Z. Phys. Chem.*, **85**, 697 (1913).
62. C.W.Oseen, *Trans. Faraday Soc.*, **29**, 883 (1933).
63. A.Saupe, *Angew. Chem.*, **80**, 99 (1968); *Angew. Chem., Int. Ed. Engl.*, **7**, 97 (1968).
64. R.Steinsträsser, *Chem. -Ztg.*, **95**, 661 (1971).
65. H.Stegemeyer and K.-J.Mainusch, *Naturwissenschaften*, **58**, 599 (1971).
66. J.L.Fergason, *Sci. Am.*, **211**, 77 (1964).
67. D.Vorländer, *Z. Phys. Chem.*, **105**, 221 (1923).
68. S.Chandrasekhar, B.K.Sadashiva and K.A.Suresh, *Pramana*, **9**, 471 (1977).
69. (a) J.D.Brooks and G.H.Taylor, *Carbon*, **3**, 185 (1965); (b) J.E.Zimmer and J.L.White, *Adv. Liq. Cryst.*, **5**, 157 (1982).
70. H.Zimmermann, D.Goldfarb, I.Belsky and Z.Luz, *J. Chem. Phys.*, **79**, 6203 (1983).
71. Calculated structure. Compare with illustration in Ref.99.
72. J.Malthete and A.Collet, *Nouv. Chim.*, **9**, 151 (1985).
73. H.Zimmermann, R.Poupko, Z.Lutz and J.Billard, *Z. Naturforsch., Teil A*, **40**, 149 (1935).
74. Structure calculated using X-ray data for 2,3,7,8,12,13-Hexamethoxy-5,10-dihydro-15*H*-tribenzo[*a,d,g*]cyclononene von S.Cerrini, E.Giglio, F.Mazza and N.V.Pavel, *Acta Crystallogr., Sect.B*, **25**, 2605 (1979).
75. R.Virchow, Virchows *Arch. Pathol. Anat. Physiol.*, **6**, 562 (1854).
76. O.Lehmann, *Ann. Physi.* **56**, 771 (1895).
77. W.Ostwald, *Z.Kristallogr.*, **79**, 222 (1931); [*Chem. Abstr.*, **26**, 23619 (1892)].
78. H.Kess, *Kolloid. Z.*, **88**, 40 (1939).
79. G.S.Hartley, *Aqueous Solutions of Paraffin-chain Salts*, Herrmann, Paris, 1936.
80. J.Stauff, *Kolloid Z.*, **89**, 224 (1939).
81. (a) F.Husson, H.Mustacchi and V.Luzzati, *Acta Crystallogr.*, **13**, 668 (1960); (b) V.Luzzati, H.Mustacchi, A.Skoulious and F.Husson, *Acta Crystallogr.*, **13**, 660 (1960).
82. W.Stoeckenius, *J. Cell Biol.*, **12**, 221 (1962), [*Chem. Abstr.*, **57**, 4135c (1962)].
83. J.F.Goodmann and J.S.Clunie, *Liq. Cryst. Plast. Cryst.*, **2**, 1 (1974) [*Chem. Abstr.*, **83**, 186432j (1975)].
84. Structure calculated using the usual bond angles and lengths.
85. F.Tokiwa and I.Kokubo, *J. Colloid Interface Sci.*, **24**, 223 (1967).
86. M.J.Vold, *J. Am. Chem. Soc.*, **63**, 1427 (1941).
87. W.Gross, *Angew Chem.*, **83**, 419 (1971); *Angew. Chem., Int. Ed. Engl.*, **10**, 388 (1979).
88. S.J.Singer and G.L.Nicolson, *Science*, **175** 720 (1972).
89. D.Vorländer, *Z. Phys. Chem.*, **126**, 449 (1927); see also ref.7.
90. C.Weygand 'Chemische Morphologie der Flüssigkeiten und Kristalle,' in A.Eucken and K.L.Wolf, *Hand- und Jahrbuch der Chemischen Physik.*, Akademische Verlagsgesellschaft Leipzig, 1941 Band 2 III C, p.21.
91. W.Kast, *Angew. Chem.*, **67**, 592 (1955).
92. G.W.Gray, *Molecular Structure and Properties of Liquid Crystals*, Academic Press,

London, 1962; G.Vertogen and W.H. de Jeu, *Thermotropic Liquid Crystals, Fundamentals*, Springer, Heidelberg, 1987.
93. D.J.Deutscher, H.M.Vorbrodt and H.Zaschke, *Z. Chem*., **21**, 9 (1981).
94. A.A.Akhrem and Y.A.Titov, *Total Steroid Synthesis*, Plenum Press, New York, 1970.
95. C.H.Carlisle and D.Crowfoot, *Proc. R. Soc. London, Ser. A*, **184**, 64, (1945).
96. M.J.S.Dewar and R.S.Goldberg, *J. Am. Chem. Soc*., **92**, 1582 (1970).
97. W.E.Bacon and G.H.Brown, *Mol. Cryst. Liq. Cryst*., **6**, 155 (1969).
98. W.R.Young, A.Aviram and R.J.Cox, *J. Am. Chem. Soc*., **94**, 3976 (1972).
99. (a) K.Tabei and E.Saitou, *Bull. Chem. Soc. Jpn*., **42**, 1440 (1969) [*Chem. Abstr*., **71**, 44203p (1969)]; (b) V.T.Nimkin, Y.A.Zhadanov, E.A.Medyantzeva and Y.A.Ostroumov, *Tetrahedron*, **23**, 3651 (1967); (c) M.A.El-Bayoumi, M.El.Aasser and F.Abdel-Halim, *J. Am. Chem. Soc*., **93**, 586 (1971).
100. J.Howard, A.J.Leadbetter and M.Sherwood, *Mol. Cryst. Liq. Cryst*., **56**, 271 (1980); G.Ferguson, *Structure Reports for 1980*, **46B**, 94 (1985); Reidel, Dordrecht.
101. D.Demus, *Z. Chem*., **15**, 1 (1975).
102. J.A.Nash and G.W.Gray, *Mol. Cryst. Liq. Cryst*., **25**, 299 (1974) [*Chem. Abstr*., **81**, 55350s (1974)].
103. G.W.Gray, K.J.Harrison and J.A.Nash, *Electron. Lett*., **9**, 130 (1973).
104. (a) R.Eidenschink, *Kontakte (Darmstadt)*, No.1, 15 (1979); (b) *Liquid Crystals, Product Survey*, E.Merck, Darmstadt, 1983.
105. G.Allen, *Chem. Ind. (London)*, **19**, 689 (1984).
106. H.Schubert, W.Schulze, H.J.Deutscher, V.Uhlig and R.Kuppe, *J. Phys. (Paris) Coll*., 379 (1975).
107. (a) W.Maier and A.Saupe, *Z. Naturforsch., Teil A*, **13**, 564 (1958); (b) W.Maier and A.Saupe, *Z. Naturforsch., Teil A*, **14**, 882 (1959); (c) W.Maier and A.Saupe, *Z. Naturforsch., Teil A*, **15**, 287 (1960).
108. B.Kohne and K.Praefcke, *Chem. -Ztg*., **109**, 121 (1985).
109. B.Böttcher, *Chem. -Ztg*., **96**, 214 (1972).
110. H.Liebig and K.Wagner, *Chem. -Ztg*., **95**, 733 (1971).
111. (a) Boeing, *U.S. Pat*., 3 439 525; (b) NCR, *Br. Pat*., 1 138 590; (c) NCR, *Br. Pat*., 1 161 039; (d) Hoffmann-LaRoche, *D.O.S*., 2 014 909.
112. J.L.Fergason, *Am. J. Phys*., **38**, 1729 (1979).
113. G.V.Lukianoff, *Mol. Cryst. Liq. Cryst*., **8**, 389 (1969).
114. (a) J.T.Crissey, J.L.Fergason and J.M.Bettenhausen, *J.Invest. Dermatol*., **45**, 329 (1965); (b) O.S.Selawry, H.S.Selawry and J.F.Holland, *Mol. Cryst*., **1**, 495 (1966) [*Chem. Abstr*., **65**, 20502d (1966)]; (c) see also literature cited in refs. 25 and 34.
115. J.Walker, *Spektrum Wiss*. 145 (1984).
116. (a) R.Williams, *J. Chem. Phys*., **39**, 384 (1963); (b) R.Williams, *U.S. Pat*., 3 322 485 (1962).
117. G.H.Heilmeier, L.A.Zanoni and L.A.Barton, *Appl. Phys. Lett*., **13**, 46 (1968).
118. R.Steinsträsser and H.Krüger, 'Flüssigkristalle,' in *Ullmanns Encyclopädie der Technischen Chemie*, 4. Aufl., Verlag Chemie, Weinheim, 1976, p.657.
119. A.Kobale and H.Krüger, *Phys. Unserer Zeit*, **6**, 66 (1975).
120. V.K.Freedericksz and V.Zolina, *Trans. Faraday Soc*., **29**, 919 (1933) [*Chem. Abstr*., **28**, 12385 (1934)].
121. R.Eidenschink, *Kontakte (Darmstadt)*, No.2, 25 (1984).
122. M.Schadt and W.Helfrich, *Appl. Phys. Lett*., **18**, 127 (1971).
123. L.Pohl, *Kontakte (Darmstadt)*, No.1, 33 (1973).
124. S.Meiboom and L.C.Snyder, *Science*, **162**, 1337 (1968); (a) S.Meiboom and L.C.Snyder, *J. Am. Chem. Soc*., **89**, 1038 (1967); (b) S.Meiboom and L.C.Snyder, *J. Am. Chem. Phys*., **52**, 3857 (1970).
125. G.R.Luckhurst, *Q. Rev. Chem. Soc*., **22**, 179 (1968).
126. L.Pohl, *Kontakte (Darmstadt)*, No.3, 27 (1973).
127. L.Pohl and R.Eidenschink, *Kontakte (Darmstadt)* No.2, 33 (1978).

128. A.Saupe, *Angew. Chem.*, **80**, 90 (1968).
129. S.Meiboom and L.C.Snyder, *Acc. Chem. Res.*, **4**, 81 (1971).
130. H.Kelker, *Z. Anal. Chem.*, **198** 254 (1963) [*Chem. Abstr.*, **60**, 3550d (1964)].
131. G.Kraus and A.Winterfeld, *Wiss. Z. Univ. Halle*, **27/M**, 83 (1978).
132. G.Solladier and R.G.Zimmermann, *Angew. Chem.*, **96**, 335 (1984); *Angew. Chem., Int. Ed. Engl.*, **23**, 348 (1984).
133. G.W.Gray and D.G.McKonnel, *Mol. Cryst. Liq. Cryst.*, **34**, 211 (1977) [*Chem. Abstr.*, **87**, 46819y (1977)].
134. P.J.Flory, *Proc. R. Soc. London, Ser. A* **234**, 60 (1956).
135. G.Rehage, *Nachr. Chem. Tech. Lab.*, **32**, 287 (1984). H.Finkelmann, *Angew. Chem.*, **99**, 840 (1987); *Angew. Chem., Int. Ed. Engl.*, **26**, 816 (1987).
136. H.Finkelmann, in *35 Jahre Fonds der Chemischen Industrie 1950–1985*, Fonds der Chemischen Industrie, Frankfurt, 1985, p.145.
137. H.Finkelmann and G.Rehage, *Makromol. Chem., Rapid Commun.*, **181**, 31 (1980).
138. H.Finkelmann, H.Ringsdorf and J.H.Wendorff, *Makromol. Chem.*, **179**, 273 (1978); cf. also M.Ballauff, *Chem. uns. Zeit*, **22**, 63 (1988); H.Ringsdorf *et al.*, *Pure Appl. Chem.*, **57**, 1009 (1985); H.Ringsdorf, R.Wüstefeld, E.Zerta, M.Ebert and J.H.Wendorff, *Angew. Chem.*, **101**, 934 (1989); *Angew. Chem., Int. Ed. Engl.*, **28**, 914 (1989); I.Cabrera, V.Krongauz and H.Ringsdorf, *Angew. Chem.*, **99**, 1204 (1987); *Angew. Chem., Int. Ed. Engl.*, **26**, 1178 (1987); W.Kranzig, B.Hüser, H.W.Spiess, W.Keuder, H.Ringsdorf and H.Zimmermann, *Adv. Mat.*, **2**, 36 (1990); W.Kreuder, H.Ringsdorf, O.Herrmann-Schönherr and J.H.Wendorff, *Angew. Chem.*, **99**, 1300 (1987); *Angew. Chem., Int. Ed. Engl.*, **26**, 1249 (1987).
139. P.Keller, *Makromol. Chem. Rapid Commun.*, **6**, 255 (1985).
140. (a) A.A.Michelson, *Philos. Mag.*, **21**, 554 (1911); (b) P.Gaubert, *C.R.Acad. Sci.*, **179**, 1148 (1924).
141. F.J.Kahn, *Appl. Phys. Lett.*, **22**, 111 (1973).
142. R.B.Meyer, L.Liebert, L.Strzelecki and P.Keller, *J. Phys. (Paris) Lett.*, **36**, L-69 (1975).
143. N.A.Clark and S.T.Lagerwall, *Appl. Phys. Lett.*, **36**, 899 (1980); *U.S. Pat.*, 4 367 924 (1980).
144. C.Escher, *Kontakte (Darmstadt)*, No.2, 3 (1986).
145. S.T.Lagerwall and I.Dahl, *Mol. Cryst. Liq. Cryst.*, **114**, 151 (1984).
146. S.M.Kelly and R.Buchecker, *Helv. Chim. Acta*, **71**, 461 (1988).
147. J.L.Serrano, P.Romero, M.Marcos and P.J.Alonso, *J. Chem. Soc., Chem. Commun.*, 859 (1990).

9 Surfactants, Micelles, Vesicles: Preorganization of Interface-active Compounds

9.1 EFFECT OF SURFACTANTS ON INTERFACES

Surfactants (surface-active agents) are compounds whose molecules are fitted with both pronounced *lipophilic* and pronounced *hydrophilic* moieties (*amphiphilic* molecules, Figure 1), and which accumulate preferentially at interfaces, intruding on local energetic conditions [1].

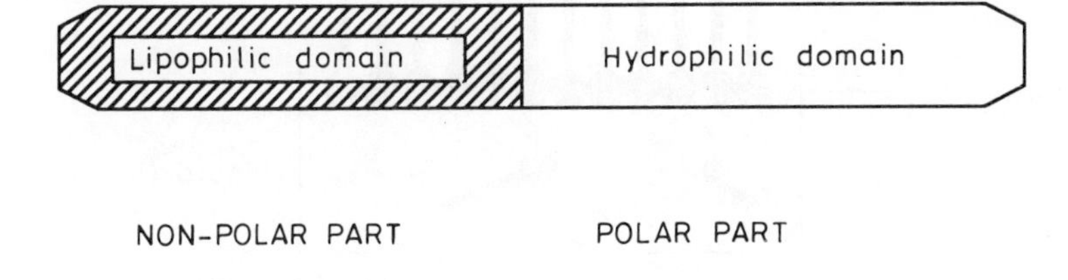

Figure 1. Schematic representation of a surfactant molecule (classic dualism of hydrophilicity and lipophilicity)

The hydrophobic parts of the molecules are generally prepared from hydrocarbon chains, and also aromatics and their perfluorinated analogues [2] and, occasionally, combinations thereof. According to the types of polar groups used, several categories of surfactants are distinguished: (1) anionic, (2) cationic, (3) amphoteric, and (4) non-ionic. In Figure 2, one characteristic example of each category is given.

The field of interactive forces emanating from such molecules, is necessarily polarized. Surfactant molecules, therefore, always accumulate in places where similar gradients are present, i.e. at interfaces of dissimilar media (e.g. water/air). In water, the polar groups of surfactants are hydrated and pulled into the aqueous phase, while the hydrophobic chains are squeezed out of it. Thus, surfactants are found at the water/air interface, where they form a monomolecular layer ('monomolecular brush', Figure 3).

Structure	Designation	Type
$H_3C{-}(CH_2)_{16}{-}COO^{\ominus}Na^{\oplus}$	Sodium stearate	Anionic
$RN^{\oplus}(CH_3)_3Cl^{\ominus}$	Quaternary ammonium salts	Cationic
$R{-}CH(N^{\oplus}(CH_3)_3){-}COO^{\ominus}$	Betaines	Amphoteric
$RO{-}(CH_2CH_2O)_n{-}H$	Fatty acid polyglycol esters	Non-ionic

Figure 2. The different types of surfactants. R = mostly long-chain alkyls

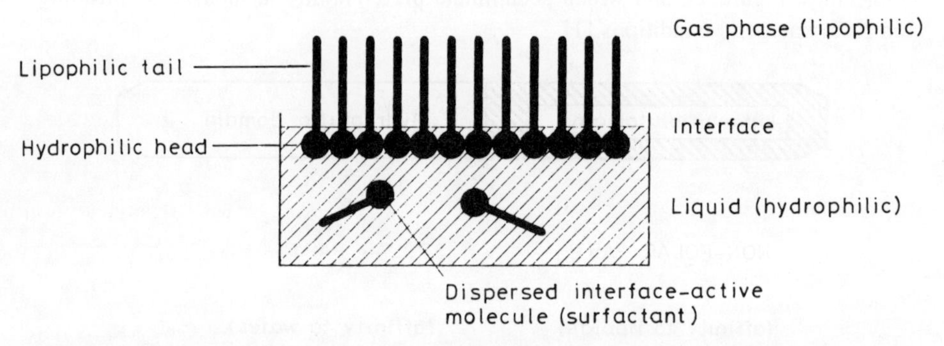

Figure 3. Accumulation and orientation of surfactant molecules at an interface (decreasing the surface tension)

As a result of their containing both non-polar and polar groups, the surfactant molecules seemingly build a bridge between lipid and aqueous phases (oriented adsorption). An accumulation of surfactant molecules at the interface only occurs if a lowering of the free surface energy accompanies it [1]. Therefore, the surface activity of a substance also depends on the solvent used.

Because of the large intermolecular forces (multiple hydrogen bonding), water has a particularly large surface tension (γ = 72.583 mN/m). However, as the attractive forces of hydrophobic hydrocarbon chains are considerably weaker, the surface tension usually decreases when surfactant molecules are added to the surface layer. As a consequence of this 'surface activation', physical effects known as spreading, wetting, solubilization, emulsion and dispersion formation, frothing, etc., are observed. They play a decisive role in the action of surfactants in washing, rinsing and cleaning processes (wettability, ability to carry dirt), and in other applications [3].

9.2 MICELLES, LAYERS, VESICLES AND OTHER ORDERED AGGREGATES

Another process whereby dissolved surfactant molecules react to the repelling action of surrounding water is aggregation to form micelles [4–6]. A rough idea of this special form of surfactant aggregation is given in the classic droplet model (spherical micelle, Figure 4), in which the hydrophobic groups (alkyl chains) of the surfactant molecules are gathered at the centre of the spherical droplet. At its surface the hydrophilic groups are found, providing contact with the surrounding water.

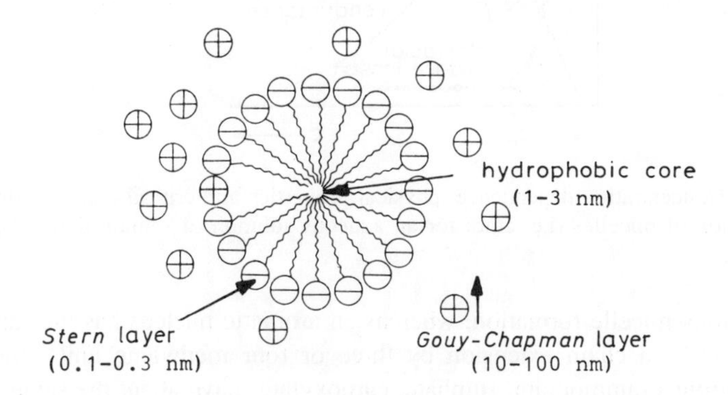

Figure 4. Cross-section of an idealized spherical anionic micelle with positive counter ions

Micellar solutions therefore represent micro-heterogeneous media, in which the lipophilic core of the surfactant micelle forms a microscopically dispersed oil-like phase in water (intramolecular emulsion). According to this model, micelles of anionic and cationic surfactants consist of three layers:

1. A hydrophobic core with a radius of 1–3 nm, roughly corresponding to the length of the hydrophobic part of the surfactant molecules;
2. the Stern layer, formed by the charged terminal groups and some of their counter ions (0.3–0.6 nm thick),
3. the Gouy–Chapman bilayer, in which the remaining hydrated counter ions are found.

The formation of micelles is only observed above a certain concentration limit of the surfactant, called the critical micellar concentration (cmc). It can be measured with relative precision, using abrupt changes in concentration-dependent physical properties of the system, such as density, conductivity, surface and interfacial tension and osmotic pressure (Figure 5), and also with light diffraction and adsorption experiments involving dyestuffs [1c].

The tendency for micelle formation is the higher and the cmc the lower, the larger is the hydrophobic part of the tenside molecule. Branching in the hydrophobic

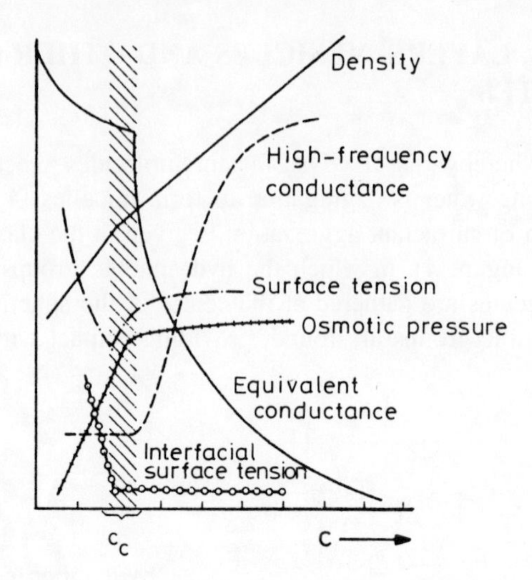

Figure 5. Concentration dependence, physical properties and critical concentration, c_c, for the formation of micelles (i.e. cmc) for an aqueous solution of sodium dodecylsulphonate [1b]

chain inhibits micelle formation, whereas an aromatic nucleus has the same effect on the cmc as a chain extension by three or four methylene units. Most ionic terminal groups (ammonium, sulphate, carboxylate) have about the same effect on the properties of their micelles.

With non-ionic surfactants the micelle formation is favoured, since electrostatic repulsion in the Stern layer is reduced, so that the cmc is lower than for ionic surfactants (ca 100-fold, Table 1). For the same reason, non-ionic surfactants form much larger micelles than do ionic surfactants (Table 1). The micelle size in each case depends on the surfactant molecule used (ionic or uncharged, bulk of the hydrophilic and hydrophobic moieties, etc.) and the experimental conditions.

If the concentration of a surfactant is increased well beyond its cmc, the originally spherical micelles will start to change into ellipsoidal and on to cylindrical aggregates (Figure 6a–c); this can be detected by small-angle scattering of X-rays in solution [6c]. This change is associated with a second cmc, e.g. 0.3 mol/l for cetyltrimethylammonium bromide [4].

Table 1. Critical micellar concentration (cmc) and average number of monomers n for several surfactants (in water at 20 °C) [7]

Surfacant	cmc (mol/l)	n
$H_3C(CH_2)_{11}N^+(CH_3)_3Br^-$	14.4	50
$H_3C(CH_2)_{11}OSO_3^-Na^+$	8.1	62
$H_3C(CH_2)_{11}COO^-K^+$	12.56	50
$H_3C(CH_2)_{11}(OCH_2CH_2)_6OH$	0.1	400

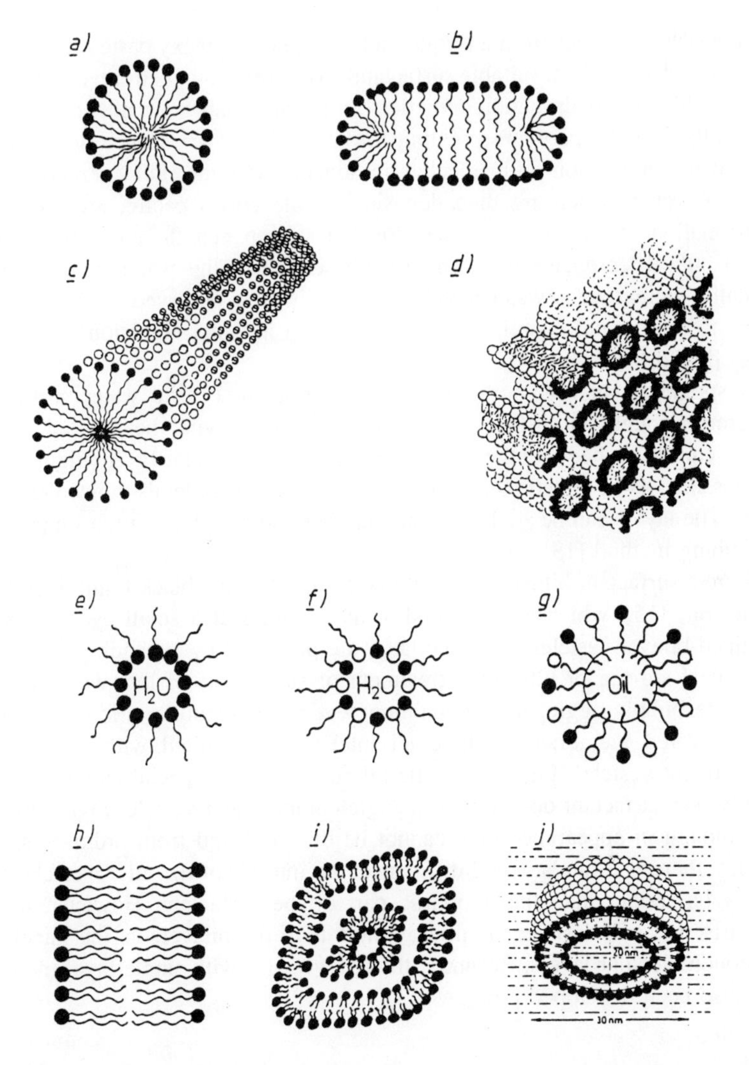

Figure 6. Examples of organized surfactant structures (schematic representation of various forms of surfactant aggregation)

At even higher surfactant concentrations, these cylindrical micelles will coagulate, and on passing a third cmc, liquid-crystalline mesophases (cf. Chapter 8) [8] with mostly nematic structures will appear (Figure 6d). With derivatives of cholesterol and other chiral molecules, the individual nematic layers are twisted at a given angle to each other, and so produce a helical structure (cholesteric liquid crystals, cf. Chapter 8).

Mesophases, in which the presence of a solvent, mostly water, in given amounts is decisive in helping produce the liquid-crystalline state, are called lyotropic. In contrast, in thermotropic systems the liquid-crystalline character only exists within a given temperature interval [9]. Lyotropic systems are strongly birefringent and their consistency varies from a turbid, mobile liquid to a waxy paste (cf. Chapter 8).

In non-polar solvents, suitable surfactants even form so-called 'inverted micelles' [10], in which the hydrophobic moieties of the amphiphiles are now on the surface and the hydrophilic parts are pointing inwards (Figure 6e). Hence the structure is inverted in comparison with micelles in aqueous solution. For energetic reasons, traces of water, which are then deposited at the core's centre, are required for the formation of inverted micelles. Such a system can therefore be treated as droplets of water enclosed in an oily phase. When the water concentration is gradually increased, a water-in-oil emulsion is first observed, until the system reverts to the normal endolipophilic form, an oil-in-water emulsion (cf. Figure 6f and g) [8a,8d,11].

On spreading an organic solution of a surfactant on the water surface of a 'Langmuir trough', a surface film is formed which, with the help of a mobile barrier, can be pushed together to an ordered monomolecular layer (Figure 3) [12]. It is possible, as in mixed membranes, to embed heterogeneous molecules in these layers. The layers can be applied in any number and orientation on a support, using the dipping method [13].

Discrete surfactant bilayers exist in two forms [14]: as 'black liquid membranes' (Figure 6h) [15], which are prepared at the opening of a small eye, and as self-contained bilayer vesicles [16]. The latter are produced by expanding, followed by mechanical stirring or ultrasonic treatment of surfactants of low water solubility. These are bubbly aggregates, containing several consecutive onion-like shells of double-layered membranes, where all interstices are filled with water ('multi-compartment vesicle', Figure 6i). The microvesicle is a special case; it is formed by only one surfactant double layer (single-compartment vesicle, Figure 6j) [10c]. Even though in appearance they cannot be differentiated from ordinary spherical vesicles, there is a fundamental difference: the interior of the microvesicle is filled with water, rather than the lipophilic part of the surfactant. However, a certain comparison is permissible: the microvesicle is to the bilayers as the micelle is to the monomolecular layers. Recently, though, vesicles with single-layer membranes have also been developed [16b].

9.3 THE CLOUD POINT

Non-ionic surfactants of the ethylene oxide adduct type (cf. Table 1) demonstrate a phenomenon known as 'reversed solubility' [1c,17]: For aqueous solutions of these compounds, there is a sharp temperature limit below which the solution is clear and homogeneous, but as soon as it is exceeded, clouding is observed and, after some time, separation into two liquid phases occurs. This transition temperature is defined as the cloud point.

Turbidity and phase separation are reversible phenomena and, on cooling below the cloud point, return to the original state. The cloud point also changes with the surfactant concentration, as well. It is only observed above the critical demixing concentration, which in accord with physical laws is always higher than the cmc, i.e. the clouding presupposes the formation of micelles [18].

Because the solubility of this type of surfactant is predominantly based on the hydration of the ethylene oxide chains [17–19], it was initially assumed that, beyond a given temperature, i.e. the cloud point, the hydrogen bonding between water and surfactant was no longer strong enough to hold the surfactant in solution [1c].

Experiments indeed proved that one of the two phases formed contains predominantly the surfactant [18]. Nevertheless, it still contains much more water, than could be expected after the dehydration of the surfactant molecules, e.g. an average of 8.7 molecules of water per ether linkage in the case of hexaethylene glycol n-octyl ester. Recent investigations showed that the effect is to be treated as any other phase separation of liquid two-component systems, using the same thermodynamic laws, but taking in account the formation of micelles [1c]. On raising the temperature, an increased amount of anisometric micelles is observed, leading to a mutual restriction of freedom of movement. For entropy reasons the system separates into two phases, the emulsified coacervate, rich with large hydrated micelles, and the equilibrium phase with only a few micelles left. Their regained freedom of movement affords the necessary increase in entropy for the total system.

This does not mean that the hydrated state of the hydrophilic ether functionalities is not involved in the phase separation process, but only insofar as favourable conditions are produced for the thermodynamic quantities which determine the lower separation limit [18]. In this respect, the non-ionic surfactants with their dual molecular structure are plainly ideal.

References

1. (a) K.Laux (Ed.) *Die Grenzflächenaktiven Stoffe*, *Chem. Technol.*, Band 4, Hauser-Verlag, Munich, 1960; (b) H.Bueren and H.Grossmann (Eds), *Grenzflächenaktive Substanzen*, Chemische Taschenbücher, Weinheim, 1974; (c) K.Schönfeldt (Ed.), *Grenzflächenaktive Äthylenoxid-Addukte*, Wissenschaftl, Verlagsgesellschaft, Stuttgart, 1976; (d) Surface-active agents, No. 14 in the series Fonds der Chemischen Industrie, Frankfurt/Main, 1987; (e) H.Hoffmann and G.Ebert, *Angew. Chem.*, **100**, 933 (1988) *Angew. Chem., Int. Ed. Engl.*, **27**, 902 (1988).
2. J.N.Meussdoerffer and H.Niederprüm, *Chem. Ztg*, **104**, 45 (1980).
3. Henkel & Cie, *Waschmittelchemie*, Hüthig, Heidelberg, 1976.
4. J.H.Fendler and E.H.Fendler (Eds). *Catalysis in Micellar and Macromolecular Systems*, Academic Press, New York, 1975.
5. C.Tanford (Ed.), *The Hydrophobic Effect: Formation of Micells and Biological Membranes*, 2nd ed., Wiley, New York, 1980.
6. (a) K.Shinoda, I.Nakagawa, B.Tamamushi and T.Isemura (Eds), *Colloidal Surfactants*, Academic Press, New York, 1963, p.1; (b) K.L.Mittel (Ed.), *Colution Chemistry of Surfactants*, Vols a and 2, Plenum Press, New York, 1979; (c) B.Lindman and H.Wennerström, *Top. Curr. Chem.*, **87**, 1 (1980); (d) F.Menger, *Acc. Chem. Res.*, **12**, 111 (1979); (e) P.Fromherz, *Nachr. Chem. Tech. Lab.*, **29**, 537 (1981); (f) H.F.Eicke, *Chimia*, **36**, 241 (1982); (g) On the subject of membranes and particularly phospholipids, compare J.Seelig, *Nachr. Chem. Tech.*, **36**, 1096 (1988).
7. J.J.Fuhrhop, *Bioorganische Chemie*, Thieme, Stuttgart, 1982.

8. (a) P.A.Winsor, *Chem. Rev.*, **68**, 1 (1968); (b) A.S.C.Lawrence, *Mol. Cryst. Liquid Cryst.*, **7**, 7 (1969); (c) B.J.Forrect and L.W.Reeves, *Chem. Rev.*, **81**, 1 (1981); (d) R.von Kleinsorgen and P.H.List, *Pharm. Unserer Zeit.*, **10**, 8 (1981); (e) E.Nürnberg and W.Pohler, *Pharm. Ztg.*, **128**, 2601 (1983).
9. (a) G.H.Brown, *Chem. Unserer Zeit*, **2**, 42 (1968); (b) R.Steinsträsser and L.Pohl, *Angew. Chem.*, **85**, 706 (1973).
10. (a) P.Ekwall, L.Mandell and K.Fontell, *Mol. Cryst. Liquid Cryst.*, **8**, 157 (1969); (b) H.F.Eicke, *Top. Curr. Chem.*, **87**, 85 (1980); (c) J.H.Fendler, *Pure Appl. Chem.*, **54**, 1809 (1982).
11. R.von Kleinsorgen and P.H.List, *Pharm. Unserer Zeit*, **9**, 109 (1980).
12. (a) D.Möbius, *Chem. Unserer Zeit*, **9**, 173 (1975); (b) G.A.Somorjai and M.A.van Hove (Eds), *Monolayers, Struct. Bonding*, Vol. 38, Springer, Berlin, 1979; (c) D.Möbius, *Acc. Chem. Res.*, **14**, 63 (1981).
13. G.Wegner, *Chimia*, **36**, 63 (1982).
14. H.Ti Tien (Ed.), *Bilayer Lipid Membranes*, Marcel Dekker, New York, 1974.
15. E.Bamberg, R.Benz, P.Läuger and G.Stark, *Chem. Unserer Zeit*, **8**, 33 (1974).
16. (a) J.H.Fendler, *Acc. Chem. Res.*, **13**, 7 (1980); (b) J.H.Fuhrhop, *Nachr. Chem. Tech. Lab.*, **28**, 792 (1980).
17. M.J.Schick (Ed.), *Nonionic Surfactants*, Marcel Dekker, New York, 1967.
18. H.Lange, *Fette Seifen Anstrichm.*, **70**, 748 (1968).
19. (a) R.Heusch, *Makromol. Chem.*, **182**, 589 (1981); (b) R.Heusch, *Tenside Detergents* **20**, 1 (1983).

10 Organic Semiconductors, Conductors and Superconductors

10.1 INTRODUCTION

In recent years, the area of research dealing with 'organic metals' has enjoyed a rapid development. Conducting, soluble and thermoplastic chemical materials are of great industrial and economic interest. Even modest conductivity produces antistatic properties, which are important in floorings, textiles, varnishes and many other things. For example, electronic instruments can be protected from electromagnetic fields by the use of casings made of conducting synthetic materials. Organic conductors are being developed as corrosion-resistant materials for electrodes, e.g. for use in novel types of batteries. In reproduction technology (electrophotography), photoconductive organic materials are already widely used. The proposed applications are so varied that the use of organic conductors would not be stopped even by high prices. This is particularly true for an (already planned) organic high-temperature superconductor.

As can be seen in Figure 1, a comparison of the conductivity in inorganic (left) and organic (right) materials shows that ordinary organic compounds are insulators. In what follows, the two groups of organic conductors indicated on the right in Figure 1 will be discussed, i.e. the organic charge-transfer complexes and the polymers with metallic conductivity.

10.2 CONDUCTING CHARGE-TRANSFER COMPLEXES

In 1973, two groups almost simultaneously discovered that the 1 : 1 charge-transfer (CT) complex of the donor tetrathiafulvalene (**1**) and the acceptor 7,7,8,8-tetracyanoquinodimethane (TCNQ, **2**), crystallized from acetonitrile, displayed metal-like electric conductivity over a wide temperature range [1].

S S S S NC CN C C NC CN

Tetrathiafulvalene (TTF) Tetracyano-*p*-quinodimethane (TCNQ)

1 2

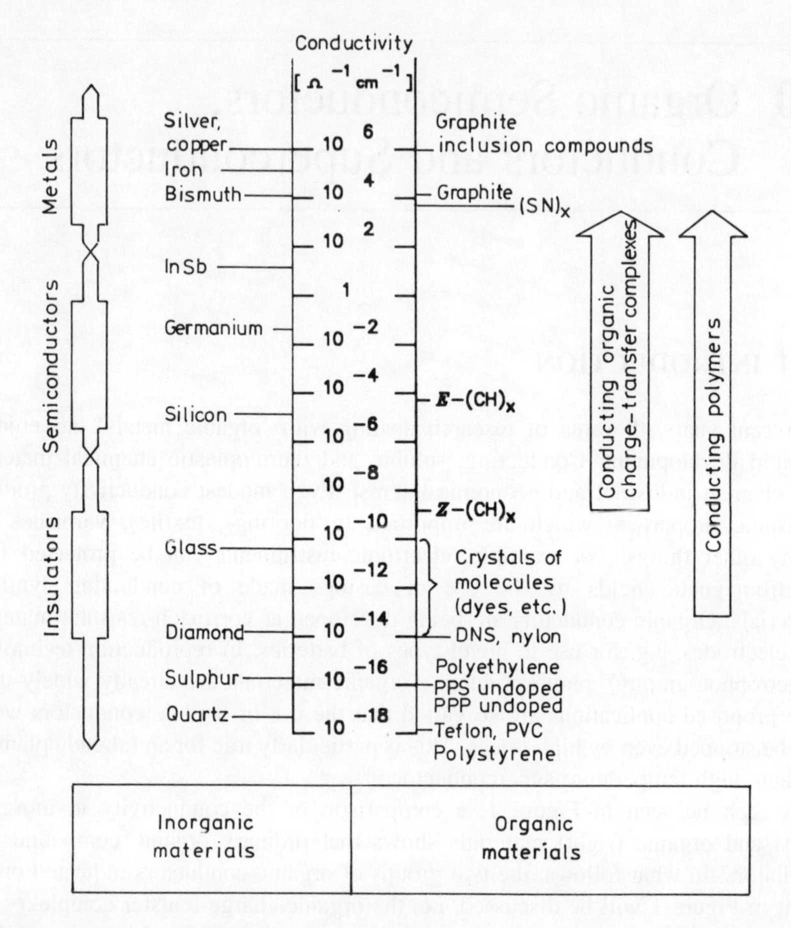

Figure 1. Electrical conductivity of inorganic (left) and organic conductors (right), semiconductors and insulators (at room temperature)

Under incident light the crystals appear black, whereas with transmitted light they are olive. They display the highest electric conductivity ($\sigma = 1.47 \times 10^4$ S/cm) at a temperature of 66 K. For the first time, a conductivity approaching that of metals was achieved (copper at 298 K: $\sigma = 6 \times 10^5$ S/cm).

It appears that, if conductivity is to be achieved in this type of crystalline organic conductor, specific geometric and electronic conditions are required. In crystalline molecular arrays, the electrons flow in 'supramolecular' orbitals, i.e. orbitals stretching over several molecules and consisting of MOs stacked in a column. For the development of a high conductivity, it is necessary that the donor and acceptor molecules crystallize into separate stacks (Figure 2).

If, on the other hand, mixed stacks are formed within the same crystal lattice, a charge-transfer during the conductivity process is almost impossible, and the material is an insulator (Figure 2a).

The separate stacking of donor and acceptor molecules alone, however, is not

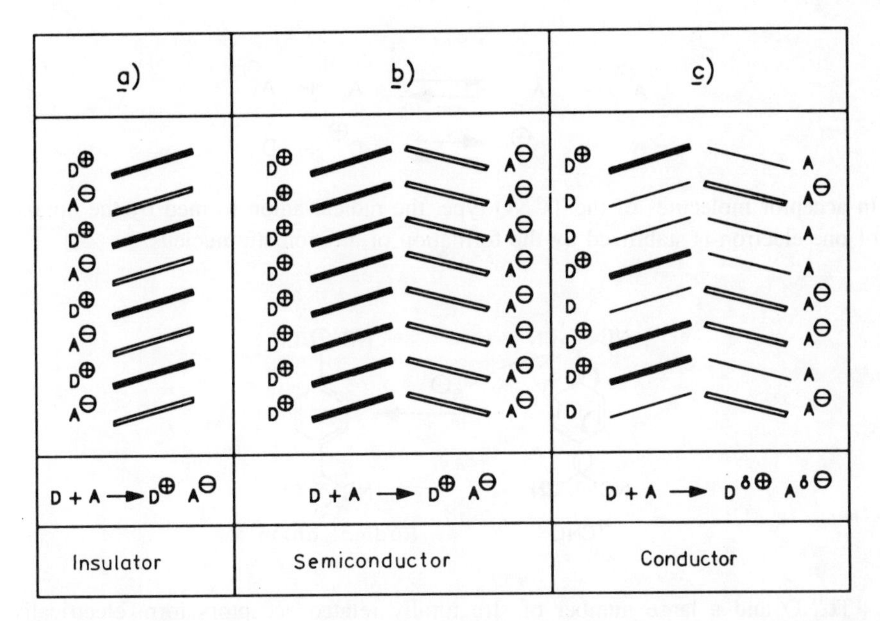

Figure 2. Crystal packing in single crystals of organic charge-transfer complexes: (*a*) mixed stacking of donor (D) and acceptor (A) molecules; (*b*) and (*c*) separate stacks of D and A

sufficient to achieve high conductivity. The electronic properties of both donor and acceptor must be tuned to each other in order to ensure that the extent of the electronic interaction, i.e. the charge-transfer process, is well balanced: if a total charge-transfer occurs (Figure 2b), CT complexes with at best semiconductor properties are obtained. The desired high conductivity, therefore, is only observed when the charge transfer is merely partial (Figure 2c).

According to the valence band theory, if a total charge-transfer occurs, all acceptor molecules will have a negative charge, and all donor molecules a positive charge. As a consequence, during the electron-transport process, intermediate, energetically unfavourable, doubly charged molecules will occur, which additionally will give rise to strong Coulomb repulsion:

$$A^{\ominus} + A^{\ominus} \rightleftharpoons A + A^{2\ominus}$$

$$D^{\oplus} + D^{\oplus} \rightleftharpoons D + D^{2\oplus}$$

Because of this, such complexes generally only afford conductivities in the range of semiconductors.

However, if the charge transfer is only partial, there will be both neutral and charged donor and acceptor molecules present. In this situation, charge transport without great energy expenditure is possible, and CT complexes with high electrical conductivity are to be expected:

$$A^{\ominus} + A \rightleftharpoons A + A^{\ominus}$$

$$D + D^{\oplus} \rightleftharpoons D^{\oplus} + D$$

In acceptor molecules of the TCNQ type, the radical anion formed by the uptake of one electron is stabilized by the formation of an aromatic nucleus:

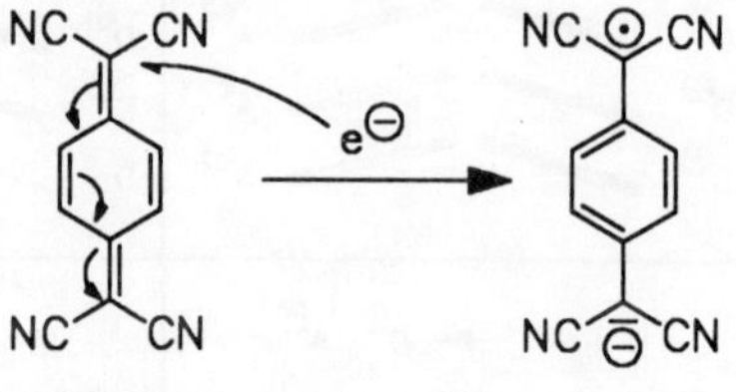

TCNQ Radical anion

TCNQ and a large number of structurally related acceptors form electrically conducting materials with many donors, e.g. alkali and alkaline earth metals, and with ammonium salts. The tetrathiafulvalene (TTF) donor can, on accepting one electron, stabilize itself by forming an aromatic sextet:

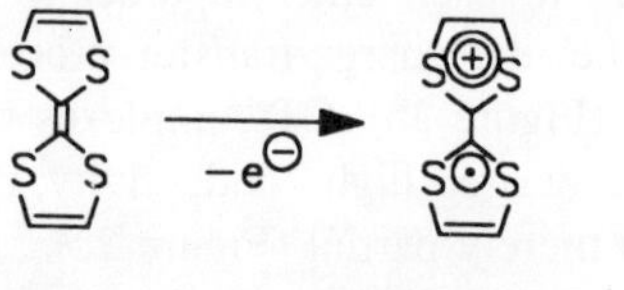

TTF Radical cation

The TTF•TCNQ charge-transfer complex belongs to the category of mixed-valent salts, as the charge transfer is not complete. Because of charge-transfer interactions between the charged and neutral species of both acceptor and donor molecules, segregated acceptor and donor stacks already form during crystallization (cf. Figure 2c). Apparently, with mixed valences this arrangement is more favourable than the alternating stacking of donor and acceptor molecules shown in Figure 2a.

When organic charge-transfer complexes are cooled below a certain temperature, the so-called Peierls transition is observed, in the TTF•TCNQ complex, for example, below 58 K. As in metals, the conductivity first increases towards lower temperatures, but then suddenly drops by three to four orders of magnitude, with a concurrent distortion of the molecular stacks taking place. This change from 'metal' to semiconductor and insulator occurs in many molecular CT complexes and radical cation salts, and only very rarely, as in tetramethyltetraselenafulvalene perchlorate, $(TMTSF)_2ClO_4$, does another phenomenon take over: superconductivity.

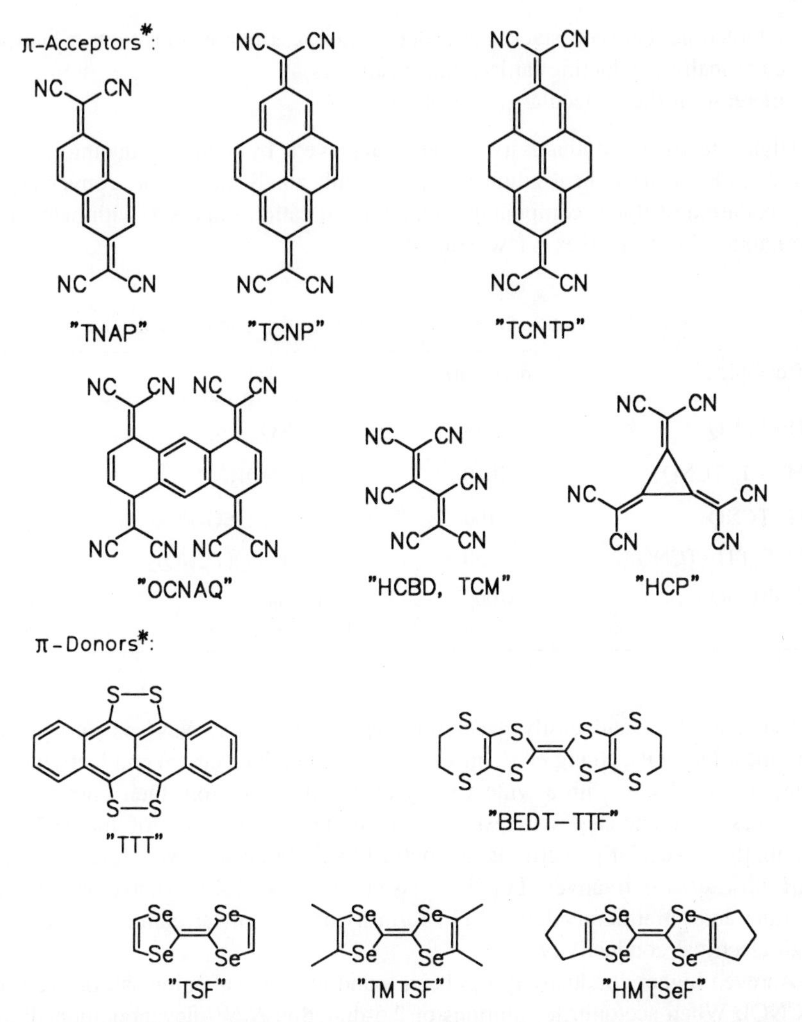

Like the Peierls transition, the latter is due, according to the BCS theory (named after the physicists Bardeen, Cooper and Schrieffer), to interactions between electrons and phonons of a solid (see below). In order to change the respective redox potentials, donor and acceptor molecules have been modified in many ways [1–3]. The aims of these optimizations of donor–acceptor pairs are as follows:

— a charge-transfer between radical species in a range from 0 (no CT) to 1 (total CT);

---

* These acronyms are often derived from names that do not conform to the IUPAC nomenclature: tetracyanonaphthoquinodimethane; tetracyanopyrenoquinodimethane; tetracyanotetrahydropyrenoquinodimethane; hexacyanobutadiene, also tetracyanomuconitrile; hexacyanotrimethylenecyclopropane; 11,11,12,12,13,13,14,14-octacyano-1,4;5,8-anthradiquinotetramethane; tetrathiatetracene; bis(ethylenedithiotetrathiafulvalene); tetraselenafulvalene; tetramethyltetraselenafulvalene; hexamethylenetetraselenafulvalene.

— interactions between stacks, in order to suppress phase transitions which turn electrically conducting lattices into insulators;
— increase in the molecular polarizability.

High electrical conductivity is also achieved by either reducing acceptor compounds to the radical anion salts, e.g. with alkali metals, or, conversely, by the oxidation of donor compounds to the radical cation salts, e.g. with halogens to the halides. Table 1 gives a few examples.

Table 1. Conductivities of some CT-complexes and radical salts

CT complex	sσ (S/cm)	Radical salt	σ (S/cm)
TTF•TCNQ	500	Li(TCNQ)	5×10^{-6}
HMTSeF•TCNQ	2000	$Cs_2(TCNQ)_3$	2×10^{-3}
TTF•$TCNQ_2$	100	$Et_3NH^+(TCNQ)_2^-$	5×10^{-2}
BEDT–TTF•TCNQ	50	Quinolinium$(TCNQ)_2$	100
TSF•TCNQ	10 000	TTF•$Cl_{0.68}$	0.6
		$(TMTSF)_2$•PF_6	1000

Using a novel and easily accessible type of acceptor, *N,N'*-dicyanoquinonediimines (DCNQI), Hünig *et al.*, in 1986 achieved high electrical conductivities [3]. They are available, with a wide variety of substituents, from their corresponding quinones in a one-step synthesis. Owing to the small angle of the =N—CN group, the molecular planarity is, in contrast to TCNQ derivatives, retained even in tetrasubstituted derivatives. For this reason, *syn-N,N'*-1,4-dicyanonaphtoquinonediimine and tetrathiafulvalene (TTF) form a charge-transfer complex with a very high electrical conductivity.

An even higher conductivity has been found in a radical anion salt derived from DCNQI. When acetonitrile solutions of 2,5-dimethyl-*N,N'*-dicyanoquinonediimine (2,5-DM-DCNQI) and copper(I) iodide are mixed, a black crystalline powder [(2,5-DM-DCNQI)$_2$Cu] precipitates which, in this form, has a conductivity of 0.4 S/cm. This limited conductivity is presumably due to contact resistance between crystallite surfaces.

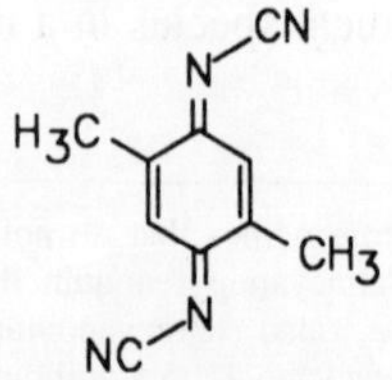

2,5–DM–DCNQI

$$2\ \text{2,5-DM-DCNQI} + \text{CuI} \xrightarrow{CH_3CN} [\text{2,5-DM-DCNQI}]_2^{\ominus}\,Cu^{\oplus} + 0.5\ I_2$$

2,5-DM-DCNQI (2,5-DM-DCNQI)$_2$Cu

Single crystals of the same composition are obtained by electrolysis of 2,5-DM-DCNQI and copper(II) bromide in acetonitrile. At 295 K they already show a conductivity of 800–1000 S/cm, which continues to increase, even below 1.3 K (Figure 3). This demonstrates the salt's metallic character, since no Peierls transition whatsoever is observed.

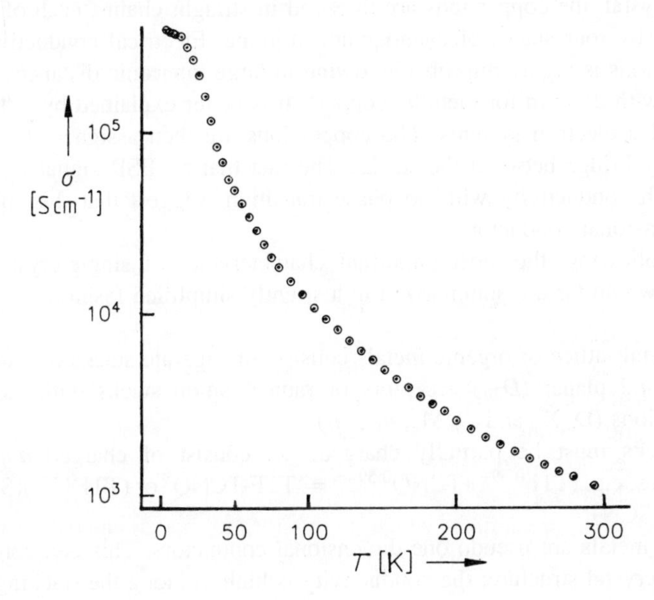

Figure 3. Temperature dependence of the electric conductivity σ of single crystals of (2,5-DM-DCNQI)$_2$Cu

At 3.5 K, particularly pure crystals of (2,5-DM-DCNQI)$_2$Cu exhibit an extremely high conductivity of ca 500 000 S/cm [the analogous TCNQ salt (TCNQ)$_2$Cu is only a semiconductor, whose conductivity drops from 1.5×10^{-5} S/cm at 530 K, to 2×10^{-6} S/cm at 290 K].

As shown by X-ray crystallography, the copper atoms are coordinated by the N atoms of four *N*-cyano groups in a slightly distorted tetrahedron; all of them are at nearly identical distances from the metal cation (Figure 4).

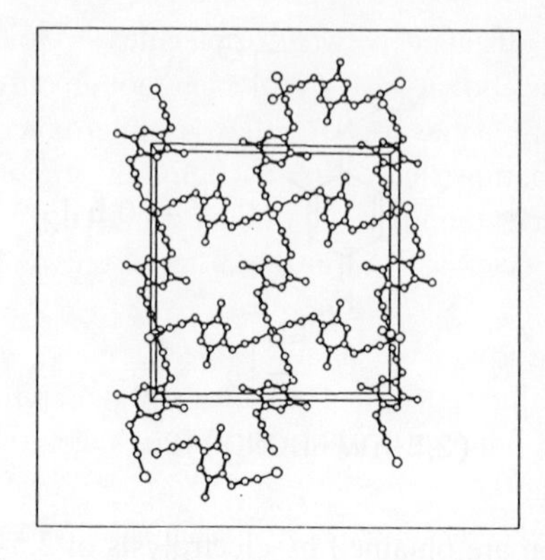

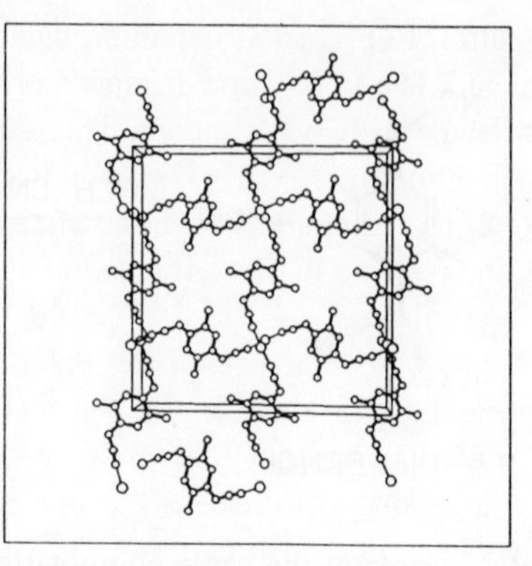

Figure 4. *a,b*-Projection of the crystal structure of $(2,5\text{-DM-DCNQI})_2Cu$ (stereoview; small circles = C, N; larger circles = Cu; for better clarity, the hydrogen atoms are not shown) [3]

In the crystal, the copper ions are threaded in straight chains, each of the latter surrounded by four stacks of cyanoquinonediimine. Electrical conductivity along the copper ions is highly improbable, owing to large interionic distances (388 pm, compared with 256 pm for metallic copper). It is better explained by a flow along the stacked π-electron systems. The copper ions are then assigned the role of a conductivity bridge between the stacks. The fact that no ESR signal is observed, and its high conductivity with no phase transition, suggest that this might be a multi-dimensional conductor.

In the following, the most important characteristics of single-crystal organic metals known so far are summarized in a slightly simplified fashion:

a. The crystal lattice of organic metals consists of separate stacks of planar (D_{2h}) donors and planar (D_{2h}) acceptors or radical anion stacks with closed-shell counter ions (D_mX_n and A_mM_n, $m > n$).
b. The stacks must be partially charged, i.e. consist of charged *and* neutral molecules, e.g. $(TTF^{0.59+})(TCNQ^{0.59-})$ = 'TTF·TCNQ' or $(TTF^{0.58+})(SCN^-)_{0.58}$ = $TTF_{12}SCN_7$.
c. Organic metals are pseudo one-dimensional conductors. This is a consequence of their crystal structure; the conductivity is highest along the stacking axes (in most cases, this is the longest axis in the crystal needles) and lowest along the remaining two crystal axes. The anisotropy in electric conductivity can be as high as a factor of 10^3.
d. Organic metals are ground-state insulators and, at a certain temperature (T_{MI}), undergo a metal-to-insulator transition. From a theoretical point of view, this temperature is a consequence of (c).
e. An exchange of sulphur for selenium in donor molecules increases the conductivity at room temperature and lowers T_{MI}.
f. Most D_mX_n crystals are highly conducting and metallic, whereas A_mM_n crystals are not.

In neutral TTF stacks, the interplanar distance between molecules is 362 pm, whereas in TTF•TCNQ this distance comes down to 347 pm, even though only one TTF molecule in two is charged. Therefore, some sort of cohesive 'intermolecular, intrastack bonding' must exist, counteracting the cationic coulombic repulsion. Because of the comparatively large distances (compared with conventional covalent bonds), the overlap in these bonds must necessarily be in the order of only 0.1 eV, or 8–13 kJ/mol.

The supramolecular orbital, or band, formed by all molecular overlaps in the stack provides a mechanism for the metallic delocalization of electrons along the stack. According to Soos [5], this type of bonding is difficult to treat theoretically, because it is at the limit of the proper applicability of VB and MO theories. So far, it has been demonstrated that the larger the heteroatom in donor molecules is, the larger is the extent of the overlap and the stronger the intermolecular bonds become, affording conductivity at room temperature.

There are several mechanistic explanations for the fact that the conductivity of an organic metal increases with falling temperature: First, the cooling of a single crystal of an organic metal produces an isotropic contraction, the one along the stack axis being more pronounced than that along the other two crystal axes. Consequently, with falling temperature the intermolecular overlap (and conductivity) increases dramatically.

Second, as the intermolecular overlap is weak, it will strongly respond to lattice vibrations, particularly along the stack axis. On cooling, the number of phonons (i.e. the amplitude of the lattice vibrations) in the crystal diminishes; the overlap is more easily maintained, and the conductivity increases. Finally, with a coherent coupling of electron movements and lattice vibrations (electron–phonon coupling, 'collective mode'), a conductivity could arise which would be higher than that expected from the narrow bandwidth in organic metals, such as TTF•TCNQ.

The electron–phonon coupling, however, is easily disturbed by either lattice defects or coulombic interactions, so that below T_{MI} the 'collective mode' waves cannot move freely through the crystal, and therefore no current is flowing. Recent spectroscopic results suggest that the 'collective mode' waves indeed contribute to the metallic conductivity of TTF•TCNQ.

As a whole, the relatively high room temperature conductivities found in organic metals can be ascribed to narrow-band formation, due to an overlap of a combination of open- and closed-shell molecular orbitals. The metallic behaviour, i.e. decreasing resistivity with falling temperature, is most probably due to a combination of lattice contractions and reduced lattice vibrations observed as the temperature drops, and perhaps to a 'flowing' collective mode contribution [1c].

10.3 THE METAL-TO-INSULATOR TRANSITION

Several mechanistic suggestions have been made for the metal → insulator transition in one- and two-dimensional solids. The phenomena responsible for the interruption of the electron flow normally originate in a combination of electronic and lattice vibration energies. In a one-dimensional array of charged molecules, subject to a periodic change in that one dimension, the crystal lattice will consist of alternating regions of higher and lower charge densities, i.e. a charge density wave (CDW). This is known as the Peierls instability.

The degree of instability of a one-dimensional metallic band depends on its degree of band filling, i.e. the number of neutral and radical ion molecules in the stack. Without neutral molecules, each has an unpaired spin, and there is an electronic driving force towards spin pairing. The molecules in the stack tend to dimerize and form a 'covalent bond', doubling the unit cell (if the original dimension of the unit cell, a, was the length of one molecule, then the new dimension of the unit cell will be $2a$).

Associated with the 'covalent bond' of the dimerized state are bonding and antibonding bands. The energy gap between bonding and antibonding bands leads to semiconductor behaviour and is responsible for the loss of free electrons and, hence, the loss of metallic conductivity. With a half-filled band, the periodicity of the charge density corresponds to that of the lattice. In this case, the CDW is fixed and the flow of electrons is inhibited: the material turns into an insulator.

When neutral molecules are introduced into the stack, the resulting band will be less than half-filled. As the CDW periodicity no longer corresponds to that of the lattice, the energy gained in the transition to a twisted structure is less favourable. The reason for this is that the build-up of charges between molecules is smaller than with a half-filled band; the transition to a twisted phase then occurs at a comparatively low temperature. In this case, a metallic state is permissible between room temperature and the lattice-distortion temperature (i.e. T_{MI}). If enough neutral molecules are built into the stack, so that half of all molecules in it are neutral, then the charge transfer again corresponds to that in the original cell, i.e. there is one electron for every two molecules. In this manner a corresponding CDW can build up, provided the molecules shift away from their uniform lattice positions to form a superlattice with a period $4a$. This is shown schematically in Figure 2c for a quarter-filled band, where the stack is made up of as many neutral as ionic molecules. In the TTF•TCNQ complex, the CDW and the underlying lattice periodicity do not really match, as the charge per stack corresponds to a ratio of only 0.59. However, as in TTF•TCNQ, a 'three-dimensional effect' is able to align the charge-density waves on different stacks; thus, at temperatures below T_{MI}, the molecules are forced to form a 'superlattice' [1c].

10.4 SUPERCONDUCTING CHARGE-TRANSFER COMPLEXES

The results so far allow one conclusion: electrons in crystalline organic metals flow along supramolecular orbitals, built from stacked molecular orbitals. The metal-like electronic band structure which relates to these giant orbitals is unstable and is subject to the formation of an energy gap by phase transition. The phenomena responsible for these phase transitions are due to electron–phonon interactions (CDW), magnetic interactions (spin-density wave, SDW) or 'mechanical' interactions (arrangement of the anion).

We shall not discuss these special effects any further. It is sufficient to point out that chemists and physicists for years concentrated their efforts on stabilizing the metallic state down to the lowest possible temperature by somehow suppressing the metal-to-insulator transition. The most surprising result in these efforts was the discovery that in certain cases, e.g. $(TMTSF)_2PF_6$, high pressure not only eliminates the metal $\rightarrow$ insulator transition, but at 0.9 K (and 12 kbar) also leads to a sudden drop in resistivity, i.e. a metal-to-superconductor transition is also observed.

In a few cases, the metallic state can be achieved through hydrostatic pressure. The best example is the family of the $(TMTSF)_2X$ salts, which generally are superconducting at low pressure. Among these there are two exceptions: the salt with the anion X = ClO_4^- is the only one to be superconducting at atmospheric pressure, and the salt with X = FSO_3^- is the only organic superconductor with T_c close to 3 K (although under pressure). In this family of compounds, T_{MI} directly relates to the anion's electronegativity, basically allowing the temperature, at which the flow of electrons is interrupted, to be controlled.

Recently, it was discovered that $(BEDT–TTF)_2ReO_4$ is a superconductor at 1.5 K and 7 kbar [BEDT–TTF = bis(ethylenedithiotetrathiafulvalene)]. This is an important result, as in the crystal structure of this salt the same chalcogen–chalcogen interactions within and between stacks occur as in $(TMTSF)_2X$ salts. Also, this demonstrates that superconductivity is not restricted to organic seleno donor compounds. Most recently, $(BEDT–TTF)_2IBr_2$ was found to have superconducting properties just above 4 K, the highest transition temperature recorded for an organic compound.

The most important problem in the research on superconductors is to raise T_c. At present, the question of whether organic superconductors with $T_c > 30$ K are possible is mostly academic. Great hopes are given to tellurium analogues. Moreover, it is as yet not clear whether the superconductivity of the compounds described above is due to a BCS mechanism, as suggested by some deuterium isotopic effects observed.

10.5 POLYMERIC CONDUCTORS

10.5.1 MATERIALS AND CONDUCTIVITY

Polymers with polyconjugated main chains are, in the ground state, insulators or, if highly crystalline, at best semiconductors with conductivities in the order of 10^{-9} S/cm. Thermal and photochemical excitation liberate only a few electrons.

The best known example of a conducting conjugated polyene is polyacetylene (PA). Acetylene is polymerized by a Ziegler–Natta procedure, modified by Shirikawa, producing fine layers with a silvery, almost metallic sheen. In this procedure, acetylene dissolved in an inert solvent is poured gently on the static surface of a concentrated solution of a Ziegler catalyst, usually titanium(IV) butoxide–triethylaluminium. A black, glistening layer of polyacetylene is formed. Depending on the amount of acetylene used, the layer's thickness can be varied between 1 μm and several mm; in this form, it is then subjected to purification and doping procedures. The all-(Z) configuration **1** first obtained can be converted to the (E)-isomer **2** by tempering:

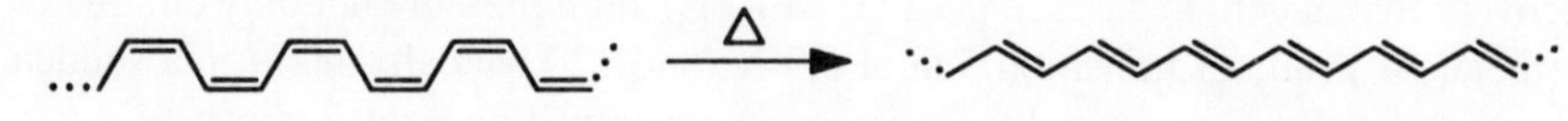

1 $\xrightarrow{\Delta}$ 2

By 'doping' with oxidizing agents (halogens, AsF_5, $FeCl_3$, $XeOF_4$) or reducing agents (e.g. alkali metals), the conductivity of polyacetylene can be improved from $\sigma = 10^{-9}$ S/cm ['cis-PA', all-(Z)-PA] and $\sigma = 10^{-5}$ S/cm ['trans'-PA, all-(E)-PA], respectively, to $\sigma = 10^{-2}$–10^{-3} S/cm. The doped polymers, now with a black metallic sheen, have the disadvantage of being unstable; in air, they lose their conductivity within hours. In spite of this, electrochemically oxidized or reduced polyacetylene has been successfully used as anodic or cathodic materials in rechargeable batteries.

On 'doping', a redox reaction between doping agent and conjugated system occurs, as a transfer of charge is observed. For example, the polymer can be oxidized to a 'polycation', the doping agent providing the counter ions. 'Doping' can be effected in solution, in the gas phase or electrochemically. Note that here 'doping' is used in a slightly different way to that in semiconductor technology.

As a mechanism for conductivity in such doped polyacetylenes, the formation of radical cations has been suggested, a formally positive charge moving along the chain (in reality, electron-deficient positions are filled with electron density). However, there are good indications that the electrons are hopping from chain to chain, as the doping has produced favourable cross-links:

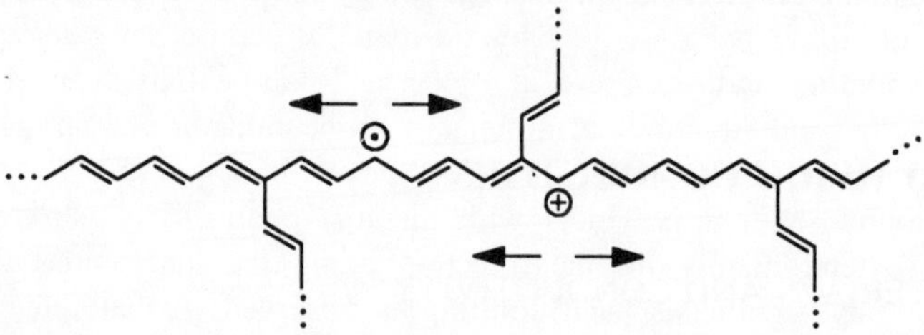

Poly-p-phenylene (PPP) and polypyrrole (PP) are further examples of polymeric conductors. The latter is a thermally stable polymer, which is prepared electrochemically from pyrrole with BF_4^- as the counter ion [11].

Poly-p-phenylene sulphide (PPS) is not conjugated and without doping is a good insulator ($\sigma = 10^{-16}$ S/cm). However, once doped with AsF_5, its conductivity increases by 16 orders of magnitude to about 1 S/cm (cf. Table 2). This value has been surpassed by some other doped polymers, but nevertheless PPS is important, since it is the first polymer with high conductivity to be soluble and thermoplastic in its doped form.

Table 2. Conductivities of doped polymeric materials

Designation	Part of polymer structure	Doping agent or counter ion	Electrical conductivity (S/cm)
(Z)-PA		I_2 AsF_5	360 560
PPP	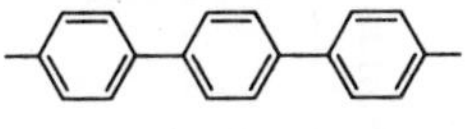	AsF_5	145
PP	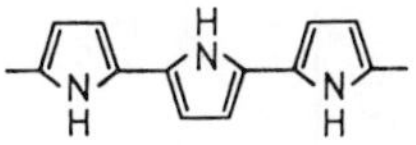	BF_4^-	100
PPS	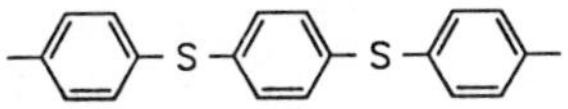	AsF_5	1

Polypyrrole is prepared in its doped form in a one-pot procedure. An aqueous solution of an electrolyte (e.g. $LiClO_4$), which also contains the pyrrole (0.06 M) and an organic solvent, is electrolysed, and the polypyrrole is deposited on the anode; if a drum-type anode is used, long sheets are obtained which are resistant to air and moisture.

10.5.2 CONDUCTIVITY AND ELECTRONIC STRUCTURE

In solids, so many atoms are participating in the bonding that electronic states group themselves into extended energy domains, called bands. Their shape, arrangement and degree of filling are described by the band structure. This determines the electronic properties and, therefore, the electrical conductivity of a given solid. In Figure 5, the band structures of a metal, a semiconductor and an insulator are shown.

In the ground state of polymers with metallic conductivity, strong bonding interactions extend mainly in one direction along the macromolecular chain. Accordingly, any significant band splitting is observed in that direction only; the distribution of the electronic states, i.e. the band structure of π-conjugated polyenes, is one-dimensional. If in polyacetylene this band was to receive electrons, it would be half-filled. Consequently, undoped polyacetylene ought to be a metallic conductor. In reality, it is an insulator. This is due to the Peierls transition, a lattice superstructure of this one-dimensional chain, in which two atoms each come together or go apart, respectively. [The Jahn–Teller effect is the molecular analogue of the Peierls transition, which, by decreasing the molecular symmetry, also leads to a gain in electronic energy (e.g. by axially stretching an octahedral complex)]. Chemically, this is connected to the localization of electrons in double and single bonds, affording a change in band structure.

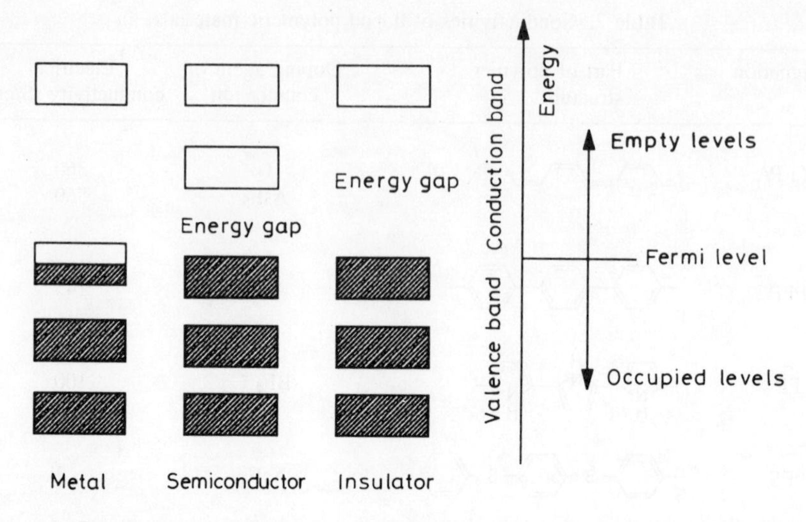

Figure 5. Band structures of a metal, a semiconductor and an insulator (band theory of solids, schematic representation). (*a*) Metals: partially filled bands close to the Fermi level allows electron mobility and charge transport in an electric field. (*b*) Semiconductors: up to the Fermi level, all energy bands are fully occupied, those above are empty. However, the energy gap being small (0.5–1.5 eV), electrons can be promoted thermally from the highest occupied band (valence band) to the lowest unoccupied band (conduction band). (*c*) Insulators: here the energy gap between the valence band and the conduction band is large (>3 eV) and cannot be crossed at normal temperatures

Near the 'Fermi level', energy levels are removed and redistributed regularly at higher and lower levels. Because only the energetically lower levels need to be filled with electrons, the energy balance of this rearrangement is positive. This leads to a gap between the fully occupied valence band and the empty conduction band, and polyacetylene, as observed, is an insulator.

The Peierls transition is due to an electron–phonon interaction and is a typically one-dimensional phenomenon. Since the corresponding lattice distortion costs elastic energy, it can only occur if, through a gain in electronic energy, the net balance is positive. In a one-dimensional situation, where electrons and phonons can only move in parallel, this is the case. By increasing the phonon energy, e.g. by raising the temperature, the Peierls transition can be suppressed, or, on the contrary, be provoked by cooling. In order to suppress it in conjugated polyenes, temperatures of several thousand degrees would be necessary, which is far above the decomposition temperature of these materials.

The band gap found in polyenes can be observed in the optical absorption spectrum of undoped polymers. A distinct absorption is, of course, only observed when the energy of the light quanta is sufficient to promote electrons from the valence band into the conduction band. The position of the absorption edge therefore corresponds to the extent of the band gap and is 1.4 eV for all-(*E*)-polyacetylene and 2.8 eV for poly-*p*-phenylene.

On doping, even with small concentrations of doping agent, strong absorptions are observed around 0.6–0.7 eV. Consequently, new electronic states must have been created in the energy gap, permitting an electronic excitation at far lower quantum energies. This is due to the charges produced on the polymer chain by the doping process. These charges, as depicted in Figure 6, are delocalized over a large section of the carbon chain's π-system. In this section, which contains approximately 15 carbon atoms, the Peierls transition is suppressed locally. As a consequence, electronic states reappear where the Peierls transition had suppressed them, midway between the valence band and the conduction band. With increasing charge density they expand and, in the metallic state, cover the whole band gap.

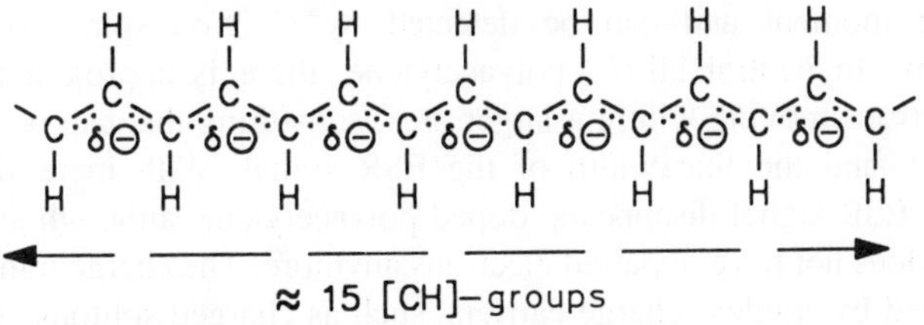

≈ 15 [CH]–groups

Figure 6. Charges produced by doping of a polymer chain [all-(E)-polyacetylene]

10.5.3 THE SOLITON CONCEPT

One physical concept which describes the charge transport along a polyene chain is that of the *solitons*. However, it can only explain the conductivity in polyacetylenes, and what is more, only in the all-(E)-transoid chain conformations. Figure 7 shows a soliton in all-(E)-polyacetylene. In the neutral state, this is a free unpaired electron (free radical), which moves in one dimension along the conjugated π-system.

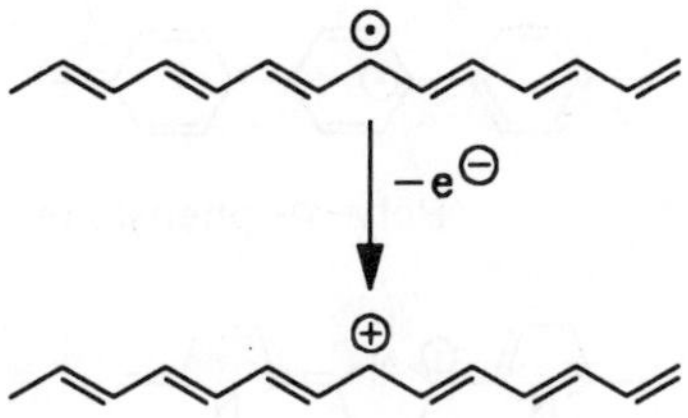

$-e^{\ominus}$

Figure 7. A neutral and a charged solitary defect in the π-conjugation system of all-(E)-polyacetylene (schematic)

The term 'soliton' refers to the mathematical description of the solution of a non-linear differential equation. One characteristic of solitons is that they do not disperse when moving. This is trivial in the case of a single electron, but not when dealing with the overlap of Bloch waves which can be used to describe electrons in a metal. A free electron in the π-orbital system is in fact delocalized over several carbon atoms. This gives rise to a local distortion of the lattice, and thus to a local suppression of the Peierls transition. The soliton creates and occupies levels, which are exactly midway between valence band and conduction band ('midgap levels'). On oxidative doping, free electrons are removed first, owing to their high energy. The soliton does not just vanish, but changes into a positive charge: along an undisturbed polyene chain, it is as delocalized and as mobile as a free electron. Such a charged soliton is spinless and can transport charges in an electric field. By changing charge and spin, it is also able to hop onto another chain.

The soliton concept has been confirmed using magnetic and optical properties of polymers. With increasing doping, further π-bonds must be broken and soliton-type charges produced, extending the midgap levels. Indeed, this can be observed in the optical absorption spectrum.

A neutral soliton carries a free electron and, therefore, a spin which has a magnetic moment and can be detected by electron spin resonance (ESR) spectroscopy. In neutral all-(*E*)-polyacetylene, there is approximately 400 ppm of free electrons moving within a larger area, something that can be deduced from the intensity and the line width of the ESR signal. With increasing degree of doping, the ESR signal disappears; doped polyacetylene, although able to conduct electricity, does not have unpaired electrons anymore. The charge transport now has to be effected by spinless charge carriers, such as charged solitons. These findings have been used as proof for the soliton mechanism. However, other physical interpretations are also possible.

In Figure 8, similar bonding defects are shown in all-(*Z*)-polyacetylene, poly-*p*-phenylene and polypyrrole. In these polymers, the soliton concept fails, since structures of different energies are isolated from each other; a solitary defect

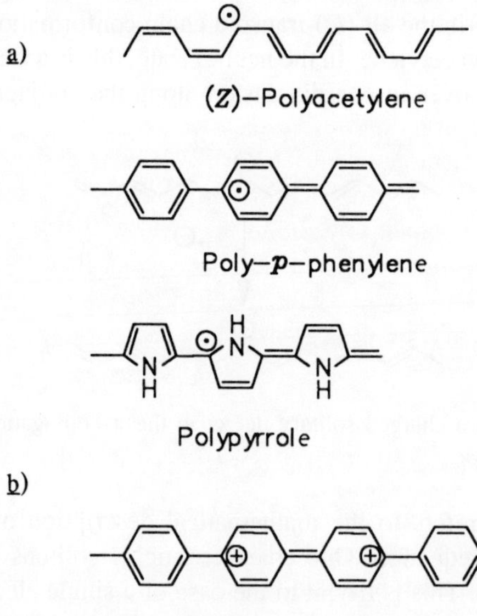

a)

(*Z*)–Polyacetylene

Poly–*p*–phenylene

Polypyrrole

b)

Figure 8. (*a*) Bonding defects in (*Z*)-polyacetylene, poly-*p*-phenylene and polypyrrole. The parts of the structure on each side of the defect have different energies. (*b*) Schematic representation of a bipolaron in poly-*p*-phenylene. The positive charges approach each other until the gain in lattice energy, obtained from the decreased number of quinoid rings, is fully compensated by the coulombic repulsion

would always run to the end of the conjugated system. Two free electrons meeting each other would recombine, forming a new chemical bond in the process. The positive charges produced on doping can only approach each other until the coulombic repulsion is equal to the lattice energy gained from the restoration of the energetically more favourable form. They then form a pair, which is held together by electronic energy and lattice forces, and is restricted to the smallest possible volume. In physics they are called bipolarons; from a chemical point of view, they could be described as dimerized radical cations. In Figure 8b such a bipolaron is shown, poly-*p*-phenylene being the example given.

10.5.4 CHARGE TRANSPORT BETWEEN CHAINS

Solitons and bipolarons can only move within a chain. Between polymer chains, 'conventional' charges must flow, hop or tunnel. There is a theory called 'inter-soliton hopping', but it is only applicable to all-(*E*)-polyacetylenes doped only slightly or not at all. Here, neutral and positively charged solitons exchange for each other when meeting on neighbouring atoms.

The measurements on polyacetylene can also be described using a general hopping theory without solitons, where only statistically distributed charges are postulated, among which there is an exchange of electrons. This hopping model also describes the conductivity in inorganic semiconductors, e.g. amorphous silicon. The model is essential for the description of charge exchanges between polymer fibrils in the polymer matrix. Soliton model and hopping theory both fail beyond a certain degree of doping, when the metallic state of the polymer is reached. Conductivity should then be described, as in a true metal, using the Drude model of a free electron gas. In a metal this is possible, irrespective of direction. Polymers have very different electronic profiles along the macromolecular chain or at right-angles to it, since the macromolecules have covalent bonds in one direction only. Longitudinally, there is a broad band width, leading to a high mobility of the electrons and metallic conductivity. In an undoped polymer, there are hardly any bonding interactions at right-angles to the chains, and therefore no energy bands. As a consequence, an anisotropic conductivity should also be observed in polymers with metallic conductivity, as well as in $(SN)_x$ and organic single-crystal charge-transfer complexes.

10.5.5 CONDUCTIVITY AND CHEMICAL STRUCTURE

In conducting organic charge-transfer complexes and radical cation salts, mostly planar organic molecules, molecular discs, are stacked one on top of another (Figure 9). The partial oxidation and close proximity of these discs leads to a charge transport and metallic conductivity (see above). Following the same principle, the structure of doped polymers with metallic conductivity would correspond to a 'molecular stack of planks', between which the counter ions and neutral molecules would be embedded. The 'planks' correspond to the planar molecular chains of the polymers with their conjugated π systems.

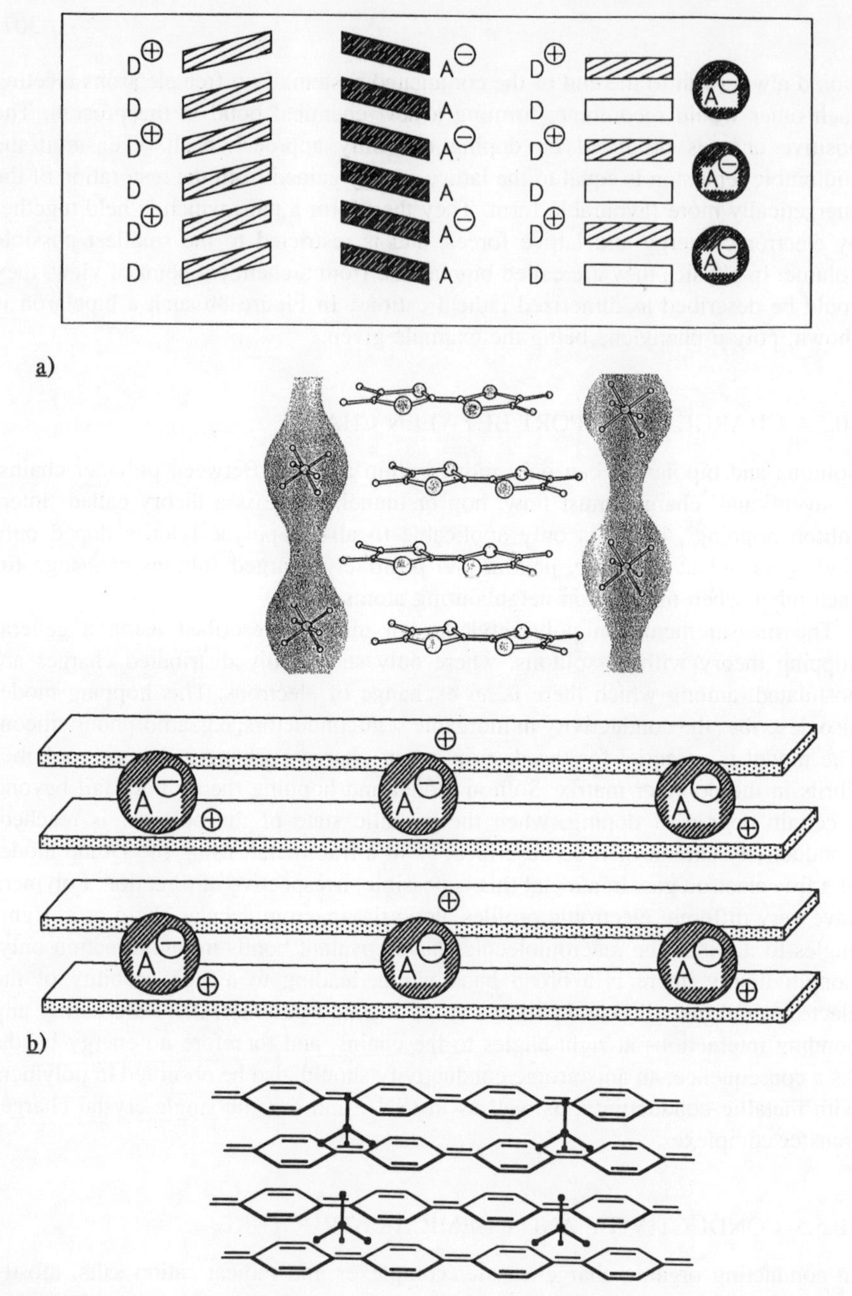

Figure 9. Structure of organic conductors. (*a*) Crystals of organic charge-transfer complexes and radical salts. In order to observe metallic conductivity along the stacks in crystals of $(TMTSF)_2PF_6$, it is essential to find one of two possible forms of stacking: either separate stacks of donor and acceptor molecules, or mixed stacks; in both cases a partial transfer of charges and an appropriate number of counter ions are required. (*b*) Comparative structural model of doped polymers with metallic conductivity (poly-*p*-phenylene with ClO_4^- as counter ion). According to this model, the molecular chains, on doping, are arranged in stacks, which allow intensive charge-transfer interactions between themselves [1h]

When doped, they can approach each other to such an extent that intensive charge-transfer interactions between them become possible; neighbouring chains now adopt an optimum configuration, i.e. they stack slightly staggered along their molecular axis. Figure 9 illustrates this principle with poly-*p*-phenylene.

10.6 CONDUCTING POLYMERIC METAL–MACROCYCLE COMPLEXES

Macrocyclic metal complexes, containing either phthalocyanine or porphyrin rings as ligands, are another group of conducting polymeric materials [1]:

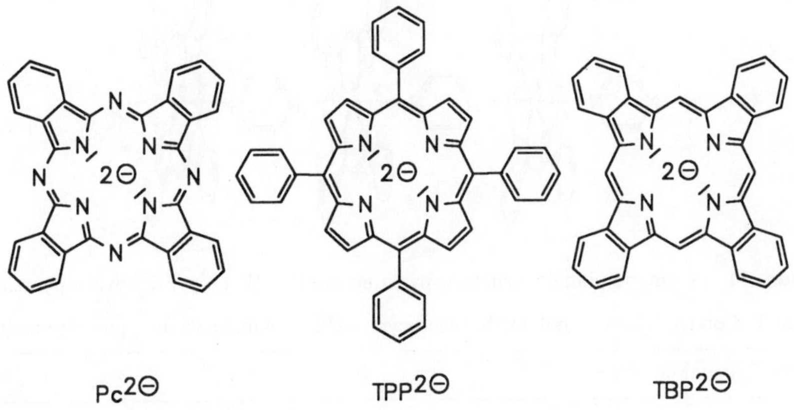

$Pc^{2\ominus}$ $TPP^{2\ominus}$ $TBP^{2\ominus}$

Starting from phthalocyanines, of which, for example, the Pb(II) complex crystallizes in a columnar structure with metal-to-metal contact, hence its conductivity, stacked, quasi-one-dimensional arrays have been produced (Figure 10).

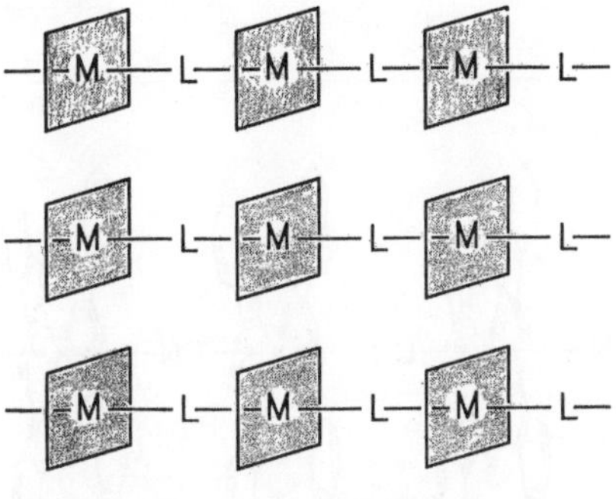

—M—L—M—L—M—L—

—M—L—M—L—M—L—

—M—L—M—L—M—L—

Figure 10. Polymeric metal macrocycles. The squares represent phthalocyanine or porphyrin; M = metal, L = ligand

Suitable central atoms M are transition metals which favour a sixfold coordination with octahedral arrangement, such as Fe, Co and Cr. Pyrazine, 4,4′-bipyridine and tetrazole proved to be good ligands L with which to link the metal ions (Figure 11).

Oxidative doping, e.g. with iodine, of such metal complexes ‘polymerized’ with two coordinative bonds, produces a dramatic increase in conductivity (Table 3). The doped materials are stable at temperatures of up to 100 °C.

Apart from these purely coordinative polymers, there are also those in which the bridging ligands form two σ bonds, or one σ-bond in one direction and a metal-to-ligand bond in the other. Polymeric complexes with CN^- as a bridging ligand belong to the latter type (Figure 12) [4].

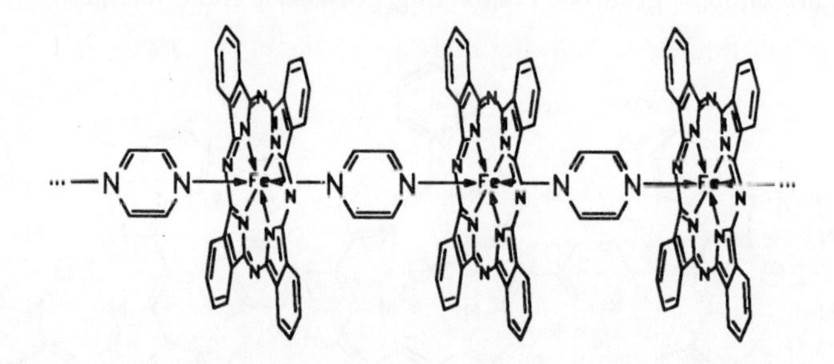

Figure 11. Polymeric phthalocyanato(μ-pyrazine)iron(II), $[PcFe(pyz)]_n$; pyz = pyrazine

Table 3. Conductivities of polymeric macrocycles (Pc = phthalocyanine; pyz = pyrazine)

Polymeric metal complex	σ (S/cm)
PcPb(II)	1×10^{-4}
$[(tert\text{-Bu})_4PcSiO]_n$	2×10^{-7}
$[(tert\text{-Bu})_4PcSiO\cdot I_x]_n$	2×10^{-3}
$[PcFe(pyz)]_n$	1×10^{-6}
$[PcFe(pyz)\cdot I_{2.6}]_n$	20
$[PcCoCN]_n$	1×10^{-2}

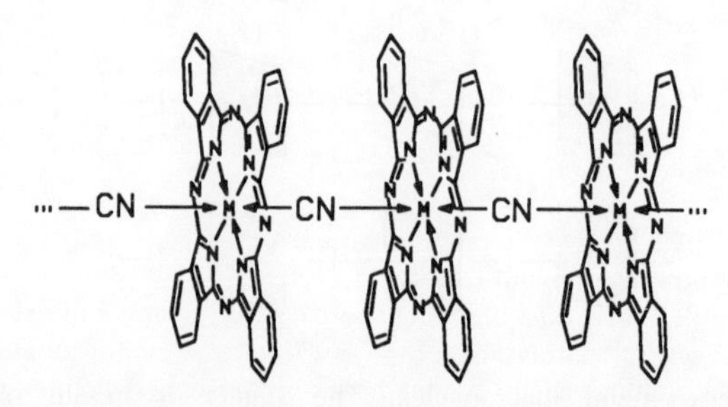

Figure 12. Polymeric phthalocyanato metal complex linked by cyanide anions, $[PcMCN]_n$

10.7 INFORMATION STORAGE ON A MOLECULAR LEVEL

Even though the mechanism of conductivity in organic metals has not yet been fully explained, concepts have already been developed where the transport of charges and spins along a polyene chain can be used for the storage of information on a molecular level. Electronic building blocks of this type would be considerably smaller than existing chips and integrated circuits, their size being on a molecular scale, i.e. 1000 times smaller.

The basic idea is to use the soliton-type defects in all-(E)-polyacetylene mentioned earlier. If the carbon atoms of a chain were numbered, one such defect would just exchange the positions of single and double bonds in the molecule.

In molecules this is only useful when functional groups or aromatic rings are present in the polyene chain, so that different positions of double bonds also lead to structures with different energies. In Figure 13 an example is given. In this particular

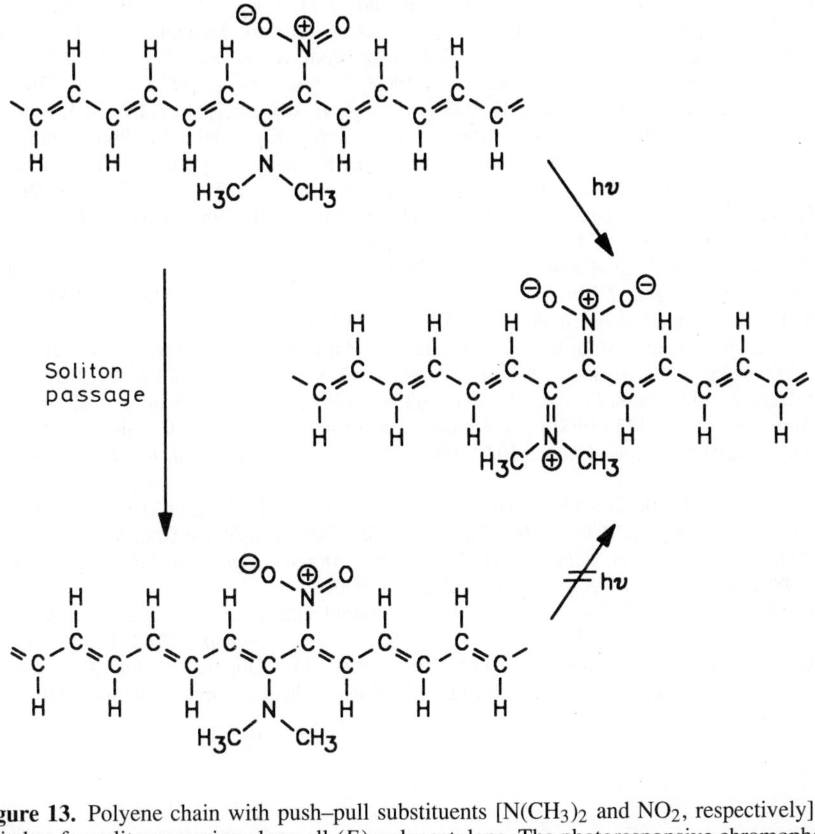

Figure 13. Polyene chain with push–pull substituents [$N(CH_3)_2$ and NO_2, respectively] as switches for solitons moving along all-(E)-polyacetylene. The photoresponsive chromophore shifts the central double bond, thus blocking the passage for any soliton moving along the polyene chain. Moreover, after the passage of a soliton, the central double bond has been shifted and the chromophore can no longer be excited. Such structures could be used as gates in 'molecular computers'

case, the higher energy state would be reached through photolytic excitation of the ground state. A soliton would only be able to pass the site if the double bond is in phase with those in the rest of the polyene chain. On the other hand, after a soliton has passed, the chromophore could no longer be excited, i.e. it would be switched off.

In a model of a molecular computer, the polyene chain corresponds to a conductor trace; the soliton is responsible for the transport of signals. The unit shown in Figure 13 represents an electronic gate. However, this model neglects the interactions of a polyene chain with its surroundings. Even if it were possible to suppress chemical reactions between polymer chains, the question would still remain of whether solitons really exist, as suggested by the model, and if they actually ‘move’.

Attempts to synthesize ferromagnetic materials have been noted [5].

References

1. For reviews, see (a) H.Meier, *Organic Semiconductors*, Verlag Chemic, Weinheim, 1974; (b) C.Hamann, J.Heim and H.Burghardt, *Organische Leiter*, Halbleiter und Photoleiter, Vieweg, Braunschweig, Wiesbaden, 1981; (c) F.Wudl, *Acc. Chem. Res.*, **17**, 227 (1984); (d) G.Wegner, *Angew. Chem.*, **93** 352 (1981); *Angew. Chem., Int. Ed. Engl.*, **20**, 361 (1981); (e) M.Hanack, *Chimia*, **37**, 238 (1983); (f) E.Amberger, H.Fuchs and K.Polborn, *Angew. Chem.*, **97**, 968 (1985); *Angew. Chem., Int. Ed. Engl.*, **24**, 968 (1985); (g) J.E.Frommer, *Acc. Chem. Res.*, **19**, 2 (1986); (h) K.Menke and S.Roth, *Chem. Unserer Zeit*, **20**, 1 (1986); **20**, 33 (1986); (i) S.V.Ley *et al.*, *Tetrahedron*, **45**, 7565 (1989); cf. G.Mahr, F.Vögtle, *J. Chem. Res. (S)*, 312 (1984); (*M*) 2901 (1984); F.Wudl, H.Yamochi, T.Suzuki, H.Isotalo, C.Fite, H.Kasmai, K.Liou, S.Srdanov, P.Coppens, K.Maly and A.Frost-Jensen, *J. Am. Chem. Soc.*, **112**, 2461 (1990); (k) T.J.Marks, *Angew. Chem.*, **102**, 886 (1990); *Angew. Chem., Int. Ed. Engl.*, **27**, 857 (1990); (l) cf. D.Ofer, R.M.Crooks and S.Wrighton, *J. Am. Chem. Soc.*, **112**, 7869 (1990).
2. (a) T.Mitsuhashi, M.Goto, K. Honda, Y.Maruyama, T.Sugawara, T.Inabe and T.Watanabe, *J. Chem. Soc., Chem. Commun.*, **1987**, 810 (1987); (b) J.Becker, J.Bernstein, S.Bittner and S.Shaik, *Pure Appl. Chem.*, **62**, 467 (1990); M.R.Bryce and A.J.Moore, *Pure Appl. Chem.*, **62**, 473 (1990); (c) Y.Yamashita, J.Eguchi, T.Suzuki, C.Kabuto, T.Miyashi and S.Tanaka, *Angew. Chem.*, **102**, 709 (1990); *Angew. Chem., Int. Ed. Engl.*, **29**, 857, (1990).
3. A.Aumüller, P.Erk, G.Klebe, S.Hünig, J.U.v.Schütz and H.P.Werner, *Angew. Chem.*, **98**, 759 (1986); *Angew. Chem., Int. Ed. Engl.*, **25**, 740 (1986); Ch.Burschka, P.Erk and S.Hünig, *et al.*, *Angew. Chem.*, **101**, 1297 (1989) *Angew. Chem., Int. Ed. Engl.*, **28**, 1245 (1989); S.Hunig, *Pure Appl. Chem.*, **62**, 395 (1990).
4. On Si–O bridge phthalocyanine crowns as supramolecules, see O.E.Sielcken, L.A.van de Kuil, W.Drenth and R.J.M.Nolte, *J. Chem. Soc., Chem. Commun.*, **1988**, 1232 (1988).
5. K.Itoh *et al.*, *J. Am. Chem. Soc.*, **112**, 4074 (1990); D.A.Dougherty, *Pure Appl. Chem.*, **62**, 519 (1990); cf. J.Thomaides, P.Maslak and R.Breslow, *J. Am. Chem. Soc.*, **110**, 3970 (1988).

11 Molecular Wires, Molecular Rectifiers and Molecular Transistors

11.1 MOLECULAR WIRES

The search for the role of biological mediators (see below), catalysts and transport and redox systems, and the mechanisms involved, introduced a new domain in chemistry, called molecular electronics [1]. Its basis is the preparation of molecules which contain a conjugated polymethine chain, as well as terminal, polar electroactive groups. They provide the necessary conductivity at the molecular level. By incorporating them into bimolecular layers of vesicles (Figure 1) (cf. Chapter 9 [2]), systems with certain capabilities for the storage, processing and transmission of signals and information might be obtained. As mediators, electrons, photons, protons or ions could play an important role.

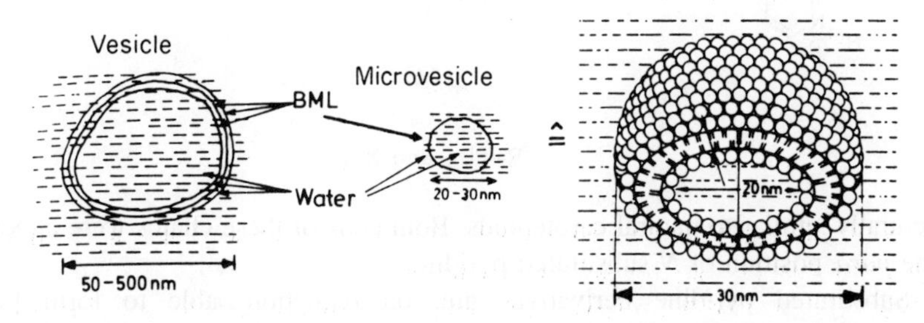

Figure 1. Vesicle consisting of several bimolecular layers (BML) with water-filled intermediate spaces and interior. Microvesicles consist of only one bimolecular layer, of which the inner surface is only about half as large as the outer, i.e. it contains about half as many molecules as the other

The assumption that in biological systems carotenoids or polyisoprene chains might be involved in the conduction of electrons prompted Lehn in 1986 to prepare four model compounds with different chain lengths, $\mathbf{1}^{2+}$– $\mathbf{4}^{2+}$, whose structures were to meet the following criteria [3a]:

1. A conjugated (polymethine) chain, permitting electrical conductivity.
2. Two terminal, polar electroactive groups for the reversible exchange of electrons.
3. A sufficient chain length, corresponding to the thickness of a double membrane.

As electron-conducting chains, conjugated double-bond systems were chosen, which, according to UV and NMR spectra, adopt an all-(E) configuration

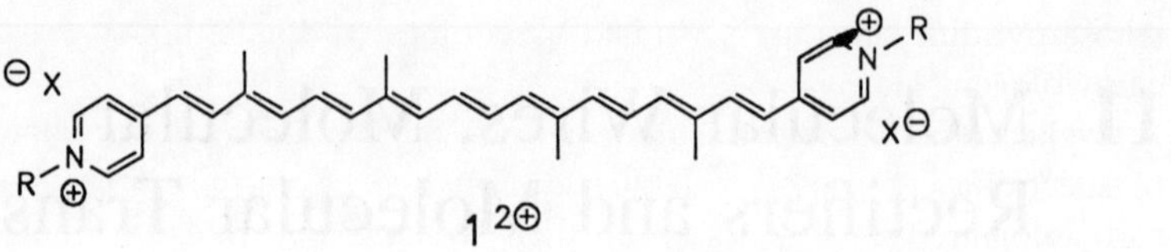

$^{\ominus}X$ R–N$^{\oplus}$ $^{\oplus}$N–R $X^{\ominus}$

1 $^{2\oplus}$

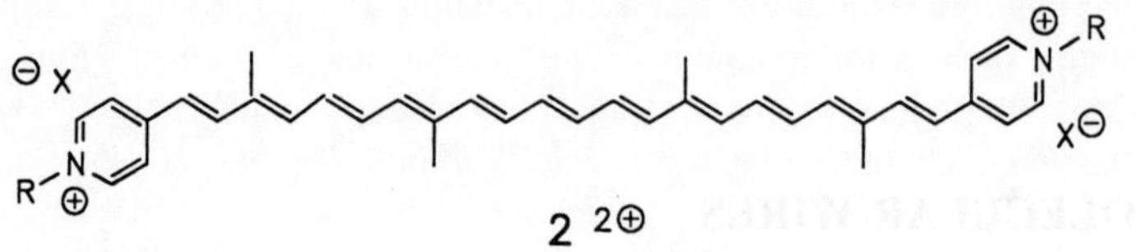

$^{\ominus}X$ R–N$^{\oplus}$ $^{\oplus}$N–R $X^{\ominus}$

2 $^{2\oplus}$

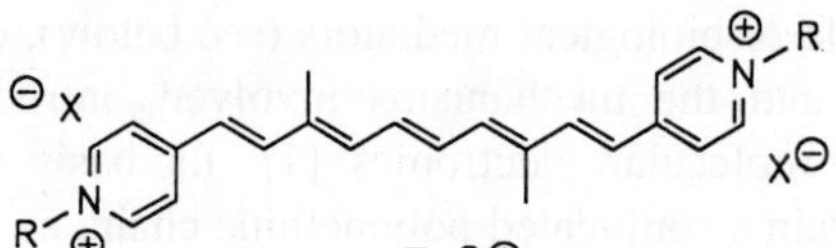

$^{\ominus}X$ R–N$^{\oplus}$ $^{\oplus}$N–R $X^{\ominus}$

3 $^{2\oplus}$

$^{\ominus}X$ R–N$^{\oplus}$ $^{\oplus}$N–R $X^{\ominus}$

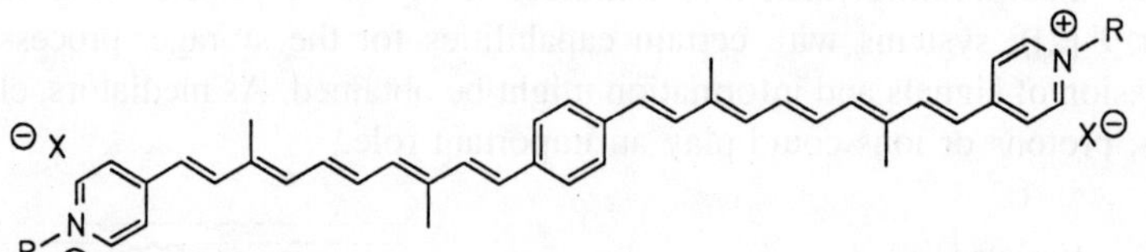

4 $^{2\oplus}$

R=CH_3 and X=I

structurally similar to natural carotenoids. Both ends of these chains were attached to the *para* position of *N*-substituted pyridines.

N-Substituted pyridine derivatives are, on reduction, able to form long-lived radicals, which easily dimerize to 'viologens' (Figure 2); after oxidation, these become even more stable viologen radical cations. In electron absorption spectroscopy, these are easily recognized by a band at about 600 nm. Viologens,

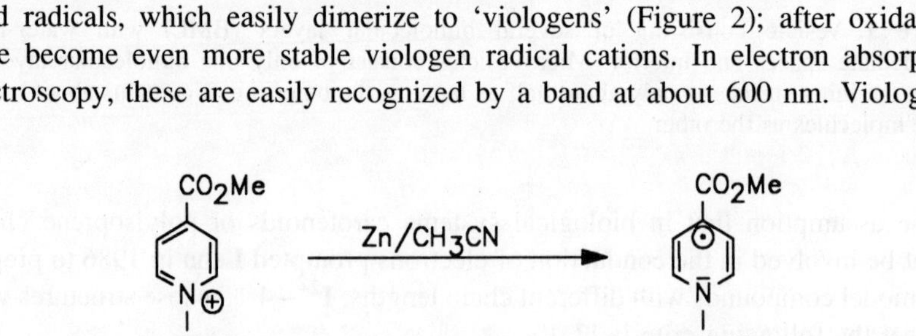

CO_2Me N$^{\oplus}$ C_2H_5 — Zn/CH_3CN → CO_2Me N C_2H_5

[R – N$^{\oplus}$ ... N – R ←→ R – N ... N$^{\oplus}$ – R]

Figure 2. Formation of a viologen from 1-ethyl-4-(methoxycarbonyl)pyridinium salts

whose pyridine rings are separated by conjugated double bonds, have been called caroviologens [1].

Cyclic voltammetric measurements, which give information on the reversibility of electronic processes, established a two-electron reduction process for the caroviologens $\mathbf{1}^{2+}$– $\mathbf{4}^{2+}$. Based on their structure and functionalities, they can be implanted into lipid bilayers, their polar groups reaching into the polar domains of bimolecular membranes. For these, among different synthetic vesicle-forming types of compounds, such as ammonium salts, phosphoric acid diesters or sulphonic acids with two oligomethylene chains, sodium dihexadecyl phosphate was chosen.

A fundamental characteristic of vesicle membranes [2] is a phase transition, which in many cases occur at a temperature close to 36 °C. Whereas below 36 °C the oligomethylene chains mostly adopt a linear conformation, giving the membrane liquid-crystalline properties, above 36 °C the crystalline domains melt, turning the hydrophobic domain of the membrane into a liquid-like melt. This behaviour was also observed in vesicles of sodium dihexadecyl phosphate, in which the caroviologens $\mathbf{1}^{2+}$, $\mathbf{2}^{2+}$ and $\mathbf{3}^{2+}$ had been incorporated.

Figure 3 shows electron spectra of $\mathbf{2}^{2+}$ in sodium dihexadecyl phosphate at temperatures of 23 °C (broken line) and 72 °C (solid line). the differences observed relate to the aggregates and monomers of $\mathbf{2}^{2+}$, respectively.

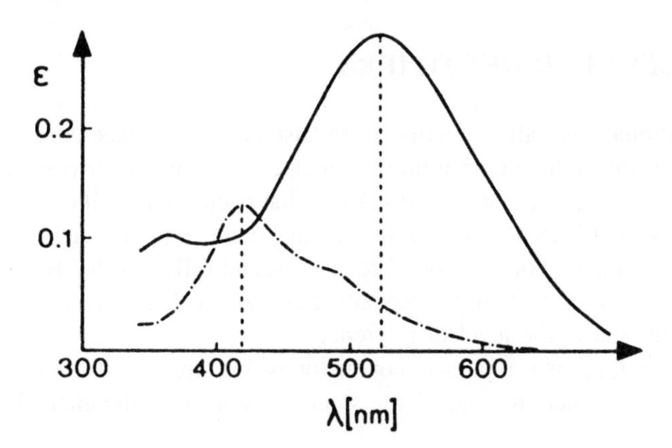

Figure 3. Electronic spectrum of $\mathbf{2}^{2+}$ in sodium dihexadecyl phosphate at 23 °C (broken line) and 72 °C (solid line)

Additionally, a phase transition of the membrane from gel to liquid crystal occurs at an average critical temperature of 59 °C. This value fits that of the undoped lipid membrane (60 °C) very well. These and other measurements led to the conclusion that the caroviologens are embedded in the lipid membrane as shown in Figure 4.

Caroviologens should therefore be considered as model compounds for electron conduction on a molecular level. Future investigations will certainly include the incorporation into complex redox systems and the investigation of the corresponding conduction phenomena, the combination of caroviologens with photoactive groups and the synthesis of push–pull carotenoids.

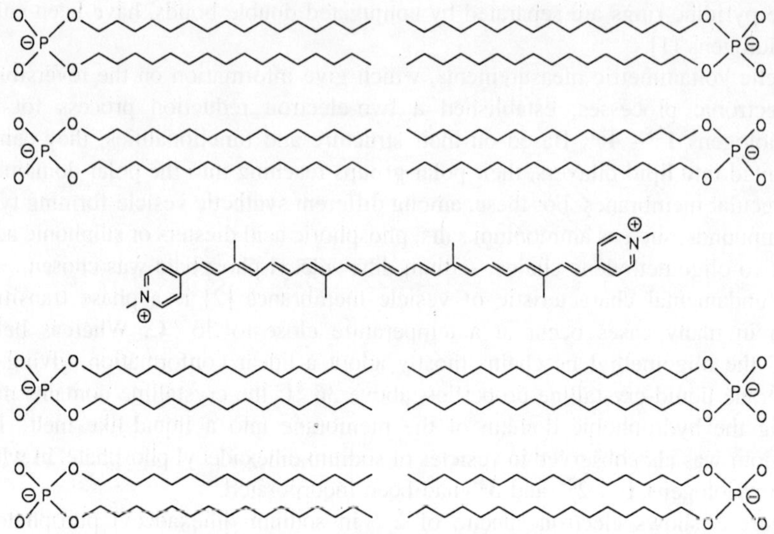

Figure 4. Caroviologen 2^{2+} embedded in dihexadecyl phosphate

11.2 MOLECULAR RECTIFIERS

The conventional, crystalline rectifiers are based on the surface contact between a metal and a semiconductor, resulting in an electric double layer (insulating layer).

For an easier understanding, the processes that occur in a rectifier are represented schematically in Figures 5 and 6 [4]. If the metal and the semiconductor are considered separately, both without free charges and fully isolated from each other, then, as seen in Figure 5, their corresponding energy levels are given by the potential of the exterior space, the level of reference.

If an excess-type or n-type semiconductor is in contact with a metal, electrons flow from the conduction band of the semiconductor into the energetically lower

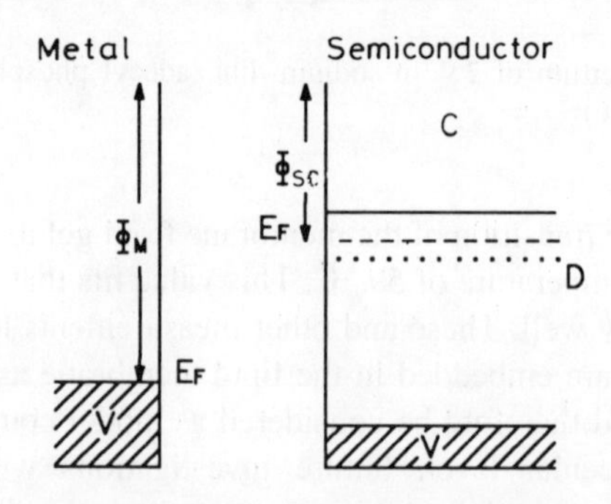

Figure 5. The potentials in metals and semiconductors, isolated from each other. V = valence band; C = conduction band; E_F = Fermi level; D = donor atom (P, As, etc); Φ_M = ionization potential for metal electrons; Φ_{SC} = ionization potential for semiconductor electrons

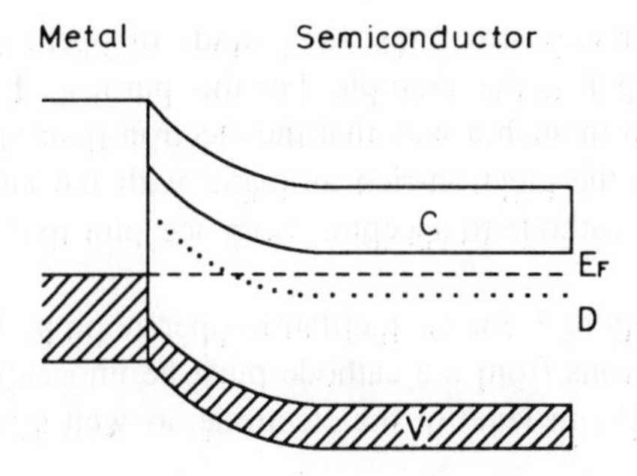

Figure 6. The potentials for an n-semiconductor in contact with a metal. Symbols as in Fig. 5

valence band of the metal (Figure 6). As a result, the metal and the semiconductor become charged, i.e. at the metal/semiconductor interface an electric double layer (insulating layer) is formed. Now, if an external voltage is applied to the rectifier, the insulating layer will grow, provided that the polarity is right, and no electric current will be able to flow through the rectifier. If the polarity is inverted, an electric current flow will become possible again. When an alternating voltage is applied to the rectifier, the insulating layer grows and shrinks according to the momentary polarity. As a consequence, alternating current is converted to direct current.

Conventional solid-state rectifiers are based on a p–n contact. For rectification to take place on a molecular level [1,5], the organic molecules involved also need a p–n contact. Aromatic systems are well suited to provide the necessary n and p building blocks, since the introduction of chosen substituents will make them electron-rich (n) or electron-poor (p). They then become electron donors and acceptors, respectively.

As in their semiconductor counterparts, they can function correctly only if the donor and acceptor moieties are separate from each other. This can be achieved by the introduction of a σ-system. Two prototypes for a molecular rectifier will now be discussed. They both have an electron-poor (acceptor) and electron-rich (donor) π-system, separated by an aliphatic bridging unit. The threefold bridging in the bicycle of **2** affords a rigid linear molecule.

Acceptor Donor Acceptor Donor

OCH_3 O OCH_3 OCH_3 O NC CN S S S S NC CN

1 2

In spite of their different structures, their mode of operation is the same, and will be explained using **1** as the example. For this purpose, **1** has to be implanted in an electronic system in such a way that the electron-poor aromatic is in contact with the cathode, and the electron-rich aromatic with the anode. Only with this polarity is a flow from cathode to acceptor, from acceptor to donor and from donor to anode possible.

As can be seen in Figure 7, for the rectifier to operate properly, those orbitals (B) which accept the electrons from the cathode must be unoccupied, and situated on the same level as the Fermi level of the electrode, as well as above the ionization level (C) of the donor.

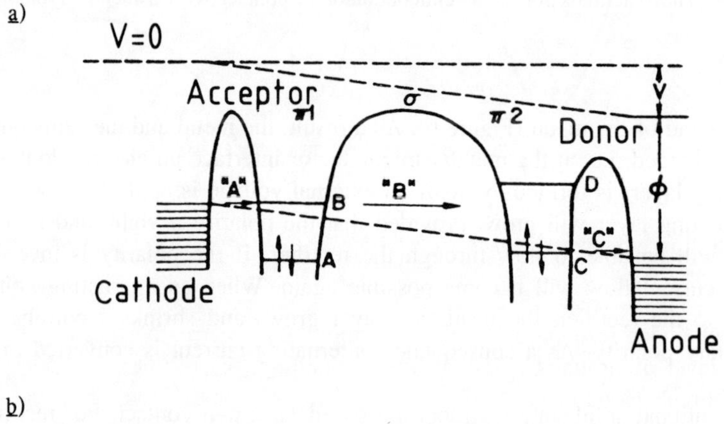

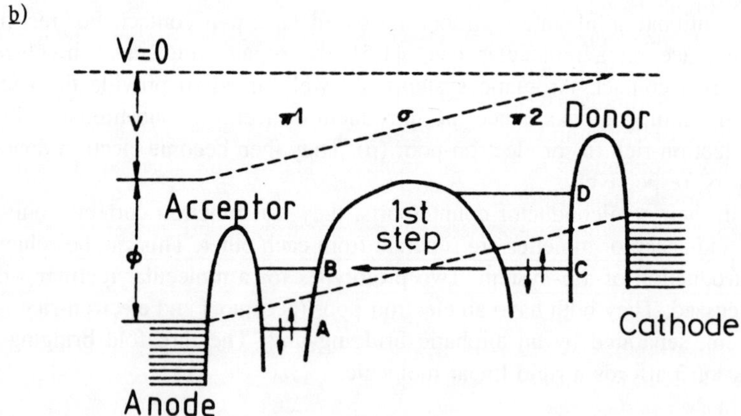

Figure 7. The mode of operation of a molecular rectifier. (*a*) Flow of electrons ('A', 'B' and 'C' are tunnelling processes); (*b*) no flow of electrons

As soon as the applied external voltage is strong enough to allow an overlap of donor and acceptor orbitals, electrons are transferred to the unoccupied orbital B of the acceptor. The energy barrier for this process depends on different factors, such as the relative positions of the energy levels and ionization potential. The same process is repeated on the other side, where an electron is transferred from the donor's C orbital to the anode. Thus, on the acceptor end a charge excess is formed, and on the donor end a charge deficiency. A transfer of charge between donor and acceptor can only occur through a tunnelling process.

The charged acceptor contains in orbital B one electron in its vibrational ground state (Figure 8). Because of the small difference in energy of B and C, it can tunnel to an unoccupied orbital C at the same energy level.

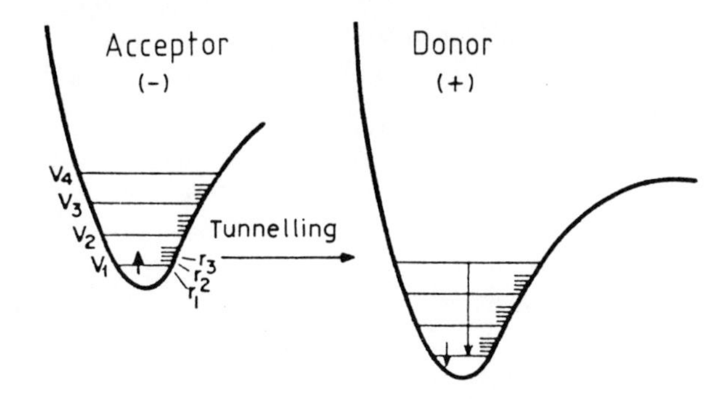

Figure 8. Tunnelling process between acceptor and donor, and their potential curves. V_n are vibrational levels

Except under resonance conditions, orbital B is normally situated above the energy level of orbital C, so that after tunnelling, the electron is occupying an excited Frank–Condon state, from where it drops without emission to the ground state. Thus, the electric circuit is closed and the overall process remains irreversible as long as orbital B is situated above orbital C.

With inverted polarity (Figure 7b), level D would have to be lowered below the Fermi level of the cathode, and the Fermi level of the anode below level A, in order to ensure a continuous flow of electrons. As shown, the necessary potential V is less favourable than in Figure 7a.

Information on molecule-based transistors can be found in literature [6].

References

1. F.L.Carter, *Molecular Electronic Devices*, Marcel Dekker, New York, Basle, 1982; see also J.M.Tour, R.Wu and J.S.Schumm, *J. Am. Chem. Soc.*, **112**, 5662 (1990).
2. J.-H.Fuhrhop, *Bioorganische Chemie*, Thieme, Stuttgart, 1982.
3. (a) J.M.Lehn, *Proc. Natl. Acad. Sci. USA*, **83**, 5355 (1986); (b) For intramolecular energy transport in polyene chains, see also F.Effenberger *et al.*, *Angew. Chem.*, **100**, 274 (1988); *Angew. Chem., Int. Ed. Engl.*, **27**, 281 (1988).
4. W.Finkelnburg, *Einführung in die Atomphysik*, Springer, Berlin, 1976.
5. A.Aviram and M.A.Ratner, *Chem. Phys. Lett.*, **29**, 277 (1974).
6. E.T.T.Jones, O.M.Chyan and M.S.Wrighton, *J. Am. Chem. Soc.*, **109**, 5526 (1987).

12 Light-induced Cleavage of Water

Plants exploit light's energy for the biosynthesis of cell components, a metabolic process of fundamental importance known as photosynthesis.

Even though only plants and a few other autotrophic organisms are able to use light's energy, all other living things indirectly obtain their energy through photosynthesis by way of the food chain. Moreover, the raw materials used for the generation of heat, light and other forms of energy, e.g. coal, mineral oil and natural gas, consist of about 90% of photosynthetic products.

The overall process of photosynthesis (equation 3) can be divided into a light reaction and a dark reaction, represented by equations 1 and 2, respectively ['CH_2O' stands for $(CH_2O)_x$].

$$2NAD^+ + 2\ H_2O \rightarrow 2\ H^+ + 2\ NADH + O_2 \quad (1)$$

$$2H^+ + CO_2 + 2\ NADH \rightarrow CH_2O + 2\ NAD^+ + H_2O \quad (2)$$

$$CO_2 + H_2O \rightarrow CH_2O + O_2 \quad (3)$$

The process mainly consists in the photolytic cleavage of water into oxygen and hydrogen, which is then trapped as NADH and H^+.

In Figure 1 the famous 'Z-scheme' of photosynthesis is shown. It is based on two photosystems (I and II) coupled together, which operate according to the same principle [1]: by absorbing a photon, an electron is promoted to a higher energy level and transferred to an acceptor (photoreduction of the acceptor). In both cases the photoactive system is a chlorophyll. In each of the photosystems, these work at different redox potentials (photosystem I ca −0.6 V; photosystem II ca −0.1 V).

The electrons excited in the chlorophyll of photosystem II are delivered to an electron transport chain and, according to the 'Förster' or 'exciton' mechanism, passed on to photosystem I. Here, a further excitation occurs, coupled with the transfer of an electron to a redox-Z system, formally ending with the reduction of CO_2. Concurrently, the electrons excited and transferred from the chlorophyll of photosystem II are replaced by the oxidation of water and a transfer from redox system Y.

Owing to the great importance of hydrogen as an alternative source of energy (e.g. the production of methane from H_2 and C), plans for the use of the photochemical cleavage of water have recently emerged [2].

A simple way of converting light into redox energy consists in the reversible photooxidation of organic dye molecules, such as metal porphyrins, embedded in bimolecular membranes (Figure 2). Shedding an electron, the dye moves across the

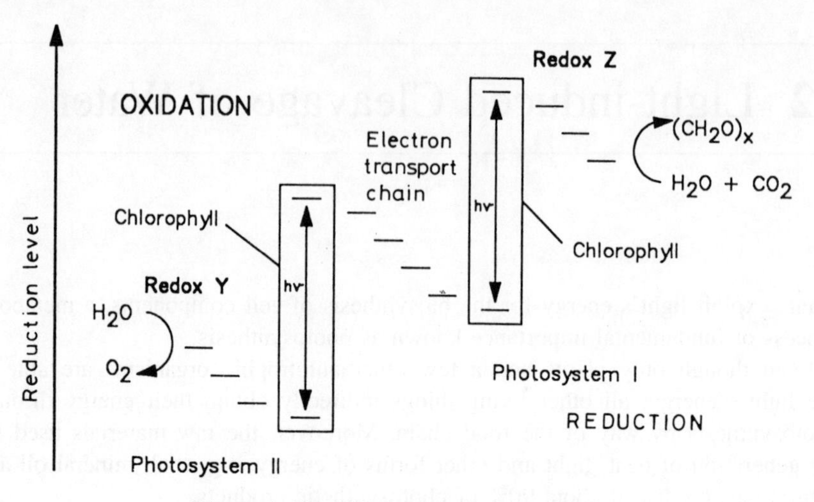

Figure 1. The 'Z-scheme' of photosynthesis

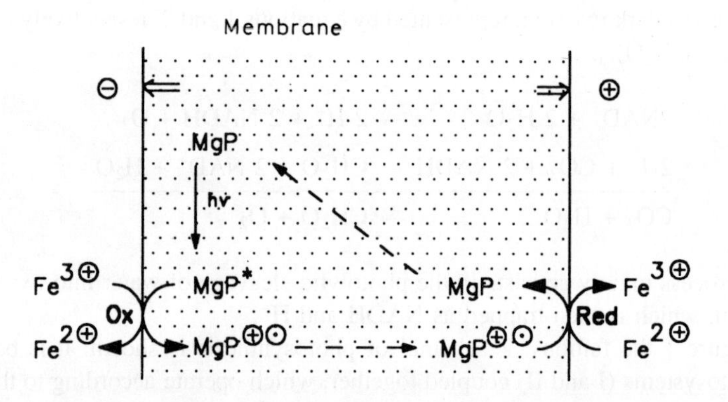

Figure 2. Conversion of light to redox energy by reversible photooxidation of organic dye complexes. P = porphyrin ligand; Mg = magnesium ions

membrane to the other side, where it is reduced to its original state. The potential produced (approximately 0.1 V), however, is not sufficient for the photolysis of water, for which a potential gradient of about 1 V is needed.

Shilov was the first to present a fairly complex system consisting of four water-soluble components:

1. either ascorbic acid or ethylenediaminetetraacetic acid;
2. a tris(2,2′-bipyridine)–ruthenium(II) complex as sensitizer [2d];

$$[Ru^{II}(bipy)_3]^{2\oplus} \equiv \left[(bipy)_2Ru^{II} - N \right]^{2\oplus}$$

3. *N,N′*-dimethylviologen as redox catalyst ('electron relay') [2e];

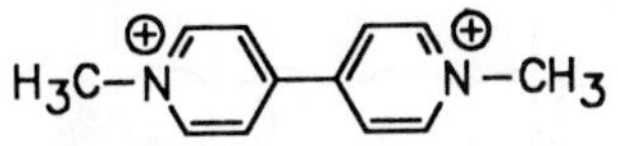

$$H_3C-\overset{\oplus}{N}\langle\;\rangle-\langle\;\rangle\overset{\oplus}{N}-CH_3$$

4. colloidal platinum (as catalyst).

In Figure 3, a schematic representation of a possible mechanism, as yet not entirely proved, is given.

Reaction pathway:

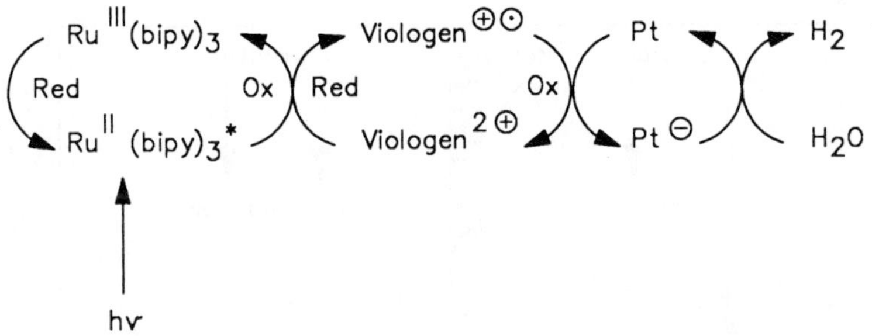

$Ru^{III}(bipy)_3$ — Red — $Ru^{II}(bipy)_3^{*}$; Ox — Red: Viologen$^{\oplus\odot}$ / Viologen$^{2\oplus}$; Ox: Pt / $Pt^{\ominus}$; H_2 / H_2O

$h\nu$

Mechanism:

$$4[Ru^{III}(bipy)_3]^{3\oplus} \xrightarrow{+4OH^{\ominus}} 4[Ru^{II}(bipy)_3]^{2\oplus} + O_2 + 2H_2O \qquad (3)$$

$$4[Ru^{II}(bipy)_3]^{2\oplus} \xrightarrow[h\nu]{+4H_2O} 4[Ru^{III}(bipy)_3]^{3\oplus} + 2H_2 + 4OH^{\ominus} \qquad (4)$$

$$2H_2O \xrightarrow[h\nu]{} 2H_2 + O_2 \qquad \text{(overall equation)} \qquad (5)$$

Figure 3. Reaction pathway and mechanism for the photochemical cleavage of H_2O in Shilov's system

The complex reaction starts with the reduction of Ru(III) to Ru(II) (equation 3). Following a photochemical excitation, an electron is then transferred to the viologen. The viologen radical cation so formed now transfers an electron to the colloidal platinum, on whose surface finally the H_3O^+ ions are reduced.

Semiconducting materials, e.g. as photoelectrochemical cells, can also be used for the photolysis of water (Figure 4). The semiconductor has a double purpose, absorbing the incident light on the one hand and its surface acting as a catalyst on the other.

Because the potential gradient necessary for the cleavage of water (1 V) can only be obtained by connecting several photoelectrochemical cells in series, a multicomponent system based on silicon semiconductors has been developed for the photochemical cleavage of HBr into H_2 and Br_2 (Figure 5). A disadvantage of such

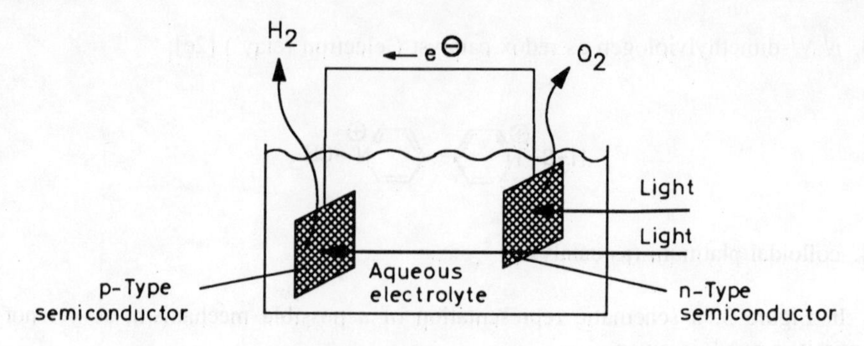

Figure 4. Semiconductor for the photochemical cleavage of water

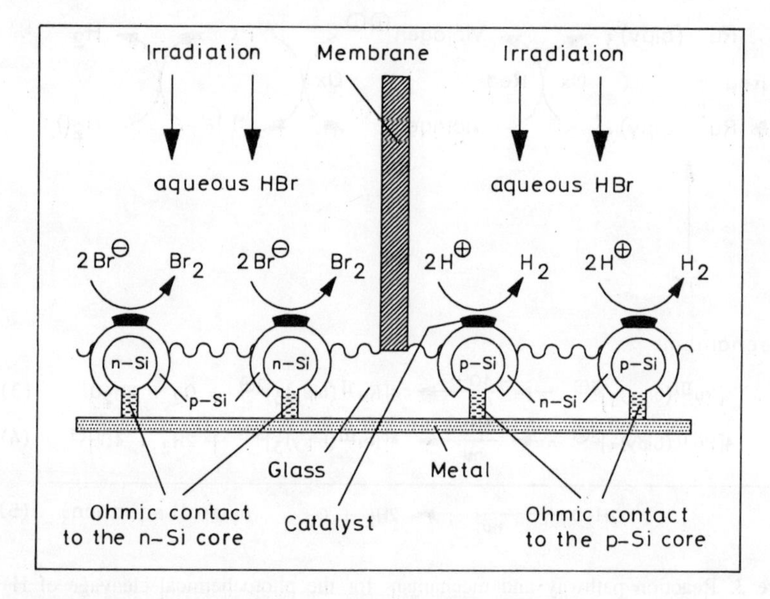

Figure 5. System for the conversion of light energy into storable chemical energy by the photolytic cleavage of 2HBr to H_2 and Br_2

photosynthetic semiconductor systems is the formation of an ohmic contact at the interface with the metal, significantly reducing the photovoltage.

Another model operating on a molecular level is presented in Figure 6. In this system, the essential building blocks consist of macromolecules, which, as in natural photosynthesis, can absorb light and then transfer the excited electrons to another component. The donor, covalently bound to the porphyrin skeleton, is coupled to an electrode which replaces the electrons that were transferred from the chromophore to the acceptor:

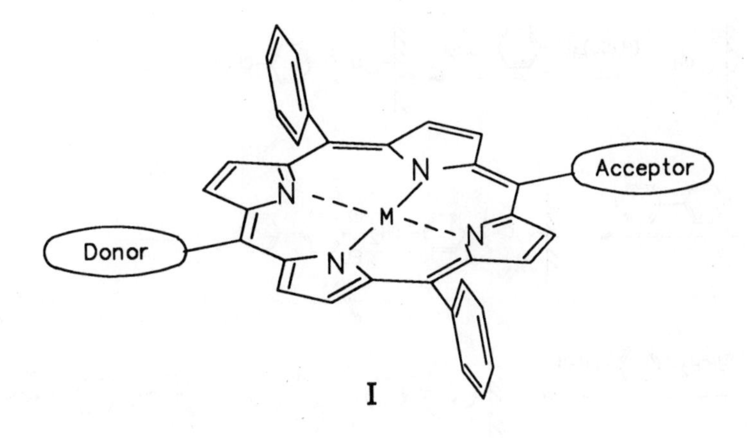

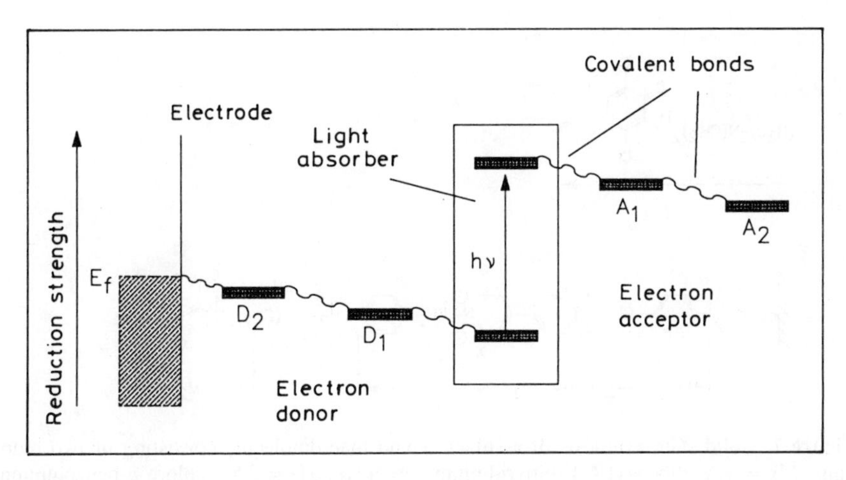

Figure 6. Layout of a molecular light absorption system that allows a light-induced electron transfer in one direction [3]

As an alternative to porphyrin derivatives, such as I, Merrifield developed a model consisting of a series of redox subunits linked to an electrochemical conductor. The preparation of such macromolecules (Figure 7) is similar to the synthesis of peptides, for which Merrifield was awarded the Nobel Prize in 1984.

The coupling of redox subunits to form a macromolecule and incorporating it into a photosystem, as shown in Figure 6, could possibly, in the future, provide an effective way of cleaving water.

The photosensitized reduction of CO_2 to CH_4 and H_2 is described elsewhere [4].

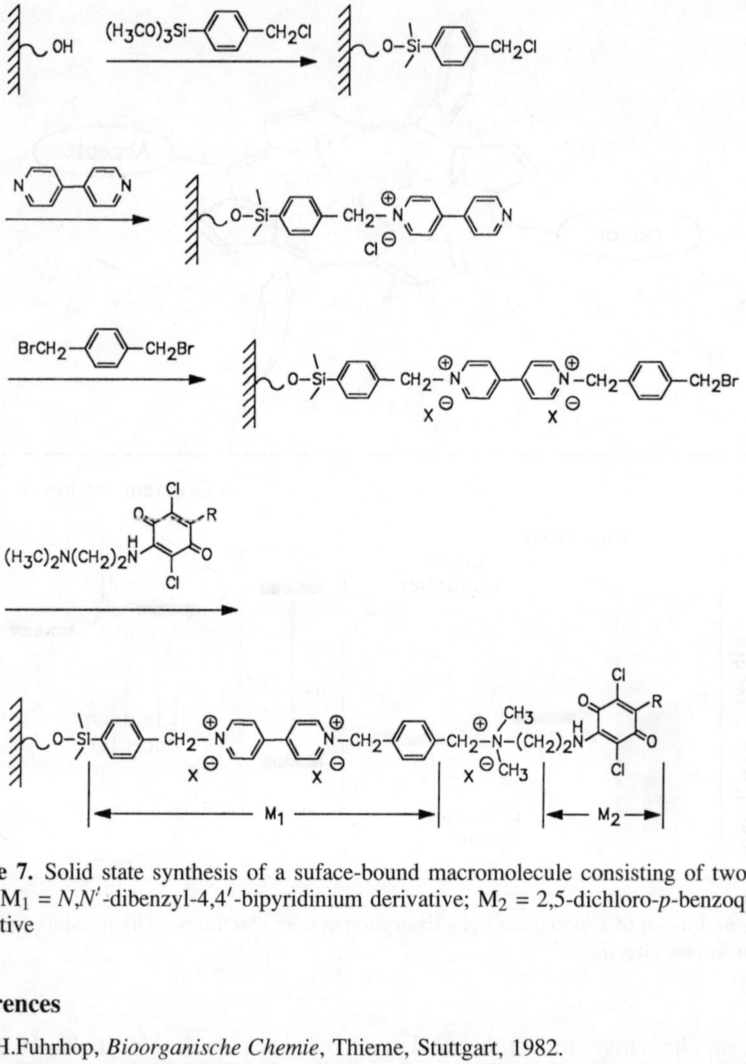

Figure 7. Solid state synthesis of a suface-bound macromolecule consisting of two redox units. M_1 = *N,N'*-dibenzyl-4,4'-bipyridinium derivative; M_2 = 2,5-dichloro-*p*-benzoquinone derivative

References

1. J.-H.Fuhrhop, *Bioorganische Chemie*, Thieme, Stuttgart, 1982.
2. (a) H.Parlar and W.Schuhmann, *Nachr. Chem. Tech. Lab.*, **36**, 1101 (1988); (b) M.S.Wrighton, *Comments Inorg. Chem.*, **9**, 269 (1985); (c) G.Renger, *Angew. Chem.*, **99** 660 (1987); *Angew. Chem., Int. Ed. Engl.*, **26**, 643 (1987); (d) for photolysis and photochemistry of Ru-complexes of bipyridine ligands, see: L.De Cola, P.Belser, F.Ebmeyer, F.Barigelletti, F.Vögtle, A.von Zelewsky and V.Balzani, *Inorg. Chem.*, **29**, 495 (1990); (e) compare, for example, H.Dürr, G.Dörr, K.Zengerle and B.Reis, *Chimia*, **37**, 245 (1983); H.Dürr *et al.*, *Nouv. J. Chim.*, **9**, 717 (1985); H.Dürr, K.Zengerle and H.P.Trierweiler, *Z. Naturforsch.*, **43b**, 361 (1988); H.Dürr, U.Thiery, P.P.Infelta and A.M.Braun, *New J. Chem.*, **13**, 575 (1989).
3. (a) Compare, for example, G.F.Strouse, L.A.Worl, J.N.Younathan and Th.J.Meyer, *J. Am. Chem. Soc.*, **11**, 9101 (1989); R.F.Beeston, S.L.Larson and M.Fitzgerald, *Inorg. Chem.*, **28**, 4189 (1989); F.Barigelletti, L.De Cola, V.Balzani, P.Belser, A.von Zelewsky, F.Ebmeyer and F.Vögtle, in *Photoconversion Processes for Energy and Chemicals* (D.O. Hall and G.Grassi, Eds), Elsevier, London, New York, 1989, p.46; E.Kimura *et al.*,

J. Chem. Soc., Chem. Commun, 1990, p.397; (b) J.v.Gersdorff, M.Huber, H.Schubert, D.Niethammer, B.Kirste, M.Plato, K.Möbius, H.Kurreck, R.Eichberger, R.Kietzmann and F.Willig, *Angew. Chem.*, **102**, 690 (1990); *Angew. Chem., Int. Ed. Engl.*, **29**, 670 (1990); (c) molecular charge-transfer relays: M.Maslak, M.P.Augustine and J.D.Burkey, *J. Am. Chem. Soc.*, **112**, 5359 (1990); (d) R.Duesing, G.Tapolski and T.J.Meyer, *J. Am. Chem. Soc.*, **112**, 5378 (1990).

4. I.Willner, R.Maidan, D.Mandler, H.Dürr, G.Dörr and K.Zengerle, *J. Am. Chem. Soc.*, **109**, 6080 (1987).

13 Final Remarks

In this book, we have looked beyond the individual molecule, to interactions between specifically designed molecules and the properties of the materials obtained. Aggregates, complexes and assemblies producing macroscopic structures with novel qualities were at the centre of our interest.

The contents had to be limited to a choice of certain subjects. There is no problem in finding other interesting supramolecular structures. The biologically important area of the self-organization of molecules (membranes, micelles, etc.) has only been touched upon. Subjects pertaining to biochemistry and polymer chemistry were hardly discussed; both can easily be found in textbooks and monographs.

In spite of these restrictions, many modern areas and methods of research have been discussed. The interdisciplinary character of supramolecular chemistry is certainly obvious. Biological and physical phenomena in many different ways link into chemistry. Many things are too recent to be classified or judged properly. There is no doubt that the subjects discussed will provide abundant results for many years to come. We are all curious to see whether it will be possible to supplement or even replace metallic conductors and magnets with organic materials.

This book teaches that chemists should stop looking at the individual molecule only, and should also consider those structures and properties arising from the interaction of several molecular units: the supramolecules [1], due to cooperative effects, offer more than the individual components [2].

There are not only attractive molecules (cf. the companion volume, *Fascinating Molecules in Organic Chemistry*), but also fascinating supramolecular structures, which are pleasing because of their microscopic or macroscopic 'architecture' or their mode of operation.

Supramolecular chemistry will gain in importance and grow even more rapidly as soon as efficient catalysts, useful active agents and materials start to leave the research laboratory and become viable in industrial applications.

References

1. First use of the designation ('Übermolekül'): K.L.Wolf, H.Frahm and H.Harms, *Z. Phys. Chem.*, **B36**, 17 (1937); K.L.Wolf, H.Dunken and K.Merkel, *Z. Phys. Chem.*, **46**, 287 (1940); K.L.Wolf and R.Wolff, *Angew. Chem.*, **61**, 191 (1949).
2. Recent overviews: (a) J.-M.Lehn, *Angew. Chem., Int. Ed. Engl.*, **29**, 1304 (1990); (b) M.Ahlers, W.Müller, A.Reichert, H.Ringsdorf and J.Venzmer, *Angew. Chem., Int. Ed. Engl.*, **29**, 1269 (1990); (c) D.Seebach, *Angew. Chem., Int. Ed. Engl.*, **29**, 1320 (1990); (d) F.Vögtle, *Cyclophan-Chemie*, Teubner, Stuttgart, 1990.

Author Index

Subject Index